'Omics' and 'Multi-Omics' Insights into Plant Responses to Abiotic Stresses

'Omics' and 'Multi-Omics' Insights into Plant Responses to Abiotic Stresses

Guest Editors

Rudo Ngara
Yuri Shavrukov

MDPI

Basel • Beijing • Wuhan • Barcelona • Belgrade • Novi Sad • Cluj • Manchester

Guest Editors
Rudo Ngara
Department of Plant Sciences
University of the Free State
Phuthaditjhaba
South Africa

Yuri Shavrukov
College of Science and Engineering
(Biological Sciences)
Flinders University
Adelaide
Australia

Editorial Office
MDPI AG
Grosspeteranlage 5
4052 Basel, Switzerland

This is a reprint of the Special Issue, published open access by the journal *Plants* (ISSN 2223-7747), freely accessible at: https://www.mdpi.com/journal/plants/special_issues/JP21M52C16.

For citation purposes, cite each article independently as indicated on the article page online and as indicated below:

Lastname, A.A.; Lastname, B.B. Article Title. *Journal Name* **Year**, *Volume Number*, Page Range.

ISBN 978-3-7258-7591-7 (Hbk)
ISBN 978-3-7258-7592-4 (PDF)
https://doi.org/10.3390/books978-3-7258-7592-4

Contents

About the Editors

Rudo Ngara

Rudo Ngara is an associate professor in the Department of Plant Sciences at the University of the Free State's Qwaqwa Campus in South Africa, where she teaches undergraduate and postgraduate modules in plant biology and plant biotechnology. Her primary research interests lie in understanding how plants modulate their physiology and biochemistry in response to single and combined abiotic stresses, such as drought, heat, and salinity. Rudo also has a keen interest in the identities and roles of secreted proteins in response to abiotic stress factors. She applies a range of plant physiology, omics, and multi-omics methods in her research.

Yuri Shavrukov

Yuri Shavrukov is Adjunct Professor in College of Science and Engineering (Biological Sciences), Flinders University, Adelaide, Australia. He is a researcher working with various projects related to plant biology, most of them in the collaboration with Kazakhstan. His research interests include plant genetics and genomics, molecular markers and plant genotyping, abiotic stress tolerance (drought, salinity and dehydration), gene identification and expression analysis in various crops and plant species. Plant 'omics' and 'multi-omics' is also important part of his research.

plants

MDPI

Editorial

Omics and Multi-Omics Insights into Plant Responses to Abiotic Stresses

Rudo Ngara [1,]* and Yuri Shavrukov [2,]*

[1] Department of Plant Sciences, University of the Free State, Qwaqwa Campus, Phuthaditjhaba 9866, South Africa
[2] College of Science and Engineering, Biological Sciences, Flinders University, Adelaide, SA 5042, Australia
* Correspondence: ngarar@ufs.ac.za (R.N.); yuri.shavrukov@flinders.edu.au (Y.S.)

Plants are often exposed to abiotic stresses such as drought, salinity, heat, cold, freezing, nutrient deficiency, and heavy metal toxicity, which limit their growth and development. In the case of crops, abiotic stresses may reduce yields, leading to insufficient food supply for the growing population. In addition, these stresses may occur concomitantly under field conditions, resulting in diverse effects on plant growth and metabolism. Whereas stress combinations such as drought and ozone may result in potentially positive interactions, others, including drought and heat or salinity with nutrient deficiency, may cause extensive plant cell damage and greater yield reductions than the individual stresses alone. Furthermore, the frequency and magnitude of stresses, such as drought and heat, will likely increase with global warming. Consequently, research interest in understanding the complex responses of plants to individual and combinations of abiotic stresses is increasing.

Omics technologies are also improving our insights into stress-responsive changes in the transcriptome, proteome, and metabolome of plants, while the sequencing of entire plant genomes is providing resources to guide crop improvement strategies. Other research groups are working on validating the roles of these 'omics-derived data' in stress adaptation using various methodologies, including transgenic plants, seed priming, and exogenous application of growth-promoting compounds.

The Special Issue "Omics and Multi-Omics Insights into Plant Responses to Abiotic Stresses" consists of 13 papers that broaden our insights into plant responses to individual abiotic stresses and their combinations using various omics approaches, including the functional validation of such data. Therefore, this Special Issue expands our understanding of plant tolerance to various abiotic stresses through omics and multi-omics approaches.

The perennial submerged aquatic plant species, *Potamogeton wrightii*, is widely distributed across China and used for medicine and as a highly nutritional feed for animals but this species is sensitive to phosphorus stress. Homeostasis of P is essential for optimal plant growth and development, with highly coordinated signaling mechanisms regulating phosphorus deficiency and excess. The authors report leaf chlorosis and chlorophyll loss in response to high phosphorus treatment, as well as increased MDA and H_2O_2 contents and higher POD, SOD, and CAT activities. The RNA-seq approach used in the study showed a thousand differentially expressed genes common across all phosphorus treatments in *P. wrightii* plants. Using an UPLCMS/MS system, 3408 metabolites arranged into 21 categories were detected in leaves. Combined transcriptomic and metabolomic analysis revealed six differentially accumulated metabolites (DAMs) and 19 differentially expressed genes (DEGs) showing down- and upregulation in glycerolipid and glycerophospholipid metabolism in response to low and high phosphorus stress in *P. wrightii* plants. The presented report is an important step towards understanding the mechanisms underlying plant responses to phosphorus stress [1].

Fungal endophytes are healthy, asymptomatic fungi living inside plants, and they are the focus of a specific area of plant phytopathology and mycology. The selected endophyte EF0801 was identified as being similar to *Sordariomycetes*, and it was used for the treatment of rice seedlings. The authors studied the benefit of endophyte infection in rice seedlings grown in a solution with sodium carbonate, 10 mM Na_2CO_3, which resulted in increased plant height but reduced root length. The application of gas chromatography and mass spectrometry revealed the enrichment of starch and sucrose metabolism, together with three amino acids (alanine, aspartate, and glutamate) in the infected rice seedlings grown under carbonate treatment compared with control plants. Together with the RNA-seq technique used, the authors found six DEGs and six DAMs involved in the response of the rice plants to Na_2CO_3 treatment. One of the most important genes identified was hexokinase, *HXK*, which encodes an enzyme that converts glucose and fructose in their respective metabolite pathways. The results of the analysis of the transcriptome and metabolome in the rice seedlings infected with the endophyte fungus provide knowledge and wide perspectives for improving rice plant tolerance to carbonate and alkaline soils [2].

An opinion paper discusses the use of polyethylene glycol (PEG), and the authors assessed 139 published papers correctly describing PEG-induced dehydration rather than drought in various studied plant species. The authors argued for the use of much clearer terminology in experiments using PEG that strictly limit discussion to "PEG-induced", "simulated" or "mimicked" dehydration, and osmotic or water stresses, with the best option being "PEG-induced dehydration". Five special cases are presented, with experiments in both soil/substrate and hydroponics in diverse plants species, including cotton, *Medicago truncatula*, wild barley, maize, and soybean. The authors concluded that the physiological parameters and gene expression profiles differed between plants grown in containers with drying soil and those subjected to dehydration using PEG-induced stress in hydroponics [3].

Plants of bread wheat cultivar Leningradskaya 6 were grown in Hoagland nutrient solution with and without 20% polyethylene glycol-6000 (PEG-6000). This treatment simulated plant dehydration with an osmotic potential of -0.75 MPa compared to controls. Proteomics analyses revealed 497 and 157 differentially abundant protein groups in the leaves and roots, respectively. Oxidative- and heat-stress-responsive proteins were identified in the leaves of the PEG-treated plants, whereas dehydration-related proteins were found in the roots of the treated plants. The authors analysed extracellular root peptides in plants after PEG-induced dehydration, leading to the identification of 16 small secreted peptides and peptides derived from proteins involved in cell wall catabolism, intercellular signaling, and stress response. The results help with better understanding the mechanisms of wheat plant adaptation to dehydration for future application in agriculture [4].

The pod dehiscence (PD) trait was studied in 170 accessions of soybean with diverse phenotypes. A genomics approach was used employing Diversity Arrays Technology (DArT) analysis covering the entire soybean genome with 60K DArT markers, including both Silico-DArT and DArT SNPs. A genome-wide association study (GWAS) identified 14 stable QTLs corresponding to PD- and pod-related traits. Two of the most stable and important genes for PD with known functions were identified from the QTLs, encoding a DNA-binding bromodomain-containing protein and a C2H2 zinc finger protein. Genomic prediction was performed helping to select PD-resistant and -sensitive soybean genotypes with estimated breeding values. The authors concluded that the identified putative candidate genes can be used to improve PD resistance in soybean genotypes using marker-assisted selection and genomic prediction [5].

 https://doi.org/10.3390/plants15050799

An in silico analysis of genome assemblies from 34 *Oryza* species, including cultivated and wild rice from available databases, was carried out targeting calcium-dependent protein kinases (CPKs) and other CPK-related kinases. Evolutionary genomics and comparative pan-genome analysis revealed distinct duplication types among the studied groups of CPK genes. Alternative splicing was identified in the transcriptome analysis as a potential mechanism contributing to the diversity of CPK protein domains. Expression studies of CPKs in multiple species of the *Oryza* genus resulted in comprehensive data provided as an atlas for comparison. It would be interesting and important to confirm the presented computer analysis with results from future experiments obtained from plants from various rice species in molecular laboratories [6].

An integrative meta-analysis of publicly available transcriptomic databases was carried out for comparative and evolutionary studies of potato and tomato, two of the most important species from the *Solanum* genus, in response to drought and heat stresses. A robust and reproducible set of stress-responsive genes was generated with reduced experiment-specific variability. Species-specific heat-responsive expression profiles were observed in tomato plants, whereas potato plants showed wide transcriptional activities under drought stress. The potato genomes exhibited lineage-specific gene duplications compared to tomato, suggesting that the duplication event in potato was relatively recent and contributed to stress adaptation. The authors concluded that the results of their meta-analysis provide an integrated view of the molecular and evolutionary mechanisms underlying drought and heat stress responses in two major Solanaceae crops [7].

Zinc is an essential micronutrient that serves as a cofactor for numerous plant enzymes. However, an overaccumulation of zinc in soils can lead to zinc stress, disrupting the normal physiological and metabolic processes of plants. In separate studies, the metabolome and proteome responses of *Amaranthus caudatus* plants to zinc stress were investigated. These studies independently highlighted organ-specific shifts in the metabolite and protein profiles of *A. caudatus* in response to the zinc stress, with the roots being the most metabolically active organ, followed by young leaves and then mature leaves. The authors attributed the high metabolic activity of the roots under zinc stress to the organ's direct role in zinc uptake and detoxification [8,9].

In the metabolomics study, low-molecular-weight metabolite analysis resulted in the detection of 419 derivatives in the samples, of which 144 could be putatively annotated. Most of the upregulated metabolites in both the leaves and roots were sugars and organic acids, with putative roles in osmotic regulation, oxidative protection, and Zn^{2+} complexation, which the authors highlighted as the probable key feature in the metabolic responses of *A. caudatus* to Zn^{2+} stress. A follow-up proteomic investigation of zinc-stressed *A. caudatus* roots and leaves was conducted using liquid chromatography–mass spectrometry (LC-MS). Overall, the authors highlighted the importance of redox homeostasis, protein synthesis, repair/folding, and degradation systems in *A. caudatus* plants under zinc stress [8,9].

A multi-omics approach integrating physiological, ultrastructural, transcriptomic, and lipidomic analyses was used to investigate the mechanisms of desiccation tolerance and senescence in the resurrection plant *Eragrostis nindensis* [10]. Non-senescent leaf tissue showed remarkable ultrastructural reorganisation, including chloroplasts with disassembled thylakoids, increased vacuolation, and densely accumulating lipid droplets as a survival mechanism under dehydration and desiccation. These features reversed upon tissue rehydration, while senescent leaf tissue cells collapsed and failed to recover. A large-scale transcriptomic analysis of the two tissue types under a range of dehydration levels from full hydration to the air-dry desiccation state further revealed a dynamic regulatory program for desiccation tolerance in *E. nindensis*. For example, genes related to seed-protective functions, autophagy, and stress-responsive pathways were upregulated in non-senescent

 https://doi.org/10.3390/plants15050799

leaves during dehydration. In contrast, photosynthesis, metabolism, and carbon fixation genes were largely downregulated under dehydration stress, which the authors attributed to the need for a metabolic shutdown and the suppression of energy-intensive processes. Overall, the study expands our understanding of the complex mechanisms of desiccation tolerance in resurrection plants, which include cellular ultrastructural modifications, maintenance of membrane and organelle integrity by lipid droplets, a complex transcriptional regulatory program, and metabolic arrest [10].

A genome-wide identification, classification, and expression study was systematically conducted of the bread wheat *Respiratory burst oxidase homologs* (*TaRboh*) gene family. In silico analysis identified 39 *TaRboh* genes, together with their chromosomal distribution patterns and gene structures. The subsequent characterisation of their *cis* elements suggested that *TaRbohs* may be involved in diverse functions, including phytohormone response, biotic and abiotic stress response, and plant growth. Quantitative expression profiling of nine candidate *TaRboh* genes under 25% PEG-6000-induced dehydration mimicking drought stress, heat, and their combined stress also revealed distinct levels of the responsiveness of the genes to the imposed stresses, thus validating their roles in drought and heat responses. Overall, the study lays a foundation for future molecular investigations into the role of *TaRboh* genes in wheat responses to drought, heat, and these stresses combined [11].

Another study evaluated the effects of chloramphenicol, a broad-spectrum antibiotic, on the germination and early seedling establishment of rice cultivar Khao Dawk Mali L 105, commonly known as Thai jasmine rice [12]. The authors rationalised that chloramphenicol persists in agricultural soils through wastewater contamination and affects growth processes in various plant species, including rice. The results showed that chloramphenicol treatment did not affect germination relative to the control but negatively affected early seedling establishment, with pronounced decreases in seedling vigor across antibiotic concentrations. A phosphoproteomic analysis using LC-MS/MS also revealed stage-specific phosphoproteome changes in 3-day- and 6-day-old geminated rice seeds in response to 15 μg/mL chloramphenicol treatment. These chloramphenicol-responsive phosphoproteins were putatively involved in metabolic- and cytokinesis-related processes during germination and hormonal signaling pathways, stress response, DNA replication, and metabolism during early seedling establishment. The authors concluded that chloramphenicol affects rice seed germination and physiological processes and suggested that future studies should compare the antibiotic's effects across different rice cultivars to expand our knowledge of plant stress responses [12].

Recent advances in crop functional genomics were discussed for their potential applications in developing stress-tolerant crops for improved global food security. The collective importance of omics-based plant research for identifying stress-responsive genes, proteins, metabolites, and/or metabolic pathways were highlighted as critical for stress tolerance. Genome editing technologies and their applications in plant stress biology were also reviewed, and the authors emphasised the need to critically evaluate the ethical and regulatory elements of these technologies to ensure their responsible and safe use. Potential challenges involved in the translation of laboratory-generated omics data to field studies were especially highlighted [13].

In the context of global warming and the onset of climate change, abiotic stresses are likely to become more intense, resulting in a decline in crop production and increasing disruptions of plant ecology. Molecular tools such as genomics, metabolomics, proteomics, other omics together with multi-omics can improve our understanding of plant responses to such stresses. The identified stress-responsive genes, proteins, and metabolites could help elucidate the biological mechanisms underlying plant tolerance to abiotic stresses. We hope that this Special Issue provides a significant contribution to this process.

 https://doi.org/10.3390/plants15050799

Author Contributions: Conceptualization, R.N. and Y.S.; writing—original draft preparation, R.N. and Y.S.; review and editing, R.N. and Y.S. All authors have read and agreed to the published version of the manuscript.

Funding: This research received no external funding.

Acknowledgments: We want to thank all authors who presented their interesting and important publications and contributed to the success of this Special Issue edition.

Conflicts of Interest: The authors declare no conflicts of interest.

References

1. Pan, C.; Zhou, B.; Tang, M.; Chen, J.; Yang, H.; Xu, X. Integrating RNA-seq and metabolomic perspectives reveals the mechanism of response to phosphorus stress of *Potamogeton wrightii*. *Plants* **2025**, *14*, 3556. [CrossRef] [PubMed]
2. Wang, X.; Li, Y.; Sun, H.; Zhang, L.; Li, X. Integrative transcriptomics and metabolomics reveal the key metabolic pathways in endophyte-infected rice seedlings resistance to Na_2CO_3 stress. *Plants* **2025**, *14*, 1524. [CrossRef] [PubMed]
3. Kylyshbayeva, G.; Bishimbayeva, N.; Jatayev, S.; Eliby, S.; Shavrukov, Y. Polyethylene glycol (PEG) application triggers plant dehydration but does not accurately simulate drought. *Plants* **2025**, *14*, 92. [CrossRef] [PubMed]
4. Azarkina, R.; Makeeva, A.; Mamaeva, A.; Kovalchuk, S.; Ganaeva, D.; Tikhonovich, I.; Fesenko, I. The proteomic and peptidomic response of wheat (*Triticum aestivum* L.) to drought stress. *Plants* **2025**, *14*, 2168. [CrossRef] [PubMed]
5. Mazkirat, S.; Bulatova, K.; Didorenko, S.; Bastaubayeva, S.; Babissekova, D.; Khalbayeva, S.; Tukenov, A.; Yespembetova, A.; Saparbayeva, N.; Shavrukov, Y. Pod dehiscence in soybean: Genome wide association study and genomic prediction. *Plants* **2025**, *14*, 3505. [CrossRef] [PubMed]
6. Shi, F.; Li, L.; Chen, M.; Chang, J.; Tu, M.; He, G.; Li, Y.; Yang, G. Genus-wide pan-genome analysis of *Oryza* calcium-dependent protein kinase genes and their related kinases highlights the complexity of protein domain architectures and expression dynamics. *Plants* **2025**, *14*, 1542. [CrossRef]
7. Bondar, E.I.; Zubairova, U.S.; Bobrovskikh, A.V.; Doroshkov, A.V. Integrative transcriptomic and evolutionary analysis of drought and heat stress responses in *Solanum tuberosum* and *Solanum lycopersicum*. *Plants* **2025**, *14*, 3851. [CrossRef] [PubMed]
8. Osmolovskaya, N.; Bilova, T.; Gurina, A.; Orlova, A.; Vu, V.D.; Sukhikh, S.; Zhilkina, T.; Frolova, N.; Tarakhovskaya, E.; Kamionskaya, A.; et al. Metabolic responses of *Amaranthus caudatus* roots and leaves to zinc stress. *Plants* **2025**, *14*, 2119. [CrossRef] [PubMed]
9. Gurina, A.; Bilova, T.; Gorbach, D.; Soboleva, A.; Stepanova, N.; Babich, O.; Ihling, C.; Kamionskaya, A.; Osmolovskaya, N.; Frolov, A. Metabolic responses to the zinc stress in the roots and leaves of *Amaranthus caudatus*: The proteomics view. *Plants* **2025**, *14*, 3315. [CrossRef] [PubMed]
10. Madden, C.F.; Williams, B.; Mundree, S.; Acket, S.; Ruelland, E.; Hilhorst, H.W.M.; Farrant, J.M. Defying death: A multi-omics approach to understanding desiccation tolerance and senescence in *Eragrostis nindensis*. *Plants* **2025**, *14*, 3360. [CrossRef] [PubMed]
11. Cao, M.; Zhang, Y.; Zou, X.; Yin, H.; Yin, Y.; Li, Z.; Xiao, W.; Liu, S.; Li, Y.; Guo, X. Genome-wide identification, classification, expression analysis, and screening of drought and heat resistance-related candidates of the *Rboh* gene family in wheat. *Plants* **2024**, *13*, 3377. [CrossRef]
12. Li, R.; Phaonakrop, N.; Roytrakul, S.; Lohmaneeratana, K.; Thamchaipenet, A. Phosphoproteomic analysis reveals impairment of rice germination by chloramphenicol. *Plants* **2025**, *14*, 2845. [CrossRef] [PubMed]
13. Goche, T.; Mavindidze, P.; Zenda, T. Advances in functional genomics for exploring abiotic stress tolerance mechanisms in cereals. *Plants* **2025**, *14*, 2459. [CrossRef] [PubMed]

https://doi.org/10.3390/plants15050799

plants

MDPI

Article

Integrating RNA-Seq and Metabolomic Perspectives Reveals the Mechanism of Response to Phosphorus Stress of *Potamogeton wrightii*

Caiyun Pan [1,2], Bing Zhou [1,2], Ming Tang [1,2], Jingan Chen [3,4], Haiquan Yang [3,4] and Xiaorong Xu [1,2,*]

1 School of Life Sciences, Guizhou Normal University, Guiyang 550025, China; pancaiyun@gznu.edu.cn (C.P.); zhoubing@gznu.edu.cn (B.Z.); mingtang@gznu.edu.cn (M.T.)
2 Key Laboratory of National Forestry and Grassland Administration on Biodiversity Conservation in Karst Mountainous Areas of Southwestern China, Guizhou Normal University, Guiyang 550025, China
3 State Key Laboratory of Environmental Geochemistry, Institute of Geochemistry, Chinese Academy of Sciences, Guiyang 550001, China; chenjingan@vip.skleg.cn (J.C.); yanghaiquan@vip.skleg.cn (H.Y.)
4 Guizhou Province Field Scientific Observation and Research Station of Hongfeng Reservoir Ecosystem, Guiyang 551499, China
* Correspondence: xxr@gznu.edu.cn

Abstract: Phosphorus is an important nutrient element in aquatic ecosystems, and its concentration directly affects the growth and ecological functions of submerged plants. However, the physiological and molecular mechanisms of *P. wrightii*'s response to phosphorus stress remain unknown. This study investigated the effects of different phosphorus concentration treatments on *P. wrightii* through physiological, RNA-seq, and metabolome analysis methods. The results indicated that phosphorus stress affected plant physiology by reducing chlorophyll content, increasing MDA and H_2O_2 accumulation, and activating the antioxidant enzyme system. Multiple phosphorus transporters (PHT, SPX, and PAP) and the transcription factor PHR1 were identified through RNA-seq and RT-qPCR analysis. The glycerol phospholipids represent a decreasing trend after low or high phosphorus stress. Through the combined analysis of RNA-seq and metabolome analysis, the response differences of 6 DAMs and 19 DEGs to the *P. wrightii* Glycerolipid metabolism and Glycerophospholipid metabolism pathways under different phosphorus stresses were revealed. Our results provide a scientific basis and guidance for restoring submerged plants in shallow lakes and for preventing and controlling eutrophication.

Keywords: *Potamogeton wrightii*; phosphorus; RNA-seq; metabolome; phosphate transporters

1. Introduction

Phosphorus (P) is an essential macronutrient for plants, serving as a fundamental component of nucleic acids, phospholipid membranes, and adenosine triphosphate (ATP) [1]. It plays a crucial role in plant growth, development, and overall metabolic regulation [2]. Moreover, phosphorus is involved in diverse physiological processes, including energy transfer, molecular transport, enzymatic reactions, and photosynthesis [3,4].

Maintaining P homeostasis is essential for optimal plant growth and development. Both P deficiency and excess can adversely affect plant physiological processes. Under low P conditions, plants such as *Arabidopsis thaliana* exhibit inhibited vegetative growth, reduced photosynthetic activity, increased anthocyanin accumulation, and altered root architecture [5,6]. Conversely, high P levels can lead to phosphorus toxicity, characterized by accelerated metabolism, premature development of reproductive organs, suppressed stem and leaf growth, and early termination of vegetative growth [7,8]. Plants respond dynamically to fluctuations in external P availability through a tightly regulated network of metabolic pathways controlled by numerous phosphorus-responsive genes. The expression of these genes varies depending on environmental phosphorus conditions. Previous studies have elucidated the molecular responses of various plant species, including *maize* [9], *soybean* [10], *rice* [11], *wheat* [12], and *Myriophyllum aquaticum* [13], under phosphorus stress.

The phosphorus signaling regulatory network is complex and highly coordinated, involving multiple components, including the phosphate starvation response transcription factor (PHR) family, SPX-domain-containing proteins (SPX family), phosphate transporters (PHT family), microRNAs, and ubiquitination-related proteins [14]. This network plays a critical role in maintaining intracellular and inter-organellar phosphorus balance [15]. Among these regulators, PHR proteins serve as central transcriptional activators in the phosphorus starvation response, directly or indirectly controlling the expression of downstream phosphate starvation-induced (*PSI*) genes. SPX-domain-containing proteins function as phosphorus sensors, modulating the activity of PHRs in response to intracellular phosphorus concentrations [16]. In *rice*, for instance, *OsSPX1* and *OsSPX2* interact with PHRs in the nucleus via their SPX domains, thereby inhibiting PHR binding to the P1BS cis-element [15,17]. Furthermore, *OsSPX4* acts as a negative regulator of OsPHR1. Under phosphorus-deficient conditions, *OsSPX4* undergoes degradation through the 26S proteasome pathway, which releases OsPHR2 to the nucleus, subsequently activating the transcription of downstream *PSI* genes [18]. The PHT family plays a pivotal role in the uptake of inorganic phosphorus (Pi) from the soil and its distribution within plant tissues. PHRs can directly bind to P1BS elements in the promoter regions of PHT genes, thereby regulating their transcription and ultimately controlling phosphorus acquisition and transport in plants [19].

With the rapid development of urbanization, industrialization, and agricultural activities worldwide, eutrophication has become a widespread environmental problem affecting many rivers and lakes. Eutrophication occurs when excessive amounts of nutrients, particularly nitrogen and phosphorus, accumulate in water bodies, resulting in degraded water quality and ecological imbalance [20,21]. Among these nutrients, phosphorus plays a critical role, as its concentration strongly influences the growth, physiological performance, and ecological functions of submerged aquatic plants [13]. Over the past few decades, extensive research has focused on strategies to restore and manage eutrophic water bodies [22,23]. These restoration approaches generally fall into three categories: physical, chemical, and bio-ecological methods. Among these, the use of aquatic plants for ecological restoration has emerged as one of the most cost-effective, environmentally friendly, and sustainable solutions [24,25]. Submerged plants introduced into eutrophic water bodies can efficiently absorb large amounts of nitrogen, phosphorus, and other dissolved nutrients, thereby lowering nutrient concentrations and improving overall water quality [21,26]. Moreover, many macrophytes have the capacity to accumulate heavy metals, absorb and settle suspended particles, and suppress algal blooms. Through these processes, they help purify the water, enhance water transparency, and contribute to the stability and resilience of aquatic ecosystems [27,28].

Potamogeton wrightii, a perennial submerged plant belonging to the genus *Potamogeton* in the family *Potamogetonaceae*, is widely distributed across China, from northern to southern provinces, including Taiwan, Guangdong, Jiangsu, Anhui, Yunnan, Guizhou, and Sichuan. Its range also extends to parts of West Asia, northwestern and northeastern India, various Southeast Asian countries, and Japan [29,30]. *P. wrightii* holds both ecological and economic value. It has long been used for medicinal purposes, as bait for herbivorous fish, and as a high-quality feed source for pigs and ducks due to its high nutritional content. Its aerial leaves are relatively thick and rich in keratin and wax, with well-developed stomata, strong tolerance to intense light, and high photochemical efficiency [31]. Moreover, *P. wrightii* exhibits strong ecological adaptability, thriving in a wide range of aquatic environments from oligotrophic to eutrophic water bodies. Its roots, stems, and leaves all possess the ability to absorb phosphorus, and the plant plays a crucial role in regulating phosphorus forms in sediments, inhibiting endogenous phosphorus release, and competing with algae for nutrients. These attributes make it highly effective in purifying various types of effluents and valuable for the prevention and control of lake eutrophication [32–34].

Currently, research on the molecular mechanisms underlying phosphorus homeostasis has primarily focused on model plants and major food crops [13,35]. However, little is known about how submerged macrophytes, such as *P. wrightii*, respond to different phosphorus stress conditions at the molecular level. To address this knowledge gap, this study integrates transcriptomic and metabolomic analyses to investigate the differentially expressed genes (DEGs), key functional pathways, and metabolic changes involved in phosphorus homeostasis in *P. wrightii* under varying phosphorus concentrations. By examining changes in gene expression, metabolite profiles, and physiological and biochemical indicators, this study aims to reveal the molecular adaptation mechanisms that enable *P. wrightii* to cope with different levels of P stress.

2. Results

2.1. Effects of Different Phosphorus Concentrations on Plant Phenotypes and Physiological Indicators

2.1.1. Morphological Differences in Plant Leaves

After 72 h of phosphorus stress treatment, clear morphological changes were observed in the leaves of *P. wrightii* (Figure 1). Compared with the control group (CK), there were no significant differences in leaf morphology under low-phosphorus (LP) conditions. However, in the high-phosphorus (HP) treatment groups (P5, P20, and P40), the leaves exhibited noticeable yellowing. This chlorosis is likely due to HP-induced interference with normal photosynthetic processes, which may have resulted in a reduction in chlorophyll content and subsequent leaf discoloration.

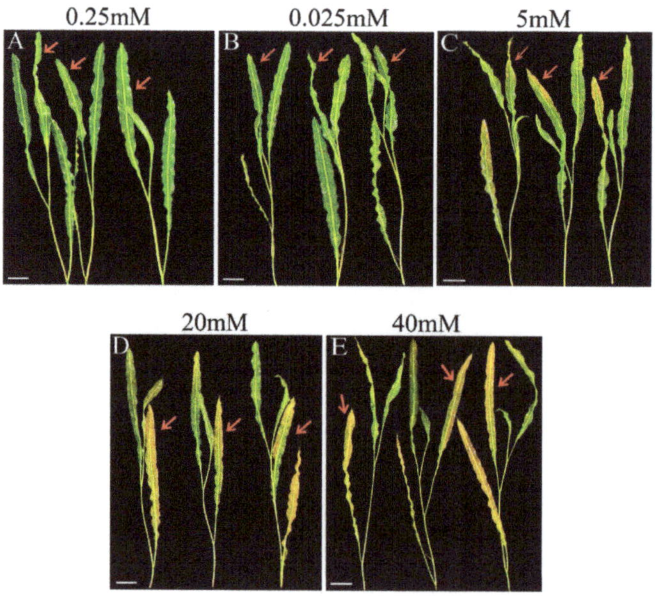

Figure 1. The effect of phosphorus stress on the leaf morphology of *P. wrightii*. Each picture includes three plants, representing three biological replicates respectively. The number at the top of the graph

represents the concentration of phosphorus treatment. The red arrows indicate the different shapes of the leaves of plants in different treatment groups. In (**C–E**), compared with CK, the leaves indicated by the red arrows are obviously yellowing. (**A**) CK (0.25 mM), (**B**) LP (0.025 mM), (**C**) P5 (5 mM), (**D**) P20 (20 mM), (**E**) P40 (40 mM). Bar = 3 cm.

2.1.2. Determination of Chlorophyll Content, Antioxidant Indices, and Inorganic Phosphorus Content

The chlorophyll content of *P. wrightii* was measured under different phosphorus treatments, and the results are presented in Figure 2A. Compared with the CK treatment group, the P20 and P40 treatment groups showed significant reductions in chlorophyll content, with the P40 treatment group exhibiting the greatest reduction. There was no significant difference between the LP and P5 treatment groups. These findings are consistent with the leaf morphological changes shown in Figure 1. In terms of oxidative stress, with the increase of phosphorus concentration, the contents of MDA, H_2O_2, and the activity of POD all showed an upward trend. Among them, the contents of MDA and the activity of POD in the P20 and P40 treatment groups were significantly enhanced. The content of H_2O_2 in the P40 treatment group was significantly increased, while the activities of SOD and CAT showed a trend of first decreasing, then increasing, and decreasing again. There was a significant improvement in both the LP and HP treatment groups. Compared with CK, there was no significant difference in proline content between the LP and P5 treatment groups; however, P20 and P40 showed significant increases. Inorganic phosphorus (Pi) measurements revealed that Pi content decreased under LP conditions but was significantly elevated in the P5 group. In contrast, no significant differences were detected between the P20, P40, and CK groups. Overall, these results indicate that phosphorus stress influences the physiological status of *P. wrightii* by reducing chlorophyll content, promoting MDA accumulation, and activating the antioxidant enzyme system. However, when phosphorus levels are excessively high, the activities of certain antioxidant enzymes decline, suggesting that high phosphorus stress has a more profound negative impact on plant physiology than low phosphorus stress (Figure 2).

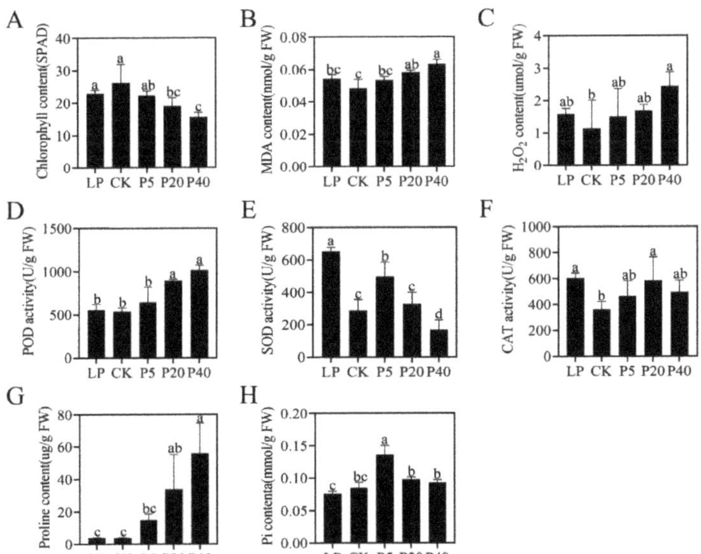

Figure 2. The effects of phosphorus stress on chlorophyll content (**A**), MDA content (**B**), H_2O_2 content (**C**), POD activity (**D**), SOD activity (**E**), CAT activity (**F**), proline content (**G**) and Pi content (**H**) in *P.*

wrightii. CK (0.25 mM), LP (0.025 mM), HP (5 mM, 20 mM, 40 mM). The different letters in the figure indicate significant differences among the various groups ($p < 0.05$), and the largest mean was marked with a. The meaning was compared with each of the following means, and those not significantly different were marked with a, until an average was found that was significantly different from it, marked b. Take the maximum average number of marks b as the standard and compare it with the following average number of non-marks, where b is not significant, and c is significant. Take the maximum average number of marks c as the criterion and compare it with the following average number of non-marks, where c is not significant, and d is significant. If two samples have the same letter, the difference is not significant.

2.2. RNA-Seq Analysis of *P. wrightii* Under Different Phosphorus Stress

2.2.1. Quality Assessment of RNA-Seq Data

To investigate the genetic mechanisms underlying *P. wrightii*'s diverse responses to LP and HP stress, RNA-seq was conducted on the plant leaves. After raw data filtering, we obtained 64.7 GB of clean bases with Q30 scores exceeding 95.85% and GC content ranging between 44.2–47.2% (Table S1). The RNA-seq data can be obtained through accession number PRJNA1335104 (https://www.ncbi.nlm.nih.gov/bioproject/PRJNA1335104 (accessed on 25 September 2025)). De novo assembly was used Trinity software and yielded a total of 149,312 Unigenes (Tables S2 and S3). rincipal component analysis (PCA) of the 15 RNA-seq samples showed that PC1 and PC2 accounted for 19.72% and 11.72% of the total variance, respectively (Figure S1A). Sample correlation heatmap analysis demonstrated high biological reproducibility among replicates (Figure S1B). These results confirm the high quality and reliability of our sequencing data for subsequent bioinformatics analyses.

2.2.2. Identification of Differentially Expressed Genes

Differentially expressed genes (DEGs) were identified by comparing RNA-seq across treatments. A total of 1006 common DEGs were detected across all four comparisons: LP vs. CK, P5 vs. CK, P20 vs. CK, and P40 vs. CK (Figure 3A). Among LP vs. CK, P5 vs. CK, P20 vs. CK, and P40 vs. CK, 5639 (3225 up-regulated and 2414 down-regulated), 7493 (5181 up-regulated and 2312 down-regulated), 6111 (3556 up-regulated and 2555 down-regulated), and 10,693 (5881 up-regulated and 4812 down-regulated) DEGs (Figure 3B). As shown in Figure 3B, the number of DEGs increased markedly with rising phosphorus concentrations. Under HP stress, the number of DEGs was substantially higher than under LP stress. Notably, the number of DEGs in the P40 vs. CK comparison was 1.9 times greater than that in the LP vs. CK comparison. These results suggest that HP stress triggers more extensive transcriptional reprogramming, indicating that these DEGs may play critical roles in the response of *P. wrightii* to varying phosphorus levels.

To further investigate the potential functions of DEGs in the stress response of *P. wrightii*, the Gene Ontology (GO) functional enrichment analysis was performed. The results showed that under different phosphorus treatments, the majority of DEGs were enriched in the biological process (BP) category, primarily involving cellular processes and metabolic processes. In the molecular function (MF) category, DEGs were mainly associated with binding and catalytic activity, while in the cellular component (CC) category, DEGs were enriched in cellular structural components and protein complexes (Figure S2). Although the overall proportion of each GO category remained relatively stable across treatments, the number of DEGs within each category varied considerably, indicating that phosphorus stress induced distinct transcriptional changes.

To identify metabolic pathways affected by phosphorus stress, DEGs were mapped to the Kyoto Encyclopedia of Genes and Genomes (KEGG) database. The top 20 significantly enriched pathways are shown in Figure S3. The metabolic pathways with the most enrichment of DEGs in LP vs. CK are Starch and sucrose metabolism, Nitrogen metabolism, Glycolysis/Gluconeogenesis, Phenylpropanoid biosynthesis, alpha-linolenic acid metabolism, and Photosynthesis antenna proteins. The metabolic pathways with more DEGs enrichment in P5 vs. CK are Photosynthesis, Photosynthesis antenna proteins, Phenylpropanoid biosynthesis, Nitrogen metabolism, Carotenoid biosynthesis, and Plant-pathogen interaction. The metabolic pathways with the most enrichment of DEGs in P20 vs. CK are Phenylpropanoid biosynthesis, Plant-pathogen interaction, Carotenoid biosynthesis, Starch and sucrose metabolism, Porphyrin metabolism, and Flavonoid biosynthesis. The metabolic pathways with the most enrichment of DEGs in P40 vs. CK are Phenylpropanoid biosynthesis, Plant-pathogen interaction, Carotenoid biosynthesis, Starch and sucrose metabolism, photosynthetic antenna proteins, and Nitrogen metabolism. These results suggest that both LP and HP stresses affect key metabolic and defense-related pathways in *P. wrightii*. In particular, pathways related to carbohydrate metabolism, secondary metabolite biosynthesis, photosynthesis, and plant defense responses were strongly influenced, highlighting their potential roles in maintaining phosphorus homeostasis and plant adaptation under stress conditions.

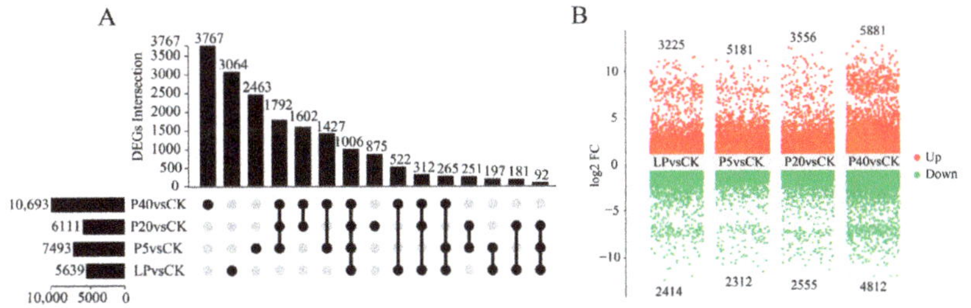

Figure 3. (**A**) UpSet plot of DEGs. The lines connecting the black dots in the Upset plot represent the intersections between sets. The columns and numbers on the left represent the quantity of DEGs contained in each of the four comparison groups. The columns and numbers above show the number of intersection DEGs with the corresponding set below; (**B**) DEGs Volcano map. The numbers in the figure represent the quantity of DEGs, namely CK (0.25 mM), LP (0.025 mM), P5 (5 mM), P20 (20 mM), and P40 (40 mM). From left to right, they are the four comparison groups: LP vs. CK, P5 vs. CK, P20 vs. CK, and P40 vs. CK. The UP (red dot) and Down (green dot) respectively represent the upward and downward adjustment of DEGs, and Log2FC is the corresponding Log2FoldChange value.

2.2.3. Expression of DEGs in P. wrightii Under Different Phosphorus Stress Conditions

Through bioinformatics analysis, the *P. wrightii* PHT family and SPX gene family were identified to contain 12 and 9 members, respectively. Gene expression levels under different phosphorus treatments are shown in Figure S4A,C. The four treatment groups were respectively compared with the control group. As long as there was a significant expression difference in any of the comparisons of genes, they were defined as DEGs. Otherwise, it is defined as non-DEGs. Based on this criterion, the PHT family included 4 DEGs (2 PHT1 and 2 PHT4), and the SPX family included 5 DEGs (1 SPX, 1 PHO1, 2 SPX-MFS, and 1 NLA). The expression profiles of these 9 DEGs under different phosphorus concentrations are shown in Figure 4. In the PHT family, Cluster-53058.0 and Cluster-75927.0 were upregulated, and Cluster-49140.1 and Cluster-63611.11 were downregulated in LP compared to CK. In P5 vs. CK, P20 vs. CK, and P40 vs. CK, Cluster-49140.1, Cluster-53058.0, and Cluster-63611.11 were consistently downregulated, while Cluster-75927.0 was upregulated. In the SPX family, Cluster-33141.7, Cluster-63729.0, and Cluster-62070.9 were upregulated, whereas Cluster-75038.0 was downregulated in LP vs. CK. For P5 vs. CK, P20 vs. CK, and P40 vs. CK, Cluster-67516.3 and Cluster-75038.0 were upregulated. In addition, a total of 15 members of the PAP (purple acid phosphatase) gene family were identified, including 5 DEGs. Among them, Cluster-45204.0, Cluster-6805.3, Cluster-55786.0, and

Cluster-74442.5 were all upregulated in the LP and HP treatment groups, while Cluster-54031.1 was all down-regulated (Figure S4B). All DEGs were predicted and analyzed, and a total of 1169 transcription factors were identified, including 55 families. Among them, WRKY had the most significant number, followed by NAC, AP2/ERF->AP2/ERF-ERF, bZIP, bHLH, C2H2, MYB, and other families (Figure S5). Further analysis of the PHR1 transcription factor family of *P. wrightii* revealed that there was a total of 5 PHR1 and homologous transcription factor PHR-Like members. Among them, two genes were significantly expressed, and Cluster-60971.0 and Cluster-69197.0 were upregulated in LP compared to CK. However, the expressions of P5 vs. CK, P20 vs. CK, and P40 vs. CK were downregulated (Figure S4D). These results indicate that LP and HP stress affect the expression of different gene families in *P. wrightii*, suggesting their roles in multiple aspects of phosphorus stress response.

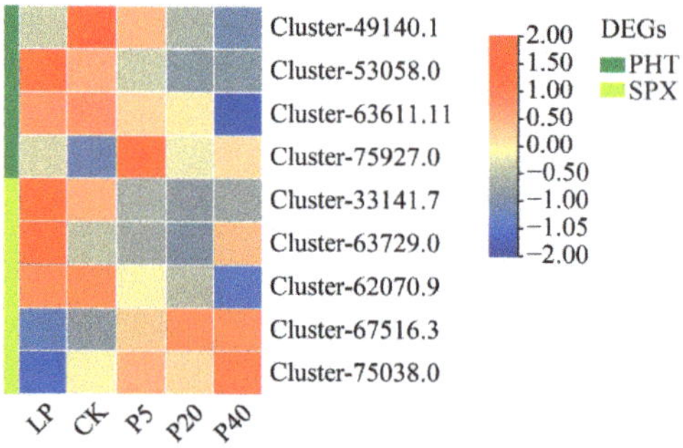

Figure 4. Heat map of DEGs related to phosphorus response. The red and blue boxes represent genes that are upregulated and downregulated under different phosphorus stress conditions. The green and yellow on the left side of the heat map represent the gene grades of the *P. wrighii* PHT family and the SPX gene family respectively. CK (0.25 mM), LP (0.025 mM), P5 (5 mM), P20 (20 mM), and P40 (40 mM). The color scale represents the gene expression levels. During the heat map drawing process, normalization processing is usually carried out, so the legend data is displayed as symmetrically distributed around 0, with red representing high expression and blue representing low expression.

2.2.4. RT-qPCR Analysis Results

To validate the RNA-seq results, 9 DEGs related to phosphorus stress response were selected for quantitative real-time PCR (RT-qPCR) analysis. The RT-qPCR results showed that the expression patterns of these genes were consistent with the FPKM values obtained from RNA-seq, confirming the reliability and accuracy of the RNA-seq data (Figure 5).

2.3. *Metabolome Analysis of P. wrightii Under Different Phosphorus Stress*

2.3.1. Qualitative and Quantitative Analysis of Etabolites

The leaves of *P. wrightii* were analyzed using the UHPLC-QTOF-MS/MS system, and total ion current (TIC) diagrams of quality control (QC) samples showed highly overlapping curves, indicating high instrument stability and data reliability (Figure S6). A total of 3408 metabolites were detected, which were classified into 21 categories, including 938 amino acids and derivatives, 510 organic acids, 330 benzene and substituted derivatives, 167 alkaloids, 157 flavonoids, 114 glycerophospholipids, 99 lipids, and 405 other compounds (Figure 6A). Principal component analysis (PCA) further revealed the overall metabolic differences among the treatment groups, with PC1 and PC2 contributing 22.46% and 11.44% of the variance, respectively, indicating significant metabolic variation under different phosphorus stress conditions (Figure 6B).

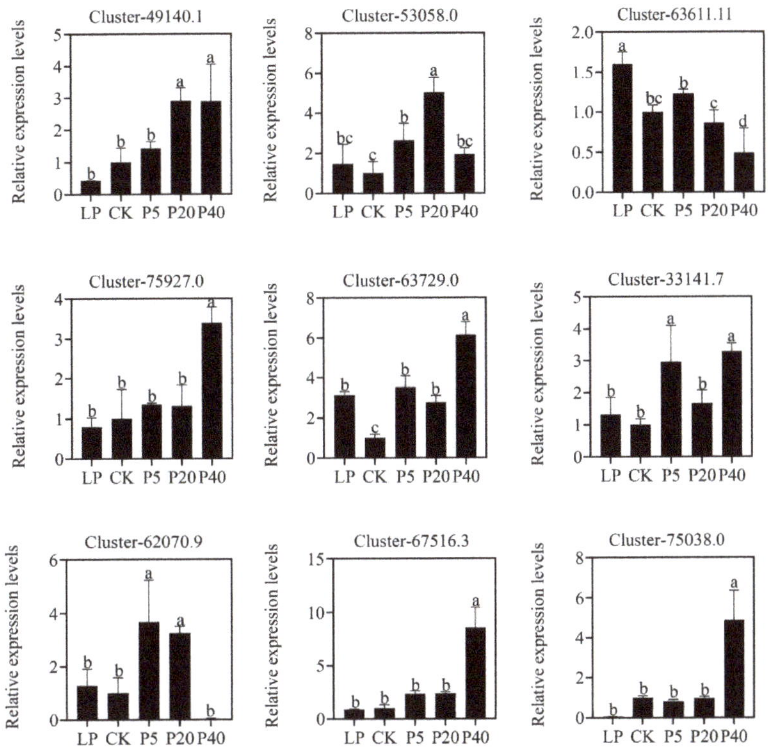

Figure 5. The expression of DEGs in the *P. wrightii* PHT family and SPX family genes under phosphorus stress was analyzed by RT-qPCR. The number directly above each figure is the ID number of DEGs. The standard errors of three biological replications and three technical replications are represented by error bars. CK (0.25 mM), LP (0.025 mM), P5 (5 mM), P20 (20 mM), and P40 (40 mM). The different letters in the figure indicate significant differences among the various groups ($p < 0.05$), and the largest mean was marked with a. The meaning was compared with each of the following means, and those not significantly different were marked with a, until an average was found that was significantly different from it, marked b. Take the maximum average number of marks b as the standard and compare it with the following average number of non-marks, where b is not significant, and c is significant. Take the maximum average number of marks c as the criterion and compare it with the following average number of non-marks, where c is not significant, and d is significant. If two samples have the same letter, the difference is not significant.

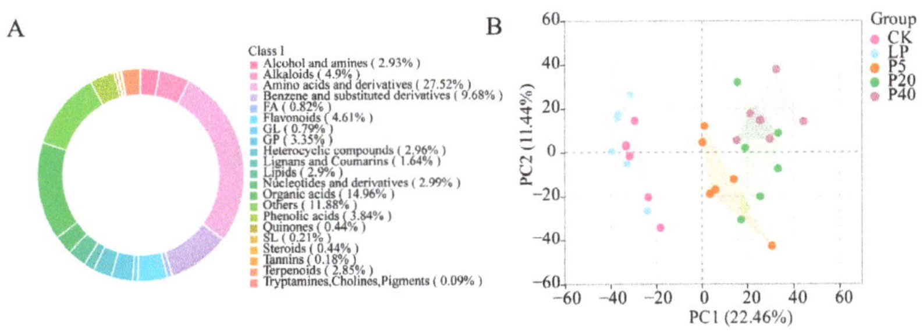

Figure 6. (**A**) Classification diagrams of metabolites identified by *P. wrightii* for different phosphorus treatments. Each color represents a category of metabolites, the area of the color block indicates the proportion of that category, and Class I represents the first-level classification of substances. (**B**) In principal component analysis (PCA), the first principal component (PC1) is represented by the *X*-axis,

and the second principal component (PC2) is represented by the *Y*-axis. Groups represent different treatment groups, namely CK (0.25 mM), LP (0.025 mM), P5 (5 mM), P20 (20 mM), and P40 (40 mM).

2.3.2. Analysis of Differentially Expressed Metabolites Among Different Treatments

To identify metabolites that enable *P. wrightii* to adapt to phosphorus stress, differential accumulation of metabolites (DAMs) was analyzed across the Four control groups, revealing a total of 1678 DAMs (Table S4). Volcano plot analysis showed 534 DAMs (121 up-regulated and 413 down-regulated) in LP vs. CK, 450 DAMs (269 up-regulated and 181 down-regulated) in P5 vs. CK, 842 DAMs (409 up-regulated and 433 down-regulated) in P20 vs. CK, and 926 DAMs (327 up-regulated and 599 down-regulated) in P40 vs. CK (Figure S7). KEGG pathway enrichment of these DAMs ($p \leq 0.05$) highlighted several key metabolic pathways, including general metabolic pathways, biosynthesis of secondary metabolites, arachidonic acid metabolism, pentose phosphate pathway, purine metabolism, glycolysis/gluconeogenesis, glycerolipid metabolism, and glycerophospholipid metabolism (Figure S8). Detailed analysis revealed that most Glycerol phospholipids (GP) contents decreased under high phosphorus (HP) stress, particularly in P40 (Figure 7), whereas flavonoid compounds such as 7-methoxyisoflavone, licoricone, irigenin, dibenzyl ether, and isobavachalcone significantly increased under HP stress. In contrast, eriodyctiol-5,3′-di-O-rutinoside, aloeresin A, and pachyrrhizone increased under low phosphorus (LP) stress (Tables S5 and S6). Given their potent antioxidant properties, flavonoids likely help *P. wrightii* eliminate free radicals and mitigate oxidative damage, suggesting that the differential accumulation of these metabolites under varying phosphorus conditions contributes to the plant's enhanced adaptability and maintenance of physiological homeostasis.

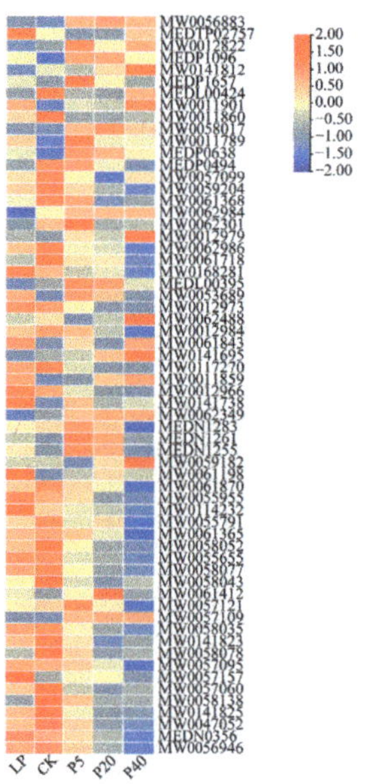

Figure 7. Accumulated heat maps of *P. wrightii* GP under different phosphorus stresses. CK (0.25 mM), LP (0.025 mM), P5 (5 mM), P20 (20 mM), and P40 (40 mM). The red and blue boxes represent DAMs that are up-regulated and down-regulated under different phosphorus stress conditions. The color scale represents the accumulation level of metabolites compared to the color scale. During the heat map drawing process, normalization processing is usually carried out, so the legend data is displayed as symmetrically distributed around 0, with red representing high expression and blue representing low expression.

2.4. Combined Transcriptomic and Metabolomic Analysis of the Response Mechanism of P. wrightii to Different Phosphorus Levels

To clarify the phosphorus stress response mechanism of *P. wrightii*, we integrated transcriptomic and metabolomics data. Through KEGG pathway enrichment analysis (Figure S9), we screened out the key pathways co-enriched by DEGs and DAMs, namely Glycerolipid metabolism and Glycerophospholipid metabolism. It is worth noting that the expression trends of some DAMs are not completely consistent with those of related DEGs, which reflects the complex post-transcriptional regulation and metabolic modification between gene transcription and metabolite generation. To further explore the correlation between the two, based on the KEGG database annotations, we identified DEGs encoding the key enzymes in these metabolic pathways and co-mapped them with DAMs to the metabolic pathways, thus constructing a gene–metabolite regulatory network.

In each difference group, genes and metabolites that meet the absolute value of the Pearson correlation coefficient greater than 0.8 and the p-value less than 0.05 are screened and displayed in the nine-quadrant graph (Figure S10). The gene–metabolite interaction network can help understand functional relationships and identify new regulatory elements. The DAMs and DEGs in the Glycerolipid metabolism and Glycerophospholipid metabolism pathways selected and screened were used to draw the gene–metabolite interaction network (Figure S11). No correlation relationship with an absolute value of the Pearson correlation coefficient greater than 0.8 and a p-value less than 0.05 was found between DEGs and DAMs in LP vs. CK. The expression of most DEGs in P5 vs. CK (such as Cluster-37182.0, Cluster-39532.3, Cluster-64735.3, etc.) is similar to that of Citicoline (MEDP1592) and Colfosceril palmitate (MW0056883), 1-Palmitoyl-2-linoleoyl-sn-glycero-3-phosphate (MW0012966) and PC (18:3(6Z,9Z,12Z)/P-16:0) (MW0057109) are related. The four key metabolites were significantly positively correlated with the expression of most DEGs (such as Cluster-39532.3, Cluster-37182.0, Cluster-64735.3, etc.). It was negatively correlated with the expression of most DEGs (such as Cluster-78097.9, Cluster-77096.14 and Cluster-12638.14). In P20 vs. CK, the expression of most DEGs (such as Cluster-61121.0, Cluster-42869.12, Cluster-62070.9, etc.) and Colfosceril palmitate (MW0056883) is associated with 1-Palmitoyl-2-linoleoyl-sn-glycero-3-phosphate (MW0012966). Two key metabolites were significantly positively correlated with the expression of most DEGs (such as Cluster-62070.9, Cluster-37182.0, Cluster-31026.5, etc.). It is negatively correlated with the expression of most DEGs (such as Cluster-72446.3, Cluster-67702.3 and Cluster-67516.3). On P40vs. The expression of most DEGs in CK (such as Cluster-62241.0, Cluster-51434.5, Cluster-77293.7, etc.) is similar to that of Choline (MEDP0125), Diethanolamine (MEDP0128), and PC(18:3, 6Z,9 Related to Z,12Z)/P-16:0 (MW0057109), Colfosceril palmitate (MW0056883), and 1-Palmitoyl-2-linoleoyl-sn-glycero-3-phosphate (MW0012966) The five key metabolites were significantly positively correlated with the expression of most DEGs (such as Cluster-51434.5, Cluster-62070.9, Cluster-37182.0, etc.). It was negatively correlated with the expression of most DEGs (such as Cluster-67702.3, Cluster-62241.0 and Cluster-67516.3). It is worth noting that, as can be found from Figure 5, the Cluster-67516.3 and Cluster-62070.9 genes involved in the Glycerolipid metabolism pathway are members of the SPX domain gene family.

To comprehensively evaluate the potential molecular mechanisms of *P. wrightii* under phosphorus stress, we plotted the Glycerolipid metabolism and Glycerophospholipid metabolism pathways by analyzing the KEGG pathways of DEGs and DAMs (Figure 7). We noticed that Citicoline accumulated similarities in the four control groups. Colfosceril palmitate, Diethanolamine, and PC (18:3(6Z,9Z,12Z)/P-16:0) accumulated increased in P5 vs. CK, P20 vs. CK, and P40 vs. CK, while accumulated decreased in LP vs. CK. The accumulation of Choline and 1-Palmitoyl-2-linoleoyl-sn-glycero-3-phosphate increased in

LP vs. CK, while the accumulation decreased in P5 vs. CK, P20 vs. CK, and P40 vs. CK. In addition, compared with LP vs. CK, more DEGs and DAMs in P5 vs. CK, P20 vs. CK and P40 vs. CK were significantly upregulated. The above results indicate that the expression of specific genes and the accumulation of metabolites under the same metabolic pathway, as well as the regulation of specific metabolic pathways, lead to the differences in *P. wrightii*'s responses to low phosphorus and high phosphorus stress (Figure 8).

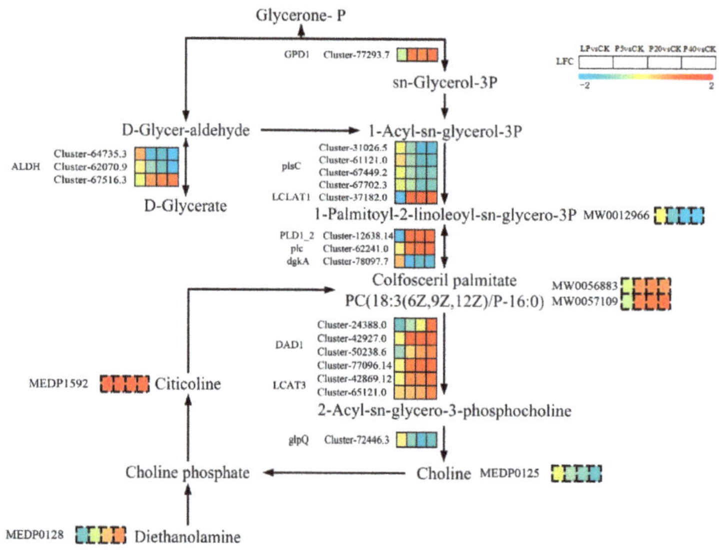

Figure 8. Response mechanisms of the *P. wrightii* Glycerolipid metabolism and Glycerophospholipid metabolism pathways under phosphorus stress. DEGs and DAMs are respectively represented by solid and dashed boxes in the path, with the boxes from left to right being LP vs. CK, P5 vs. CK, P20 vs. CK, and P40 vs. CK, and LFC being the corresponding Log2FoldChange value. Red and blue, respectively, represent the upward and downward adjustments of DEGs and DAMs. CK (0.25 mM), LP (0.025 mM), P5 (5 mM), P20 (20 mM), and P40 (40 mM). plsC:1-acyl-sn-glycerol-3-phosphate acyltransferase [EC:2.3.1.51]; GPD1:glycerol-3-phosphate dehydrogenase (NAD+) [EC:1.1.1.8]; LCLAT1:lysocardiolipin and lysophospholipid acyltransferase [EC:2.3.1.- 2.3.1.51]; PLD1-2:phospholipase D1/2 [EC:3.1.4.4]; plc:phospholipase C [EC:3.1.4.3]; dgkA:diacylglycerol kinase (ATP) [EC:2.7.1.107]; DAD1:phospholipase A1 [EC:3.1.1.32]; LCAT3:phospholipase A1 [EC:3.1.1.32]; qlpQ:glycerophosphoryl diester phosphodiesterase [EC:3.1.4.46]; ALDH:aldehyde dehydrogenase (NAD+) [EC:1.2.1.3].

3. Discussion

Research on *P. rightii*'s phosphorus removal and absorption has been reported [36,37], but the molecular mechanism of its response to phosphorus stress remains unknown. The morphological manifestations of plants vary with different phosphorus treatments. Low phosphorus stress inhibits the photosynthetic activity of plants, resulting in a slower metabolism [38]. High phosphorus stress leads to the loss of leaf greenness and phosphorus poisoning in plants [7], thereby affecting plant growth and development, which is consistent with the results of this study. Under high phosphorus stress, compared with the control group, the leaves of *P. wrightii* turned yellow and the chlorophyll content decreased significantly, while there was no significant difference under low phosphorus stress. Reactive oxygen species (ROS) are key signaling molecules that enable cells to respond rapidly to various stimuli [39]. This study found that under high phosphorus stress (P40) conditions, the contents of MDA and H_2O_2 in plant leaves accumulated significantly. This result clearly indicates that phosphorus stress disrupts the dynamic balance between the production and clearance of reactive oxygen species (ROS) within plants, leading to a significant accumulation of ROS and subsequently triggering severe oxidative stress, which damages the structural and functional integrity of cell membranes. To cope with oxidative stress, plants activate their defense systems through the use of antioxidant enzymes. In this experiment, the enzymatic activity of POD increased with the rise in phosphorus concentration. In contrast, the dynamic changes in SOD and CAT activities were more complex, exhibiting a trend of first decreasing, then increasing, and then decreasing again with increasing phosphorus concentration. This result reflects the different antioxidant strategies that plants may adopt under different degrees of phosphorus stress to maintain cellular homeostasis. Under high phosphorus stress, plants accumulate a large amount of proline. Proline accumulation usually enhances the tolerance to osmotic stress, maintains the water absorption capacity of cells, and thereby alleviates osmotic dysregulation caused by stress [40]. The results of inorganic phosphorus detection in plant leaves showed that the inorganic phosphorus content in the LP treatment group decreased, while that in the P5 group significantly increased, indicating that LP stress limited phosphorus absorption. In contrast, moderate phosphorus increase (P5) promoted phosphorus accumulation. However, there was no difference between the P20 and P40 groups and CK. Considering their growth inhibition and changes in the antioxidant system, it is speculated that high phosphorus may cause metabolic disorders, leading to a decrease in phosphorus absorption and utilization efficiency. Overall, the negative impact of HP stress on plants is more severe than that of LP stress, as evidenced by a reduction in chlorophyll content, the induction of reactive oxygen species accumulation, and membrane lipid peroxidation, ultimately leading to oxidative damage in cells. In response, plants enhance their ROS clearance capacity by increasing the activity of certain antioxidant enzymes and accumulating substances such as proline to stabilize the intracellular environment, thereby forming a complex defense network against phosphorus stress.

Maintaining the homeostasis of phosphorus within plant cells is an important biological mechanism, and phosphate transporters (PHT) play a key role in Pi acquisition and in vivo balance [41]. Based on protein sequences, structural features, subcellular localization and functions, the PHT family is divided into four subfamilies: PHT1 (Sugar (and other) transporter), PHT2 (Phosphate transporter), PHT3 (Mitochndrial carrier pretein), and PHT4 (Major Facility (Superfamily) [42]. In model plants such as *Arabidopsis thaliana* and *Oryza sativa*, most high-affinity *PHT* genes are predominantly expressed in roots, although some can also be induced in leaves under Pi starvation [43,44]. In *O. sativa*, *OsPHT2* and *OsPHT6* are primarily expressed in roots, but their expression increases in leaves during phosphorus (Pi) deficiency [45]. Based on phylogenetic, sequence, and functional analyses, the *PHT* gene family of *P. wrightii* was identified, and the expression of its differentially expressed genes (DEGs) under low phosphorus (LP) and high phosphorus (HP) stress was analyzed. Some members of the PHT1 and PHT4 subfamilies were upregulated under LP stress and downregulated under HP stress, whereas other members exhibited the opposite trend. The SPX domain encoding protein plays a significant role in Pi homeostasis and signaling pathways. Proteins containing the SPX domain are classified into four families based on whether other domains exist in their structure: SPX (SPX-Single domain), SPX-MFS (SPX-Major Facility Superfamily) (also known as VPTs or PHT5) SPX-EXS (SPX-ERD1/XPR1/SYG1) (also known as PHO1) and SPX-RING (SPX-really Interesting New Gene) (also called NLA) [46]. In *A. thaliana* and *O. sativa*, most *SPX* genes are Pi-responsive, except *OsSPX4* and AtSPX4 [47]. *NLA*, *OsSPX-MFS1*, and *OsSPX-MFS3* are repressed under Pi deficiency, while *OsSPX-MFS2* is induced [46,48]. In *P. wrightii*, members of the SPX and PHO1 subfamilies were upregulated under LP treatment and downregulated under HP treatment, while NLA subfamily members showed downregulation under LP and upregulation under HP. The two DEGs of SPX-MFS members exhibited opposite expression patterns, consistent with observations in *O. sativa*. PAP not only release phosphorus from the soil but also regulate its distribution throughout plant growth [49]. In this study, five PAP family DEGs were identified, of which four were upregulated under both LP and HP treatments, while one was downregulated, suggesting potential functional diversification. AtPHR1 is a MYB transcription factor that plays a key role in Pi starvation signaling [50]. PHR1 affects the expression of PSI genes, including key components of Pi signaling such as SPX, PHT, PAP, miRNA399s, and IPS1 [51]. Two PHR-like genes in *P. wrightii* were upregulated under LP stress but downregulated under HP stress, indicating their potential role in regulating SPX, *PHT*, and *PAP* to maintain phosphorus homeostasis. Collectively, these genes play key roles in the plant's adaptation to varying phosphorus conditions, facilitating phosphorus uptake, transport, and physiological balance.

The cell membrane is mainly composed of lipids, proteins and carbohydrates, among which the phospholipid bilayer is its basic framework. The growth, development and physiological and biochemical functions of plants can affect the structure and composition of membrane lipids [52]. Phosphorus stress induces the expression of plant phospholipases (such as PLC and PLD), accelerating Phosphatidylcholine (PC) and Phosphatidylethanolamine (PE) is hydrolyzed into Phosphatidate (PA) and diacylglycerol (DAG), and then DAG is used as the substrate. Under the catalysis of enzyme transformation sulfoquinovosyldiacylglycerol (SQDG) [53]. Pi reduces the relative amount of total phosphatide, hunger and added some of phospholipids, especially monogalactosyldiacylglycerol (MGDG), digalactosyldiacylglycerol (DGDG) and (SQDG) [54]. Ding et al. [55] determined that in the presence of both phosphorus deficiency and excess, the synthesis of phosphorylated metabolites was reduced. *P. wrightii* responds to changes in phosphorus nutritional status by precisely regulating the turnover of membrane phospholipids. We observed that under low phosphorus stress, the expression of phospholipase-coding genes (such as PLD1-2, plc, DAD1, LCAT3) was down-regulated, and the corresponding degradation products The accumulation levels of Colfosceril palmitate and PC (18:3(6Z,9Z,12Z)/P-16:0) also decreased simultaneously. On the contrary, under high-phosphorus conditions, the expression of these genes and metabolites is upregulated. This synergistic change in the expression of genes and metabolites indicates that *P. wrightii* may maintain the integrity of the cell membrane structure by inhibiting the degradation of membrane phospholipids under low phosphorus stress. Under high phosphorus stress, the degradation pathway is activated to release inorganic phosphorus for other life activities. In addition, PC can be hydrolyzed by DAD1 and LCAT3 to form lysophospholipids. glpQ, as an important hydrolase in the phospholipid metabolic pathway, is responsible for catalyzing the hydrolysis of glycerophosphate diesters to generate glycerophosphate and corresponding alcohols (such as choline), which are used for the next round of phospholipid synthesis (such as PC). In this study, the expression of glpQ was upregulated under low phosphorus stress and downregulated under high phosphorus stress. Therefore, glpQ is involved in *P. wrightii*'s adaptation to phosphorus stress and plant growth and development. SQDG is widely present in plant chloroplasts, participates in the function and evolution of photosynthetic membranes, and is of great significance to plant photosynthesis [56]. However, there was no significant difference in the content of SQDG in *P. wrightii* under different phosphorus stresses. Therefore, SQDG may not play a major role in the photosynthesis, growth or development of *P. wrightii*, which is basically consistent with the research of Yu et al. And wang et al. [8,57]. It is worth noting that in our study, it was found that some genes encoding the ALDH enzyme are members of the SPX domain gene family. This gene may play a key role in regulating the Glycerolipid metabolism and Glycerophospholipid metabolism pathways. In particular, it affects the activity of related enzymes by responding to the homeostasis of phosphorus within cells. The above-mentioned genes require further functional verification.

4. Materials and Methods

4.1. Materials and Sample Preparation

P. wrighti seedlings were collected from the Huaxi River, Guiyang City, Guizhou Province, China (106°27′ E, 26°11′ N) and propagated under laboratory conditions. The plants were cultivated in a controlled light incubator at 24 ± 2 °C with a 12-h light/12-h dark photoperiod (Figure S12). The phosphorus stress test selected newly propagated plants after two weeks of culture, with a plant length of approximately 35 to 40 cm. For the experiment, seedlings were placed in transparent columns (15 cm in diameter and 50 cm in height), each containing 6.5 L of either the standard 10% solution or a phosphorus-deficient Hoagland nutrient solutions (Hunan Hoagland Biotechnology Co., Ltd., Xiangtan, China). Six seedlings were placed in each column. After a 7—day acclimation period, phosphorus treatments were applied using solutions with different phosphorus concentrations: CK (control, 0.25 mM), LP (low phosphorus, 0.025 mM), and HP (high phosphorus, 5 mM, 20 mM, and 40 mM), with the pH adjusted to 6.0. Each treatment included three independent biological replicates. The nutrient solution was refreshed once during the treatment period. After 72 h of exposure, plants were harvested and rinsed thoroughly three times with sterile water to remove residual nutrient solutions. The leaves were then immediately frozen in liquid nitrogen and stored at −80 °C for subsequent analyses.

4.2. Determination of Chlorophyll, Antioxidant Indicators, and Phosphorus Content in Plant Leaves

Chlorophyll content was measured using a SPAD-502 Plus chlorophyll meter (Konica Minolta, Tokyo, Japan). The activities of antioxidant enzymes in plant leaves, including superoxide dismutase (SOD), peroxidase (POD), and catalase (CAT), as well as the content of malondialdehyde (MDA), Hydrogen Peroxide (H_2O_2), Proline (Pro), inorganic phosphorus content (Pi) were determined using commercially available kits (Sod No: BC0170, Pod No:BC0090, Cat No:BC0200, Mda No:BC0025, H_2O_2 No:BC3590, Pro No:BC0290, and Pi No:BC2840; Solarbio Science & Technology Co., Ltd., Beijing, China) following the manufacturer's instructions and previous reports [58–62]. Each treatment group included three independent biological replicates.

4.3. Leaf RNA-Seq and Data Analysis

A reference genome-free RNA-seq strategy was adopted [13]. RNA extraction, quality assessment, library construction, and high-throughput sequencing were performed by Metware Biotechnology Co., Ltd. (Wuhan, China), with three independent biological replicates per treatment group. Sequencing was conducted on the Illumina NovaSeq^{TM} X Plus platform (Illumina, CA, USA) using the sequencing-by-synthesis method.

The original data was filtered using fastp 0.23.2 [63], the clean reads were assembled using Trinity v2.15.1 [64], and the CDS of the assembled transcripts was predicted using TransDecoder v5.3.0 [65]. Obtain the amino acid sequence corresponding to the transcript. The Unigene sequence was compared with the KEGG, NR, Pfam, Swiss-Prot, GO, COG/KOG and Trembl databases using DIAMOND v2.0. [66] After predicting the amino acid sequence of Unigene, the HMMER 3.2 [67] package software was used to compare it with the Pfam database to obtain the annotation information of Unigene. The parameters of the seven major databases are shown in Table S7. The expression levels of transcripts were calculated using RSEM v1.3.1 software [68], and then the FPKM of each transcript was calculated based on its length. According to the expression levels of each sample, DESeq2 v1.22.2 software [69,70] was used to conduct differential analysis of gene expression. Based on the corrected P-values, If the log2Fold Change is greater than or equal to 1 and the FDR is less than 0.05, the genes are screened as significantly differentially expressed genes (DEGs). Upset [71] of DEGs analysis figure and volcanic figure drawing through online website https://cnsknowall.com/#/Home/Contain/BottomContainAll (accessed on 7 November 2025) and https://www.bioinformatics.com.cn/ (accessed on 6 June 2025). The heat map was drawn after normalizing the DEGs data using TBtools-II v2.136. software [72]. The software parameters are shown in Table S8.

4.4. Leaf Metabolome and Data Analysis

Metabolome analysis was conducted by Metware Biotechnology Co., Ltd. (Wuhan, China), with six independent biological replicates in each group. The analysis was conducted using a TripleTOF 6600+ mass spectrometer (SCIEX, Foster City, CA, USA) and an LC-30A ultra-high performance liquid chromatograph (Shimadzu, Kyoto, Japan), namely an ultra-high performance liquid chromatography-tandem mass spectrometry (UPLC-MS/MS) system. The raw data were converted to mzXML format using ProteoWizard [73]. Peak extraction, alignment, and retention time correction were carried out using the XCMS program. Principal component analysis (PCA) was conducted with the prcomp function in R v4.1.2 [74,75] to assess the overall variation and grouping of samples. The orthogonal partial least squares discriminant analysis (OPLS-DA) model was applied to identify key metabolites, and the variable importance in projection (VIP) score was calculated to evaluate the contribution of each metabolite to sample classification. Differentially accumulated metabolites (DAMs) were screened based on the criteria of VIP > 1 and $|\log_2 \text{fold change}| \geq 1$. Identified metabolites were functionally annotated using the Kyoto Encyclopedia of Genes and Genomes (KEGG) database [76], and then mapped to KEGG pathways to explore metabolic changes associated with phosphorus stress. The heat map was drawn after normalizing the DAMs data using TBtools-II v2.136 software.

4.5. Combined RNA-Seq and Metabolome Analyses

The Pearson correlation coefficient between genes and metabolites was calculated using the cor function of R v2.2.5. Use Rv3.3.6 and R v1.3.4 to draw the nine-quadrant graph and network graph for correlation analysis. Based on the Kyoto Encyclopedia of Genes and Genomes (KEGG) database (https://www.genome.jp/kegg/ (accessed on 11 November 2025)), correlation analysis was conducted between differentially expressed genes (DEGs) and differentially accumulated metabolites (DAMs) within the same metabolic pathways. The correlated DEGs and DAMs were then mapped to KEGG pathway charts to reveal their functional relationships. Final visualizations were created using Microsoft Excel 2021 and Microsoft PowerPoint 2021 (Microsoft, Redmond, WA, USA).

4.6. Analysis of the Expression Characteristics of Key Genes in the Phosphorus Pathway by RT-qPCR

To verify the reliability of RNA-seq data, nine genes related to phosphate transporters were selected from DEGs for RT-qPCR analysis. The primers are listed in Table S9. The relative expression levels of each gene were calculated using the $2^{-\Delta\Delta Ct}$ method [77].

5. Conclusions

This study elucidated the physiological and molecular regulatory mechanisms underlying the response of *P. wrightii* to phosphorus stress. Under HP stress, plants exhibited visible leaf yellowing and a marked decrease in chlorophyll content, while no significant changes were observed under LP stress. In terms of oxidative stress, phosphorus stress induces the accumulation of reactive oxygen species, as evidenced by a significant increase in the contents of MDA and H_2O_2, accompanied by the activation of the antioxidant defense system. The activities of POD, SOD, and CAT all increased, with SOD and CAT displaying a dynamic pattern of initially increasing and then decreasing under HP stress. At the molecular level, several DEGs belonging to the PHT, SPX, and PAP gene families, as well as the transcription factor *PHR1*, were identified and analyzed under varying P conditions. Through joint analysis, the study revealed the response differences of 6 DAMs and 19 DEGs to the *P. wrightii* Glycerolipid metabolism and Glycerophospholipid metabolism pathways under different phosphorus stresses. Overall, this research deepens our understanding of the mechanisms by which submerged plants respond to phosphorus stress, providing a theoretical foundation for ecological restoration and eutrophication management in shallow lake ecosystems.

Supplementary Materials: The following supporting information can be downloaded at https://www.mdpi.com/article/10.3390/plants14233556/s1, Figure S1: (A) Sample PCA, the first principal component (PC1) is represented by the *X*-axis, and the second principal component (PC2) is represented by the *Y*-axis. (B) Sample correlation heatmap, Pearson correlation coefficient (R^2) > 0.8 between biological duplicate samples; Figure S2: GO Function Annotation of *P. wrightii* DEGs under phosphorus stress. (A) LP vs. CK, (B) P5 vs. CK, (C) P20 vs. CK, (D) P40 vs. CK. The horizontal axis represents the secondary GO entry, and the vertical axis represents the number of genes on the comment of this GO entry; Figure S3: KEGG enrichment analysis of *P. wrightii* DEGs under phosphorus stress. (A) LP vs. CK, (B) P5 vs. CK, (C) P20 vs. CK, (D) P40 vs. CK. The vertical axis represents the KEGG pathway, and the horizontal axis represents the Rich factor. The greater the Rich factor, the greater the degree of enrichment; Figure S4: Expression heat maps of members of the (A) PHT family, (B) PAP gene family, (C) SPX gene family, and (D) PHR1 transcription factor family. LP (0.025 mM), CK (0.25 mM), P5 (5 mM), P20 (20 mM), P40 (40 mM). Red and blue represent genes that are upregulated and downregulated under different phosphorus stresses; Figure S5: Analysis of the *P. wrightii* transcription factor family under phosphorus stress; Figure S6. (A): The total ion flow (TIC) shows the positive ion modes, (B): The total ion flow (TIC) shows the negative ion modes. Figure S7: Volcanic map of *P. wrightii* DAMs under phosphorus stress. (A) LP vs. CK, (B) P5 vs. CK, (C) P20 vs. CK, (D) P40 vs. CK. The green dots represent the down-regulated differential metabolites, the red dots represent the up-regulated differential metabolites, and the gray dots represent the metabolites detected but with insignificant differences. The horizontal axis represents the logarithm of the multiple of the relative content difference in a certain metabolite in the two groups of samples, and the vertical axis represents the level of significance of the difference; Figure S8: KEGG enrichment analysis of *P. wrightii* DAMs under phosphorus stress. (A) LP vs. CK, (B) P5 vs. CK, (C) P20 vs. CK, (D) P40 vs. CK. The vertical axis represents the KEGG pathway, and the horizontal axis represents the Rich factor. The greater the Rich factor, the greater the degree of enrichment.; Figure S9: Bar chart of *P. wrightii* KEGG enrichment analysis under phosphorus stress. (A) LP vs. CK, (B) P5 vs. CK, (C) P20 vs. CK, (D) P40 vs. CK. The horizontal axis represents the P-value of the significance test enriched to this pathway, and the red and green axes represent the metabolome and RNA-seq respectively; Figure S10: Correlation analysis and nine-quadrant chart. (A) LP vs. CK, (B) P5 vs. CK, (C) P20 vs. CK, (D) P40 vs. CK. Each point represents a pair of correlations, with the horizontal axis representing the Log2FC of genes and the vertical axis representing the Log2FC of metabolites. The black dotted line is divided into quadrants 1–9 from left to right and from topto bottom. For quadrant 5, the genes, metabolites, and differently grouped genes and metabolites are not differentially expressed. For quadrants 3 and 7, the genes and metabolites have the same differential expression patterns. For genes and metabolites with consistent expression level trends, the genes may positively regulate the changes in the metabolites. For quadrants 1 and 9, the genes and metabolites have the opposite differential expression patterns. For genes and metabolites with inconsistent expression level trends, the genes may negatively regulate the changes in the metabolites. For quadrants 2, 4, 6, 8, when the metabolites are unchanged, the genes are upregulated or downregulated and when the genes are unchanged, the metabolites are upregulated or downregulated; Figure S11: The gene–metabolite interaction network diagrams of the Glycerolipid metabolism and Glycerophospholipid metabolism pathways. In the figure, metabolites are marked with green squares and genes with red circles. Solid lines represent positive correlation, while dashed lines represent negative correlation. ko00561 represents the Glycerolipid metabolism pathway and ko00564 represents the Glycerophospholipid metabolism pathway. Figure S12: Diagram of the experimental setup; Table S1: RNA-seq data of 15 samples; Table S2: Assembly Result Statistics Table; Table S3: Annotation statistics table of each database; Table S4: According to the statistics table of DAMs for the four processing groups; Table S5: DAMs statistics Table of the four processing group GP; Table S6: Accumulation of flavonoid DAMs in LP vs. CK, P5 vs. CK, P20 vs. CK, and P40 vs. CK; Table S7: Software and parameters of the gene function annotation database; Table S8: Analyze the list of software and their versions; Table S9: Primers used for RT-qPCR analysis in this study.

Author Contributions: Software, validation, writing—original draft preparation and visualization: C.P. and B.Z.; formal analysis, investigation and data curation: J.C. and H.Y.; conceptualization, resources, writing—review and editing: M.T.; supervision, project administration and funding acquisition: X.X. All authors have read and agreed to the published version of the manuscript.

Funding: This research was funded by the National Natural Science Foundation of China (42263014); Guizhou Provincial Major Scientific and Technological Program (Oian Ke He Maior Proiect [2024]009).

Data Availability Statement: The RNA-seq data presented in this study are openly available; the processed data can be obtained through accession number PRJNA1335104 (https://www.ncbi.nlm.nih.gov/bioproject/PRJNA1335104 (accessed on 25 September 2025)). Accession numbers for DEGs: PwACTIN, PwNLA, PwPHO1, PwPHT1, PwPHT1-Like, PwPHT4, PwPHT4-Like, PwSPX, PwSPX-MFS, PwSPX-MFS-Like: C_AA126357.1—C_AA126366.1.

Acknowledgments: We thank Xia Huang (State Key Laboratory of Environmental Geochemistry, Institute of Geochemistry) for field investigation work. We also thank Yao Chen, Wenxin Xu, Wencui Wang and Jing Guo, and Yin Yi (School of Life Sciences, Guizhou Normal University) for all the help on this study.

Conflicts of Interest: The authors declare no conflicts of interest.

References

1. Wang, Z.; Kuo, H.F.; Chiou, T.J. Intracellular phosphate sensing and regulation of phosphate transport systems in plants. *Plant Physiol.* **2021**, *187*, 2043–2055. [CrossRef] [PubMed]
2. Carstensen, A.; Herdean, A.; Schmidt, S.B.; Sharma, A.; Spetea, C.; Pribil, M.; Husted, S. The impacts of phosphorus deficiency on the photosynthetic electron transport chain. *Plant Physiol.* **2018**, *177*, 271–284. [CrossRef]
3. Puga, M.I.; Rojas-Triana, M.; De Lorenzo, L.; Leyva, A.; Rubio, V.; Paz-Ares, J. Novel signals in the regulation of Pi starvation responses in plants: Facts and promises. *Curr. Opin. Plant Biol.* **2017**, *39*, 40–49. [CrossRef]
4. Dissanayaka, D.M.S.B.; Ghahremani, M.; Siebers, M.; Wasaki, J.; Plaxton, W.C. Recent insights into the metabolic adaptations of phosphorus-deprived plants. *J. Exp. Bot.* **2020**, *72*, 199–223. [CrossRef]
5. Péret, B.; Clément, M.; Nussaume, L.; Desnos, T. Root developmental adaptation to phosphate starvation: Better safe than sorry. *Trends Plant Sci.* **2011**, *16*, 442–450. [CrossRef]
6. Li, H.; He, K.; Zhang, Z.; Hu, Y. Molecular mechanism of phosphorous signaling inducing anthocyanin accumulation in *Arabidopsis*. *Plant Physiol. Biochem.* **2023**, *196*, 121–129. [CrossRef]
7. Sun, T.; Zhang, J.; Zhang, Q.; Li, X.; Li, M.; Yang, Y.; Zhou, J.; Wei, Q.; Zhou, B. RNA-seq and metabolome analyses revealed the response mechanism of apple to different phosphorus stresses. *Plant Physiol. Biochem.* **2021**, *167*, 639–650. [CrossRef]
8. Wang, Q.; Guo, Y.; Huang, T.; Zhang, X.; Zhang, P.; Xie, H.; Liu, J.; Li, L.; Kong, Z.; Qin, P. RNA-seq and metabolome analyses revealed the response mechanism of quinoa seedlings to different phosphorus stresses. *Int. J. Mol. Sci.* **2022**, *23*, 4704. [CrossRef] [PubMed]
9. Du, Q.; Wang, K.; Xu, C.; Zou, C.; Xie, C.; Xu, Y.; Li, W.X. Strand-specific RNA-Seq RNA-seq analysis of genotypes with and without low-phosphorus tolerance provides novel insights into phosphorus-use efficiency in maize. *BMC Plant Biol.* **2016**, *16*, 222. [CrossRef] [PubMed]
10. Zeng, H.Q.; Wang, G.P.; Zhang, Y.Q.; Hu, X.Y.; Pi, E.X.; Zhu, Y.Y.; Wang, H.Z.; Du, L.Q. Genome-wide identification of phosphate-deficiency-responsive genes in soybean roots by high-throughput sequencing. *Plant Soil.* **2016**, *398*, 207–227.
11. Deng, Q.W.; Luo, X.D.; Chen, Y.L.; Zhou, Y.; Zhang, F.T.; Hu, B.L.; Xie, J.K. RNA-seq analysis of phosphorus stress responsiveness in the seedlings of Dongxiang wild rice (*Oryza rufipogon* Griff.). *Biol. Res.* **2018**, *51*, 7. [CrossRef] [PubMed]

12. Wang, J.; Qin, Q.; Pan, J.; Sun, L.; Sun, Y.; Xue, Y.; Song, K. RNA-seq analysis in roots and leaves of wheat seedlings in response to low-phosphorus stress. *Sci. Rep.* **2019**, *9*, 19802. [CrossRef] [PubMed]
13. Jiang, C.; Xu, S.; Wang, R.; Sun, Q.; Zuo, J.; Zhuang, X. Transcriptomics Insights into Phosphorus Stress Response of *Myriophyllum aquaticum*. *Int. J. Mol. Sci.* **2023**, *24*, 4874. [CrossRef]
14. Wu, P.; Shou, H.; Xu, G.; Lian, X. Improvement of phosphorus efficiency in rice on the basis of understanding phosphate signaling and homeostasis. *Curr. Opin. Plant Biol.* **2013**, *16*, 205–212. [CrossRef]
15. Wang, Z.; Ruan, W.; Shi, J.; Zhang, L.; Xiang, D.; Yang, C.; Li, C.; Wu, Z.; Liu, Y.; Yu, Y.; et al. Rice SPX1 and SPX2 inhibit phosphate starvation responses through interacting with PHR2 in a phosphate-dependent manner. *Proc. Natl. Acad. Sci. USA* **2014**, *111*, 14953–14958. [CrossRef]
16. Wild, R.; Gerasimaite, R.; Jung, J.Y.; Truffault, V.; Pavlovic, I.; Schmidt, A.; Saiardi, A.; Jessen, H.J.; Poirier, Y.; Hothorn, M.; et al. Control of eukaryotic phosphate homeostasis by inositol polyphosphate sensor domains. *Science* **2016**, *352*, 986–990. [CrossRef]
17. Puga, M.I.; Mateos, I.; Charukesi, R.; Wang, Z.; Franco-Zorrilla, J.M.; de Lorenzo, L.; Irigoyen, M.L.; Masiero, S.; Bustos, R.; Rodríguez, J.; et al. SPX1 is a phosphate-dependent inhibitor of Phosphate Starvation Response 1 in *Arabidopsis*. *Proc. Natl. Acad. Sci. USA* **2014**, *111*, 14947–14952. [CrossRef]
18. Lv, Q.; Zhong, Y.; Wang, Y.; Wang, Z.; Zhang, L.; Shi, J.; Wu, Z.; Liu, Y.; Mao, C.; Yi, K.; et al. SPX4 negatively regulates phosphate signaling and homeostasis through its interaction with PHR2 in rice. *Plant Cell.* **2014**, *26*, 1586–1597. [CrossRef]
19. Bustos, R.; Castrillo, G.; Linhares, F.; Puga, M.I.; Rubio, V.; Pérez-Pérez, J.; Solano, R.; Leyva, A.; Paz-Ares, J. A central regulatory system largely controls transcriptional activation and repression responses to phosphate starvation in *Arabidopsis*. *PLoS Genet.* **2010**, *6*, e1001102. [CrossRef]
20. Wu, Z.; Liu, Y.; Liang, Z.; Wu, S.; Guo, H. Internal cycling, not external loading, decides the nutrient limitation in eutrophic lake: A dynamic model with temporal Bayesian hierarchical inference. *Water Res.* **2017**, *116*, 231–240. [CrossRef] [PubMed]
21. Wang, D.; Gan, X.; Wang, Z.; Jiang, S.; Zheng, X.; Zhao, M.; Zhang, Y.; Fan, C.; Wu, S.; Du, L. Research status on remediation of eutrophic water by submerged macrophytes: A review. *Process Saf. Environ.* **2023**, *169*, 671–684. [CrossRef]
22. Fang, Y.Y.; Babourina, O.; Rengel, Z.; Yang, X.E.; Pu, P.M. Ammonium and nitrate uptake by the floating plant *Landoltia punctata*. *Ann. Bot.* **2007**, *99*, 365–370. [CrossRef] [PubMed]
23. Zhang, H.; Zhang, S.; Peng, Y.; Li, Y.; Chen, Z.; Xu, H.; Yu, Z.; Zheng, W.; Zheng, T. Effects of marine actinomycete on the removal of a toxicity alga *Phaeocystis globose* in eutrophication waters. *Front. Microbiol.* **2015**, *6*, 474. [CrossRef]
24. Bakker, E.S.; Van Donk, E.; Declerck, S.A.J.; Helmsing, N.R.; Hidding, B.; Nolet, B.A. Effect of macrophyte community composition and nutrient enrichment on plant biomass and algal blooms. *Basic Appl. Ecol.* **2010**, *11*, 432–439. [CrossRef]
25. Gao, H.; Qian, X.; Wu, H.; Li, H.; Pan, H.; Han, C. Combined effects of submerged macrophytes and aquatic animals on the restoration of a eutrophic water body—A case study of Gonghu Bay, Lake Taihu. *Ecol. Eng.* **2017**, *102*, 15–23. [CrossRef]
26. Chao, C.; Wang, L.; Li, Y.; Yan, Z.; Liu, H.; Yu, D.; Liu, C. Response of sediment and water microbial communities to submerged vegetations restoration in a shallow eutrophic lake. *Sci. Total Environ.* **2021**, *801*, 149701. [CrossRef]
27. Keskinkan, O.; Goksu, M.Z.L.; Yuceer, A.; Basibuyuk, M.; Forster, C.F. Heavy metal adsorption characteristics of a submerged aquatic plant (*Myriophyllum spicatum*). *Process Biochem.* **2003**, *39*, 179–183. [CrossRef]
28. Zhou, Y.; Zhou, X.; Han, R.; Xu, X.; Wang, G.; Liu, X.; Bi, F.; Feng, D. Reproduction capacity of *Potamogeton crispus* fragments and its role in water purification and algae inhibition in eutrophic lakes. *Sci. Total Environ.* **2017**, *580*, 1421–1428. [CrossRef] [PubMed]
29. Dong, B.; Qin, B.; Gao, G.; Cai, X. Submerged macrophyte communities and the controlling factors in large, shallow Lake Taihu (China): Sediment distribution and water depth. *J. Great Lakes Res.* **2014**, *40*, 646–655. [CrossRef]
30. Peng, K.; Luo, C.; Lou, L.; Li, X.; Shen, Z. Bioaccumulation of heavy metals by the aquatic plants *Potamogeton pectinatus* L. and *Potamogeton malaianus* Miq. and their potential use for contamination indicators and in wastewater treatment. *Sci. Total Environ.* **2008**, *392*, 22–29. [CrossRef]
31. Han, S.; Xing, Z.; Jiang, H.; Li, W.; Huang, W. Biological adaptive mechanisms displayed by a freshwater plant to live in aquatic and terrestrial environments. *Environ. Exp. Bot.* **2021**, *191*, 104623. [CrossRef]
32. Xiao, C.; Wang, X.; Xia, J.; Liu, G. The effect of temperature, water level and burial depth on seed germination of *Myriophyllum spicatum* and *Potamogeton malaianus*. *Aquat. Bot.* **2010**, *92*, 28–32. [CrossRef]
33. Zuo, S.; Wang, H.; Gan, L.D.; Shao, M. Allelopathy appraisal of worm metabolites in the synergistic effect between *Limnodrilus hoffmeisteri* and *Potamogeton malaianus* on algal suppression. *Ecotoxicol. Environ. Saf.* **2019**, *182*, 109482. [CrossRef]
34. Gao, Y.; Zhang, Y.; Wei, Q.; Qi, X.; Yin, Q.; Liu, B.; He, K. Response and synergistic effect of microbial community to submerged macrophyte in restoring urban black and smelly water bodies. *J. Water Process Eng.* **2023**, *53*, 103906. [CrossRef]
35. Yang, S.Y.; Lin, W.Y.; Hsiao, Y.M.; Chiou, T.J. Milestones in understanding transport, sensing, and signaling of the plant nutrient phosphorus. *Plant Cell.* **2024**, *36*, 1504–1523. [CrossRef]
36. Xu, J.; Liu, J.; Hu, J.; Wang, H.; Sheng, L.; Dong, X.; Jiang, X. Nitrogen and phosphorus removal in simulated wastewater by two aquatic plants. *Environ. Sci. Pollut. Res.* **2021**, *28*, 63237–63249. [CrossRef] [PubMed]

37. Huang, W.; Yang, X.; Chen, H.; Lu, C.; Che, F. Promoting effect of submerged plants on phosphorus source and sink in sediments: Response of phosphorus release and microbial community to *Vallisneria natans* and *Potamogeton malaianus* growth. *Environ. Technol. Innov.* **2024**, *36*, 103899. [CrossRef]
38. Zhang, K.; Liu, H.; Song, J.; Wu, W.; Li, K.; Zhang, J. Physiological and comparative proteome analyses reveal low-phosphate tolerance and enhanced photosynthesis in a maize mutant owing to reinforced inorganic phosphate recycling. *BMC Plant Biol.* **2016**, *16*, 129. [CrossRef] [PubMed]
39. Mittler, R.; Zandalinas, S.I.; Fichman, Y.; Van Breusegem, F. Reactive oxygen species signalling in plant stress responses. *Nat. Rev. Mol. Cell Biol.* **2022**, *23*, 663–679. [CrossRef]
40. Furlani, A.L.; Bianuccii, E.; Giordanoi, W.; Castroi, S.; Beckeri, D.F. Proline metabolic dynamics and implications in drought tolerance of peanut plants. *Plant Physiol. Biochem.* **2020**, *151*, 566–578. [CrossRef]
41. Lv, S.; Wang, D.; Jiang, P.; Jia, W.; Li, Y. Variation of PHT families adapts salt cress to phosphate limitation under salinity. *Plant Cell Environ.* **2021**, *44*, 1549–1564. [CrossRef]
42. Sun, T.; Li, M.; Shao, Y.; Yu, L.; Ma, F. Comprehensive genomic identification and expression analysis of the phosphate transporter (PHT) gene family in apple. *Front. Plant Sci.* **2017**, *8*, 426. [CrossRef] [PubMed]
43. Paszkowski, U.; Kroken, S.; Roux, C.; Briggs, S.P. Rice phosphate transporters include an evolutionarily divergent gene specifically activated in arbuscular mycorrhizal symbiosis. *Proc. Natl. Acad. Sci. USA* **2002**, *99*, 13324–13329. [CrossRef]
44. Mudge, S.R.; Rae, A.L.; Diatloff, E.; Smith, F.W. Expression analysis suggests novel roles for members of the Pht1 family of phosphate transporters in *Arabidopsis*. *Plant J.* **2002**, *31*, 341–353. [CrossRef] [PubMed]
45. Ai, P.H.; Sun, S.B.; Zhao, J.N.; Fan, X.R.; Xin, W.J.; Guo, Q.; Yu, L.; Shen, Q.R.; Wu, P.; Miller, A.J.; et al. Two rice phosphate transporters, OsPht1; 2 and OsPht1; 6, have different functions and kinetic properties in uptake and translocation. *Plant J.* **2009**, *57*, 798–809. [CrossRef] [PubMed]
46. Secco, D.; Wang, C.; Arpat, B.A.; Wang, Z.; Poirier, Y.; Tyerman, S.D.; Wu, P.; Shou, H.; Whelan, J. The emerging importance of the SPX domain-containing proteins in phosphate homeostasis. *New Phytol.* **2012**, *193*, 842–851. [CrossRef]
47. Wang, Z.; Hu, H.; Huang, H.; Duan, K.; Wu, Z.; Wu, P. Regulation of OsSPX1 and OsSPX3 on expression of OsSPX domain genes and Pi-starvation signaling in rice. *J. Integr. Plant Biol.* **2009**, *51*, 663–674. [CrossRef]
48. Wang, C.; Huang, W.; Ying, Y.; Li, S.; Secco, D.; Tyerman, S.; Whelan, J.; Shou, H. Functional characterization of the rice SPX-MFS family reveals a key role of OsSPX-MFS1 in controlling phosphate homeostasis in leaves. *New Phytol.* **2012**, *196*, 139–148. [CrossRef]
49. Liu, H.C.; Rao, L.; Meng, J.H.; Zuo, W.T.; Sun, T.T. Genome-Wide Analysis, Identification, and Transcriptional Profile of the Response to Abiotic Stress of the Purple Acid Phosphatases (PAP) Gene Family in Apple. *Int. J. Mol. Sci.* **2025**, *26*, 1011. [CrossRef]
50. Rubio, V.; Linhares, F.; Solano, R.; Martín, A.C.; Iglesias, J.; Leyva, A.; Paz-Ares, J.A. Conserved MYB transcription factor involved in phosphate starvation signaling both in vascular plants and in unicellular algae. *Genes Dev.* **2001**, *15*, 2122–2133. [CrossRef]
51. Wang, Z.; Zheng, Z.; Zhu, Y.; Kong, S.; Liu, D. PHOSPHATE RESPONSE 1 family members act distinctly to regulate transcriptional responses to phosphate starvation. *Plant Physiol.* **2023**, *191*, 1324–1343. [CrossRef] [PubMed]
52. Harayama, T.; Riezman, H. Understanding the diversity of membrane lipid composition. *Nat. Rev. Mol. Cell Biol.* **2018**, *19*, 281–296. [CrossRef]
53. Mühlroth, A.; Winge, P.; El Assimi, A.; Jouhet, J.; Maréchal, E.; Hohmann-Marriott, M.F.; Vadstein, O.; Bones, A.M. Mechanisms of phosphorus acquisition and lipid class remodeling under P limitation in a marine microalga. *Plant Physiol.* **2017**, *175*, 1543–1559. [CrossRef]
54. Yu, L.; Zhou, C.; Fan, J.; Shanklin, J.; Xu, C. Mechanisms and functions of membrane lipid remodeling in plants. *Plant J.* **2021**, *107*, 37–53. [CrossRef]
55. Ding, Z.; Jia, S.; Wang, Y.; Xiao, J.; Zhang, Y. Phosphate stresses affect ionome and metabolome in tea plants. *Plant Physiol. Biochem.* **2017**, *120*, 30–39. [CrossRef]
56. Sato, N. Roles of the acidic lipids sulfoquinovosyl diacylglycerol and phosphatidylglycerol in photosynthesis: Their specificity and evolution. *J. Plant Res.* **2004**, *117*, 495–505. [CrossRef] [PubMed]
57. Yu, B.; Xu, C.; Benning, C. *Arabidopsis* disrupted in SQD2 encoding sulfolipid synthase is impaired in phosphate-limited growth. *Proc. Natl. Acad. Sci. USA* **2002**, *99*, 5732–5737. [CrossRef] [PubMed]
58. Zhang, D.; Liu, J.; Zhang, Y.; Wang, H.; Wei, S.; Zhang, X.; Zhang, D.; Ma, H.; Ding, Q.; Ma, L. Morphophysiological.; proteomic and metabolomic analyses reveal cadmium tolerance mechanism in common wheat (*Triticum aestivum* L.). *J. Hazard. Mater.* **2023**, *445*, 130499. [CrossRef]
59. Xiao, S.; Song, W.; Xing, J.; Su, A.; Zhao, Y.; Li, C.; Shi, Z.; Li, Z.; Wang, S.; Zhang, R.; et al. ORF355 confers enhanced salinity stress adaptability to S-type cytoplasmic male sterility maize bymodulating the mitochondrial metabolic homeostasis. *J. Integr. Plant Biol.* **2023**, *65*, 656–673. [CrossRef]
60. Ma, Y.; Liu, L.; Ye, Z.; Xu, L.; Li, Y.; Liu, S.; Song, G.; Zhang, X.B. Engineering of cyanine-based nanoplatform with tunable response toward reactive species for ratiometric NIR-II fluorescent imaging in mice. *Sci. Bull.* **2023**, *68*, 2382–2390. [CrossRef]

61. Liu, X.; Matsumoto, H.; Lv, T.; Zhan, C.; Fang, H.; Pan, Q.; Xu, H.; Fan, X.; Chu, T.; Chen, S.; et al. Phyllosphere microbiome induces host metabolic defence against rice false-smut disease. *Nat. Microbiol.* **2023**, *8*, 1419–1433. [CrossRef]
62. Li, C.; Li, K.; Liu, X.; Ruan, H.; Zheng, M.; Yu, Z.; Gai, J.; Yang, S. Transcription Factor GmWRKY46 Enhanced Phosphate Starvation Tolerance and Root Development in Transgenic Plants. *Front. Plant Sci.* **2021**, *12*, 700651. [CrossRef] [PubMed]
63. Chen, S.; Zhou, Y.; Chen, Y.; Gu, J. Fastp: An ultra-fast all-in-one FASTQ preprocessor. *Bioinformatics* **2018**, *34*, i884–i890. [CrossRef] [PubMed]
64. Grabherr, M.G.; Haas, B.J.; Yassour, M.; Levin, J.Z.; Thompson, D.A.; Amit, I.; Regev, A. Full-length RNA-seq assembly from RNA-Seq data without a reference genome. *Nat. Biotechnol.* **2011**, *29*, 644–652. [CrossRef]
65. Haas, B.J.; Papanicolaou, A.; Yassour, M.; Grabherr, M.; Blood, P.D.; Bowden, J.; Couger, M.B.; Eccles, D.; Li, B.; Lieber, M.; et al. De novo transcript sequence reconstruction from RNA-seq using the Trinity platform for reference generation and analysis. *Nat. Protoc.* **2013**, *8*, 1494–1512. [CrossRef] [PubMed]
66. Buchfink, B.; Xie, C.; Huson, D.H. Fast and sensitive protein alignment using DIAMOND. *Nat Methods.* **2015**, *12*, 59–60. [CrossRef]
67. Mistry, J.; Chuguransky, S.; Williams, L.; Qureshi, M.; Salazar, G.A.; Sonnhammer, E.L.L.; Tosatto, S.C.E.; Paladin, L.; Raj, S.; Richardson, L.J.; et al. Pfam: The protein families database in 2021. *Nucleic Acids Res.* **2021**, *49*, D412–D419. [CrossRef]
68. Dewey, C.N.; Li, B. RSEM: Accurate transcript quantification from rna-seq data with or without a reference genome. *BMC Bioinf.* **2011**, *12*, 323.
69. Love, M.I.; Huber, W.; Anders, S. Moderated estimation of fold change and dispersion for RNA-seq data with DESeq2. *Genome Biol.* **2014**, *15*, 550. [CrossRef]
70. Varet, H.; Brillet-Guéguen, L.; Coppée, J.-Y.; Dillies, M.-A. SARTools: A DESeq2- and EdgeR-based r pipeline for comprehensive differential analysis of RNA-seq data. *PLoS ONE* **2016**, *11*, e0157022. [CrossRef]
71. Connell, J.; Bates, H.J.; Geoghegan, I.; Wilson, F.; Harrison, R.J.; Price, R.J. Mutation of the LRG1 Rho-GAP gene is responsible for the hyper branching C-variant phenotype in the quorn mycoprotein fungus Fusarium venenatum A3/5. *Fungal Biol. Biotechnol.* **2025**, *12*, 3. [CrossRef] [PubMed]
72. Chen, C.; Wu, Y.; Li, J.; Wang, X.; Zeng, Z.; Xu, J.; Liu, Y.; Feng, J.; Chen, H.; He, Y.; et al. TBtools-II: A "one for all, all for one" bioinformatics platform for biological big-data mining. *Mol. Plant.* **2023**, *16*, 1733–1742. [CrossRef]
73. French, W.R.; Zimmerman, L.J.; Schilling, B.; Gibson, B.W.; Miller, C.A.; Townsend, R.R.; Sherrod, S.D.; Goodwin, C.R.; McLean, J.A.; Tabb, D.L. Wavelet-based peak detection and a new charge inference procedure for MS/MS implemented in ProteoWizard's msConvert. *J. Proteome Res.* **2015**, *14*, 1299–1307. [CrossRef]
74. Chong, J.; Xia, J. MetaboAnalystR: An R package for flexible and reproducible analysis of metabolomics data. *Bioinformatics* **2018**, *34*, 4313–4314. [CrossRef]
75. Jiang, P.; Zhang, Z.; Yu, Q.; Wang, Z.; Diao, L.; Li, D. ToxDAR: A Workflow Software for Analyzing Toxicologically Relevant Proteomic and Transcriptomic Data, from Data Preparation to Toxicological Mechanism Elucidation. *Int. J. Mol. Sci.* **2024**, *25*, 9544. [CrossRef] [PubMed]
76. Kanehisa, M.; Goto, S. KEGG: Kyoto encyclopedia of genes and genomes. *Nucleic Acids Res.* **2000**, *28*, 27–30. [CrossRef] [PubMed]
77. Livak, K.J.; Schmittgen, T.D. Analysis of relative gene expression data using real-time quantitative PCR and the $2^{-\Delta\Delta CT}$ Method. *Methods* **2001**, *25*, 402–408. [CrossRef]

plants

MDPI

Article

Integrative Transcriptomics and Metabolomics Reveal the Key Metabolic Pathways in Endophyte-Infected Rice Seedlings Resistance to Na_2CO_3 Stress

Xinnan Wang [1], Yanan Li [1], Hefei Sun [1], Lihong Zhang [2,*] and Xuemei Li [1,*]

1 College of Life Science, Shenyang Normal University, Shenyang 110034, China; wangxinnan0917@163.com (X.W.); 15042129995@163.com (Y.L.); sunhefei99@163.com (H.S.)
2 School of Environmental Science, Liaoning University, Shenyang 110036, China
* Correspondence: lihongzhang132@163.com (L.Z.); lxmls132@163.com (X.L.)

Abstract: Soil saline-alkalization is a key factor affecting rice growth and physiological metabolism, which leads to reduced yields. Endophyte EF0801 significantly promoted growth and improved its saline-alkali resistance. We investigated growth parameters and physiological indices of endophyte EF0801-infected and control rice seedlings under sodium carbonate (Na_2CO_3) stress. The results showed that endophyte-infected rice seedlings showed plant height increase by 1.25-fold, root length shortening by 0.79-fold, sucrose synthase (SS), sucrose phosphosynthase (SPS), hexokinase (HXK), and α-glucosidase (α-GC) activities increased by 0.15-fold, 0.29-fold, 0.06-fold, and 1.45-fold, respectively, and β-glucosidase (β-GC) activity decreased by 0.12-fold. Utilizing gas chromatography and mass spectrometry (GC-MS) technology and RNA sequencing (RNA-seq) technology, we identified 419 differentially expressed genes (DEGs) and 37 differentially accumulated metabolites (DAMs). Comprehensive enrichment analysis of DAMs and DEGs showed that 6 DEGs and 6 DAMs were strongly correlated with the mitigating effects of endophytes on rice leaves under Na_2CO_3 treatment, highlighting the co-enrichment in starch and sucrose metabolism, as well as alanine, aspartate, and glutamate metabolism. The gene encoding HXK was found to be upregulated in endophyte-infected rice seedlings under Na_2CO_3 stress. HXK plays a key role in the conversion of fructose and glucose to fructose 6-phosphate (F-6-P) and glucose 6-phosphate (G-6-P), which are important intermediates in cellular energy metabolism and glycolytic pathways, providing energy and biosynthesis of precursor substances. Our findings provide a potential perspective for unraveling the molecular response of endophyte-mediated saline-alkali resistance in rice leaves and a theoretical rationale for exploring the mechanisms of growth-promoting effects by endophytes.

Keywords: *Oryza sativa* L.; Na_2CO_3 stress; endophyte-plant interaction; RNA-seq; GC-MS analysis

1. Introduction

Soil saline-alkalization critically affects global food crop growth and production [1]. Reports indicate that an estimated 113 million hectares of land are affected by salinization globally, about 25% of the global land area [2,3]. Generally, saline-alkali causes osmotic stress and high pH in plants and disrupts ionic balance [4,5], ultimately affecting plant growth and production [6,7]. Rice, the cereal crop with the widest cultivation area in China, is particularly sensitive to saline-alkali, especially at its seedling stage and growth period [8,9]. Therefore, it is necessary to study the damage caused by Na_2CO_3 treatment to plants to improve their resistance to saline-alkali.

Endophytes are microorganisms that live in normally developing plant organs and tissues and do not negatively affect the host [10]. Numerous studies have found that endophytes can promote plant nutrient uptake [11,12], enhance defense-related gene expression [13], increase endogenous hormone levels [14], and positively alter related metabolite production [15] under abiotic stress. Based on previous studies, it was found that a fungal endophyte EF0801 significantly increased the stability of photosystem II in rice seedlings [16], regulated mineral element accumulation and organic acid metabolism [17], and increased antioxidant enzyme activities [18] under Na_2CO_3 treatment.

Metabolomes are used to observe a variety of metabolites in plants at specific times and environments [19], which is important for studying plant metabolism pathways and improving plant response to adverse stress [20]. Currently, metabolomics has been applied to study the reaction of barley to Na_2CO_3 treatment under endophyte infection [21]. The transcriptome is used to analyze plant gene expression levels [22] and alleviation mechanisms of endophytes in rice seedlings under Na_2CO_3 treatment [23]. However, integrative transcriptomic and metabolomic analyses provide a novel and formidable tool for elucidating the relationship between genotype and phenotype. Li et al. [24] demonstrated how ecological adaptations develop in two distinct ecotypes of wild soybeans by examining gene expression and metabolic variations in a thorough and detailed study of the transcriptome and metabolome. Shao et al. [25] combined transcriptomic and metabolomic analyses to analyze the effects of the biocontrol endophyte R31 on sweet corn and found that DEGs and DAMs were co-enriched for flavonoid biosynthesis. Therefore, we performed metabolomics and transcriptomics of endophyte-infected rice leaves under saline-alkali stress using GC-MS and RNA-Seq to identify and screen differentially accumulated metabolites (DAMs) and differentially expressed genes (DEGs). The common key metabolic pathways of DAMs and DEGs were further explored through co-enrichment analysis. This study focuses on revealing the mechanism of enhancing saline-alkali stress tolerance in rice seedlings by endophytes at the molecular level, screening potential resistance genes and metabolites, and providing theoretical support for agricultural production and scientific research.

2. Results

2.1. Growth Indexes of Rice Seedlings

After 7 d of treatment, significant differences in phenotype and growth were observed between endophyte-infected and control rice seedlings (Figure 1A,B). Compared with control rice seedlings, endophyte-infected rice seedlings showed that plant height increased by 1.25-fold (Figure 1C) and root length decreased by 0.79-fold (Figure 1D).

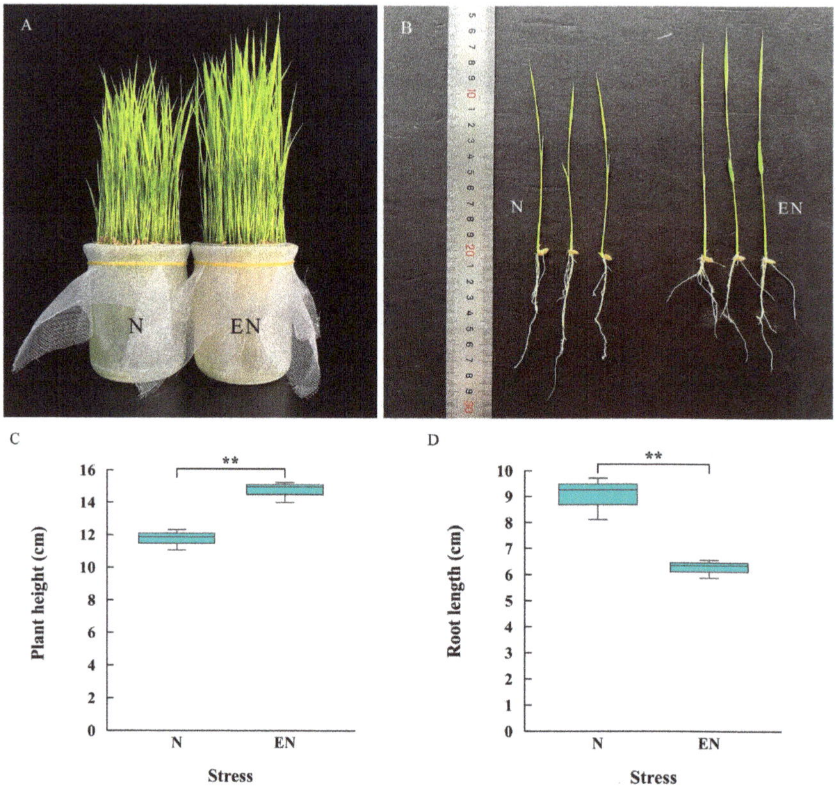

Figure 1. The growth of endophyte-infected and control rice seedlings under Na_2CO_3 stress. Phenotypes (**A**,**B**), plant height (**C**), and root length (**D**). ** indicated the significant differences of $p < 0.01$.

2.2. Changes and Validation of Sucrose Metabolism-Related Enzyme Activities

Under Na_2CO_3 treatment, sucrose synthase (SS), sucrose phosphate synthase (SPS), α-Glucosidase (α-GC), and hexokinase (HXK) activities were increased by 0.15-fold, 0.29-fold, 0.06-fold, and 1.45-fold, respectively, and β-Glucosidase (β-GC) was decreased by 0.12-fold in rice leaves inoculated with endophytes (Figure 2A). To test the accuracy of sucrose metabolism-related enzymes' (SS, SPS, HXK, α-GC, and β-GC) activity data, qRT-PCR was performed to verify the enzyme-regulated related genes. The validation results showed that the trend of sucrose metabolism-related enzyme activities was consistent with the qRT-PCR validation results (Figure 2B).

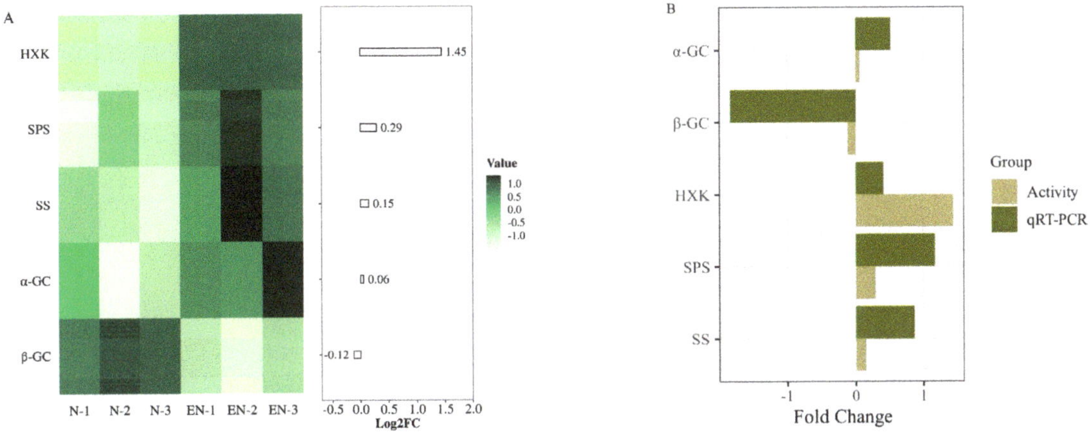

Figure 2. Changes in activities and expressions of sucrose metabolism-related enzymes in endophyte-infected versus control rice seedlings under Na_2CO_3 stress. Heatmap analysis of sucrose metabolism-related enzyme activities in endophyte-infected versus control rice seedlings under Na_2CO_3 stress (**A**). Bar graphs were plotted using log_2FC values of enzyme activity and qRT-PCR relative expression for fold change comparison (**B**).

2.3. Analysis and Validation of RNA-Seq Data

Clean reads of 44.32 gigabytes (Gb) were obtained for six cDNA libraries. The sample Q30% was more than 93.87%, and the GC content was between 50.76% and 51.88%. To check transcriptome data similarity and credibility, we performed Spearman correlation analysis and PCA on the sequencing data, and the results obtained suggested that the data were reliable and ready for subsequent analysis (Figure 3A,B). To further determine the RNA-seq data reliability, we randomly picked six genes to perform qRT-PCR validation. The validation results showed a consistent trend between RNA-seq data and qRT-PCR validation (R^2 = 0.9165) (Figure 4).

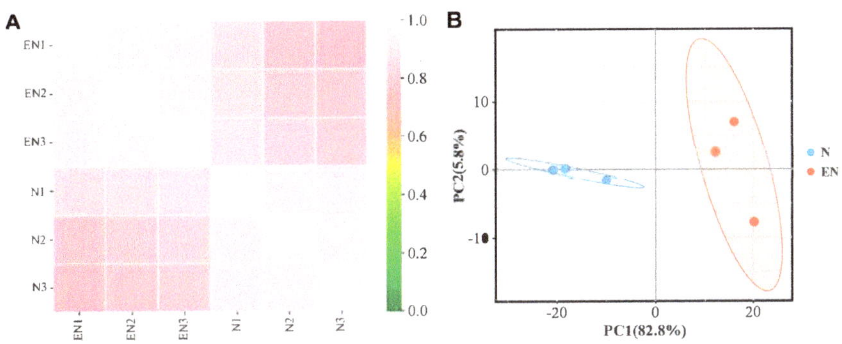

Figure 3. RNA-Seq analysis of endophyte-infected versus control rice seedlings under Na_2CO_3 stress. Spearman correlation analysis (**A**) and PCA (**B**).

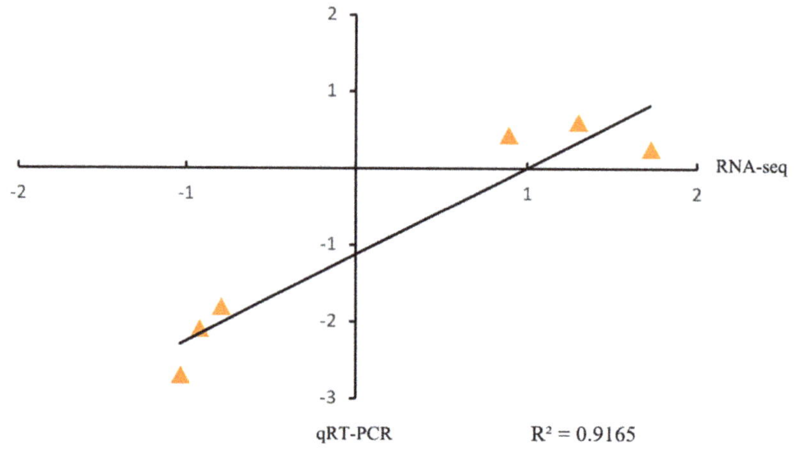

Figure 4. Validation of the expression level of genes from RNA-seq using qRT-PCR. A comparison of fold change (FC) was done using scatter plots using $\log_2$FC values obtained from RNA-seq and qRT-PCR.

2.4. Screening and Functional Analysis of DEGs

Based on the FDR < 0.05 and $|\log_2 FC| \geq 0.585$ of the screened DEGs, 419 DEGs were detected, including 229 up- and 190 down-regulated DEGs (Figure 5A). KEGG enrichment showed that 69 DEGs were enriched in 53 metabolic pathways (Table S2). Among them, eight significantly enriched pathways were starch and sucrose metabolism (ko00500), monoterpenoid biosynthesis (ko00902), cysteine and methionine metabolism (ko00270), nitrogen metabolism (ko00910), galactose metabolism (ko00052), cyanoamino acid metabolism (ko00460), carotenoid biosynthesis (ko00906) and alanine, aspartate, and glutamate metabolism (ko00250) (Figure 5B).

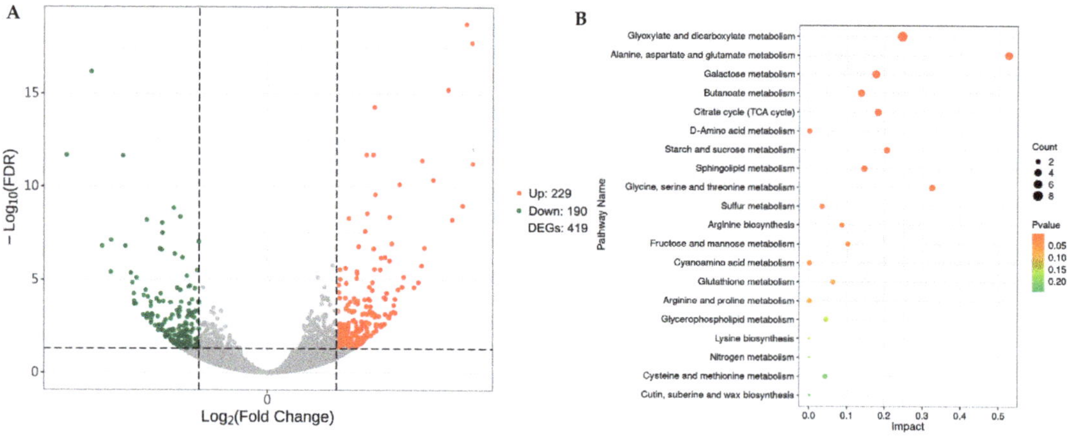

Figure 5. Transcriptome analysis of endophyte-infected versus control rice seedlings under Na_2CO_3 stress. Volcano plot of DEGs (**A**) and KEGG enrichment analysis (**B**). Significantly up- and down-regulated genes are indicated by red and green dots, respectively.

2.5. Assessment and Bioinformatics Analysis of DAMs

In this experiment, we analyzed the metabolites of endophyte-infected and control rice seedlings under Na_2CO_3 treatment. PC1 (first principal component) was 45.2%, and PC2 (second principal component) was 11.9%, and the results showed significantly different data for the two treatment groups (Figure 6A). PLS-DA results showed that components contributing to PC1 were oxalic acid, L-malic acid, pyroglutamic acid, and D-fructose; components contributing to PC2 were citric acid, quinic acid, and stearic acid (Figure 6B, Table S3). Screening of metabolites according to VIP > 1 and $p < 0.05$ revealed 37 metabolites with significant changes between the N and EN groups, including 10 amino acids, 6 glycolysis and TCA cycles, 9 organic acids, 12 sugars and polyols (Table 1). Fold change analysis showed that 21 and 16 DAMs were up- and down-regulated (Figure 6C). Hierarchical clustering analysis could more directly show the fold change of metabolites between E and EN groups (Figure 6D).

To elucidate the role of DAMs in the N and EN groups, KEGG pathway analysis was performed on DAMs. The results show that 38 metabolic pathways were enriched and 12 metabolic pathways were significantly enriched (Table S4). Among them, DAMs were most enriched in carbohydrate- and amino acid-related metabolic pathways (Figure 7).

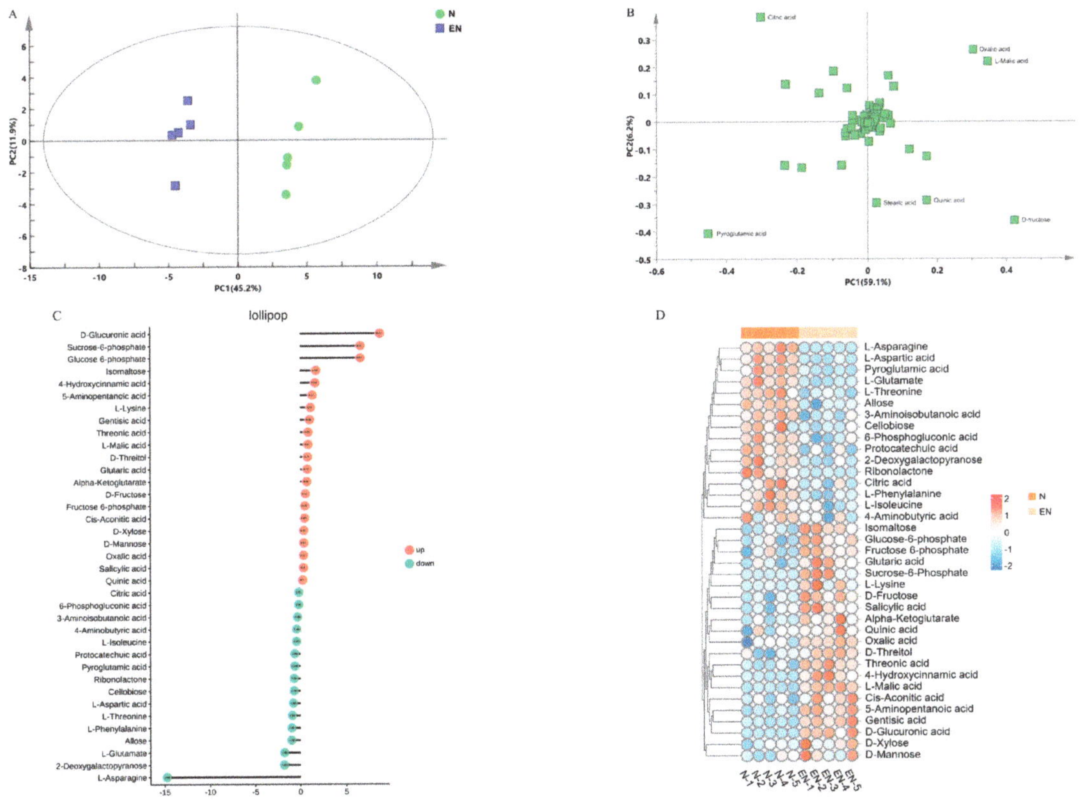

Figure 6. Metabolome analysis of endophyte-infected versus control rice seedlings under Na_2CO_3 stress. PCA (**A**), PLS-DA (**B**), fold changes (**C**), and hierarchical clustering heat map (**D**).

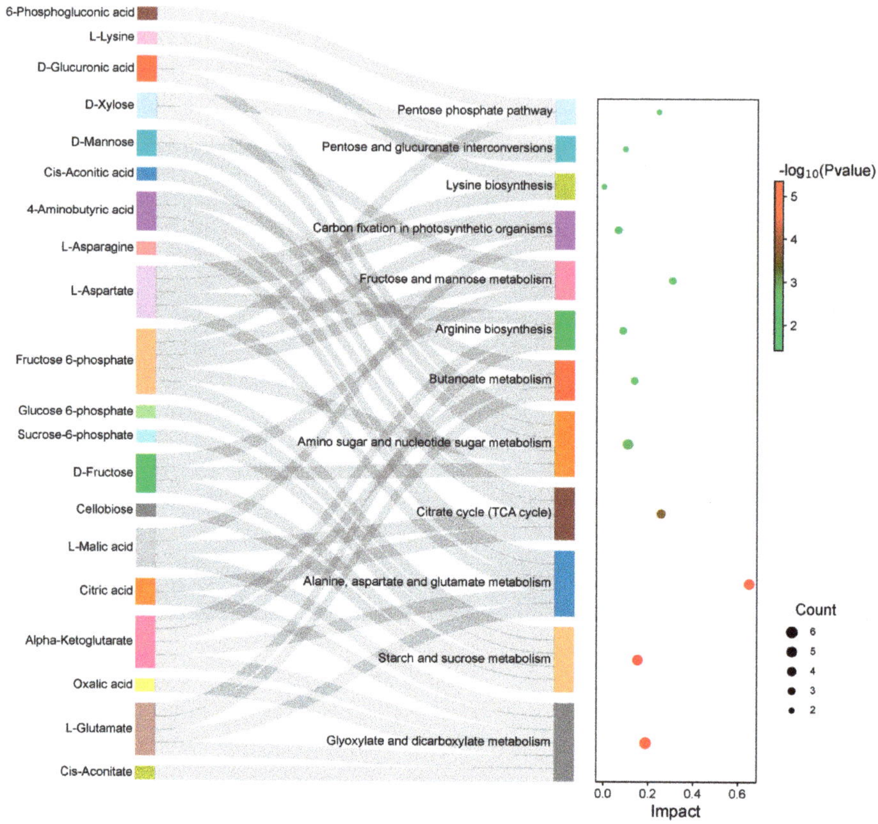

Figure 7. KEGG enrichment analysis of endophyte-infected versus control rice seedlings under Na_2CO_3 stress.

Table 1. Changes in metabolite contents of endophyte-infected versus control rice seedlings under Na_2CO_3 stress.

Metabolite	Name	N	EN	FC
Glycolysis and TCA cycles	L-Malic acid	123.43 ± 8.37	216.20 ± 15.91	0.81
	Glucose 6-phosphate	0.49 ± 0.13	0.76 ± 0.10	6.52
	Fructose 6-phosphate	1.81 ± 0.43	2.49 ± 0.21	0.46
	Cis-Aconitic acid	0.15 ± 0.03	0.21 ± 0.03	0.43
	Alpha-Ketoglutarate	25.09 ± 5.94	40.39 ± 11.26	0.68
	Citric acid	729.92 ± 66.67	627.08 ± 59.73	−0.21
Amino acid	L-Asparagine	3.18 ± 0.74	0.00 ± 0.00	−14.8
	L-Glutamate	38.99 ± 6.88	11.07 ± 3.21	−1.82
	L-Aspartic acid	4.43 ± 0.89	3.36 ± 1.96	−0.83
	5-Aminopentanoic acid	0.16 ± 0.03	0.71 ± 0.23	1.27
	Pyroglutamic acid	445.79 ± 43.75	281.20 ± 22.41	−0.66
	L-Threonine	95.25 ± 14.58	49.21 ± 11.62	−0.95
	L-Phenylalanine	3.47 ± 0.62	1.73 ± 0.45	−1.01
	3-Aminoisobutanoic acid	0.42 ± 0.02	0.33 ± 0.04	−0.36
	L-Isoleucine	7.95 ± 1.35	5.72 ± 1.32	−0.47
	L-Lysine	2.20 ± 0.26	4.69 ± 2.39	1.09

Table 1. *Cont.*

Metabolite	Name	N	EN	FC
Sugars and polyols	D-Glucuronic acid	0.23 ± 0.03	0.31 ± 0.07	8.62
	Sucrose-6-phosphate	0.00 ± 0.00	0.09 ± 0.06	6.52
	Threonic acid	1.21 ± 0.22	2.15 ± 0.23	0.83
	Allose	0.48 ± 0.03	0.23 ± 0.08	−1.03
	D-Mannose	146.01 ± 25.61	187.22 ± 43.21	0.34
	D-Fructose	242.96 ± 42.73	357.43 ± 57.83	0.53
	6-Phosphogluconic acid	0.91 ± 0.03	0.78 ± 0.07	−0.22
	2-Deoxygalactopyranose	7.60 ± 2.92	2.12 ± 1.84	−1.84
	D-Threitol	0.10 ± 0.04	0.17 ± 0.03	0.76
	Cellobiose	0.90 ± 0.28	0.54 ± 0.11	−0.74
	D-Xylose	0.09 ± 0.01	0.11 ± 0.02	0.34
	Isomaltose	0.21 ± 0.12	0.66 ± 0.17	1.66
Organic acid	4-Hydroxycinnamic acid	0.10 ± 0.02	0.29 ± 0.09	1.58
	Oxalic acid	368.55 ± 54.49	456.90 ± 13.54	0.31
	Quinic acid	2.62 ± 0.38	3.02 ± 0.30	0.2
	Gentisic acid	0.24 ± 0.02	0.47 ± 0.06	0.98
	Salicylic acid	5.98 ± 0.43	7.11 ± 0.58	0.3
	Protocatechuic acid	0.15 ± 0.03	0.04 ± 0.06	−0.66
	Ribonolactone	00.14 ± 0.03	0.08 ± 0.00	−0.74
	Glutaric acid	0.09 ± 0.03	0.13 ± 0.03	0.72
	4-Aminobutyric acid	95.74 ± 27.83	69.45 ± 23.55	−0.46

Note: Relative concentration and standard deviation increased by 100 times. FC: fold change. The fold changes of metabolites were represented as $\log_2$(EN/N).

2.6. Integrated Transcriptome and Metabolome Analysis

To study the metabolic changes in endophyte-infected and control rice under Na_2CO_3 treatment, we performed KEGG joint enrichment analysis of DEGs and DAMs. Data indicated that the co-enrichment pathways were starch and sucrose metabolism as well as alanine, aspartate, and glutamate metabolism pathways (Figure 8). Construction of the gene and metabolite networks of starch and sucrose metabolic pathways revealed that the gene encoding hexokinase (HXK) [EC:2.7.1.1] (*Os01g0940100*) was highly correlated with f-6-p and g-6-p; the genes encoding α-glucosidase (α-GC) [EC:3.2.1.20] (*Os07g0656200*, *Os03g0749500*, and *Os03g0749650*) were positively correlated with cellobiose (Figure 9). Construction of gene and metabolite networks for alanine, aspartate, and glutamate metabolic pathways revealed that the gene encoding asparagine synthase (ASNS) [EC: 6.3.5.4] (*Os03g0291500*) was positively correlated with L-aspartic acid and L-asparagine; the gene encoding glutamate dehydrogenase (GDH) [EC. 1.4.1.3] gene (*Os04g0543900*) was positively correlated with L-glutamate and pyroglutamic acid (Figure 10). The final six genes were found to be associated with starch and sucrose metabolic pathways and alanine, aspartate, and glutamate metabolic pathways. It was demonstrated that starch and sucrose metabolic pathways and alanine, aspartate, and glutamate metabolic pathways are crucial in the mechanism of endophyte-mediated saline-alkali resistance in rice.

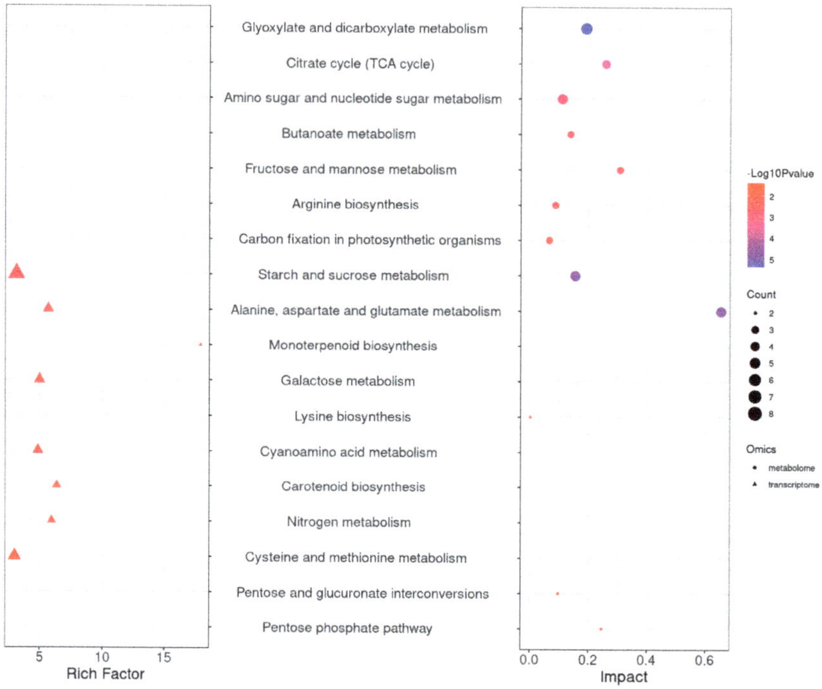

Figure 8. Integrated analysis of transcriptome and metabolome pathway enrichment in endophyte-infected versus control rice seedlings under Na_2CO_3 stress.

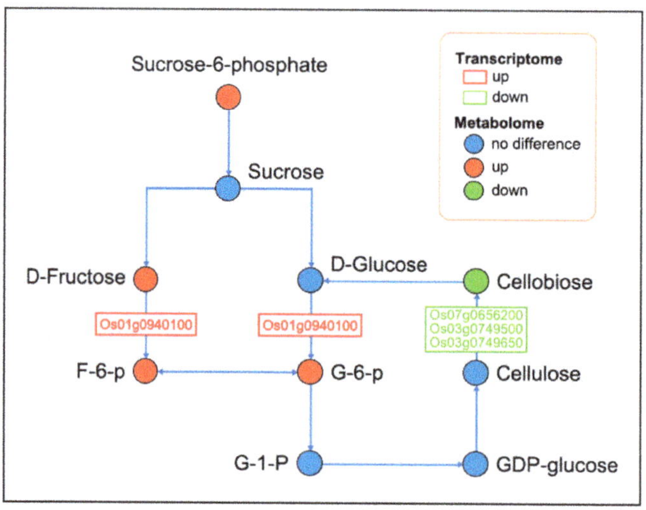

Figure 9. Changes in DAMs and DEGs in starch and sucrose metabolism pathways in endophyte-infected versus control rice seedlings under Na_2CO_3 stress.

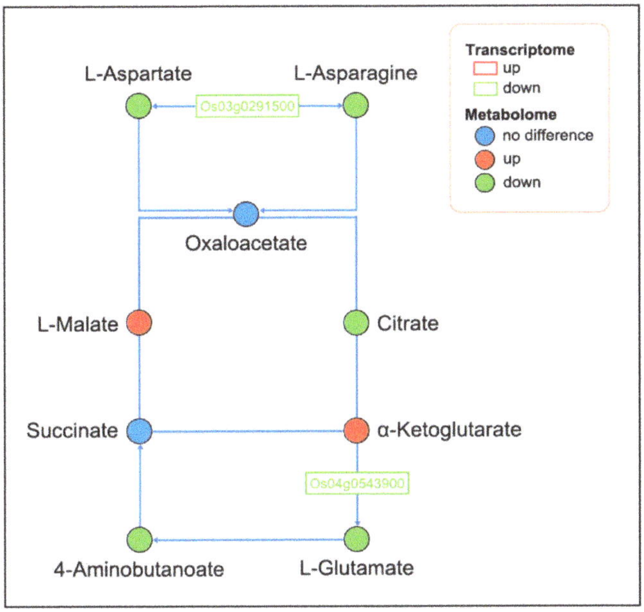

Figure 10. Changes in DAMs and DEGs in alanine, aspartate, and glutamate metabolism pathways in endophyte-infected versus control rice seedlings under Na_2CO_3 stress.

3. Discussion

Soil saline-alkalization markedly affects rice growth, development, and ultimately yields. Therefore, improving rice tolerance to sustain rice production requires a comprehensive characterization of the multiple biological processes under Na_2CO_3 treatment. Analysis of endophyte-host interactions revealed that the endophyte significantly promotes plant growth and maintains homeostatic balance [26,27], and based on our previous studies [28], we found that endophyte infection significantly increased the mineral element content of rice seedlings. Studies found that rice seedlings were significantly taller but had reduced root length after endophyte infection compared to the Na_2CO_3-treated group. Consistent with our previous descriptions [16,18], rice seedlings were sensitive to Na_2CO_3 treatment, and endophyte infection resulted in significant changes in phenotype and growth parameters. Wu et al. [27] found that inoculation with the dark septate endophyte S16 promoted sweet cherry growth and development.

Genes regulate various metabolic processes in plants, gene expression changes affect metabolite levels to some extent, and these metabolites are associated with potential enzymes or genes that can be directly involved in plant physiological, biochemical, and metabolic processes [29]. In this study, KEGG co-enrichment analysis of screened DAMs and DEGs by integrating transcriptomic and metabolomic data revealed starch and sucrose metabolism, as well as alanine, aspartate, and glutamate metabolism as co-enrichment pathways.

Starch and sucrose metabolism are the main carbohydrate metabolism pathways in plants, providing an adequate carbon source for plant growth and development, whereas carbohydrates, as the main products of photosynthesis, not only provide energy for plant growth but also regulate osmotic homeostasis thus maintaining the dynamic balance of cells under various stress states [30–33]. In addition, carbohydrates are key molecules and reactants that regulate various metabolic processes [34]. It was found that five DAMs (including f-6-p, g-6-p, sucrose-6-phosphate, fructose, and cellobiose) were involved in the starch and sucrose metabolism pathway. In this study, we determined the activities of SS, SPS, HXK, α-GC, and β-GC in endophyte-infected and control rice seedlings under Na_2CO_3 treatment. SS and SPS are the main enzymes involved in sucrose metabolism, and they catalyze the production of UDP-glucose, fructose, and sucrose-6-phosphate, respectively [35]. α-GC, also known as α-D-glucoside hydrolase, is a carbohydrate-digesting enzyme that catalyzes the hydrolysis of α-1,4-glycosidic bonds, or transfers free glucose residues to another carbohydrate substrate to form α-1,6-glycosidic bonds and form isomaltose, while β-GC is an enzyme that hydrolyzes β-glycosidic bonds and participates in the hydrolysis of cellobiose [17,36]. HXK catalyzes the production of f-6-p from fructose, which is further phosphorylated during glycolysis by phosphofructokinase, and glucose is catalyzed by HXK to produce g-6-p, which is essential for polysaccharide biosynthesis [37], as well as providing sufficient substrates for a variety of metabolic pathways.

Metabolome data showed that the majority of DAMs participating in starch and sucrose metabolism were up-regulated, implying that endophyte-infected rice seedlings mitigated the inhibitory effect of saline-alkali by accumulating carbohydrates. Richardson et al. [38] found that the content of carbohydrates in fungal endophyte- (*Acremonium coenophialum* Morgan-Jones & Gains) infected leaves of tall fescue increased when grown under drought stress compared to uninfected tall fescue, which is consistent with the results of the present study. Combined with the metabolome data and sucrose metabolism-related enzyme activity assay data, the results showed that the increase of SS, SPS, HXK, and α-GC activities promoted the accumulation of fructose, g-6-p, f-6-p, and isomaltose, and the decrease of β-GC activity inhibited the synthesis of cellobiose, indicating that sucrose metabolism-related enzymes also actively participated in the synthesis and decomposition of metabolites in starch and sucrose metabolism pathways. Transcriptome data showed that β-GC and HXK were partially induced in endophyte-infected rice seedlings under Na_2CO_3 treatment, indicating activation of starch and sucrose metabolism. SkZ et al. [39] have found that maize seedlings infected with *Pseudomonas putida* showed a strengthened tolerance to drought treatment due to the ability of *Pseudomonas putida* to participate in the regulation of carbohydrate metabolic pathway genes. In the combined analysis of DEGs and DAMs, only f-6-p, g-6-p, and cellobiose content were detected. It was found that the expression of the gene encoding HXK was up-regulated with the accumulation of f-6-p and g-6-p, while the expression of the gene encoding β-GC was down-regulated with the decrease of cellobiose. In addition, combined transcriptome and metabolome data determined that the expression of DEGs involved in starch and sucrose metabolism followed the same trend as the content of DAMs.

Amino acid metabolism is one of the key pathways in plant response to a variety of abiotic stresses [40]. Amino acids are critical in modulating plant growth, physiological metabolism, maintaining intracellular osmoregulation, and protein structural integrity [41,42], and they are an important indicator for assessing crop tolerance to growth, development, and various stresses [43]. Alanine, aspartate, and glutamate metabolism plays important roles in participating in the regulation of phytohormones, thus directly or indirectly regulating crop growth and stress tolerance [44]. GDH acts as a reactive enzyme that catalyzes pyroglutamic acid generation from L-glutamate and assimilates $NH4^+$-N [1]. ASNS is a metabolic enzyme that catalyzes the catabolism of glutamine and aspartate into asparagine [45]. Integration of transcriptome and metabolome data found that down-regulation expression of GDH (*Os04g0543900*) and ASNA *(Os03g0291500)* led to a decrease in the corresponding downstream products L-aspartate, L-glutamate, and L-asparagine, and a decrease in amino acid content, which may indicate that the presence of the endophyte altered the pathway of sugar synthesis and depletion, thereby affected the TCA cycle, changed the composition of the small-molecule osmotic-regulating substances in the plant from amino acids to sugar and sugar alcohols, and ultimately affected the pathway of energy production under Na_2CO_3 treatment. The results showed that the expression of relevant DEGs involved in the alanine, aspartate, and glutamate metabolism was consistent with the trend of DAMs content. Wang et al. [46] reported that genes involved in the regulation of amino acid metabolism were detected to be down-regulated in *Achnatherum inebrians* inoculated with *Epichloë gansuensis* under salt treatment, which is the same as the results of the present experiment.

Given that our results indicate that *Os04g0543900*, *Os03g0749500*, and *Os03g0749650*, which were screened under alanine, aspartate, and glutamate metabolic pathways and starch and sucrose metabolic pathways, have not yet been cloned from plants but may also be involved in the regulation of endophyte to alleviate Na_2CO_3 treatment in rice seedlings, there is a need for further research on these DEGs that are involved in plant–endophyte interactions.

Combining all the results of the studies, it was found that Na_2CO_3-treated rice seedlings showed some changes in gene expression and metabolite levels after endophyte infection; thus, the functions of these metabolites and genes are crucial for studying the intricate interactions between rice and endophytes. Some studies have revealed the saline-alkali resistance mechanism in rice [47,48], but few studies have focused on the mechanism of Na_2CO_3 treatment resistance in rice under endophyte infection. We believe that it is valuable to analyze and study the transcriptome and metabolome of endophyte-infected rice leaves under Na_2CO_3 treatment. In this experiment, DEGs are involved in starch and sucrose metabolic pathways, and alanine, aspartate, and glutamate metabolic pathways have rarely been reported in previous studies. We believe that these DEGs and DAMs will help us better study and understand the molecular mechanisms underlying the reactions of endophyte-infected rice seedlings to Na_2CO_3 treatment.

4. Materials and Methods

4.1. Endophyte Culture and Rice Treatment

Endophyte EF0801 is a fungus provided by the Microbiology Laboratory, College of Life Sciences, Shenyang Normal University (Shenyang, China). Endophyte EF0801 was isolated and screened from *Suaeda salsa* leaves growing on Panjin's Red Beach, which can improve plant saline-alkali tolerance and was identified to be 99% similar to *Sordariomycetes* sp. [28]. Endophyte EF0801 was transferred to 125 mL of potato dextrose agar (PDA) medium and placed on the shaker at 26 °C and 120 rpm for 9 days [49]. The culture can be used to treat rice seedlings after successful cultivation.

The experimental material was Liaoxing No. 1 rice, a major cultivar in Liaoning Province, provided by the Rice Crops Research Institute of Shenyang Agricultural University (Shenyang, China). First, healthy seeds were chosen and soaked in 1% sodium hypochlorite for 0.5 h, then washed with purified water. The 100 seeds were cultivated in a beaker containing 750 mL of Hoagland nutrient solution and incubated in a light incubator (28 °C for 16 h day/26 °C for 8 h night, light intensity 10,000 Lux, and relative humidity 80%). According to our previous research [16,18], 10 mM Na_2CO_3 concentration was selected to treat rice seedlings in this study. Rice seedlings were cultured for five days and then divided into two groups. Na_2CO_3 stressed group (N): cultured in Hoagland solution containing 10 mM Na_2CO_3; Na_2CO_3 stressed and endophyte EF0801-infected group (EN): cultured in Hoagland solution containing 5% fermentation broth and 10 mM Na_2CO_3. Based on the methodology of Liu and Chen [50], it was determined that more than 90% of the rice seedlings were endophyte-infected. On day 7 after the stress treatment, rice seedlings were sampled for further experiments.

4.2. Determination of Growth Parameters

Three replicates were established for both the N and EN groups, with 10 seedlings randomly selected from each replicate to measure plant height and root length.

4.3. Determination of Sucrose Metabolism-Related Enzyme Activities

SS and SPS activities were determined using the method of Zhang et al. [51]. The leaf sample was homogenized in HEPES-NaOH buffer (50 mM, pH 7.5) and then centrifuged at 14,340× *g* for 10 min at 4 °C to obtain the enzyme extract. The mixture contained supernatant, HEPES-NaOH buffer (50 mM; pH 7.5), $MgCl_2$ (50 mM), Fructose (10 mM), and UDPG (10 mM), which was incubated at 37 °C for 30 min. The reaction was stopped by adding 30% NaOH. The activity of SS was determined spectrophotometrically at 480 nm. SPS activity was determined in the same way as SS, except that fructose 6-phosphate was used instead of fructose in the mixture. The activity of SPS was determined spectrophotometrically at 480 nm.

α-Glucosidase (α-GC), β-Glucosidase (β-GC), and Hexokinase (HXK) activities were determined by using α-Glucosidase, β-Glucosidase, and Hexokinase kits (Suzhou Grace Biotechnology Co., Ltd. Suzhou, China)

4.4. Metabolome Analysis

Rice leaves of 30 mg were transferred to a 2 mL Eppendorf (EP) tube and extracted with 0.9 mL of extraction solution ($V_{Methanol}$:V_{H2O} = 3:1). In addition, 0.02 mL of ribitol was added as an internal quantitative standard. The sample was ground in a grinder at 50 Hz for 4 min, and then sonicated in an ice water bath for 10 min, and this process was repeated 3 times. Then, the sample was centrifuged at 14,340× *g* for 15 min at 4 °C. Next, 200 μL of the supernatant was transferred into a 1.5 mL EP tube, 40 μL of methoxyamine hydrochloride was added, and the sample was dried under vacuum at 80 °C for 2 h. After that, 60 μL of BSTFA reagent (1% TMCS, *v*/*v*) was added to the sample aliquots and incubated for 1.5 h at 70 °C. All samples were analyzed using an Agilent 7890 GC-MS (Santa Clara, CA, USA). The mass spectrometry data were acquired in full-scan mode with the *m*/*z* range of 50–500 at a rate of 20 spectra per second after a solvent delay of 6.27 min.

The data were processed using ChromaTOF software (V 4.3x, LECO, St. Joseph, MI, USA) and subsequently analyzed for metabolite data using SMICA 14.1 software, including principal component analysis (PCA), partial least squares discriminant analysis (PLS-DA), and orthogonal partial least squares discriminant analysis (OPLS-DA). DAMs were screened for $Log_2FC \geq 1$, $p < 0.05$ and VIP > 1. DAMs were projected to the MetaboAnalyst 6.0 websites (https://www.metaboanalyst.ca/, accessed on 18 March 2024) and KEGG (https://www.kegg.jp/, accessed on 10 March 2024) database to identify the metabolic pathways involved.

4.5. Transcriptome Analysis

Total RNA was extracted from frozen rice leaves of the N and EN groups using the RNA Simple Kit (TIANGEN Bio Beijing Co., Ltd. Beijing, China). The purity, concentration, RNA integrity, and quality inspection were examined using a Qubit RNA IQ and Qubit Flex fluorometer (Thermo Fisher Scientific Inc., Waltham, MA, USA). A cDNA library was obtained by PCR enrichment, and the 6 gene expression libraries were named N1, N2, N3, EN1, EN2, and EN3. Finally, sequencing was performed with the Illumina HiSeq platform (Biomarker Technologies Co. Ltd, Beijing, China).

Mapping high-quality clean reads to the *Japonica* rice genome database (https://rapdb.dna.affrc.go.jp/, accessed on 13 January 2024) was conducted using HISAT2 (version 2.0.4). String Tie (FPKM = [total exon fragments/mapped reads (millions) × exon length (kb)]) was used to calculate the mRNA expression level. DESeq2 (version 1.6.3) was used for differential expression analysis, and genes with a false discovery rate (FDR) < 0.05 and $|Log_2FC| \geq 0.585$ were considered DEGs. GO enrichment and KEGG enrichment analyses of DEGs were performed using the GOseq R software package (version 1.34.1), KOBAS website, and KEGG database. Statistical tests for GO and KEGG analyses were performed using Fisher's method, and $p \leq 0.05$ was considered a significant change.

4.6. Quantitative Real-Time PCR Analysis

Expression levels of endophyte-infected and control rice leaves under Na_2CO_3 treatment were examined using qRT-PCR to verify the accuracy of the transcriptome data and the indices of sucrose metabolism-related enzyme activities. Specific primers (Tables S5 and S6) were designed through the Primer3plus website (https://www.primer3plus.com/, accessed on 25 April 2024) (Tables S1 and S2). Total RNA was synthesized into complementary DNA (cDNA) using a PrimeScript RT kit (Takara Bio Inc., Shiga, Japan) containing gDNA Eraser. The reaction mixture contained 4 μL of 5× PrimeScript Buffer, 1 μL of PrimeScript RT Enzyme Mix I, and 1 μL of gDNA Eraser. Reaction conditions are as follows: 37 °C for 15 min and 85 °C for 5 s to inactivate the reverse transcriptase. The qRT-PCR run was conducted on the LightCycler96 PCR system (Roche Co., Ltd., Basel, Switzerland), using SYBR Green I as the fluorescent dye. Each 20 μL reaction mixture contained 10 μL of 2× SYBR Green I Master Mix, 2 μL of diluted cDNA, 0.8 μL of PCR forward primer (10 μM), 0.8 μL of PCR reverse primer, and 6.4 μL nuclease-free water. The qRT-PCR cycling conditions were as follows: initial denaturation at 95 °C for 10 min, 40 cycles at 95 °C for 15 s each, and 60 °C for 1 min. The specificity of the amplification was confirmed by melting curve analysis. In addition, 18s rRNA was quantified as an internal control, and $2^{-\Delta\Delta CT}$ was used to calculate the relative expression of genes.

4.7. Data Analysis

Significance was analyzed using a two-tailed Student's *t*-test. Statistical analysis was performed using SPSS 29.0, and $p < 0.05$ was considered significant. Results were expressed as mean ± standard deviation (SD).

Supplementary Materials: The following supporting information can be downloaded at: https://www.mdpi.com/article/10.3390/plants14101524/s1, Table S1: Mapping of RNA-Seq data of rice seedlings to the reference genome. Table S2: KEGG enrichment analysis of DEGs in endophyte-infected versus control rice seedlings under Na_2CO_3 stress. Table S3: Loading of metabolites to the first principal component (PC1) and the second principal component (PC2). Table S4: KEGG enrichment analysis of endophyte-infected versus control rice seedlings under Na_2CO_3 stress. Table S5: Primers used for the qRT-PCR. Table S6: Primers for qRT-PCR of sucrose metabolism-related enzymes.

Author Contributions: Conceptualization, X.W.; Data curation, X.W. and H.S.; Formal analysis, Y.L.; Funding acquisition, X.L.; Investigation, X.W. and H.S.; Methodology, X.W. and Y.L.; Project administration, L.Z. and X.L.; Resources, X.L.; Software, X.W. and Y.L.; Supervision, L.Z. and X.L.; Validation, X.W.; Visualization, L.Z.; Writing—original draft, X.W.; Writing—review and editing, L.Z. and X.L. All authors have read and agreed to the published version of the manuscript.

Funding: This research was funded by the National Natural Science Foundation (31270369) and the Department of Education of Liaoning Province (LZD202004).

Data Availability Statement: Data will be made available on request.

Conflicts of Interest: The authors have declared that no conflicts of interest exist.

References

1. Shahid, M.A.; Balal, R.M.; Khan, N.; Zotarelli, L.; Liu, G.D.D.; Sarkhosh, A.; Fernández-Zapata, J.C.; Martínez, N.J.J.; Garcia-Sanchez, F. Selenium impedes cadmium and arsenic toxicity in potato by modulating carbohydrate and nitrogen metabolism *Ecotoxicol. Environ. Saf.* **2019**, *180*, 588–599. [CrossRef] [PubMed]
2. Morton, M.J.L.; Awlia, M.; Al-Tamimi, N.; Saade, S.; Pailles, Y.; Negrão, S.; Tester, M. Salt stress under the scalpel-dissecting the genetics of salt tolerance. *Plant J.* **2019**, *97*, 148–163. [CrossRef] [PubMed]
3. Zahra, N.; Uzair, M.; Zaid, I.U.; Attia, K.A.; Inam, S.; Fiaz, S.; Abdallah, R.M.; Naeem, M.K.; Farooq, U.; Rehman, N.; et al. The comparative transcriptome analysis of two green super rice genotypes with varying tolerance to salt stress. *Mol. Biol. Rep.* **2024**, *51*, 22. [CrossRef]
4. Liu, X.L.; Xie, X.Z.; Zheng, C.K.; Wei, L.X.; Li, X.W.; Jin, Y.Y.; Zhang, G.H.; Jiang, C.J.; Liang, Z.W. RNAi-mediated suppression of the abscisic acid catabolism gene *OsABA8ox1* increases abscisic acid content and tolerance to saline-alkaline stress in rice (*Oryza sativa* L.). *Crop. J.* **2022**, *10*, 354–367. [CrossRef]
5. Shaheen, S.; Baber, M.; Aslam, S.; Aslam, S.; Shaheen, M.; Waheed, R.; Seo, H.; Azhar, M.T. Effect of NaCl on morphophysiological and biochemical responses in *Gossypium hirsutum* L. *Agronomy* **2023**, *13*, 1012. [CrossRef]
6. Wang, H.; Wu, Z.; Chen, Y.; Yang, C.; Shi, D. Effects of salt and alkali stresses on growth and ion balance in rice (*Oryza sativa* L.). *Plant Soil Environ.* **2011**, *57*, 286–294. [CrossRef]
7. Amirbakhtiar, N.; Ismaili, A.; Ghaffari, M.-R.; Mirdar, M.R.; Sanjari, S.; Shobbar, Z.-S. Transcriptome analysis of bread wheat leaves in response to salt stress. *PLoS ONE* **2021**, *16*, e0254189. [CrossRef] [PubMed]
8. Gupta, P.; De, B. Metabolomics analysis of rice responses to salinity stress revealed elevation of serotonin and gentisic acid levels in leaves of tolerant varieties. *Plant Signal. Behav.* **2017**, *12*, e1335845. [CrossRef]
9. Sun, J.; Xie, D.W.; Zhang, E.Y.; Zheng, H.L.; Wang, J.G.; Liu, H.L.; Yang, L.M.; Zhang, S.L.; Wang, L.; Zou, D.T. QTL mapping of photosynthetic-related traits in rice under salt and alkali stresses. *Euphytica* **2019**, *215*, 147. [CrossRef]
10. Gond, S.K.; Bergen, M.S.; Torres, M.S.; White, J.J. Endophytic *Bacillus* spp. produce antifungal lipopeptides and induce host defence gene expression in maize. *Microbiol. Res.* **2015**, *172*, 79–87. [CrossRef]
11. Usuki, F.; Narisawa, K. A mutualistic symbiosis between a dark septate endophytic fungus, *Heteroconium chaetospira*, and a nonmycorrhizal plant, Chinese cabbage. *Mycologia* **2007**, *99*, 175–184. [CrossRef] [PubMed]
12. Yadav, V.; Kumar, M.; Deep, D.K.; Kumar, H.; Sharma, R.; Tripathi, T.; Tuteja, N.; Saxena, A.K.; Johri, A.K. A phosphate transporter from the root endophytic fungus *Piriformospora indica* plays a role in phosphate transport to the host plant. *J. Biol. Chem.* **2010**, *285*, 26532–26544. [CrossRef] [PubMed]
13. Jiang, C.H.; Yao, X.F.; Mi, D.D.; Li, Z.J.; Yang, B.Y.; Zheng, Y.; Qi, Y.J.; Guo, J.H. Comparative transcriptome analysis reveals the biocontrol mechanism of *Bacillus velezensis* F21 against *Fusarium* wilt on watermelon. *Front. Microbiol.* **2019**, *10*, 652. [CrossRef] [PubMed]
14. Waqas, M.; Khan, A.L.; Shahzad, R.; Ullah, I.; Khan, A.R.; Lee, I.J. Mutualistic fungal endophytes produce phytohormones and organic acids that promote japonica rice plant growth under prolonged heat stress. *J. Zhejiang Univ. Sci. B* **2015**, *16*, 1011–1018. [CrossRef]
15. Khan, A.L.; Lee, I.J. Endophytic *Penicillium funiculosum* LHL06 secretes gibberellin that reprograms *Glycine max* L. growth during copper stress. *BMC Plant Biol.* **2013**, *13*, 86. [CrossRef]
16. Li, X.M.; Chen, M.J.; Li, J.; Ma, L.J.; Li, Y.Y.; Bu, N.; Li, Y.Y.; Zhang, L.H. Effect of endophyte infection on chlorophyll *a* fluorescence in salinity stressed rice. *Biol. Plant.* **2014**, *58*, 589–594. [CrossRef]
17. Li, P.; Liang, H.; Lin, W.T.; Feng, F.; Luo, L.X. Microbiota dynamics associated with environmental conditions and potential roles of cellulolytic communities in traditional Chinese cereal starter solid-state fermentation. *Appl. Environ. Microbiol.* **2015**, *81*, 5144–5156. [CrossRef]
18. Ren, X.N.; Shan, Y.; Li, X.; Wang, L.L.; Li, Y.Y.; Ma, L.J.; Li, X.M. Endophytic infection programs the ascorbate-glutathione cycle in rice (*Oryza sativa* L.) under Na_2CO_3 stress. *Appl. Ecol. Environ. Res.* **2021**, *19*, 1895–1907. [CrossRef]
19. Ott, K.H.; Aranibar, N.; Singh, B.; Stockton, G.W. Metabonomics classifies pathways affected by bioactive compounds. Artificial neural network classification of NMR spectra of plant extracts. *Phytochemistry* **2003**, *62*, 971. [CrossRef]
20. Sekiyama, Y.; Okazaki, K.; Kikuchi, J.; Ikeda, S. NMR-Based metabolic profiling of field-grown leaves from sugar beet plants harbouring different levels of resistance to cercospora leaf spot disease. *Metabolites* **2017**, *7*, 4. [CrossRef]
21. Ghaffari, M.R.; Ghabooli, M.; Khatabi, B.; Hajirezaei, M.R.; Schweizer, P.; Salekdeh, G.H. Metabolic and transcriptional response of central metabolism affected by root endophytic fungus *Piriformospora indica* under salinity in barley. *Plant Mol. Biol.* **2016**, *90*, 699–717. [CrossRef] [PubMed]
22. Shi, Y.L.; Sheng, Y.Y.; Cai, Z.Y.; Yang, R.; Li, Q.S.; Li, X.M.; Li, D.; Guo, X.Y.; Lu, J.L.; Ye, J.H.; et al. Involvement of salicylic acid in anthracnose infection in tea plants revealed by transcriptome profiling. *Int. J. Mol. Sci.* **2019**, *20*, 2439. [CrossRef] [PubMed]
23. Ren, X.N.; Shan, Y.; Li, X.; Fan, J.H.; Li, Y.Y.; Ma, L.J.; Wang, L.L.; Li, X.M. Application of RNA sequencing to understand the benefits of endophytes in the salt-alkaline resistance of rice seedlings. *Environ. Exp. Bot.* **2022**, *196*, 104820. [CrossRef]

24. Li, M.; Zhou, J.; Lang, X.; Han, D.; Hu, Y.; Ding, Y.; Wang, G.; Guo, J.; Shi, L. Integrating transcriptomic and metabolomic analysis in roots of wild soybean seedlings in response to low-phosphorus stress. *Front. Plant Sci.* **2022**, *13*, 1006806. [CrossRef]
25. Shao, M.; Chen, Y.; Gong, Q.; Miao, S.; Li, C.; Sun, Y.; Qin, D.; Guo, X.; Yan, X.; Cheng, P.; et al. Biocontrol endophytes *Bacillus subtilis* R31 influence the quality; transcriptome and metabolome of sweet corn. *PeerJ* **2023**, *11*, e14967. [CrossRef] [PubMed]
26. Malik, A.; Butt, T.A.; Naqvi, S.; Yousaf, S.; Iqbal, M. Lead tolerant endophyte *Trametes hirsuta* improved the growth and lead accumulation in the vegetative parts of *Triticum aestivum* L. *Heliyon* **2020**, *6*, e04188. [CrossRef]
27. Wu, F.; Qu, D.; Tian, W.; Wang, M.; Chen, F.; Li, K.; Sun, Y.; Su, Y.; Yang, L.; Su, H.; et al. Transcriptome analysis for understanding the mechanism of dark septate endophyte S16 in promoting the growth and nitrate uptake of sweet cherry. *J. Integr. Agric.* **2021**, *20*, 1819–1831. [CrossRef]
28. Li, X.M.; Ma, L.J.; Bu, N.; Li, Y.Y.; Zhang, L.H. Endophytic infection modifies organic acid and mineral element accumulation by rice under Na_2CO_3 stress. *Plant Soil* **2017**, *420*, 93–103. [CrossRef]
29. Rodriguez-Garcia, A.; Sola-Landa, A.; Barreiro, C. RNA-Seq-Based Comparative Transcriptomics, RNA Preparation and Bioinformatics. In *Microbial Steroids, Methods and Protocol*; Humana Press: New York, NY, USA, 2017; pp. 59–72.
30. Coueé, I.; Sulmon, C.; Gouesbet, G.; Amrani, A.E. Involvement of soluble sugars in reactive oxygen species balance and responses to oxidative stress in plants. *J. Exp. Bot.* **2006**, *3*, 449–459. [CrossRef]
31. Jammer, A.; Gasperl, A.; Luschin-Ebengreuth, N.; Heyneke, E.; Chu, H.; Cantero-Navarro, E.; Großkinsky, K.D.; Albacete, A.; Stabentheiner, E.; Franzaring, J.; et al. Simple and robust determination of the activity signature of key carbohydrate metabolism enzymes for physiological phenotyping in model and crop plants. *J. Exp. Bot.* **2015**, *66*, 5531–5542. [CrossRef]
32. Krasensky, J.; Jonak, C. Drought, salt and temperature stress-induced metabolic rearrangements and regulatory networks. *J. Exp. Bot.* **2012**, *63*, 1593–1608. [CrossRef] [PubMed]
33. Nemati, F.; Ghanati, F.; Gavlighi, H.A.; Sharifi, M. Comparison of sucrose metabolism in wheat seedlings during drought stress and subsequent recovery. *Biol. Plant.* **2018**, *62*, 595–599. [CrossRef]
34. Xu, Z.; Chen, X.; Lu, X.; Zhao, B.; Yang, Y.; Liu, J. Integrative analysis of transcriptome and metabolome reveal mechanism of tolerance to salt stress in oat (*Avena sativa* L.). *Plant Physiol. Biochem.* **2021**, *160*, 315–328. [CrossRef] [PubMed]
35. Fu, J.; Huang, B.; Fry, J. Osmotic potential, sucrose level, and activity of sucrose metabolic enzymes in tall fescue in response to deficit irrigation. *J. Am. Soc. Hortic. Sci.* **2010**, *135*, 506–510. [CrossRef]
36. Yan, H.; Nie, Y.; Cui, K.; Sun, J. Integrative transcriptome and metabolome profiles reveal common and unique pathways involved in seed initial imbibition under artificial and natural salt stresses during germination of halophyte quinoa. *Front. Plant Sci.* **2022**, *13*, 853326. [CrossRef]
37. Granot, D.; Kelly, G.; Stein, O.; David-Schwartz, R. Substantial roles of hexokinase and fructokinase in the effects of sugars on plant physiology and development. *J. Exp. Bot.* **2014**, *65*, 809–819. [CrossRef]
38. Richardson, M.D.; Chapman, G.W.; Hoveland, C.S.; Bacon, C.W. Sugar alcohols in endophyte-infected tall fescue under drought. *Crop Sci.* **1992**, *32*, 1060–1061. [CrossRef]
39. SkZ, A.; Vardharajula, S.; Vurukonda, S.S.K.P. Transcriptomic profiling of maize (*Zea mays* L.) seedlings in response to *Pseudomonas putida stain FBKV2* inoculation under drought stress. *Ann. Microbiol.* **2018**, *68*, 331–349. [CrossRef]
40. Heinemann, B.; Hildebrandt, T.M. The role of amino acid metabolism in signaling and metabolic adaptation to stress-induced energy deficiency in plants. *J. Exp. Bot.* **2021**, *72*, 4634–4645. [CrossRef]
41. Huang, T.; Jander, G. Abscisic acid-regulated protein degradation causes osmotic stress-induced accumulation of branched-chain amino acids in *Arabidopsis thaliana*. *Planta* **2017**, *246*, 737. [CrossRef]
42. Zhao, X.; Huang, L.J.; Sun, X.F.; Zhao, L.L.; Wang, P.C. Transcriptomic and metabolomic analyses reveal key metabolites; pathways and candidate genes in *Sophora davidii* (franch.) skeels seedlings under drought stress. *Front. Plant Sci.* **2022**, *13*, 785702. [CrossRef] [PubMed]
43. Dey, S.; Kundu, R.; Gopal, G.; Mukherjee, A.; Nag, A.; Paul, S. Enhancement of nitrogen assimilation and photosynthetic efficiency by novel iron pulsing technique in *Oryza sativa* L. var Pankaj. *Plant Physiol. Biochem.* **2019**, *144*, 207–221. [CrossRef] [PubMed]
44. Less, H.; Galili, G. Principal transcriptional programs regulating plant amino acid metabolism in response to abiotic stresses. *Plant Physiol.* **2008**, *147*, 316. [CrossRef] [PubMed]
45. Rashmi, D.; Barvkar, V.T.; Nadaf, A.; Mundhe, S.; Kadoo, N.Y. Integrative omics analysis in *Pandanus odorifer* (Forssk.) Kuntze reveals the role of asparagine synthetase in salinity tolerance. *Sci. Rep.* **2019**, *9*, 932. [CrossRef]
46. Wang, C.; Huang, R.; Wang, J.; Jin, J.; Malik, K.; Niu, X.; Tang, R.; Hou, W.; Cheng, C.; Liu, Y.; et al. Comprehensive analysis of transcriptome and metabolome elucidates the molecular regulatory mechanism of salt resistance in roots of *Achnatherum inebrians* mediated by *Epichloë gansuensis*. *J. Fungi* **2022**, *8*, 1092. [CrossRef]
47. Ren, X.N.; Fan, J.H.; Li, X.; Shan, Y.; Wang, L.L.; Ma, L.J.; Li, Y.Y.; Li, X.M. Application of RNA sequencing to understand the response of rice seedlings to salt-alkali stress. *BMC Genom.* **2023**, *24*, 21. [CrossRef]
48. Zhou, Z.; Liu, J.; Meng, W.; Sun, Z.; Tan, Y.; Liu, Y.; Tan, M.; Wang, B.; Yang, J. Integrated analysis of transcriptome and metabolome reveals molecular mechanisms of rice with different salinity tolerances. *Plants* **2023**, *12*, 3359. [CrossRef]

49. Li, X.M.; Bu, N.; Li, Y.Y.; Ma, L.J.; Xin, S.G.; Zhang, L.H. Growth, photosynthesis and antioxidant responses of endophyte infected and non-infected rice under lead stress conditions. *J. Hazard. Mater.* **2012**, *213*, 55–61. [CrossRef]
50. Liu, R.J.; Chen, Y.L. *Mycorrhizology*; Science Press: Beijing, China, 2007.
51. Zhang, Z.L.; Qu, W.J.; Li, X.F. *Plant Physiology Laboratory Guide*; Higher Education Press: Beijing, China, 2009.

plants

MDPI

Opinion

Polyethylene Glycol (PEG) Application Triggers Plant Dehydration but Does Not Accurately Simulate Drought

Gulnar Kylyshbayeva [1], Nazira Bishimbayeva [2], Sativaldy Jatayev [3], Serik Eliby [4] and Yuri Shavrukov [5,*]

1 Faculty of Natural Sciences, Central Asian Innovation University, Shymkent 160000, Kazakhstan; kuntun-gulnar@mail.ru
2 Research Institute for Biology and Biotechnology Problems, Al-Farabi Kazakh National University, Almaty 050040, Kazakhstan
3 Faculty of Agronomy, S. Seifullin Kazakh AgroTechnical Research University, Astana 010011, Kazakhstan; s.jatayev@kazatu.edu.kz
4 Australian Centre for Plant Functional Genomics, University of Adelaide, Adelaide, SA 5064, Australia; selib1961@gmail.com
5 College of Science and Engineering, Biological Sciences, Flinders University, Adelaide, SA 5042, Australia
* Correspondence: yuri.shavrukov@flinders.edu.au

Abstract: Polyethylene glycol (PEG), especially at high molecular weights, is highly soluble in water, and these solutions have reduced water potential. It is convenient to use PEG in hydroponics (liquid nutrient solution) for experiments with plants. However, some authors have been found to describe the application of PEG to plants incorrectly, such as drought, dehydration, osmotic, or water stresses, which can mislead readers. The presented opinion paper shows our arguments for a terminology in such experiments that is strictly limited to 'PEG-induced' or 'simulated' or 'mimicked' dehydration, and osmotic or water stresses, with the best option being 'PEG-induced dehydration'. The most popular term, 'drought', is inappropriate to be used for hydroponics at all, with or without PEG. Traditionally, drought stress study was related to only plants in soil or other substrates mixed with soil. Based on 139 published papers, the examples presented in our opinion paper can demonstrate differences in gene expression between plants grown in containers with soil and under PEG-induced stress in hydroponics. Researchers can carry out any type of experiments suitable for the purposes of their study. However, clear and correct description of experiments and careful interpretation of the results are strongly required, especially with PEG, to avoid incorrect information. In all cases, at the final stage, results of experiments in controlled conditions have to be verified in field trials with naturally occurring drought.

Keywords: containers with soil; dehydration; drought; field trial; gene expression; hydroponics; polyethylene glycol (PEG)

1. Introduction

Readers cannot imagine the many thousands or even millions of papers in the literature that begin with a phrase along the lines of 'Drought is the biggest threat ...' or similar. This indicates the extent to which drought is regarded as both an extremely important abiotic stress and one of the biggest challenges for living organisms, and plants especially, in conditions of global climate changes. Drought is defined as a single stretch of water deficit, lack, or cessation [1]. However, drought stress affecting crops and natural plant species has extremely complicated effects, where many factors, including the nature of plant species, developmental stage, geographic location, climate, soil, and many others,

https://doi.org/10.3390/plants14010092

are involved [2]. Dehydration is an unavoidable component of drought and means loss of water from plant cells as a result of drought, heat, or salinity stress [3,4]. Dehydration is generally triggered by osmotic signals [5] and is an important part of plant reactions to drought. Osmotic stress occurs due to an imbalance of the osmotic gradient inside and outside of cells and usually requires rapid cellular responses to adjust and maintain a stable osmotic balance and adapt to challenging environmental conditions [6,7].

The study of natural drought is possible only in field trials, where plants grow in soil and rain-fed water is somehow stopped or limited. However, such natural environments are unpredictable due to fluctuations of climate and natural precipitation [8]. Therefore, for experiments studying plants under drought stress, artificially controlled and stable conditions with drying soil are much more suitable, particularly for reproducibility and repetition of experiments, including high-throughput phenotyping technologies [9]. Various bins, buckets, pots, and any other types of containers filled with soil are typically used in controlled greenhouses, chambers, cabinets, and even on laboratory benchtops to study the effects of drought on plants. However, in strong terms, such experiments must be classified as 'artificially controlled drought' compared to a 'natural' one. Nevertheless, this designation is very rarely used, perhaps because the description of 'research field' or 'non-field' conditions is obvious and clear in plant study.

In contrast, a dramatic difference can be made when artificial chemical compounds are used to absorb water and limit its availability to studied plants. Polyethylene glycol (PEG), usually with high molecular weight (6000 or 8000), remains the most popular reagent for such experiments, usually used at a concentration of 10–30% [10]. Liquid PEG is used as a supplement to growth solutions containing the nutrients required for plant growth, and despite quite variable protocols, descriptions, and applications, all experiments fall under the umbrella term of 'hydroponics'. It was perfectly described on the web site 'Use of PEG' [11].

Only one paper has been found [12], where a PEG solution was applied directly to pots with transgenic *Arabidopsis thaliana* plants grown in Sunshine Mix #5 soil (Sungro, Chesapeake, VI, USA) mixed with vermiculite (3:1). The authors described it as 'Chemical drought stress' but it is hard to predict how PEG interacts with soil particles, organic and non-organic matter in soil. Nevertheless, it seems this method of PEG application is a very rare case [12]. Additionally, PEG can be added to agar-based tissue culture media in Petri dishes, which very convenient for model plant analyses or germinated seeds and young seedlings [4]. These methods of PEG treatments, together with soil and hydroponic strategies, were comprehensively reviewed with many examples [13]. However, the authors concluded that this method of using solid agar media with added PEG is very far from a direct extrapolation from laboratory models to field trials in dry conditions, but it remains very suitable for pre-screening of certain compounds in initial studies.

The aim of the presented paper was to justify our opinion that the application of PEG in any type of hydroponics is very often incorrectly described in publications as 'drought' or 'PEG-drought'. Instead, the best option is to define it more accurately as 'PEG-induced dehydration'. The assessment is provided for 139 published papers where both terms 'PEG' and 'drought' were used, and five cases were analysed in more detail where both drought in soil and PEG-induced dehydration were studied simultaneously.

2. Water Potential in Experiments with Soil and in Hydroponics

Experiments in containers with soil and in hydroponics represent two different types of study. Soil is a very complicated mix composed of inorganic and organic products, along with diverse communities of microorganisms existing in and around plant roots [14,15].

In contrast, hydroponics was designed for quick delivery of water and nutrients from the solution to plant roots [16,17].

Water potential Ψ_W, or water available for plant uptake [4], can be reduced progressively in experiments with soil after watering is withheld, and this process is very dependent on soil type and its water capacity, volume, and overall density, which all affect the speed of water loss [18]. In some cases, water field capacity in soil or permanent wilting point was used in experiments with plants grown in containers with soil [19]. However, this required additional calculations to estimate water potential. There are some difficulties in precise measurement of Ψ_W in soil-based experiments, which is easier to do in hydroponics with the addition of PEG [20]. To compare different experiments with plants grown in soil and in solutions with PEG, it is important to arrange a similar level of water potential in leaves, Ψ_{leaf}, which could be the most suitable indicator showing that plants have similar levels of water limitation in different conditions of growing (containers with soil or hydroponics). Two excellent examples of such a comparison are presented in greater detail in Section 4 below as special cases [21,22].

However, these two cases providing evidence of similar levels of measured water potential Ψ_{leaf} in plants grown in soil without watering and in hydroponics with PEG are very rare. The majority of all other studied publications only indicate the percentage of PEG added and include no study of plants in soil, since researchers are primarily interested in producing consistent and reproducible results. Therefore, concentrations of applied PEG and duration of treatments were determined empirically to reach an adequate level of stress [23] or were based on previous publications and initial tests [24,25].

The estimated reduction of water potential was present in experiments with the gradual addition of 10% PEG-6000 for plant treatments in hydroponics, where water (or osmotic) potential was reduced to reach levels of −1.48 MPa [24] or −1.50 MPa [26]. However, Ψ_W in solution is not the same as that within leaves and, therefore, such results cannot be directly compared with other data, particularly in experiments with soil.

3. Correct and Incorrect Description of Experiments with PEG Application to Plants

A portion of published papers was searched in the Scopus database using the terms 'PEG', 'drought', and 'dehydration' in the Title, Abstract, and Keywords. This yielded 143 published papers suitable for individual checking. Four papers were eliminated in the initial stage as studies involving tissue culture, calli, and embryonic cells, which are not suitable for the subject of the presented analysis. Results of the remaining 139 published papers are present in Table 1, with full details in Supplementary Table S1.

The largest number of papers (40%) described their experiments using PEG solution as a supplement in hydroponics as 'PEG-treatment' or sometimes as 'PEG stress'. This is a very neutral description avoiding any question about the type of stress, which authors do not want to discuss as perhaps beyond the scope of their studies (Table 1).

The four most critical groups of publications described the conditions used to grow plants in hydroponics with PEG as follows: drought (26%), dehydration (18%), osmotic stress (7%), and water stress (9%). With some variation, about half of the published papers in each of these four groups described PEG application or supplements to hydroponics using the simple terms 'drought, dehydration, osmotic and water stress', respectively. This incorrect use of these terms for hydroponics accounts for more than $\frac{1}{4}$ of the studied papers in total (27% or 38 papers in Table 1). There is also a very interesting observation that each author's team interpreted the application of PEG as they wished, depending on the focus of their research. For example, if the authors studied drought, their descriptions and discussions of PEG treatment experiments were only related to drought using exclusively this term. Similarly, if the topics of papers concerned dehydration, osmotic or water stress, only these terms were used, respectively.

Table 1. Analysis overview of 139 published papers containing PEG application to plants in hydroponics and author descriptions of their experiments as either PEG treatment, drought, dehydration, osmotic or water stress. Special terms are indicated in bold.

<table>
<tr><th>Description</th><th>Sub-Description</th><th>No. of Papers</th><th>%</th><th>Description in Abstracts or in the Text of Papers</th><th>Ref.</th></tr>
<tr><td colspan="2">PEG treatment</td><td>55</td><td>40</td><td>PEG treatment; PEG stress</td><td>[27,28]</td></tr>
<tr><td rowspan="2">Drought</td><td>Drought</td><td>17</td><td rowspan="2">26</td><td>Drought stress; drought (PEG treatment)</td><td>[29,30]</td></tr>
<tr><td>PEG-induced drought</td><td>19</td><td>PEG-induced drought; PEG treatments to simulate drought stress; drought stress induced by PEG</td><td>[31,32]</td></tr>
<tr><td rowspan="2">Dehydration</td><td>Dehydration</td><td>12</td><td rowspan="2">18</td><td>Dehydration stress; dehydration tolerance in response to PEG treatment</td><td>[33,34]</td></tr>
<tr><td>PEG-induced dehydration</td><td>13</td><td>PEG-induced dehydration; PEG-simulated dehydration; PEG-mimic dehydration</td><td>[35,36]</td></tr>
<tr><td rowspan="2">Osmotic stress</td><td>Osmotic stress</td><td>4</td><td rowspan="2">7</td><td>Osmotic stress; PEG osmotic stress</td><td>[37,38]</td></tr>
<tr><td>PEG-induced osmotic stress</td><td>6</td><td>PEG-induced osmotic stress; PEG-simulated osmotic stress</td><td>[39,40]</td></tr>
<tr><td rowspan="2">Water-stress</td><td>Water-stress</td><td>5</td><td rowspan="2">9</td><td>Water stress; water-deficit stress</td><td>[41,42]</td></tr>
<tr><td>PEG-induced water- stress</td><td>8</td><td>PEG-induced water deficiency; PEG-mediated water stress; PEG-simulated water deficit; water stress induced by PEG; water stress mimicked by PEG</td><td>[25,43]</td></tr>
<tr><td colspan="2">Total</td><td>139</td><td>100</td><td>PEG supplement (application) in hydroponics</td><td></td></tr>
</table>

The biggest problem we see for these authors is that the application of PEG is non-direct and artificially causes effects in plants, which differ from the traditionally described phenomenon of drought, dehydration, osmotic or water stress. Therefore, authors from the latter groups showed much better descriptions, using the terms 'PEG-induced', 'simulated', 'mediated', or 'mimicked' (33% or 46 papers in Table 1, indicated in bold). These authors emphasized that PEG application causes indirect effects and needs to be described with additional words like 'induced, simulated, mediated', and so on. Such a decision is the best option since it clearly indicates correct use of the scientific terms for the description of their experiments with PEG treatment (Table 1).

4. Analysis of Special Cases with Experiments Both in Soil/Substrate and Hydroponics

Below, five papers were selected as examples where authors studied plants grown in both soil/substrate and hydroponics with PEG simultaneously. For these cases we compared the published results for gene expression analyses when it was possible. All presented cases are very different, and it was particularly of interest to show studies and results with very diverse plant species or different stages of plants and the specific experimental designs. Our attention was especially directed to the Materials and Methods section and how the authors described the conditions they used to grow plants in soil or substrates and PEG-treated hydroponics and what kind of results were reported.

4.1. Case 1: Cotton Plants in Hydroponics with Sand and PEG Treatment

The first selected paper was a study of the comparison of cotton plants grown "under stress of PEG-drought compared with Soil-drought" [21]. Despite the authors' statement, there was no soil but only sand as the substrate with added growth solution. So, in fact, it was a study of a "culture matrix" with or without the addition of PEG solution. Nevertheless, the authors in this study were particularly focused on the identical water potential in leaves, Ψ_{leaf}, as described in Section 2 above [21].

Three-leaf stage young plants of the upland cotton accession Zhong H177 were grown in sand as the culture substrate. For PEG treatment, growth solution with 5% PEG was used for growing plants for 6 h. For dehydration without PEG, growth solution was withdrawn for three days, and plants were exposed to slowly drying sand. The very important step was to choose conditions of treatments and to make identical water potential in leaves of plants in both treatments, $\Psi_{leaf} = -2.07$ MPa. It made possible the near-perfect comparison of plants from different experiments but with the same water potential [21].

In the results, the authors concluded that the physiological and biochemical characteristics in the studied cotton plants were similar in both treatments, with and without PEG. However, significant differences were found in the expression of seven genes related to photosynthesis, which were up-regulated during slow dehydration without PEG, but the expression of the same genes was steady or down-regulated in the presence of 5% PEG. The similar increased expression levels were reported for eight glucose metabolite-related genes, three sugar transporter genes, and some other genes in slowly drying sand without PEG compared to PEG treatment but with the same level of water potential in leaves [21]. Therefore, PEG application and slowly drying sand had different effects on plants with the same water potential.

4.2. Case 2: Medicago Truncatula Plants in Substrate and PEG Treatment

In the second case, seedlings of *Medicago truncatula* were grown for 11 weeks in a mix of perlite–vermiculite (1:3), irrigated with growth solutions. The substrate mix used is artificial and commonly used in laboratory experiments where absorption and retention of moisture for a certain time is required, but this is still different from real soil. In this experiment, no genes were analysed but only physiological and biochemical parameters [22].

After 11 weeks of growth, *M. truncatula* plants were treated for 7 days either with a 2/3 reduction of irrigation in the substrate mix, indicated as 'No-W (no watering)', or with the addition of 25% PEG-6000 solution. Leaf water potential was measured and confirmed as identical in both treatments, $\Psi_{leaf} = -1.68$ MPa, making equal dehydration pressure, corresponding to a moderate water-deficit level [22].

Based on the comparative analysis, the authors concluded that some physiological parameters, including stomatal conductance and leaf transpiration, were the same in both the No-W and the PEG studies. However, accumulation and content of water were significantly higher in roots and lower in the leaves of *M. truncatula* plants grown in PEG compared to No-W treatment. Similar differences were found for carbohydrate accumulation in leaves, significantly higher for sucrose content and a remarkable 'zero' accumulation of starch in leaves after PEG treatment compared to exactly opposite trends in No-W plants. The authors reported on many other studied compounds and traits showing big differences between No-W and PEG treatments, and they finally summarised that PEG is "a false mimetic of drought". It is important to emphasize that leaf water potential was identical in the presented experiments [22].

4.3. Case 3: Wild Barley Plants in Soil, Leaf Dehydration, and PEG Treatment

The next case was related to gene analysis, but water potential was not measured or reported in this study. Young barley plants (four-leaf stage) were studied in several conditions, including the following: (1) gradual drought in pots with soil (local soil, nutrient soil, and vermiculite in a ratio of 7:2:1); (2) rapid dehydration of detached leaves on a bench at room temperature (called 'dehydration shock'); and (3) 20% PEG 6000 supplied to 1/8 liquid MS medium (hydroponics) [44]. Relative expression of the *HvLRX* (light-responsive X), described by the authors as a new dehydration- and light-responsive gene, was analysed in two drought-tolerant and sensitive accessions.

Under drought stress, the *HvLRX* expression gradually increased during 5 and 7 days of drought, reaching about three units in leaves of both accessions. In contrast, under rapid dehydration, time points and magnitudes of *HvLRX* expression were sharply upregulated over 1 and 5 h to about 17–22 units. However, results of *HvLRX* expression in hydroponics with PEG in leaves of the same barley accessions was dissimilar to both rapid dehydration and gradual drought, instead fluctuating between about 2.2 and 3.2 units with the maximum after 8–12 h of PEG treatment [44]. The results of this study indicated that PEG treatment showed the induction of plant reaction to stress; however, the simulation of real drought or dehydration conditions in this case was very limited.

4.4. Case 4: Breeding Lines of Maize Grown in Soil, Dehydrated Plants, and PEG Treatment

Young maize plants (three-leaf stage) were studied in the following conditions: (1) gradual drought in plastic containers filled with a mixture of nutrient soil and vermiculite (1:1); (2) rapid dehydration treatment with plants pulled from pots and air-dried on paper; and (3) 20% PEG treatment. Plants grown in vermiculite were rinsed and immersed in Hoagland solution containing PEG. Relative expressions of NAC transcription factor genes, *ZmNAC87* and *ZmJUB1*, were analysed in maize inbred lines Lv28 and H082183 [45].

In an experiment with moderate and severe drought (several days of treatment), expressions of both genes were mixed in both roots and shoots in the studied breeding lines, reaching maximums of 8–12 unit levels for gene *ZmNAC87* and 6–8 unit levels for *ZmJUB1*. Very different results were reported for rapid dehydration, where expression of *ZmNAC87* was highest in shoots (30 units, 3 h) and in roots (35–55 units, 3–24 h). The expression of *ZmJUB1* was recorded at levels 5–35 in shoots and 1–5 units in roots. Importantly, some similarity was found in *ZmJUB1* expression during PEG treatment, which varied between 3–15 units in shoots and 1–5 in roots. It was particularly clear for breeding line Lv28 but more variable in breeding line H082183. In contrast, expressions of *ZmNAC87* in PEG-treated plants were very low and quite different from those in the dehydration experiment [45]. The presented results indicated relative similarity between experiments with rapid dehydration and PEG treatment in hydroponics for *ZmJUB1* expression but not for the *ZmNAC87* gene. However, it is important to note that water potential was not measured in this study, and, therefore, we cannot exclude the possibility that the differences observed between treatments may be more related to variations in water potential than to the treatments themselves.

4.5. Case 5: Soybean Plants in Soil and PEG Treatment at Different Stages

For the gradual drought experiment, water was withheld from young soybean plants (25 days after seed sowing, V4 stage) grown in pots with a mixture of soil and vermiculite (2:1). PEG treatment (20%) was applied to other soybean plants in hydroponics at the V1 stage. Expression of 15 genes from the *GmLAX* (like Auxin) family genes was measured in one soybean cultivar, Tianlong 1 [46].

All studied genes were down-regulated in response to either drought or PEG treatment. However, a big difference was observed between the control and drought-exposed plants whereas the difference was much smaller in the PEG-supplement experiment. The transcript level of *GmLAX3* was very similar in both treatments, but other genes, such as *GmLAX2* and *GmLAX13*, showed significant differences in their expressions under drought compared to experiments with PEG treatment [46]. In addition to the absence of water potential measurement, a mixed conclusion can be made for the analysis of the present paper since it remains unclear what role the different stages of plant development in the studied soybean—V1 stage in PEG treatment and V4 stage in drought—played in the final analysis. Nevertheless, there were big differences between drought and PEG treatments, and the authors attempted to separate the results from the 'drought in soil' experiment and 'PEG-treatment in hydroponics'.

5. Conclusions

Researchers are permitted to use any type of experiments suitable for the purposes of their study. However, clear and correct description is required to avoid inferring incorrect information and potentially misleading their readers. It is recommended to use the terms gradual or slowly 'induced drought' or 'simulated drought' in containers with drying soil. At the same time, the best option for experiments with hydroponics supplemented with PEG solutions are the terms 'PEG-induced dehydration' or 'PEG induced dehydration simulating drought'. Alternatively, 'PEG-treatment' experiments can be used and described properly. However, the authors must be especially careful in the interpretation of their results since experiments with PEG cannot be considered as true 'drought' studies. In all cases, results of experiments in controlled conditions must ultimately be verified in field trials with naturally occurring drought.

Supplementary Materials: The following supporting information can be downloaded at https://www.mdpi.com/article/10.3390/plants14010092/s1: Table S1: Summary of 143 published paper analysis, where both 'drought' and PEG' were used in Titles, Abstracts, and Keywords, based on search in Scopus database.

Author Contributions: Conceptualization, Y.S. and S.E.; writing—original draft preparation, S.J. and Y.S.; writing—review and editing, G.K. and N.B.; visualization, S.E.; supervision, S.J. All authors have read and agreed to the published version of the manuscript.

Funding: This research received no external funding.

Data Availability Statement: The original contributions presented in this study are included in the opinion paper and in the Supplementary Material. Further inquiries can be directed to the corresponding author.

Acknowledgments: We want to thank Carly Schramm for critical comments during editing of the manuscript. Additionally, special thanks to the staff and students of our Universities for their support in this research and help with the manuscript preparation.

Conflicts of Interest: The authors declare no conflicts of interest.

References

1. Lawlor, D.W. Genetic engineering to improve plant performance under drought: Physiological evaluation of achievements, limitations, and possibilities. *J. Exp. Bot.* **2013**, *64*, 83–108. [CrossRef]
2. Langridge, P.; Reynolds, M. Breeding for drought and heat tolerance in wheat. *Theor. Appl. Genet.* **2021**, *134*, 1753–1769. [CrossRef]
3. Bartels, D.; Sunkar, R. Drought and salt tolerance in plants. *Crit. Rev. Plant Sci.* **2005**, *24*, 23–58. [CrossRef]
4. Verslues, P.E.; Agarwal, M.; Katiyar-Agarwal, S.; Zhu, J.; Zhu, J.K. Methods and concepts in quantifying resistance to drought, salt and freezing, abiotic stresses that affect plant water status. *Plant J.* **2006**, *45*, 523–539. [CrossRef]

5. Martignago, D.; Rico-Medina, A.; Blasco-Escámez, D.; Fontanet-Manzaneque, J.B.; Caño-Delgado, A.I. Drought resistance by engineering plant tissue-specific responses. *Front. Plant Sci.* **2020**, *10*, 1676. [CrossRef] [PubMed]
6. Haswell, E.S.; Verslues, P.E. The ongoing search for the molecular basis of plant osmosensing. *J. Gen. Physiol.* **2015**, *145*, 389–394. [CrossRef]
7. Yu, B.; Chao, D.Y.; Zhao, Y. How plants sense and respond to osmotic stress. *J. Integr. Plant Biol.* **2024**, *66*, 394–423. [CrossRef] [PubMed]
8. Tuberosa, R. Phenotyping for drought tolerance of crops in the genomics era. *Front. Physiol.* **2012**, *3*, 347. [CrossRef] [PubMed]
9. Negin, B.; Moshelion, M. The advantages of functional phenotyping in pre-field screening for drought-tolerant crops. *Funct. Plant Biol.* **2017**, *44*, 107–118. [CrossRef] [PubMed]
10. Zivcak, M.; Brestic, M.; Sytar, O. Osmotic adjustment and plant adaptation to drought stress. In *Drought Stress Tolerance in Plants*; Hossain, M.A., Wani, S.H., Bhattacharjee, S., Burritt, D.J., Tran, L.S.P., Eds.; Springer International Publishing: Berlin/Heidelberg, Germany, 2016; Volume 1, pp. 105–143. [CrossRef]
11. Use of PEG. Available online: https://plantstress.com/use-of-peg (accessed on 10 November 2024).
12. Shin, S.Y.; Kim, H.Y.; Kim, I.S.; Kim, J.J.; Kim, Y.S.; Yoon, H.S. The dehydrin gene of the Arctic plant *Cerastium arcticum*, *CaDHN*, increases tolerance to multiple stresses in *Arabidopsis thaliana*. *Plant Biotechnol. Rep.* **2020**, *14*, 387–395. [CrossRef]
13. Osmolovskaya, N.; Shumilina, J.; Kim, A.; Didio, A.; Grishina, T.; Bilova, T.; Keltsieva, O.A.; Zhukov, V.; Tikhonovich, I.; Tarakhovskaya, E.; et al. Methodology of drought stress research: Experimental setup and physiological characterization. *Int. J. Mol. Sci.* **2018**, *19*, 4089. [CrossRef]
14. Gamalero, E.; Bona, E.; Todeschini, V.; Lingua, G. Saline and arid soils: Impact on bacteria, plants, and their interaction. *Biology* **2020**, *9*, 116. [CrossRef] [PubMed]
15. Gu, Z.; Hu, C.; Gan, Y.; Zhou, J.; Tian, G.; Gao, L. Role of microbes in alleviating crop drought stress: A review. *Plants* **2024**, *13*, 384. [CrossRef]
16. Shavrukov, Y.; Genc, Y.; Hayes, J. The use of hydroponics in abiotic stress tolerance research. In *Hydroponics. A Standard Methodology for Plant Biological Researches*; Asao, T., Ed.; InTech Open Access Publisher: Rijeka, Croatia, 2012; pp. 39–66.
17. Rajendran, S.; Domalachenpa, T.; Arora, H.; Li, P.; Sharma, A.; Rajauria, G. Hydroponics: Exploring innovative sustainable technologies and applications across crop production, with emphasis on potato mini-tuber cultivation. *Heliyon* **2024**, *10*, e26823. [CrossRef]
18. Landsberg, J.; Waring, R. Water relations in tree physiology: Where to from here? *Tree Physiol.* **2017**, *37*, 18–32. [CrossRef]
19. Dutta, M.; Moin, M.; Saha, A.; Dutta, D.; Bakshi, A.; Kirti, P.B. Gain-of-function mutagenesis through activation tagging identifies *XPB2* and *SEN1* helicase genes as potential targets for drought stress tolerance in rice. *Theor. Appl. Genet.* **2021**, *134*, 2253–2272. [CrossRef] [PubMed]
20. Lawlor, D.W. Absorption of polyethylene glycols by plants and their effects on plant growth. *New Phytol.* **1970**, *69*, 501–513. [CrossRef]
21. Dai, M.; Zhang, Y.; Lu, X.; Wang, D.; Zhang, Y.; Zhou, N.; Wu, Z.; Liu, L.; Rui, C.; Wang, X.; et al. Starch granules of the sugar-pathway were eliminated under the stress of PEG-drought compared with Soil-drought. *Ind. Crops Prod.* **2023**, *193*, 116158. [CrossRef]
22. Castañeda, V.; González, E.M. Strategies to apply water-deficit stress: Similarities and disparities at the whole plant metabolism level in *Medicago truncatula*. *Int. J. Mol. Sci.* **2021**, *22*, 2813. [CrossRef]
23. Youssef, M.S.; Renault, S.; Hill, R.D.; Stasolla, C. The soybean *Phytoglobin1* (*GmPgb1*) is involved in water deficit responses through changes in ABA metabolism. *J. Plant Physiol.* **2021**, *267*, 153538. [CrossRef] [PubMed]
24. Sahay, S.; Robledo-Arratia, L.; Glowacka, K.; Gupta, M. Root *NRT, NiR, AMT, GS, GOGAT* and *GDH* expression levels reveal NO and ABA mediated drought tolerance in *Brassica juncea* L. *Sci. Rep.* **2021**, *11*, 7992. [CrossRef] [PubMed]
25. Mira, M.M.; Ibrahim, S.; So, K.; Kowatsch, R.; Duncan, R.W.; Hill, R.D.; Stasolla, C. Specificity in root domain accumulation of Phytoglobin1 and nitric oxide (NO) determines meristematic viability in water-stressed *Brassica napus* roots. *Ann. Bot.* **2023**, *131*, 475–490. [CrossRef] [PubMed]
26. Pashkovskiy, P.P.; Vankova, R.; Zlobin, I.E.; Dobrev, P.; Ivanov, Y.V.; Kartashov, A.V.; Kuznetsov, V.V. Comparative analysis of abscisic acid levels and expression of abscisic acid-related genes in Scots pine and Norway spruce seedlings under water deficit. *Plant Physiol. Biochem.* **2019**, *140*, 105–112. [CrossRef]
27. Zhang, Y.; He, Y.; Zhao, H.; Zhang, Y.; Yang, J.; Ou, X.; Zhang, J.; Zhu, Q. A 14-3-3 protein-encoding gene, *BdGF14g*, confers better drought tolerance by regulating ABA biosynthesis and signaling. *Plants* **2023**, *12*, 3975. [CrossRef] [PubMed]
28. Xu, J.; Zhou, Y.; Xu, Z.; Chen, Z.; Duan, L. Combining physiological and metabolomic analysis to unravel the regulations of coronatine alleviating water stress in tobacco (*Nicotiana tabacum* L.). *Biomolecules* **2020**, *10*, 99. [CrossRef] [PubMed]
29. Mohamed, E.A.; Ahmed, A.A.M.; Schierenbeck, M.; Hussein, M.Y.; Baenziger, P.S.; Börner, A.; Sallam, A. Screening spring wheat genotypes for *TaDreb-B1* and *Fehw3* genes under severe drought stress at the germination stage using KASP technology. *Genes* **2023**, *14*, 373. [CrossRef]

30. Ding, Z.T.; Li, C.; Shi, H.; Wang, H.; Wang, Y. Pattern of *CsICE1* expression under cold or drought treatment and functional verification through analysis of transgenic *Arabidopsis*. *Genet. Mol. Res.* **2015**, *14*, 11259–11270. [CrossRef]
31. Jaybhaye, S.G.; Deshmukh, A.S.; Chavhan, R.L.; Patade, V.Y.; Hinge, V.R. GA3 and BAP phytohormone seed priming enhances germination and PEG induced drought stress tolerance in soybean by triggering the expression of osmolytes, antioxidant enzymes and related genes at the early seedling growth stages. *Environ. Exp. Bot.* **2024**, *226*, 105870. [CrossRef]
32. Ibrahime, M.; Kibar, U.; Kazan, K.; Özmen, C.Y.; Mutaf, F.; Aşçı, S.D.; Aydemir, B.Ç.; Ergül, A. Genome-wide identification of the LEA protein gene family in grapevine (*Vitis vinifera* L.). *Tree Genet. Genomes* **2019**, *15*, 55. [CrossRef]
33. Allagulova, C.; Avalbaev, A.; Lubyanova, A.; Plotnikov, A.; Yuldashev, R.; Lastochkina, O. Nitric oxide (NO) improves wheat growth under dehydration conditions by regulating phytohormone levels and induction of the expression of the *TaDHN* dehydrin gene. *Plants* **2023**, *12*, 4051. [CrossRef]
34. Geng, L.; Su, L.; Fu, L.; Lin, S.; Zhang, J.; Liu, Q.; Jiang, X. Genome-wide analysis of the rose (*Rosa chinensis*) NAC family and characterization of *RcNAC091*. *Plant Mol. Biol.* **2022**, *108*, 605–619. [CrossRef] [PubMed]
35. Rani, V.; Rana, S.; Muthamilarasan, M.; Joshi, D.C.; Yadav, D. Expression profiling of Nuclear Factor-Y (NF-Y) transcription factors during dehydration and salt stress in finger millet reveals potential candidate genes for multiple stress tolerance. *Planta* **2024**, *259*, 136. [CrossRef] [PubMed]
36. Verma, G.; Dhar, Y.V.; Srivastava, D.; Kidwai, M.; Chauhan, P.S.; Bag, S.K.; Asif, M.H.; Chakrabarty, D. Genome-wide analysis of rice dehydrin gene family: Its evolutionary conservedness and expression pattern in response to PEG induced dehydration stress. *PLoS ONE* **2017**, *12*, e0176399. [CrossRef]
37. Pakzad, R.; Fatehi, F.; Kalantar, M.; Maleki, M. Proteomics approach to investigating osmotic stress effects on pistachio. *Front. Plant Sci.* **2023**, *13*, 1041649. [CrossRef] [PubMed]
38. Fu, Y.; Poli, M.; Sablok, G.; Wang, B.; Liang, Y.; La Porta, N.; Velikova, V.; Loreto, F.; Li, M.; Varotto, C. Dissection of early transcriptional responses to water stress in *Arundo donax* L. by unigene-based RNA-seq. *Biotechnol. Biofuels* **2016**, *9*, 54. [CrossRef]
39. Liu, T.; Wang, W.; He, J.; Wang, X.; Dong, K.; Hao, Y.; Wang, M.; Zhang, L.; Ren, R.; Yang, P.; et al. Transcriptome profiling and TCP family analysis of broomcorn millet (*Panicum miliaceum* L.) seedlings under hyperosmotic stress. *Plant Mol. Biol. Rep.* **2023**, *41*, 277–291. [CrossRef]
40. Zhao, M.R.; Han, Y.Y.; Feng, Y.N.; Li, F.; Wang, W. Expansins are involved in cell growth mediated by abscisic acid and indole-3-acetic acid under drought stress in wheat. *Plant Cell Rep.* **2012**, *31*, 671–685. [CrossRef]
41. Gajurel, G.; Hasan, R.; Medina-Bolivar, F. Water-deficit stress induces prenylated stilbenoid production and affects biomass in peanut hairy roots: Exploring the role of stilbenoid prenyltransferase downregulation. *Plant Physiol. Biochem.* **2024**, *210*, 108596. [CrossRef]
42. Zia, R.; Nawaz, M.S.; Yousaf, S.; Amin, I.; Hakim, S.; Mirza, M.S.; Imran, A. Seed inoculation of desert-plant growth-promoting rhizobacteria induce biochemical alterations and develop resistance against water stress in wheat. *Physiol. Plant.* **2021**, *172*, 990–1006. [CrossRef]
43. Ambrosone, A.; Batelli, G.; Bostan, H.; D'Agostino, N.; Chiusano, M.L.; Perrotta, G.; Leone, A.; Grillo, S.; Costa, A. Distinct gene networks drive differential response to abrupt or gradual water deficit in potato. *Gene* **2017**, *597*, 30–39. [CrossRef] [PubMed]
44. Liang, J.; Zhang, H.; Yi, L.; Tang, Y.; Long, H.; Yu, M.; Deng, G. Identification of *HvLRX*, a new dehydration and light responsive gene in Tibetan hulless barley (*Hordeum vulgare* var. *nudum*). *Genes Genom.* **2021**, *43*, 1445–1461. [CrossRef] [PubMed]
45. Ding, N.; Zhao, Y.; Wang, W.; Liu, X.; Shi, W.; Zhang, D.; Chen, J.; Ma, S.; Sun, Q.; Wang, T.; et al. Transcriptome analysis in contrasting maize inbred lines and functional analysis of five maize NAC genes under drought stress treatment. *Front. Plant Sci.* **2023**, *13*, 1097719. [CrossRef] [PubMed]
46. Yang, H.; Guo, W.; Cao, D.; Chen, H.; Chen, S.; Yang, Z.; Shan, Z.; Chen, L.; Zhou, X. Expression pattern of *GmLAX* genes under different stresses in soybean drought sensitive cultivar and tolerant cultivar. *Oil Crop Sci.* **2021**, *6*, 58–65. [CrossRef]

plants

MDPI

Article

The Proteomic and Peptidomic Response of Wheat (*Triticum aestivum* L.) to Drought Stress

Regina Azarkina [1,2,*,†], Arina Makeeva [1,†], Anna Mamaeva [1], Sergey Kovalchuk [1], Daria Ganaeva [1], Igor Tikhonovich [3,4] and Igor Fesenko [5]

1 Shemyakin-Ovchinnikov Institute of Bioorganic Chemistry of the Russian Academy of Sciences, Moscow 117997, Russia; aryamakeeva@gmail.com (A.M.); physcomitrella.proteomics@gmail.com (A.M.); xerx222@gmail.com (S.K.)

2 Laboratory of System Analysis of Proteins and Peptides, Shemyakin and Ovchinnikov Institute of Bioorganic Chemistry, Russian Academy of Sciences, Moscow 117997, Russia

3 All-Russia Research Institute for Agricultural Microbiology, Podbelscogo hwy 3, Pushkin, St. Petersburg 196608, Russia

4 Department of Genetics and Biotechnology, Faculty of Biology, Saint Petersburg State University, 7-9 Universitetskaya Embankment, Saint Petersburg 199034, Russia

5 Independent Researcher, Moscow 117997, Russia; fesigor@gmail.com

* Correspondence: khazigaleeva.regina@gmail.com

† These authors contributed equally to this work.

Abstract: Drought conditions impact plants at the morphological, physiological, and molecular levels. Plant tolerance to drought conditions is frequently associated with maintaining proteome stability, highlighting the significance of proteomic analysis in understanding the mechanisms underlying plant resilience. Here, we performed proteomic and peptidomic analysis of spring wheat (*Triticum aestivum* L.) under drought stress conditions. Using isobaric tags for relative and absolute quantitation (iTRAQ), we identified 497 and 157 differentially abundant protein (DAP) groups in leaves and roots, respectively. The upregulated DAP groups in leaves were primarily involved in stress responses, such as oxidative stress and heat response, whereas those in roots were associated with responses to water deprivation and sulfur compound metabolic processes. The analysis of the extracellular root peptidome revealed 2294 native peptides, including members of small secreted peptide (SSP) families. In the peptidomes of stress-induced plants, we identified 16 SSPs as well as peptides derived from proteins involved in cell wall catabolism, intercellular signaling, and stress response. These peptides represent potential candidates as regulators of drought responses. Our results help us to understand adaptation mechanisms and develop new agricultural technologies to increase productivity.

Keywords: proteome; peptidome; secretome; drought; LC-MS/MS; iTRAQ; wheat

1. Introduction

Drought is one of the major factors contributing to global yield losses, accounting for up to 50% of crop reductions worldwide [1]. The impact of drought is particularly severe during specific developmental stages, such as fruit formation, often resulting in substantial yield losses. Historically, severe drought conditions have repeatedly resulted in widespread famine and continue to pose a significant threat to global food security and agricultural sustainability [2,3].

Drought affects plant tissues and organs by membrane disintegration, protein degradation, and growth suppression, primarily due to inhibited cell elongation [4]. To maintain cellular osmotic balance and counteract turgor loss, plants accumulate osmoprotectants, such as proline, glycine, betaine, and various sugars [5]. These molecules not only stabilize proteins and membranes but also scavenge reactive oxygen species (ROS). Indeed, drought-induced oxidative stress results from excessive ROS production [6]. In response, plants upregulate antioxidant defense systems, including enzymatic components such as superoxide dismutase (SOD), catalase (CAT), and various peroxidases (POD) [7]. Plants also induce the expression of Late Embryogenesis Abundant (LEA) proteins and dehydrins (DHNs), which function as molecular chaperones, protecting macromolecules and cellular structures from damage by stabilizing proteins and membranes [4,8].

Drought-tolerant cultivars elevated levels of LEA proteins, DHNs, heat shock proteins, and enzymes linked to antioxidant defense and osmotic adjustment under water deficit [9]. These changes are the result of both transcriptional reprogramming and selective accumulation and stabilization of proteins that safeguard cellular structures and metabolic homeostasis [9,10].

Besides proteins, Small Secreted Peptides (SSPs) were shown to play an important role as key signaling molecules under drought stress [11–17]. SSPs are derived from protein precursors that contain a signal peptide. They have been shown to play a role as mediators of intercellular communication [18–20]. Exogenous application of SSPs from specific families has been shown to enhance drought tolerance. For example, treatment with synthetic CLAVATA3 (CLV3)/Embryo Surrounding Region 25 (CLE25), C-terminally Encoded Peptide 5 (CEP5), and overexpression of peptide hormone *CLAVATA3/ESR-RELATED 9* (*CLE9*) in *Arabidopsis thaliana* increases drought tolerance [21–23]. Also, the drought-induced peptide Rapid Alkalinization Factor 8 (RALF8) in *A. thaliana* facilitates cell wall remodeling under water deficit [24]. The premature drought-induced leaf drop is shown to be regulated by phytosulfokine (PSK) and Inflorescence Deficient in Abscission (IDA) peptides in tomatoes and *A. thaliana*, respectively [25,26]. Also, some members of the Epidermal Patterning Factor-Like (EPFL) peptide family modulate drought response [27–29].

In addition to SSPs, peptides, which are released from functional proteins, can be involved in the plant stress response [13,30–34]. In contrast to animals, the functional roles of such peptides are mostly unknown [35]. However, isolation and identification of both SSPs and degradome-derived peptides are difficult due to their very low abundance in vivo [36].

Wheat is one of the most important cereal crops in the world [37]. The development of proteomic, genomic, and transcriptomic technologies led to the identification of new drought-responsive wheat proteins that have potential for use in marker-assisted selection [38]. In particular, analysis of the developing grains' transcriptome revealed a large number of drought-responsive genes associated with water regulation, osmotic stress, and stress signaling [39]. The transcriptome analysis of wheat roots under drought conditions revealed significant upregulation of genes encoding DHNs, aquaporins, and various transcriptional factors [40,41]. In addition, 15 calmodulin (*TaCAM*) and 113 calmodulin-like (*TaCML*) genes were upregulated under drought and salt stresses [42]. Long-term water deprivation leads to transcriptional changes of thousands of genes, most of which belong to flavonoid biosynthesis, plant hormone signaling, and the antioxidant pathway in wheat [43].

Numerous proteomic studies have investigated wheat responses to drought stress [44–52]. Most of these focused on individual organs such as leaves, grains, or roots. These studies identified drought-responsive proteins involved in various metabolic pathways, including sugar and amino acid synthesis, antioxidant defense, cytoskeletal stabilization, and signal transduction [45–52]. For example, proteins involved in stress defense and detoxification, such as heat shock proteins (HSPs) and ROS scavenging proteins (ascorbate peroxidase, glutathione S-transferase (GST), thiol-specific antioxidant protein), are upregulated under drought conditions in wheat [49,53]. In addition, chaperones and enzymes involved in proline biosynthesis also show increased abundance. Furthermore, proteins associated with cell biogenesis and degradation are upregulated in the leaves of stress-tolerant wheat genotypes [54].

The studies on wheat grains have focused on developmental and storage processes affected by water deficit [46]. It has been demonstrated that albumin and gliadin are significantly upregulated under drought stress [46]. Proteomic analysis of wheat seed germination under drought stress conditions revealed significant effects of pathways involved in phenylpropanoid metabolism and fatty acid biosynthesis [55]. Although the wheat root proteome response to drought conditions has been less extensively studied, available data highlight the importance of proteins involved in water uptake, carbon metabolism, and nutrient metabolism [47,49]. The first comprehensive root proteome study showed that HSPs, O-methyltransferase, and 2-caffeoyl CoA-methyltransferase were upregulated in tolerant genotypes [56]. Despite extensive research on proteins, the role of bioactive peptides, such as SSPs, remains relatively understudied [57].

In this study, we performed a quantitative analysis of the leaf and root proteomes as well as the extracellular root peptidomes of spring wheat under osmotic stress. Our study provided a better understanding of the drought adaptation mechanisms in wheat and the potential development of technologies to improve drought tolerance.

2. Results

2.1. Morphological Changes in Wheat Plants Under Drought Stress

To analyze proteomic and peptidomic changes under drought conditions in wheat, we selected the commonly used early-maturing spring wheat cultivar Leningradskaya 6, which is adapted to the growing conditions in the northwest regions of Russia [58] and has shown the strongest response to inoculation with plant growth-promoting *Bacillus* sp. *V2026* [59]. In plants, drought stress can be induced through various methods [60]. Due to our interest in the analysis of the extracellular wheat peptidome, wheat plants were grown in a hydroponic system, and the polyethylene glycol (PEG-6000) was used as an osmotic stressor [49,61]. To simulate drought stress, wheat plants were incubated with 20% PEG-6000 for 4 and 8 days in the hydroponic system.

The most pronounced reduction in root and leaf length was observed on day 8 (Figure 1); therefore, further analyses were conducted on plants incubated under drought conditions for 8 days.

2.2. Drought Conditions Result in Significant Proteome Changes

At first, quantitative proteomic analysis using iTRAQ labeling was performed on leaves collected after 4 and 8 days of treatment with 20% PEG-6000. We identified 41 differentially abundant proteins (DAPs) concordantly changed across these time points, further supporting the reproducibility of our analysis (Tables S1–S3). Since the most pronounced changes in protein abundance were observed on day 8, further analysis focused on plants incubated under drought conditions for 8 days (Figure 2B, Tables S1 and S2).

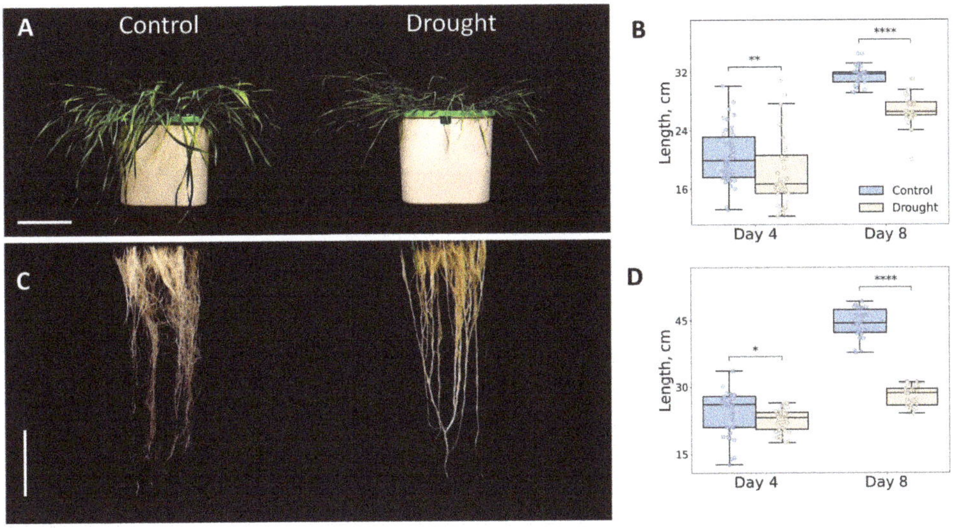

Figure 1. Morphological changes in seedlings of wheat (*Triticum aestivum* L., Leningradskaya 6 cultivar) under 8-day treatment with PEG 20%. (**A**) Shoots of plants under 8-day treatment with PEG 20% and control plants. Scale bar = 10 cm. (**B**) Changes in leaf length after 4 and 8 days of treatment with 20% PEG. Error bars indicate the standard error of three biological replicates. (**C**) Roots of plants under 8-day treatment with PEG 20% and control plants. Scale bar = 10 cm. (**D**) Changes in root length under 4 and 8 days treatment with PEG-20%. Error bars indicate the standard error of three biological replicates. Statistical significance was assessed using a two-tailed Mann–Whitney U test. Asterisks indicate significance levels: * $p < 0.05$, ** $p < 0.01$, **** $p < 0.0001$.

In wheat leaves, we identified 857 DAPs, which were assigned to 497 (239 up- and 258 down-regulated) DAP groups (Figure 2B, Table S2) under 8 days of drought stress. The proteomic changes in roots were less pronounced and included 256 DAPs, which were assigned to 157 (102 up- and 55 down-regulated) DAP groups (Figure 2B; Table S4) under 8 days of drought stress. Since a protein group may contain multiple putative proteins, we performed a comparative analysis between all DAPs in roots and leaves. However, only 27 DAPs were shared between roots and leaves, indicating minimal overlap (Figure 2C). Among these, 4 proteins were down-regulated in both organs, 20—upregulated, and 2 proteins showed opposite regulation patterns between leaves and roots (Table S5).

PCA demonstrated a high degree of consistency among the biological replicates in two groups (Figure 2D,E), indicating that drought stress caused a significant difference in protein expression.

The GO term analysis provides insights into the molecular pathways that enable plants to adapt to environmental stresses. We next performed GO term annotation of DAP groups using the bioinformatic tool ShinyGO 0.80 [62]. Among the 497 DAP groups, 339 (69%) were successfully annotated at the GO level in leaves and 99 (63%) of the 157 DAP groups in roots. We found that upregulated DAP groups in leaves were enriched in such GO terms as the responses to different stimuli and stress conditions, response to oxygen-containing compounds (GO:1901700), obsolete response to inorganic substances (GO:0010035), and response to temperature stimulus (GO:0009266; Figure 3). The upregulated DAP groups in leaves were enriched in such KEGG pathways as oxidative phosphorylation and nitrogen metabolism (Figure 4). The down-regulated DAP groups were enriched in such GO terms as catabolic process (GO:0009056), cellular macromolecule biosynthetic process (GO:0034645), and peptide metabolic process (GO:0006518) in Figure 3, indicating a decline in cytoskeletal dynamics under drought.

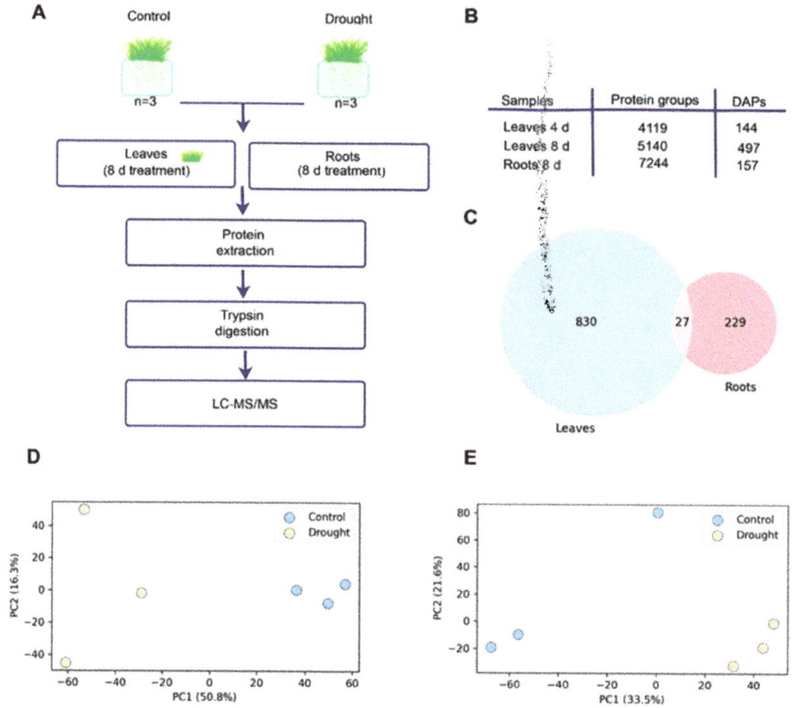

Figure 2. Analysis of the proteome. (**A**) Overview of the experimental workflow. (**B**) Quantitative proteome analysis of wheat under 8-day drought stress. (**C**) Venn diagram showing the overlap of DAPs between the leaves and roots' samples. The intersection represents proteins commonly regulated in both tissues. (**D**) PCA of wheat under drought for 8 days and controls in leaves. (**E**) PCA of wheat under drought for 8 days and controls in the roots.

In leaves, the most strongly upregulated proteins belonged to well-known stress-responsive groups, including DHNs, RNA-recognition motif (RRM) domain-containing proteins, non-specific lipid-transfer proteins (nsLTPs), and water stress and hypersensitive response domain-containing proteins (Figure 5A; Table S2). We also identified seven significantly upregulated nsLTPs (Figure 5A) exclusively in leaves but not in roots. The identification of DAP groups associated with lignin biosynthesis (dirigent protein (DIR)) and stress-related functions (nsLTPs) shows that plants employ a multifaceted strategy to manage drought stress.

The upregulated DAP groups in roots were enriched in such GO terms as response to chemical stimulus (GO:0042221), water deprivation (GO:0009414), and sulfur compound metabolic process (GO:0006790) (Figure 3). The most enriched KEGG pathways that involved upregulated proteins were starch and sucrose metabolism, biosynthesis of secondary metabolites, biosynthesis of nucleotide sugars, and galactose metabolism (Figure 4). The changes in the starch and sucrose metabolism pathways indicate that the plants increase sugar biosynthesis in response to drought stress.

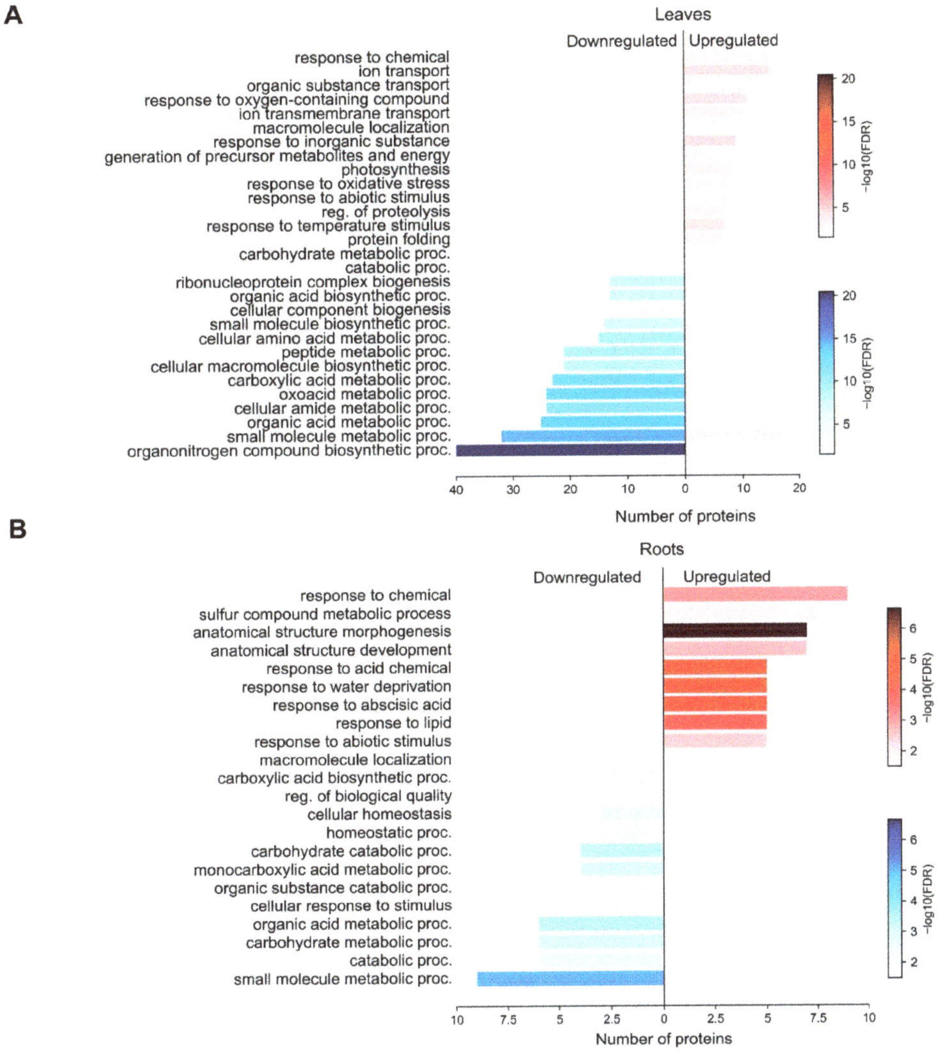

Figure 3. The GO term analysis of wheat. (**A**) GO terms of DAP groups involved in biological processes in leaves after the 8 d drought. (**B**) GO terms of DAP groups involved in biological processes in roots after the 8 d drought. The X-axis represents the number of proteins in each GO Term. The Y-axis is GO Terms. Enrichment significance is indicated by $-\log_{10}$(FDR) values displayed next to each bar. GO analysis was carried out by ShinyGO 0.80.

Root proteome analysis revealed pronounced changes in proteins involved in stress adaptation and redox regulation under drought conditions. Upregulated proteins included DHNs, which are well-established protective molecules stabilizing cellular structures during dehydration (Figure 5B; Table S4). Notably, malic enzyme (ME), an important player in carbon metabolism and NADPH production, was markedly increased. LEA proteins and SMP domain-containing proteins, associated with osmoprotection and membrane stabilization, were also significantly enriched, underscoring multifaceted protective strategies employed by wheat roots in drought (Figure 2E; Table S4).

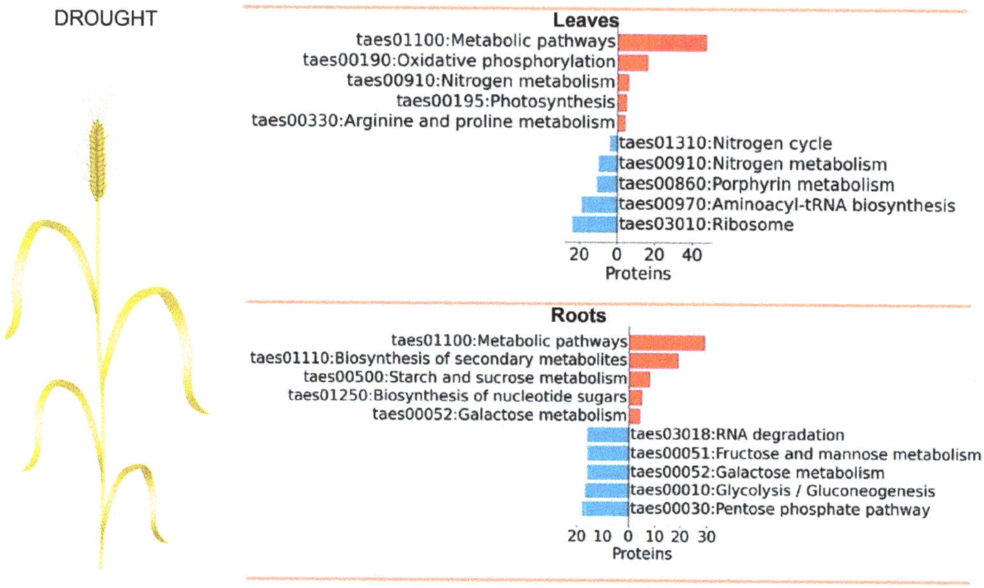

Figure 4. The KEGG annotation for the DAP groups under drought stress. Red bars indicate KEGG pathways enriched among upregulated DAP groups, while blue bars represent those enriched among down-regulated DAP groups. The number of associated proteins is shown along the x-axis. Enrichment significance is indicated by $-\log_{10}$(FDR) values displayed next to each bar.

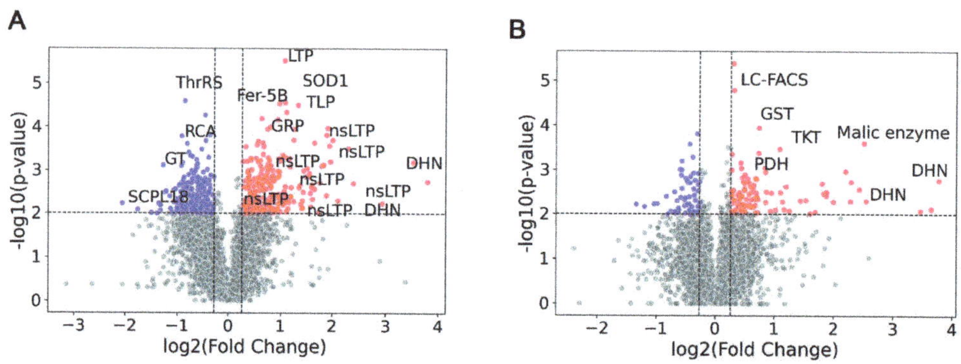

Figure 5. (**A**) Volcano plot indicating DAP groups present in leaves. (**B**) Volcano plot indicating DAP groups present in roots. Each spot represents one protein. On the y-axis, $-\text{Log}_{10}$ (*p*-value) is indicated. On the x-axis, Log_2 (fold-change) is shown; negative values indicate proteins that were down-regulated under drought conditions, while positive values indicate proteins that were upregulated. Blue dots correspond to proteins that were significantly down-regulated, red dots represent significantly upregulated proteins, and grey dots represent non-regulated proteins.

Several proteins were consistently upregulated under drought conditions in both leaves and roots. These included multiple dehydrins (A0A3B6QLW3, A0A3B6LSK1, A0A3B6LUF1), a salt-induced YSK2 dehydrin (W5GW81), a LEA protein (A0A024CKY0), an S-like RNase (Q6R326), and an uncharacterized protein (A0A3B6QLV9) (Figure 2C, Table S5). Additionally, EF-hand and PH domain-containing proteins were commonly regulated across both tissues, pointing to potential roles in calcium signaling and membrane-associated stress responses. This overlap highlights a conserved core of drought-responsive proteins in wheat, irrespective of tissue type.

A small set of proteins was consistently down-regulated in both roots and leaves under drought conditions. These included a 14-3-3 protein (G5DFC5), tripeptidyl-peptidase II (A0A3B6NRW7), lysine–tRNA ligase (A0A3B6MZJ5), and a LURP-one-related protein (A0A1D6B0F9). Their consistent repression suggests that drought stress may suppress specific components of protein synthesis, proteolysis, and signaling, potentially as part of an energy-saving or growth-inhibitory response.

2.3. Extracellular Peptides in Root Secretome Under Drought Conditions

Peptides are key signaling molecules in the communication between roots and leaves, thereby activating plant defenses and physiological changes that promote tolerance to stress [22,25,63,64]. We suggested that the stress peptidome, which includes cryptic bioactive peptides released during the degradation of functionally active proteins [35], might be a part of the plant's drought response. It has been shown previously that secreted peptides from roots would diffuse in the culture medium [65]. Therefore, to identify biologically active secreted peptides, wheat plants were grown in a hydroponic system under normal and drought conditions for 8 days (Figure 6A), and native peptides were extracted from the culture medium. Using top-down mass spectrometry analysis, we identified 1718 peptides derived from 551 protein precursors in normal conditions (Figure 6B; Table S6). In addition, we identified 2294 peptides derived from 721 protein precursors under PEG treatment (Figure 6B,C; Table S7). The GO term analysis showed that the majority of the protein precursors participate in cell wall metabolism, peptide metabolism, and purine ribonucleotide metabolism (Figure S1).

SSPs are known to play a critical role in intercellular communication and drought stress response [66,67]. Therefore, we additionally searched for evidence of SSPs translation in our peptidomic data. For this analysis, protein precursor sequences of identified peptides (Tables S6 and S7) were re-annotated using the Small Secreted Peptide Database [68]. Under normal conditions, we obtained the evidence of translation for 4 SSPs from four families—Plantacyanin (PCY), nsLTP, Subtilisin inhibitor (SubIn), and Nodule-specific Glycine-rich Protein (NodGRP) (Table S8). In drought stress, we obtained the evidence of translation for 16 SSPs from 4 SSP families, including phytocystatins (PhyCys), PCY, CAP-derived peptide (CAPE), and SubIn (Figure 6E; Table S8). Furthermore, we detected an increased number of Plantacyanin/Chemocyanin (PCY) families under drought stress. Additionally, our proteomic data revealed the presence of PhyCys and PCY families under drought stress (Table S8).

Following the identification of SSPs in the extracellular peptidome, we re-analyzed publicly available RNA-seq data on drought-stressed wheat roots of two *T. aestivum* cultivars—ZM366 and CM42 [41] to assess the expression patterns of SSP-encoding genes. Our analysis identified 22 transcripts from 9 SSP families that were upregulated under drought conditions ($log_2FC > 1$, adjusted p-value < 0.05) and 14 down-regulated transcripts from 6 SSP families ($log_2FC < -1$, adjusted p-value < 0.05) in the CM42 cultivar (Table S9). In the ZM366 cultivar, drought stress resulted in upregulation of 39 transcripts from 14 SSP families and down-regulation of 12 transcripts from 7 SSP families (Table S10).

In both lines, there were significant changes in the transcription of cysteine-rich SSP genes, including *PCY*, *CAP-derived peptide* (*CAPE*), *nsLTP*, *Hevein*, and *Bougainvillea bract protein inhibitor* (*BBPI*). Thus, we observed a notable increase in the transcriptional level of cysteine-rich SSP families, particularly PCY, that were also identified in extracellular peptidomes (Figure 6E; Table S9). We also observed an upregulation of protease inhibitor PhyCys at both transcriptome, proteome, and peptidome levels. Additionally, the ZM366 line showed a significantly increased expression of 17 members of the nsLTP family, which is known to facilitate lipid transport. Therefore, we detected changes in the abundance of nsLTPs across both the transcriptomic and proteomic datasets, which indicates their relevance in the plant drought stress response. The difference in SSPs' expression at the transcriptomic level between the two cultivars suggests potential molecular mechanisms that enable different wheat lines to withstand such stresses.

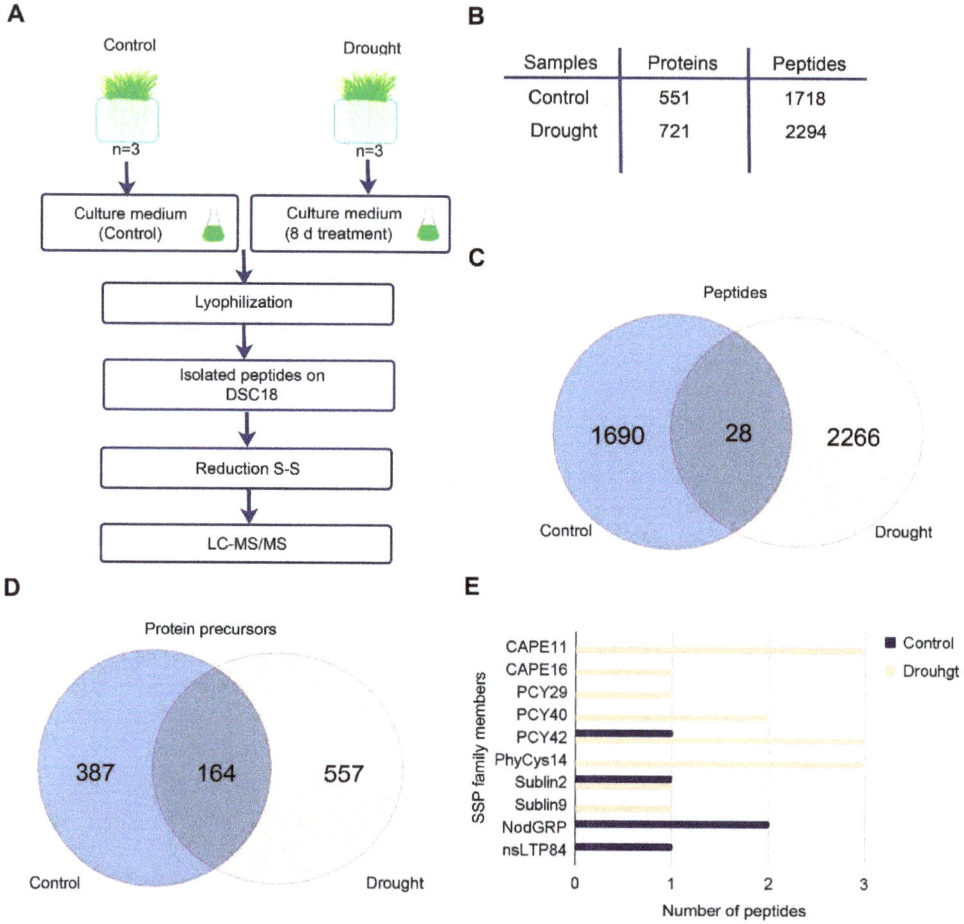

Figure 6. Analysis of secretome peptides. (**A**) Overview of the experimental workflow. (**B**) Characteristics of the peptidome dataset after PEG treatment for 8 days. (**C**) Venn diagram showing comparison of identified endogenous peptides between control and PEG 20% treatment. (**D**) Venn diagram showing comparison of identified protein precursors of peptides between control and PEG 20% treatment. (**E**) The families of SSP are identified in the extracellular peptidome.

3. Discussion

In this study, we performed proteomic and peptidomic analysis of the early-maturing spring wheat cultivar Leningradskaya 6 under drought stress conditions. Our analysis revealed DAP groups involved in stress responses, such as oxidative stress and heat response in leaves, and associated with responses to water deprivation and sulfur compound metabolic processes in roots. Notably, only a small subset of proteins was shared between the two organs, reflecting distinct tissue-specific drought adaptation strategies.

Several proteomic studies have previously addressed drought responses in wheat, but these typically focused on individual organs and early stress stages. For example, Ford et al.

(2011) conducted a shotgun proteomic analysis of leaves and identified 159 DAP groups related to photosynthesis and ROS scavenging [45]. Zhang et al. (2014) compared leaf and grain tissues, highlighting tissue-specific responses, but did not examine roots [46].

In contrast to previous studies, we utilized iTRAQ-based quantitative proteomic analysis to investigate the late-phase adaptive responses in both leaf and root proteomes, which had not been described previously. The depth of our dataset also allowed us to perform a more detailed comparison of proteomic changes between different organs [69]. Consistent with earlier findings [51], we observed clear differences in drought responses between roots and leaves. Notably, only a small subset of proteins was shared between the two organs, highlighting distinct tissue-specific adaptation strategies to drought stress. Furthermore, the overall magnitude of proteomic changes was smaller in roots compared to leaves.

Consistent with prior research, we identified DHNs as the most upregulated proteins in both leaves and roots, confirming their role in enhancing drought tolerance by stabilizing proteins and membranes during desiccation [70]. Similarly, the upregulation of SODs, GSTs and peroxidases corroborates previous findings about the activation of antioxidant systems protecting cells from oxidative damage [71,72].

Previous studies have shown that overexpression of the Photosystem II reaction center Psb28 subunit impaired drought tolerance in transgenic *Arabidopsis thaliana* [73]. In our analysis, the Photosystem II reaction center Psb28 protein was significantly upregulated in wheat leaves under drought stress, suggesting its involvement in the drought response.

In leaves, we observed increased levels of DIR proteins (A0A3B6FG70; A0A341W7H8; A0A3B6SD56), which regulate lignin biosynthesis and contribute to cell wall reinforcement, as well as RRM domain-containing proteins (A0A3B6REA1, A0A3B6KQ24, A0A3B6MJX3), likely involved in post-transcriptional regulation under stress [74,75].

Our study highlights the upregulation of seven non-specific nsLTPs in leaves, which have been implicated in stress responses, symbiosis establishment, and immune reactions [76–78]. The fact that two of these proteins (A0A3B6GKQ2; W5D2I6) have also been identified in wheat in other studies acknowledges their importance and suggests a conserved role in the plant's defense mechanisms against environmental stressors [79]. The tissue-specific presence of these nsLTPs, supported by increased transcript levels, suggests that they have a specific function in protecting leaf tissues from drought-induced damage. Furthermore, we detected differential regulation of calcium-binding proteins, such as calreticulin, which decreased under drought, and calcium-dependent protein kinases, some of which were upregulated, indicating a complex modulation of calcium signaling pathways [80].

Another intriguing aspect of our data is the involvement of sulfur metabolism in drought adaptation. Upregulated proteins linked to sulfur-containing compounds, such as glutathione transferase (A0A3B6GSQ6) and 5-methyltetrahydropteroyltriglutamate–homocysteine S-methyltransferase (A0A3B6MJZ2), point to enhanced antioxidant capacity and metabolic adjustments, consistent with previous reports on the importance of sulfur in stress adaptation [81,82]. Meanwhile, the down-regulation of sulfotransferase (A0A3B5Y3I8) in roots may reflect organ-specific metabolic shifts.

In addition to the proteomic analysis, we examined extracellular peptide pools under control and drought conditions, identifying over 2000 endogenous peptides, including 16 SSPs in the root secretome. Although secreted peptides are increasingly recognized as mediators of stress responses, such data for wheat under drought conditions have not been reported previously.

Intercellular communication between roots and leaves is crucial for the proper activation of plant defenses and physiological changes that promote tolerance to stress.

Plantacyanins are components of the plant cell wall and play various roles in plant development and stress response [83,84]. It is shown that PCY is involved in the regulation of seed germination, anther development, and pollen tube chemotropism [84–86]. Moreover, drought-induced expression of *PCY* in the stomatal guard cells increases the ROS level, leading to stomatal closure and enhanced drought tolerance [83]. Our peptidomic analysis and publicly available RNA-seq data of drought-stressed wheat roots both revealed a notable upregulation of the PCY family peptides under drought stress.

We also observed the consistent presence and upregulation of PhyCys family members both in the extracellular root peptidomes and transcriptome in response to drought stress. While their classical function involves inhibiting papain-like cysteine proteases (PLCPs), several family members have been implicated in responses to abiotic stresses such as salt, drought, and heat [87–89]. Previous transcriptomic analyses have shown that *Oryza sativa* PhyCys14 is induced under drought conditions, yet its specific functional role remains poorly characterized [90,91].

Among the peptides identified in the extracellular root secretome under drought stress, we also detected CAPE11 and CAPE9—members of the CAPE family derived from the precursor pathogenesis-related protein 1 (PR-1). CAPEs have been previously implicated in immune signaling; for instance, CAPE1 functions as a damage-associated molecular pattern (DAMP) that activates jasmonic acid and ethylene signaling pathways to promote immune responses. Recent work has shown that CAPE1 expression is down-regulated under salt stress, and its reduced levels are associated with diminished salt tolerance [92]. In our study, however, CAPE9 and CAPE11 were consistently detected in the drought-stressed wheat root secretome, suggesting a possible role in the drought response that has not been previously described.

Taken together, these results highlight the potential significance of specific SSP families in increased plant resistance to abiotic stress, suggesting a need for further research to understand their functions and mechanisms in drought tolerance.

4. Materials and Methods

4.1. Plant Growth Conditions and Treatments

The seed of the commonly used early-maturing spring wheat cultivar Leningradskaya 6, which is adapted to the growing conditions in the northwest regions of Russia [58], has shown the strongest response to inoculation with plant growth-promoting *Bacillus* sp. *V2026* (https://doi.org/10.3390/plants11141817) was obtained from the All-Russia Research Institute of Agricultural Microbiology, St. Petersburg, Russia [58,93]. The seeds were sterilized by 0.1% sodium hypochlorite (NaClO) solution for 10 min, washed 7 times, and then germinated for 24 h on wet filter paper at room temperature in the dark. The germinated seeds were transferred to a cultivation plastic container (190 × 190 × 175 cm) with 3 L full-strength Hoagland solution [94] in growth chambers (*Panasonic* PHCBI *MLR-352H-PE* Plant Growth Humidity Light Klima Schrank Cabinet, Tokyo, Japan) with a temperature of 22 ± 2 °C under 16 h photoperiod under photosynthetically active light (250 µmole m^{-2} s^{-1}). The pH (6.0–6.5) was optimized by adding NaOH. Fresh air was supplied in solution by an aeration pump (220 V). The nutritional solution was changed every 3 days. Drought stress was imposed on seedlings starting from the three-leaf stage by adding 20% *w*/*v* PEG-6000 (polyethylene glycol-6000) to the Hoagland solution to obtain an osmotic potential (ψ s) of –0.75 MPa [49]. Samples were collected after 8 days of drought treatment. Plant materials were stored at –80 °C prior to use. The experiments were conducted in three biological replicates.

4.2. Protein Extraction

Protein extraction from leaves and roots and following trypsin digestion were conducted as previously described [92], with three biological replicates for each treatment. Briefly, proteins were extracted using the phenol extraction method [95], dissolved in Sample buffer (8 M urea, 2 M thiourea, 10 mM Tris), quantified by Bradford protein assays (Bio-Rad), reduced in 5 mM dithiothreitol for 30 min at 50 °C, and then alkylated by 10 mM iodoacetamide for 20 min. Obtained proteins were dissolved four times in 40 mM ammonium bicarbonate and digested by incubating with trypsin (Promega) overnight at 37 °C.

4.3. iTRAQ Labeling and Cation Exchange Fractionation

Proteins were labeled with the iTRAQ tags as follows: leaves, control—113, 115, and 119 tags, 8 days of drought stress—114, 116, and 121 tags; roots, control—115, 117, and 119 tags, 8 days of drought stress—116, 118, and 121 tags. Samples were mixed, vacuum-dried, and dissolved in 1% trifluoroacetic acid (TFA). To increase the number of identified proteins, labeled peptides were fractionated by cation exchange chromatography on strong cation exchange (SCX) extraction disks (Supelco, Bellefonte, PA, USA). Peptides were eluted successively by 50, 75, 125, 200, and 300 mM ammonium acetate in 0.5% formic acid and 20% acetonitrile; 5% NH_4OH in 80% acetonitrile; 10% NH_4OH in 60% acetonitrile. This procedure has been shown not to significantly affect protein quantification by iTRAQ (Hao et al. 2015 [49]). A total of 6 fractions were collected, and each fraction was dried in a vacuum concentrator for the next step. Peptide fractions were resuspended with 100 μL solvent (water with 0.1% formic acid) and desalted on DSC-C18 extraction disk columns and were washed with 0.1% TFA and eluted with 80% acetonitrile with 0.1% TFA, and each fraction was dried in a vacuum concentrator for the next step.

4.4. LC-MS/MS Analysis

LC-MS analysis was carried out using an UltiMate 3000 RSLCnano HPLC system connected to an Orbitrap Fusion Lumos mass spectrometer (ThermoFisher Scientific, Markham, ON, Canada). Samples were loaded to a home-made trap column 40 × 0.1 mm, packed with Reprosil Pur C18 5 um (MZ-ANALYSENTECHNIK GmbH, Mainz, Germany) in the loading buffer (2% ACN, 98% H_2O, 0.1% TFA) at 4.5 μL/min flow and separated at RT in a fused-silica column 300 × 0.1 mm packed with Reprosil PUR C18AQ 1.9 (MZ-ANALYSENTECHNIK GmbH, Mainz, Germany) into the emitter prepared with P2000 Laser Puller (AutoMate Scientific, San Diego, CA, USA) [96]. Fractionated samples were eluted with a linear gradient of 80% ACN, 19.9% H_2O, 0.1% FA (buffer B) in 99.9% H_2O, and 0.1% FA (solvent A) from 8 to 45% of solvent B in 2 h at 0.5 μL/min flow at RT. MS data were collected in DDA mode with 2 s cycle time. MS1 parameters were as follows: 120K resolution, 350–1500 scan range, injection time, and AGC target set to auto. Ions were isolated with a 1.4 m/z window, charge +2 to +7, and 5×10^4 intensity threshold. Dynamic exclusion was set to 40 s. MS2 fragmentation was carried out in HCD mode at 15K resolution with stepped CE 30, 35, and 40%. The initial mass was set to 100. AGC (Automatic Gain Control) target and injection time set to auto.

4.5. Extracellular Peptides Extraction

For the peptide assays, wheat plants were grown in a hydroponic system under normal and drought conditions for 8 days. Then the wheat seedlings were cultured with Hoagland nutrient solution without PEG-6000 for an additional day. The Hoagland solution was sterilized by filtration through a 0.22 μm membrane filter (Millipore, Burlington, MA, USA). This incubation aimed to establish a controlled environment for the effective extraction of extracellular peptides. Following this, 400 mL of wheat culture medium was collected for peptides' extraction. The culture medium was filtered through a 0.22 μm membrane filter (Millipore), lyophilized, and isolated on reversed-phase SPE (Solid-Phase Extraction) cartridges DSC-18 (Discovery DSC-18, Supelco, USA) using 500 μL 50% acetonitrile with 0.1% TFA solutions for the elution. The eluted peptides were concentrated in a SpeedVac to 5 μL and followed by the reduction of disulfide bonds as previously described [97]. For this peptide, samples were resuspended in 40 μL disulfide reduction solution (0.02 M CAA, 0.025 M TCEP, and 0.1 M 8.5 pH Tris–HCl) and incubated for 10 min at 90 °C. The reduced pool of peptides was desalted by solid-phase extraction on reversed-phase SPE cartridges DSC-18 in the microcentrifugation column using a 100 μL 50% acetonitrile with 0.1% TFA solution for the elution. Eluted peptides were concentrated in a SpeedVac and resuspended in 15 μL 5% acetonitrile with 0.1% TFA, followed by mass-spectrometry analysis.

4.6. Peptide and Protein Identification

Tandem mass spectra were analyzed by PEAKS Studio version 8.0 software (Bioinformatics Solutions Inc., Waterloo, ON, Canada). The custom database was built from the UniProt database *T. aestivum* (UP000019116) combined with chloroplast and mitochondrial proteins (130,673 records). The database search was performed with the following parameters: a fragmentation mass tolerance of 0.05 Da; parent ion tolerance of 10 ppm; fixed modification—carbamidomethylation; variable modifications—oxidation (M) and acetylation (Protein N-term). The resulting protein list was filtered by a 1% false discovery rate (FDR). The parameter "Digestion Mode" was set to "unspecific" for extracellular peptide identification. For the mass spectrometry analysis, the peptide size was limited to a maximum of 65 amino acids, as defined by the settings in PEAKS Studio version 8.0.

PEAKS Q was used for iTRAQ quantification "https://www.bioinfor.com/quantification/ (accessed on 2 June 2024)". Normalization was performed by averaging the abundance of all peptides. Median values were used for averaging. Given that iTRAQ quantification typically underestimates the degree of real fold changes between two samples, differential protein screening was performed using a fold change ratio ≥ 1.20 and significance ≥ 20 on the basis of iTRAQ studies [98–100]. The fold threshold for changed protein fold-change ratios was set at ± 1.20, which covers a 95% quantification area based on the normal distribution of two samples in all biological replicates [98,101]. Protein significance analysis was performed in PEAKS 8.0 (Bioinformatics Solutions Inc., Waterloo, Canada). A two-tailed t-test was used for the calculation of significant differences in protein ratios ($p < 0.05$).

4.7. Statistical Analysis

Two-component principal component analysis (PCA) was performed using the *scikit-learn* Python package (version 1.6) based on standardized iTRAQ intensity values of each protein group across the samples. Standardization and dimensionality reduction were performed in the *scikit-learn* package with default parameters. The columns "Coverage (%)", "Avg. Mass", and "#Peptides" from the PEAKS result tables were used for analysis of quality control metrics of the proteomic datasets. Peptide lengths were calculated as the difference between the start and stop positions of each peptide. Data visualization was made in Python using the seaborn package (v. 0.11.1) [102].

4.8. Analysis of Transcriptomic Data

Publicly available transcriptomic data of root tissues from two *T. aestivum* cultivars—ZM366 and CM42 [41]—exposed to drought stress were analyzed (NCBI BioProject accession: PRJNA950485). Quality control of the raw data was performed using FastQC v. 0.12.1 [103]. Trimmomatic v. 0.39 [104] was used to clip adapters and nucleotides of low quality. Reference genome and genome annotation were accessed via Ensembl Plants [105], and IWGSC v.60 was accessed. Illumina short reads were mapped to the reference genome using HISAT2 v. 2.2.1 [106]. Transcripts were de novo assembled using Stringtie v2.2 [107] with the -e mode applied for quantification. Next, orfipy [108] was used to predict open reading frames (ORFs) from the AUG start codon and to further translate ORFs to sequences of length over 30 amino acid residues. The longest ORF per transcript was selected. Differential gene expression analysis was performed using the DESeq2 package [109]. Transcripts with logFC > 1 and adjusted p-value (FDR) < 0.05 were considered upregulated, with logFC < -1 and adjusted p-value < 0.05—down-regulated. Translated ORFs of length under 200 amino acids were considered small ORFs (smORFs). Finally, SSP Prediction Tool [68] was used to determine smORFs encoding precursors of known SSP families based on homology to SSPs from other plants.

4.9. Annotation Methods and Functional Enrichment

The gene ontology (GO) enrichment analysis was conducted using the ShinyGO 0.80 tool "https://bioinformatics.sdstate.edu/go80/ (accessed on 2 June 2024)". Representative proteins from each protein group were used as input, and GO terms with a false discovery rate (FDR) ≤ 0.05 were considered significantly enriched. In addition, Kyoto Encyclopedia of Genes and Genomes (KEGG) pathway enrichment analysis was performed using the DAVID tool "https://davidbioinformatics.nih.gov/ (accessed on 10 June 2024)". A significance threshold of $p < 0.05$ was applied to identify enriched pathways [62,110].

5. Conclusions

In this study, morphological changes, proteome, and peptidome alterations were determined in the roots and the leaves of the control and the drought-treated wild wheat seedlings to reveal proteomic responses to long-term drought stress in the two organs. Comparative proteomic analysis identified 497 and 157 DAP groups in leaves and roots, respectively. The upregulated DAP groups in leaves were primarily involved in stress responses, such as oxidative stress and heat response, whereas those in roots were associated with responses to water deprivation and sulfur compound metabolic processes. In leaves, the most strongly upregulated proteins belonged to well-known stress-responsive groups, including DHNs, RRM domain-containing proteins, nsLTPs, and water stress and hypersensitive response domain-containing proteins. Root proteome analysis revealed pronounced changes in proteins involved in stress adaptation and redox regulation under drought conditions. Upregulated proteins in roots included DHNs, ME, and LEA under drought stress. Notably, only a small subset of proteins was shared between the two organs, reflecting distinct tissue-specific drought adaptation strategies. Additional analysis of the extracellular root peptidome revealed 2294 native peptides, including members of small secreted peptide (SSP) families. In the peptidomes of stress-induced plants, we identified 16 SSPs as well as peptides derived from proteins involved in cell wall catabolism, intercellular signaling, and stress response. These peptides represent potential candidates as regulators of drought responses. Our results help us to understand adaptation mechanisms and develop new agricultural technologies to increase productivity.

Supplementary Materials: The following supporting information can be downloaded at https://www.mdpi.com/article/10.3390/plants14142168/s1, Figure S1: The GO term analysis of the extracellular peptides. A. Functional analysis of precursor proteins of secretome peptides in control. B. Functional analysis of precursor proteins of secretome peptide under drought stress. X-axis represents the number of proteins in each GO Term. Y-axis is GO Terms. Table S1: Differentially expressed leaf proteins under 4 days of drought stress. Table S2: Differentially expressed leaf proteins under 8 days of drought stress. Table S3: Common significantly regulated proteins expressed in leaves under 4- and 8-day drought stress. Table S4: Differentially expressed root proteins under 8 days of drought stress. Table S5: Common significantly regulated proteins expressed in both roots and leaves under 8 days of drought stress. Table S6: List of native extracellular root peptides under normal conditions. Table S7: List of native extracellular root peptides under drought stress. Table S8: List of predicted SSP families identified in root peptidomes. Table S9: List of predicted SSP families identified in CM42 transcriptome data. Table S10: List of predicted SSP families identified in ZM366 transcriptome data.

Author Contributions: Conceptualization: R.A., A.M. (Anna Mamaeva), A.M. (Arina Makeeva), I.T. and I.F.; investigation, R.A., A.M. (Arina Makeeva), A.M. (Anna Mamaeva) and S.K.; formal analysis, R.A., A.M. (Arina Makeeva) and D.G. All authors have read and agreed to the published version of the manuscript.

Funding: This work was supported by the Russian Science Foundation (project no. 23-66-10013).

Data Availability Statement: The datasets that we have presented in this study are available in public repositories. The mass spectrometry proteomics data have been deposited in the ProteomeXchange Consortium via the PRIDE partner repository with the dataset identifier PXD057751. Reviewer access details. Log in to the PRIDE website using the following details: Project accession: PXD057751. Token: uCyx7erSjr4r. Alternatively, the reviewer can access the dataset by logging in to the PRIDE website using the following account details: Username: reviewer_pxd057751@ebi.ac.uk. Password: mdojSHEC9vVI.

Conflicts of Interest: The authors declare no conflicts of interest.

Abbreviations

The following abbreviations are used in this manuscript:

ABA	Abscisic acid
CAPE	Cap-derived peptide
CAT	Catalase
DAPs	Differentially abundant proteins
DAPgroups	Differentially abundant protein groups
DHN	Dehydrin
DIR	Dirigent protein
iTRAQ	Isobaric tag for relative and absolute quantitation
IDA	Inflorescence deficient in abscission
JA	Jasmonic acid
LC/MS	Liquid chromatography–mass spectrometry
LEA	Late embryogenesis abundant
ME	Malic enzyme
NodGRP	Nodule-specific glycine-rich protein
nsLTP	Non-specific lipid-transfer protein
ORF	Open reading frame
PEG	Polyethylene glycol
POD	Peroxidase
PSK	Phytosulfokine
RuBisCO	Ribulose-1,5-bisphosphatecarboxylase/oxygenase
ROS	Reactive oxygen species
SA	Salicylic acid
SOD	Superoxide dismutase
SOT	Sulfotransferase
SSP	Small secreted peptide
SubIn	Subtilisin inhibitor
TCA	Tricarboxylic acid cycle
TCEP	Tris(2-carboxyethyl)phosphine

TFA	Trifluoroacetic acid
CAA	Chloroacetamide

References

1. Notaguchi, M.; Okamoto, S. Dynamics of Long-Distance Signaling via Plant Vascular Tissues. *Front. Plant Sci.* **2015**, *6*, 161. [CrossRef] [PubMed]
2. Gebremeskel Haile, G.; Tang, Q.; Sun, S.; Huang, Z.; Zhang, X.; Liu, X. Droughts in East Africa: Causes, Impacts and Resilience. *Earth-Sci. Rev.* **2019**, *193*, 146–161. [CrossRef]
3. Raza, A.; Mubarik, M.S.; Sharif, R.; Habib, M.; Jabeen, W.; Zhang, C.; Chen, H.; Chen, Z.-H.; Siddique, K.H.M.; Zhuang, W.; et al. Developing Drought-Smart, Ready-to-Grow Future Crops. *Plant Genome* **2023**, *16*, e20279. [CrossRef] [PubMed]
4. Oguz, M.C.; Aycan, M.; Oguz, E.; Poyraz, I.; Yildiz, M. Drought Stress Tolerance in Plants: Interplay of Molecular, Biochemical and Physiological Responses in Important Development Stages. *Physiologia* **2022**, *2*, 180–197. [CrossRef]
5. Hasan, H.; Uzma; Gul, A.; Amir, R.; Ali, M.; Kubra, G.; tuz Zahra Khan, F.; Yousaf, S.; Ajmal, K.B.; Naseer, H.; et al. Chapter 13—Role of Osmoprotectants and Drought Tolerance in Wheat. In *Climate Change and Food Security with Emphasis on Wheat*; Ozturk, M., Gul, A., Eds.; Academic Press: Cambridge, MA, USA, 2020; pp. 207–216. ISBN 978-0-12-819527-7.
6. Samanta, S.; Seth, C.S.; Roychoudhury, A. The Molecular Paradigm of Reactive Oxygen Species (ROS) and Reactive Nitrogen Species (RNS) with Different Phytohormone Signaling Pathways During Drought Stress in Plants. *Plant Physiol. Biochem.* **2024**, *206*, 108259. [CrossRef]
7. Reactive Oxygen Species and Antioxidant Defense in Plants Under Abiotic Stress: Revisiting the Crucial Role of a Universal Defense Regulator. Available online: https://www.mdpi.com/2076-3921/9/8/681 (accessed on 2 July 2025).
8. Hara, M. The Multifunctionality of Dehydrins: An Overview. *Plant Signal. Behav.* **2010**, *5*, 503–508. [CrossRef]
9. Fang, Y.; Xiong, L. General Mechanisms of Drought Response and Their Application in Drought Resistance Improvement in Plants. *Cell. Mol. Life Sci.* **2015**, *72*, 673–689. [CrossRef] [PubMed]
10. Wheat Drought Tolerance: Unveiling a Synergistic Future with Conventional and Molecular Breeding Strategies. Available online: https://www.mdpi.com/2223-7747/14/7/1053 (accessed on 2 July 2025).
11. Okamoto, S.; Shinohara, H.; Mori, T.; Matsubayashi, Y.; Kawaguchi, M. Root-Derived CLE Glycopeptides Control Nodulation by Direct Binding to HAR1 Receptor Kinase. *Nat. Commun.* **2013**, *4*, 2191. [CrossRef]
12. Imin, N.; Mohd-Radzman, N.A.; Ogilvie, H.A.; Djordjevic, M.A. The Peptide-Encoding CEP1 Gene Modulates Lateral Root and Nodule Numbers in Medicago Truncatula. *J. Exp. Bot.* **2013**, *64*, 5395–5409. [CrossRef]
13. Chen, Y.-L.; Lee, C.-Y.; Cheng, K.-T.; Chang, W.-H.; Huang, R.-N.; Nam, H.G.; Chen, Y.-R. Quantitative Peptidomics Study Reveals that a Wound-Induced Peptide from PR-1 Regulates Immune Signaling in Tomato. *Plant Cell* **2014**, *26*, 4135–4148. [CrossRef]
14. Doblas, V.G.; Smakowska-Luzan, E.; Fujita, S.; Alassimone, J.; Barberon, M.; Madalinski, M.; Belkhadir, Y.; Geldner, N. Root Diffusion Barrier Control by a Vasculature-Derived Peptide Binding to the SGN3 Receptor. *Science* **2017**, *355*, 280–284. [CrossRef] [PubMed]
15. Haruta, M.; Sabat, G.; Stecker, K.; Minkoff, B.B.; Sussman, M.R. A Peptide Hormone and Its Receptor Protein Kinase Regulate Plant Cell Expansion. *Science* **2014**, *343*, 408–411. [CrossRef]
16. Kumpf, R.P.; Shi, C.-L.; Larrieu, A.; Stø, I.M.; Butenko, M.A.; Péret, B.; Riiser, E.S.; Bennett, M.J.; Aalen, R.B. Floral Organ Abscission Peptide IDA and Its HAE/HSL2 Receptors Control Cell Separation During Lateral Root Emergence. *Proc. Natl. Acad. Sci. USA* **2013**, *110*, 5235–5240. [CrossRef]
17. Stahl, Y.; Wink, R.H.; Ingram, G.C.; Simon, R. A Signaling Module Controlling the Stem Cell Niche in *Arabidopsis* Root Meristems. *Curr. Biol.* **2009**, *19*, 909–914. [CrossRef] [PubMed]
18. Xu, K.; Tian, D.; Wang, T.; Zhang, A.; Elsadek, M.A.Y.; Liu, W.; Chen, L.; Guo, Y. Small Secreted Peptides (SSPs) in Tomato and Their Potential Roles in Drought Stress Response. *Mol. Hortic.* **2023**, *3*, 17. [CrossRef]
19. Marshall, E.; Costa, L.M.; Gutierrez-Marcos, J. Cysteine-Rich Peptides (CRPs) Mediate Diverse Aspects of Cell–Cell Communication in Plant Reproduction and Development. *J. Exp. Bot.* **2011**, *62*, 1677–1686. [CrossRef] [PubMed]
20. Murphy, E.; Smith, S.; De Smet, I. Small Signaling Peptides in Arabidopsis Development: How Cells Communicate Over a Short Distance. *Plant Cell* **2012**, *24*, 3198–3217. [CrossRef]
21. Takahashi, F.; Suzuki, T.; Osakabe, Y.; Betsuyaku, S.; Kondo, Y.; Dohmae, N.; Fukuda, H.; Yamaguchi-Shinozaki, K.; Shinozaki, K. A Small Peptide Modulates Stomatal Control via Abscisic Acid in Long-Distance Signalling. *Nature* **2018**, *556*, 235–238. [CrossRef]
22. Zhang, L.; Shi, X.; Zhang, Y.; Wang, J.; Yang, J.; Ishida, T.; Jiang, W.; Han, X.; Kang, J.; Wang, X.; et al. CLE9 Peptide-Induced Stomatal Closure Is Mediated by Abscisic Acid, Hydrogen Peroxide, and Nitric Oxide in Arabidopsis Thaliana. *Plant Cell Environ.* **2019**, *42*, 1033–1044. [CrossRef]
23. Smith, S.; Zhu, S.; Joos, L.; Roberts, I.; Nikonorova, N.; Vu, L.D.; Stes, E.; Cho, H.; Larrieu, A.; Xuan, W.; et al. The CEP5 Peptide Promotes Abiotic Stress Tolerance, As Revealed by Quantitative Proteomics, and Attenuates the AUX/IAA Equilibrium in *Arabidopsis*. *Mol. Cell. Proteom.* **2020**, *19*, 1248–1262. [CrossRef]

24. Atkinson, N.J.; Lilley, C.J.; Urwin, P.E. Identification of Genes Involved in the Response of Arabidopsis to Simultaneous Biotic and Abiotic Stresses. *Plant Physiol.* **2013**, *162*, 2028–2041. [CrossRef] [PubMed]
25. Reichardt, S.; Piepho, H.-P.; Stintzi, A.; Schaller, A. Peptide Signaling for Drought-Induced Tomato Flower Drop. *Science* **2020**, *367*, 1482–1485. [CrossRef]
26. Patharkar, O.R.; Walker, J.C. Core Mechanisms Regulating Developmentally Timed and Environmentally Triggered Abscission *Plant Physiol.* **2016**, *172*, 510–520. [CrossRef] [PubMed]
27. Jiang, Q.; Yang, J.; Wang, Q.; Zhou, K.; Mao, K.; Ma, F. Overexpression of *MdEPF2* Improves Water Use Efficiency and Reduces Oxidative Stress in Tomato. *Environ. Exp. Bot.* **2019**, *162*, 321–332. [CrossRef]
28. Genome-Wide Identification of Epidermal Pattern Factor (EPF) Gene Family in Potato and Functional Characterization of StEPF4 in Regulating Drought Stress. Available online: https://www.mdpi.com/2073-4395/14/12/2948 (accessed on 2 July 2025).
29. Liu, S.; Wang, C.; Jia, F.; An, Y.; Liu, C.; Xia, X.; Yin, W. Secretory Peptide PdEPF2 Enhances Drought Tolerance by Modulating Stomatal Density and Regulates ABA Response in Transgenic *Arabidopsis thaliana*. *Plant Cell Tissue Organ Cult.* **2016**, *125*, 419–431. Available online: https://link.springer.com/article/10.1007/s11240-016-0957-x (accessed on 2 July 2025). [CrossRef]
30. Mamaeva, A.; Taliansky, M.; Filippova, A.; Love, A.J.; Golub, N.; Fesenko, I. The Role of Chloroplast Protein Remodeling in Stress Responses and Shaping of the Plant Peptidome. *New Phytol.* **2024**, *227*, 1326–1334. [CrossRef]
31. Fesenko, I.; Azarkina, R.; Kirov, I.; Kniazev, A.; Filippova, A.; Grafskaia, E.; Lazarev, V.; Zgoda, V.; Butenko, I.; Bukato, O.; et al. Phytohormone Treatment Induces Generation of Cryptic Peptides with Antimicrobial Activity in the Moss Physcomitrella Patens. *BMC Plant Biol.* **2019**, *19*, 9. [CrossRef] [PubMed]
32. Fesenko, I.A.; Arapidi, G.P.; Skripnikov, A.Y.; Alexeev, D.G.; Kostryukova, E.S.; Manolov, A.I.; Altukhov, I.A.; Khazigaleeva, R.A.; Seredina, A.V.; Kovalchuk, S.I.; et al. Specific Pools of Endogenous Peptides Are Present in Gametophore, Protonema, and Protoplast Cells of the Moss Physcomitrella Patens. *BMC Plant Biol.* **2015**, *15*, 87. [CrossRef]
33. Filippova, A.; Lyapina, I.; Kirov, I.; Zgoda, V.; Belogurov, A.; Kudriaeva, A.; Ivanov, V.; Fesenko, I. Salicylic Acid Influences the Protease Activity and Posttranslation Modifications of the Secreted Peptides in the Moss Physcomitrella Patens. *J. Pept. Sci.* **2024**, *25*, e3138. [CrossRef]
34. Khazigaleeva, R.A.; Vinogradova, S.V.; Petrova, V.L.; Fesenko, I.A.; Arapidi, G.P.; Kamionskaya, A.M.; Govorun, V.M.; Ivanov, V.T. Antimicrobial Activity of Endogenous Peptides of the Moss Physcomitrella Patens. *Russ. J. Bioorg. Chem.* **2017**, *43*, 248–254. [CrossRef]
35. Lyapina, I.; Filippova, A.; Fesenko, I. The Role of Peptide Signals Hidden in the Structure of Functional Proteins in Plant Immune Responses. *Int. J. Mol. Sci.* **2019**, *20*, 4343. [CrossRef]
36. Patel, N.; Mohd-Radzman, N.A.; Corcilius, L.; Crossett, B.; Connolly, A.; Cordwell, S.J.; Ivanovici, A.; Taylor, K.; Williams, J.; Binos, S.; et al. Diverse Peptide Hormones Affecting Root Growth Identified in the Medicago Truncatula Secreted Peptidome. *Mol. Cell. Proteom.* **2018**, *17*, 160–174. [CrossRef] [PubMed]
37. Meena, R.; Ramadas, S.; Tripathi, S.; Chander, S.; Chhokar, R.; Sharma, R. Hydro-Priming of Seed Improves the Water Use Efficiency, Grain Yield and Net Economic Return of Wheat Under Different Moisture Regimes. *SAARC J. Agric.* **2013**, *11*, 149–159. [CrossRef]
38. Ghatak, A.; Chaturvedi, P.; Weckwerth, W. Cereal Crop Proteomics: Systemic Analysis of Crop Drought Stress Responses Towards Marker-Assisted Selection Breeding. *Front. Plant Sci.* **2017**, *8*, 757. [CrossRef]
39. Yu, Y.; Zhu, D.; Ma, C.; Cao, H.; Wang, Y.; Xu, Y.; Zhang, W.; Yan, Y. Transcriptome Analysis Reveals Key Differentially Expressed Genes Involved in Wheat Grain Development. *Crop J.* **2016**, *4*, 92–106. [CrossRef]
40. Iquebal, M.A.; Sharma, P.; Jasrotia, R.S.; Jaiswal, S.; Kaur, A.; Saroha, M.; Angadi, U.B.; Sheoran, S.; Singh, R.; Singh, G.P.; et al. RNAseq Analysis Reveals Drought-Responsive Molecular Pathways with Candidate Genes and Putative Molecular Markers in Root Tissue of Wheat. *Sci. Rep.* **2019**, *9*, 13917. [CrossRef] [PubMed]
41. Xi, W.; Hao, C.; Li, T.; Wang, H.; Zhang, X. Transcriptome Analysis of Roots from Wheat (*Triticum Aestivum* L.) Varieties in Response to Drought Stress. *Int. J. Mol. Sci.* **2023**, *24*, 7245. [CrossRef]
42. Mao, H.; Li, S.; Wang, Z.; Cheng, X.; Li, F.; Mei, F.; Chen, N.; Kang, Z. Regulatory Changes in TaSNAC8-6A Are Associated with Drought Tolerance in Wheat Seedlings. *Plant Biotechnol. J.* **2024**, *18*, 1078–1092. [CrossRef]
43. Niu, Y.; Li, J.; Sun, F.; Song, T.; Han, B.; Liu, Z.; Su, P. Comparative Transcriptome Analysis Reveals the Key Genes and Pathways Involved in Drought Stress Response of Two Wheat (*Triticum aestivum* L.) Varieties. *Genomics* **2023**, *115*, 110688. [CrossRef]
44. Halder, T.; Choudhary, M.; Liu, H.; Chen, Y.; Yan, G.; Siddique, K.H.M. Wheat Proteomics for Abiotic Stress Tolerance and Root System Architecture: Current Status and Future Prospects. *Proteomes* **2022**, *10*, 17. [CrossRef]
45. Ford, K.L.; Cassin, A.; Bacic, A. Quantitative Proteomic Analysis of Wheat Cultivars with Differing Drought Stress Tolerance. *Front. Plant Sci.* **2011**, *2*, 44. [CrossRef] [PubMed]
46. Zhang, Y.; Huang, X.; Wang, L.; Wei, L.; Wu, Z.; You, M.; Li, B. Proteomic Analysis of Wheat Seed in Response to Drought Stress. *J. Integr. Agric.* **2014**, *13*, 919–925. [CrossRef]

47. Cheng, L.; Wang, Y.; He, Q.; Li, H.; Zhang, X.; Zhang, F. Comparative Proteomics Illustrates the Complexity of Drought Resistance Mechanisms in Two Wheat (*Triticum aestivum* L.) Cultivars under Dehydration and Rehydration. *BMC Plant Biol.* **2016**, *16*, 188. [CrossRef]
48. Deng, X.; Liu, Y.; Xu, X.; Liu, D.; Zhu, G.; Yan, X.; Wang, Z.; Yan, Y. Comparative Proteome Analysis of Wheat Flag Leaves and Developing Grains Under Water Deficit. *Front. Plant Sci.* **2018**, *9*, 425. [CrossRef]
49. Hao, P.; Zhu, J.; Gu, A.; Lv, D.; Ge, P.; Chen, G.; Li, X.; Yan, Y. An Integrative Proteome Analysis of Different Seedling Organs in Tolerant and Sensitive Wheat Cultivars Under Drought Stress and Recovery. *Proteomics* **2015**, *15*, 1544–1563. [CrossRef]
50. Peng, Z.; Wang, M.; Li, F.; Lv, H.; Li, C.; Xia, G. A Proteomic Study of the Response to Salinity and Drought Stress in an Introgression Strain of Bread Wheat. *Mol. Cell. Proteom.* **2009**, *8*, 2676–2686. [CrossRef] [PubMed]
51. Liu, H.; Sultan, M.A.R.F.; Liu, X.L.; Zhang, J.; Yu, F.; Zhao, H.X. Physiological and Comparative Proteomic Analysis Reveals Different Drought Responses in Roots and Leaves of Drought-Tolerant Wild Wheat (*Triticum boeoticum*). *PLoS ONE* **2015**, *10*, e0121852. [CrossRef] [PubMed]
52. Zhang, Y.; Lou, H.; Guo, D.; Zhang, R.; Su, M.; Hou, Z.; Zhou, H.; Liang, R.; Xie, C.; You, M.; et al. Identifying Changes in the Wheat Kernel Proteome Under Heat Stress Using iTRAQ. *Crop J.* **2018**, *6*, 600–610. [CrossRef]
53. Zhang, H.; Zhang, L.; Lv, H.; Yu, Z.; Zhang, D.; Zhu, W. Identification of Changes in *Triticum aestivum* L. Leaf Proteome in Response to Drought Stress by 2D-PAGE and MALDI-TOF/TOF Mass Spectrometry. *Acta Physiol. Plant* **2014**, *36*, 1385–1398. [CrossRef]
54. Faghani, E.; Gharechahi, J.; Komatsu, S.; Mirzaei, M.; Khavarinejad, R.A.; Najafi, F.; Farsad, L.K.; Salekdeh, G.H. Comparative Physiology and Proteomic Analysis of Two Wheat Genotypes Contrasting in Drought Tolerance. *J. Proteom.* **2015**, *114*, 1–15. [CrossRef]
55. Yan, M.; Xue, C.; Xiong, Y.; Meng, X.; Li, B.; Shen, R.; Lan, P. Proteomic Dissection of the Similar and Different Responses of Wheat to Drought, Salinity and Submergence During Seed Germination. *J. Proteom.* **2020**, *220*, 103756. [CrossRef]
56. Alvarez, S.; Hicks, L.M.; Pandey, S. ABA-Dependent and -Independent G-Protein Signaling in Arabidopsis Roots Revealed Through an iTRAQ Proteomics Approach. *J. Proteome Res.* **2011**, *10*, 3107–3122. [CrossRef] [PubMed]
57. Tian, D.; Xie, Q.; Deng, Z.; Xue, J.; Li, W.; Zhang, Z.; Dai, Y.; Zheng, B.; Lu, T.; De Smet, I.; et al. Small Secreted Peptides Encoded on the Wheat (*Triticum Aestivum* L.) Genome and Their Potential Roles in Stress Responses. *Front. Plant Sci.* **2022**, *13*, 1000297. [CrossRef]
58. Tokbaeva, A.A.; Satsyuk, K.A.; Cheng, X.; Ustinova, A.S.; Barakova, N.V. The Prospects of Application of Leningradskaya 6 Wheat in Ethyl Alcohol Technology. *Process. Food Prod. Equip.* **2019**, *4*, 95–101. [CrossRef]
59. Mirskaya, G.V.; Khomyakov, Y.V.; Rushina, N.A.; Vertebny, V.E.; Chizhevskaya, E.P.; Chebotar, V.K.; Chesnokov, Y.V.; Pishchik, V.N. Plant Development of Early-Maturing Spring Wheat (*Triticum aestivum* L.) Under Inoculation with *Bacillus* sp. *V2026. Plants* **2022**, *11*, 1817. [CrossRef]
60. Osmolovskaya, N.; Shumilina, J.; Kim, A.; Didio, A.; Grishina, T.; Bilova, T.; Keltsieva, O.A.; Zhukov, V.; Tikhonovich, I.; Tarakhovskaya, E.; et al. Methodology of Drought Stress Research: Experimental Setup and Physiological Characterization. *Int. J. Mol. Sci.* **2018**, *19*, 4089. [CrossRef] [PubMed]
61. Zhang, X.; Li, C.; Lu, W.; Wang, X.; Ma, B.; Fu, K.; Li, C.; Li, C. Comparative Analysis of Combined Phosphorus and Drought Stress-Responses in Two Winter Wheat. *PeerJ* **2022**, *10*, e13887. [CrossRef]
62. Ge, S.X.; Jung, D.; Yao, R. ShinyGO: A Graphical Gene-Set Enrichment Tool for Animals and Plants. *Bioinformatics* **2020**, *36*, 2628–2629. [CrossRef]
63. Kuromori, T.; Fujita, M.; Takahashi, F.; Yamaguchi-Shinozaki, K.; Shinozaki, K. Inter-Tissue and Inter-Organ Signaling in Drought Stress Response and Phenotyping of Drought Tolerance. *Plant J.* **2022**, *109*, 342–358. [CrossRef]
64. Stührwohldt, N.; Bühler, E.; Sauter, M.; Schaller, A. Phytosulfokine (PSK) Precursor Processing by Subtilase SBT3.8 and PSK Signaling Improve Drought Stress Tolerance in Arabidopsis. *J. Exp. Bot.* **2021**, *72*, 3427–3440. [CrossRef]
65. Ohyama, K.; Ogawa, M.; Matsubayashi, Y. Identification of a Biologically Active, Small, Secreted Peptide in Arabidopsis by in Silico Gene Screening, Followed by LC-MS-Based Structure Analysis. *Plant J.* **2008**, *55*, 152–160. [CrossRef]
66. Sauter, M. Phytosulfokine Peptide Signalling. *J. Exp. Bot.* **2015**, *66*, 5161–5169. [CrossRef]
67. Nakayama, T.; Shinohara, H.; Tanaka, M.; Baba, K.; Ogawa-Ohnishi, M.; Matsubayashi, Y. A Peptide Hormone Required for Casparian Strip Diffusion Barrier Formation in Arabidopsis Roots. *Science* **2017**, *355*, 284–286. [CrossRef]
68. Boschiero, C.; Dai, X.; Lundquist, P.K.; Roy, S.; Christian de Bang, T.; Zhang, S.; Zhuang, Z.; Torres-Jerez, I.; Udvardi, M.K.; Scheible, W.-R.; et al. MtSSPdb: The Medicago Truncatula Small Secreted Peptide Database1. *Plant Physiol.* **2020**, *183*, 399–413. [CrossRef]
69. Casado-Vela, J.; Martínez-Esteso, M.J.; Rodriguez, E.; Borrás, E.; Elortza, F.; Bru-Martínez, R. iTRAQ-Based Quantitative Analysis of Protein Mixtures with Large Fold Change and Dynamic Range. *Proteomics* **2010**, *10*, 343–347. [CrossRef] [PubMed]
70. Riyazuddin, R.; Nisha, N.; Singh, K.; Verma, R.; Gupta, R. Involvement of Dehydrin Proteins in Mitigating the Negative Effects of Drought Stress in Plants. *Plant Cell Rep.* **2022**, *41*, 519–533. [CrossRef] [PubMed]

71. Wang, Y.; Zhang, X.; Huang, G.; Feng, F.; Liu, X.; Guo, R.; Gu, F.; Zhong, X.; Mei, X. iTRAQ-Based Quantitative Analysis of Responsive Proteins Under PEG-Induced Drought Stress in Wheat Leaves. *Int. J. Mol. Sci.* **2019**, *20*, 2621. [CrossRef] [PubMed]
72. Sharma, M.; Gupta, S.K.; Majumder, B.; Maurya, V.K.; Deeba, F.; Alam, A.; Pandey, V. Salicylic Acid Mediated Growth, Physiological and Proteomic Responses in Two Wheat Varieties Under Drought Stress. *J. Proteom.* **2017**, *163*, 28–51. [CrossRef]
73. Overexpression of the Wheat TaPsb28 Gene Enhances Drought Tolerance in Transgenic Arabidopsis. Available online: https://www.mdpi.com/1422-0067/24/6/5226 (accessed on 2 July 2025).
74. Paniagua, C.; Bilkova, A.; Jackson, P.; Dabravolski, S.; Riber, W.; Didi, V.; Houser, J.; Gigli-Bisceglia, N.; Wimmerova, M.; Budínská, E.; et al. Dirigent Proteins in Plants: Modulating Cell Wall Metabolism During Abiotic and Biotic Stress Exposure. *J. Exp. Bot.* **2017**, *68*, 3287–3301. [CrossRef]
75. Wang, S.; Wang, R.; Liang, D.; Ma, F.; Shu, H. Molecular Characterization and Expression Analysis of a Glycine-Rich RNA-Binding Protein Gene from Malus Hupehensis Rehd. *Mol. Biol. Rep.* **2012**, *39*, 4145–4153. [CrossRef]
76. Gao, S.; Guo, W.; Feng, W.; Liu, L.; Song, X.; Chen, J.; Hou, W.; Zhu, H.; Tang, S.; Hu, J. LTP3 Contributes to Disease Susceptibility in Arabidopsis by Enhancing Abscisic Acid (ABA) Biosynthesis. *Mol. Plant. Pathol.* **2016**, *17*, 412–426. [CrossRef]
77. Missaoui, K.; Gonzalez-Klein, Z.; Pazos-Castro, D.; Hernandez-Ramirez, G.; Garrido-Arandia, M.; Brini, F.; Diaz-Perales, A.; Tome-Amat, J. Plant Non-Specific Lipid Transfer Proteins: An Overview. *Plant Physiol. Biochem.* **2022**, *171*, 115–127. [CrossRef] [PubMed]
78. Li, H.; Liu, Y.; Gao, W.; Zhu, J.; Zhang, H.; Wang, Z.; Liu, C.; Li, X. Genome-Wide Characterization of Small Secreted Peptides in Nicotiana Tabacum and Functional Assessment of NtLTP25 in Plant Immunity. *Physiol. Plant* **2024**, *176*, e14436. [CrossRef] [PubMed]
79. Xiao, Y.; Xiao, C.; He, X.; Yang, X.; Tong, Z.; Wang, Z.; Sun, Z.; Qiu, W. A Novel Non-Specific Lipid Transfer Protein Gene, CmnsLTP6.9, Enhanced Osmotic and Drought Tolerance by Regulating ROS Scavenging and Remodeling Lipid Profiles in Chinese Chestnut (*Castanea mollissima* Blume). *Plants* **2023**, *12*, 3916. [CrossRef]
80. Zhang, H.; Zeng, Y.; Seo, J.; Kim, Y.-J.; Kim, S.T.; Kwon, S.-W. Global Identification and Characterization of C2 Domain-Containing Proteins Associated with Abiotic Stress Response in Rice (*Oryza sativa* L.). *Int. J. Mol. Sci.* **2022**, *23*, 2221. [CrossRef]
81. Chen, J.; Gao, L.; Baek, D.; Liu, C.; Ruan, Y.; Shi, H. Detoxification Function of the Arabidopsis Sulphotransferase AtSOT12 by Sulphonation of Xenobiotics. *Plant Cell Environ.* **2015**, *38*, 1673–1682. Available online: https://onlinelibrary.wiley.com/doi/10.1111/pce.12525 (accessed on 28 November 2024). [CrossRef]
82. Qi, J.; Song, C.P.; Wang, B.; Zhou, J.; Kangasjärvi, J.; Zhu, J.K.; Gong, Z. Reactive Oxygen Species Signaling and Stomatal Movement in Plant Responses to Drought Stress and Pathogen Attack. *J. Integr. Plant Biol.* **2018**, *60*, 805–826. Available online: https://onlinelibrary.wiley.com/doi/full/10.1111/JIPB.12654 (accessed on 28 November 2024). [CrossRef]
83. Yang, Y.; Xu, L.; Hao, C.; Wan, M.; Tao, Y.; Zhuang, Y.; Su, Y.; Li, L. The microRNA408–Plantacyanin Module Balances Plant Growth and Drought Resistance by Regulating Reactive Oxygen Species Homeostasis in Guard Cells. *Plant Cell* **2024**, *36*, 4338–4355. [CrossRef] [PubMed]
84. Dong, J.; Kim, S.T.; Lord, E.M. Plantacyanin Plays a Role in Reproduction in Arabidopsis. *Plant Physiol.* **2005**, *138*, 778–789. [CrossRef]
85. Kim, S.; Mollet, J.-C.; Dong, J.; Zhang, K.; Park, S.-Y.; Lord, E.M. Chemocyanin, a Small Basic Protein from the Lily Stigma, Induces Pollen Tube Chemotropism. *Proc. Natl. Acad. Sci. USA* **2003**, *100*, 16125–16130. [CrossRef]
86. Jiang, A.; Guo, Z.; Pan, J.; Yang, Y.; Zhuang, Y.; Zuo, D.; Hao, C.; Gao, Z.; Xin, P.; Chu, J.; et al. The PIF1-miR408-PLANTACYANIN Repression Cascade Regulates Light-Dependent Seed Germination. *Plant Cell* **2021**, *33*, 1506–1529. [CrossRef] [PubMed]
87. Moloi, S.J.; Ngara, R. The Roles of Plant Proteases and Protease Inhibitors in Drought Response: A Review. *Front. Plant Sci.* **2023**, *14*, 1165845. [CrossRef]
88. Chojnacka, M.; Szewińska, J.; Mielecki, M.; Nykiel, M.; Imai, R.; Bielawski, W.; Orzechowski, S. A Triticale Water-Deficit-Inducible Phytocystatin Inhibits Endogenous Cysteine Proteinases In Vitro. *J. Plant Physiol.* **2015**, *174*, 161–165. [CrossRef] [PubMed]
89. Groten, K.; Dutilleul, C.; van Heerden, P.D.; Vanacker, H.; Bernard, S.; Finkemeier, I.; Dietz, K.J.; Foyer, C.H. Redox Regulation of Peroxiredoxin and Proteinases by Ascorbate and Thiols during Pea Root Nodule Senescence. *FEBS Lett.* **2006**, *580*, 1269–1276. Available online: https://febs.onlinelibrary.wiley.com/doi/10.1016/j.febslet.2006.01.043 (accessed on 28 November 2024). [CrossRef]
90. Quain, M.D.; Makgopa, M.E.; Márquez-García, B.; Comadira, G.; Fernandez-Garcia, N.; Olmos, E.; Schnaubelt, D.; Kunert, K.J.; Foyer, C.H. Ectopic Phytocystatin Expression Leads to Enhanced Drought Stress Tolerance in Soybean (*Glycine max*) and Arabidopsis Thaliana Through Effects on Strigolactone Pathways and Can Also Result in Improved Seed Traits. *Plant Biotechnol. J.* **2014**, *12*, 903–913. [CrossRef]
91. Mangena, P. Phytocystatins and Their Potential Application in the Development of Drought Tolerance Plants in Soybeans (*Glycine max* L.). *Protein Pept. Lett.* **2020**, *27*, 135–144. [CrossRef] [PubMed]
92. Chien, P.-S.; Nam, H.G.; Chen, Y.-R. A Salt-Regulated Peptide Derived from the CAP Superfamily Protein Negatively Regulates Salt-Stress Tolerance in Arabidopsis. *J. Exp. Bot.* **2015**, *66*, 5301–5313. [CrossRef]

93. Pishchik, V.N.; Filippova, P.S.; Mirskaya, G.V.; Khomyakov, Y.V.; Vertebny, V.E.; Dubovitskaya, V.I.; Ostankova, Y.V.; Semenov, A.V.; Chakrabarty, D.; Zuev, E.V.; et al. Epiphytic PGPB Bacillus Megaterium AFI1 and Paenibacillus Nicotianae AFI2 Improve Wheat Growth and Antioxidant Status Under Ni Stress. *Plants* **2021**, *10*, 2334. [CrossRef]
94. Hoagland, D.R.; Arnon, D.I. The Water-Culture Method for Growing Plants Without Soil. 1950. Available online: https://openlibrary.org/books/OL25240089M/The_water-culture_method_for_growing_plants_without_soil (accessed on 1 July 2025).
95. Faurobert, M.; Pelpoir, E.; Chaïb, J. Phenol Extraction of Proteins for Proteomic Studies of Recalcitrant Plant Tissues. *Methods Mol. Biol.* **2007**, *355*, 9–14. [CrossRef]
96. Kovalchuk, S.I.; Ziganshin, R.; Shelukhina, I. Simple In-House Ultra-High Performance Capillary Column Manufacturing with the FlashPack Approach. *J. Vis. Exp.* **2021**, *178*, e62522. [CrossRef]
97. Lyapina, I.; Fesenko, I. Intracellular and Extracellular Peptidomes of the Model Plant, Physcomitrium Patens. *Methods Mol. Biol.* **2024**, *2758*, 375–385. [CrossRef]
98. Kumari, M.; Kapoor, R.; Devanna, B.N.; Varshney, S.; Kamboj, R.; Rai, A.K.; Sharma, T.R. iTRAQ Based Proteomic Analysis of Rice Lines Having Single or Stacked Blast Resistance Genes: Pi54/Pi54rh During Incompatible Interaction with Magnaporthe Oryzae. *Physiol. Mol. Biol. Plants* **2023**, *29*, 871–887. [CrossRef]
99. Yang, M.; Dong, J.; Zhao, W.; Gao, X. Characterization of Proteins Involved in Early Stage of Wheat Grain Development by iTRAQ. *J. Proteom.* **2016**, *136*, 157–166. [CrossRef] [PubMed]
100. Yang, R.; Li, M.; Harrison, M.T.; Fahad, S.; Wei, M.; Li, X.; Yin, L.; Sha, A.; Zhou, M.; Liu, K.; et al. iTRAQ Proteomic Analysis of Wheat (*Triticum aestivum* L.) Genotypes Differing in Waterlogging Tolerance. *Front. Plant Sci.* **2022**, *13*, 890083. [CrossRef] [PubMed]
101. Unwin, R.D.; Griffiths, J.R.; Whetton, A.D. Simultaneous Analysis of Relative Protein Expression Levels Across Multiple Samples Using iTRAQ Isobaric Tags with 2D Nano LC–MS/MS. *Nat. Protoc.* **2010**, *5*, 1574–1582. [CrossRef]
102. Waskom, M.L. Seaborn: Statistical Data Visualization. *J. Open Source Softw.* **2021**, *6*, 3021. [CrossRef]
103. LaMar, D. FastQC. 2015. Available online: https://qubeshub.org/resources/fastqc (accessed on 1 December 2024).
104. Bolger, A.M.; Lohse, M.; Usadel, B. Trimmomatic: A flexible trimmer for Illumina sequence data. *Bioinformatics* **2014**, *30*, 2114–2120. [CrossRef]
105. Ensembl Plants. Available online: https://plants.ensembl.org/index.html (accessed on 2 July 2025).
106. Kim, D.; Langmead, B.; Salzberg, S.L. HISAT: A Fast Spliced Aligner with Low Memory Requirements. *Nat. Methods* **2015**, *12*, 357–360. Available online: https://www.nature.com/articles/nmeth.3317 (accessed on 2 July 2025). [CrossRef]
107. Pertea, M.; Pertea, G.M.; Antonescu, C.M.; Chang, T.C.; Mendell, J.T.; Salzberg, S.L. StringTie Enables Improved Reconstruction of a Transcriptome from RNA-Seq Reads. *Nat. Biotechnol.* **2015**, *33*, 290–295. Available online: https://www.nature.com/articles/nbt.3122 (accessed on 2 July 2025). [CrossRef]
108. Singh, U.; Wurtele, E.S. Orfipy: A Fast and Flexible Tool for Extracting ORFs. *Bioinformatics* **2021**, *37*, 3019–3020. [CrossRef]
109. Love, M.I.; Huber, W.; Anders, S. Moderated Estimation of Fold Change and Dispersion for RNA-Seq Data with DESeq2. *Genome Biol.* **2014**, *15*, 550. [CrossRef] [PubMed]
110. Sherman, B.T.; Hao, M.; Qiu, J.; Jiao, X.; Baseler, M.W.; Lane, H.C.; Imamichi, T.; Chang, W. DAVID: A Web Server for Functional Enrichment Analysis and Functional Annotation of Gene Lists (2021 Update). *Nucleic Acids Res.* **2022**, *50*, W216–W221. [CrossRef] [PubMed]

plants

MDPI

Article

Pod Dehiscence in Soybean: Genome Wide Association Study and Genomic Prediction

Shynar Mazkirat [1], Kulpash Bulatova [1,*], Svetlana Didorenko [1], Sholpan Bastaubayeva [1], Dilyara Babissekova [1], Sholpan Khalbayeva [1], Azamat Tukenov [1], Akzhan Yespembetova [1], Nurgul Saparbayeva [1] and Yuri Shavrukov [2,*]

1 Kazakh Research Institute of Agriculture and Plant Growing, Almalybak 040909, Almaty District, Kazakhstan; shynarbek.mazkirat@gmail.com (S.M.); svetl_did@mail.ru (S.D.); sh.bastaubayeva@kazapg.kz (S.B.) dilyara280188@mail.ru (D.B.); sholpan_2706@mail.ru (S.K.); tukenov97@mail.ru (A.T.); akzhanes@mail.ru (A.Y.); n.saparbaeva@mail.ru (N.S.)

2 College of Science and Engineering, Biological Sciences, Flinders University, Adelaide, SA 5042, Australia

* Correspondence: bulatova_k@rambler.ru (K.B.); yuri.shavrukov@flinders.edu.au (Y.S.)

Abstract: Pod dehiscence is one of the main factors which play a vital role on the final yield of many crops including soybean and, therefore, it is important to elucidate genetic mechanisms associated with this trait. In this study, morphological, physiological and biochemical analysis was conducted for pod and pod-related traits on 170 soybean genotypes with diverse origins. Subsequently, a genome-wide association study (GWAS) was performed using Silico-DArT and DArT SNPs markers. In total, 48 QTLs were identified with 14 stable QTLs, mostly located on chromosomes 6, 13 and 16, corresponding to pod dehiscence and pod-related traits. From putative candidate genes, two most stable and important genes for pod dehiscence with known functions were emphasised from the QTLs: Glyma.13G184500 and Glyma.16G141100, encoding transcription factors DNA-binding bromodomain-containing protein and C2H2 zinc finger protein, respectively. Finally, a genomic prediction approach was implemented to select genotypes most resistant to pod dehiscence. GWAS-derived markers confirmed the stable prediction of pod dehiscence in studied accessions from different populations and the best non-dehiscent soybean genotypes were successfully selected.

Keywords: DArT markers; genome-wide association study (GWAS); genomic prediction; pod dehiscence; soybean

1. Introduction

A major source of high-quality protein and oil for human diets and animal feed, soybean [*Glycine max* (L.) Merr.] plays a crucial role in ensuring global food and nutritional security. The demand for soybeans has increased due to the world's expanding population, underscoring the necessity of developing high-yielding, stress tolerant and nutritionally improved cultivars [1]. Soybean can grow in a variety of climatic and agronomic conditions and its total cultivated area is about 122 million hectares, producing over 353 million metric tons annually. It is mainly grown in Brazil, the USA, Argentina, China and India [2–4].

In Kazakhstan, the area under soybean cultivation is increasing progressively due to rising commercial interest. For example, since 2000 in Kazakhstan, the soybean total growing area increased from 3500 ha to nearly 128000 ha in 2022 [5]. The main soybean sowing area in Kazakhstan includes the Almaty and Zhetysu regions [6]. However, the harvesting period in these regions overlaps with strong drought conditions, increasing seed loss by pod dehiscence.

Pod dehiscence (PD), or pod shattering, in wild soybean or landraces is an important trait for plant populations in natural environments [7,8]. In domesticated crops, PD is an entirely unfavourable trait due to the detrimental impact it has on seed yield during harvest [9]. Therefore, the production of genotypes with no or minimal PD is strongly emphasised in all soybean breeding programs. Resistance to PD is a complex of physical, physiological and biochemical processes [10], and is associated with the pod-related traits: pod length, width and thickness as well as ratios, including length/width, length/thickness and width/thickness. Usually, PD occurs when pods reach the stage of physiological maturity. One of the reasons relates to decreased humidity in pod walls and reduced relative water content in the pod [11–15]. However, there are few investigated cases of pod opening, which occurs when the pods are still immature and green. Water deficit can be a trigger to immature pod opening in soybean, as reported in experiments in greenhouse conditions studying the effect of drought on plants during pod formation [16].

The chlorophyll content is an important trait to estimate crop growth, and photosynthetic rate [17,18]. Using a Minolta 502 SPAD meter (Konica Minolta, Tokyo, Japan) for rapid measurement of chlorophyll content in soybean plant leaves, it was found that the chlorophyll content remained stable from stage R2 (full flower) till stage R5 (beginning of seed formation) but declined after that [19]. The content of mineral elements Mg and Fe in the pod valves of grain soybeans at the R5 and R6 stages negatively correlates with PD. These elements participate directly in photosynthesis and chlorophyll synthesis, increasing their level and pod resistance to dehiscence. Presumably, the high concentration of chlorophyll in soybean plants at the pod filling stage may be an indicator for resistance to PD [20].

Numerous yield parameters can be associated with PD. For example, a soybean genotype with small pods was shown to have less seeds and non-dehiscent pods [21]. Additionally, the percentage of PD was positively correlated with the number of productive branches, 100-seed weight and some pod-related traits [20]. Plant height was also shown to be associated with PD. For example, taller and stronger plants from the genus *Brassica* were more resistant to dehiscence and shattering of siliques [22,23]. Soybean cultivars resistant to PD showed a tendency to have a longer stem with more nodes, lower height to first pod and smaller seed size [24].

Heat stress causes excessive generation of reactive oxygen species (ROS), which harms intracellular machinery [25,26]. Antioxidant enzymes, including superoxide dismutase (SOD), catalase (CAT), peroxidase (POX), and ascorbate peroxidase (APX), play an important role in neutralizing excess ROS and reducing their detrimental effects [27,28]. It was confirmed that drought stress during the flowering and pod development stages had a severe effect on PD in *Brassica napus* L. [29], and a positive correlation between PD and antioxidant enzymes such as SOD and CAT were reported recently [30]. Additionally, the activity of antioxidant enzymes SOD, POD and CAT was shown to strongly correlate with each other [31]. Among these enzymes, SOD has the greatest importance for active oxygen scavenging [32,33]. Therefore, that could be sufficient to determine the changes in the activity of one of these enzymes, for example SOD, and similar trend in the activity of other antioxidant enzymes is expected. Therefore, increased activity of antioxidant enzymes in plants can help to support tolerance of plants to stress factors.

The study of Quantitative trait loci (QTL) is often used in plant biology to identify genes of interest [34,35]. For PD in soybean, the first QTL was reported in chromosome 16 using Restriction fragment length polymorphism (RFLP) markers [36]. This QTL was later identified and mapped in the genetic region between SSR markers Sat_093 and Sat_366, and it was named qPDH [37]. Several other unique QTLs were discovered on chromosomes 1, 5, 8 and 14 using the Specific-locus amplified fragment sequencing method

(SLAF) [38]. The major QTL, *pdh1* (Glyma.16G141400) linked to PD was revealed in cultivated soybean [9]. In the same chromosome 16, the *SHAT1-5* gene, a NAC Transcription Factor (Glyma.16G019400) was identified as an activator of secondary wall biosynthesis, promoting thicker fiber cap cells in wild soybean accessions [39]. Bi-parental populations with various genetic origins were used in several studies making the linkage map of the *qPDH1* locus. Additionally, several minor QTLs were also found in chromosomes 2, 5, 8, 10, 14 and 19 [36,40]. An RNA-sequencing approach was successfully used for QTL identification in soybean chromosome 19 [41].

Genome wide association study (GWAS) represents another advanced method of QTL identification based on the comparison of genotyping and phenotyping results in diverse sets of germplasms in different environments rather than segregating hybrid populations [42,43]. In soybean, GWAS application resulted in the discovery of several more QTLs for PD dispersed throughout the genome [44]. In total, 163 SNPs were identified and most of them (136 SNPs) were linked to the known qPDH1 region. Other QTLs were found on chromosomes 1, 4, 6, 8, 9, 11, 16, 17, 18 and 20, and among them, Glyma09G06290 corresponding to a bHLH family transcription factor was suggested to be the most important putative candidate gene linked to PD. Most recently, GWAS has helped to identify 86 QTLs associated with PD on all chromosomes across four *G. max* populations. Among them, eight important QTLs linked to PD were revealed on chromosomes 1, 3, 4, 10, 11, 17 and 19. These eight genes were homologous to *Arabidopsis thaliana* genes which played important roles in PD through the lignification, breakdown, and biosynthesis of the cell wall [45]. Special attention was paid to soybean chromosome 16, where one main QTL was identified and titled qPS-DS16-1 with the candidate gene Glyma.16G076600, encoding Cytochrome P450, which contributed to PD resistance in soybean [46]. Recently, gene *Sh1* (Glyma.16G141100), the repressor of *SHAT1-5*, was found in the same chromosome 16. This *Sh1* gene encoded a C2H2-like zinc finger transcription factor and was found to cooperate with another gene *Pdh1*, which encodes a dirigent-like protein. The partner genes, *Sh1* and *Pdh1*, were reported to strongly regulate *SHAT1-5*, making them major controllers of PD in soybean [47].

As mentioned above, PD is correlated with many other traits and, therefore, controlled by multiple structural and regulatory genes, which play significant roles in the development of cell-walls in pods. Numerous minor loci along with major QTLs are also involved in the control of PD. Genomic prediction (GP) is a powerful method to reveal valuable information in complex traits rather than marker-assisted selection (MAS) [48,49]. GP utilizing high density markers across the genome can accelerate the breeding process and enable the integration of multiple environment data for predicting crop traits [50]. GP has been used widely to improve the genetic gain for various traits in many crops including soybean. For instance, GP was implemented for enhancing yield in soybean [51,52], disease resistance [53] and tolerance to abiotic stresses [54]. These findings emphasize the potential of GP in speeding up crop genetic improvement.

Despite some attempts of genetic analyses conducted to identify the genes controlling resistance to PD, the fundamental biological mechanisms of this trait remain unclear. Some studies used very limited genotypes, while others used low-density molecular maps for the analysis. Therefore, it is important to conduct Diversity Arrays Technology analysis (DArT) using GWAS with high-density SNP panels in sufficient and diverse soybean germplasm accessions.

The aims in this study were: (1) Analysis of genomic regions corresponding to PD based on GWAS with DArT markers in 170 soybean genotypes with diverse origins; (2) Identification of potential candidate genes linked to PD; (3) Prediction and selection of genotypes resistant to PD for soybean breeding programs using genomic prediction.

2. Materials and Methods

2.1. Plant Material and Field Trials

In the current study, 170 soybean accessions with different levels of PD and diverse origins were studied. The population size of 170 accessions is considered as moderate for plant research, but still acceptable for GWAS. This point can be supported by genetically very diverse accessions mentioned above, high-quality DArT markers that passed strict quality-control filters, described in Section 2.4 below, and rigorous logarithm of odds threshold used in this study and described in the following Section 2.6. The major germplasm collection of soybean cultivars originated from the Kazakh Research Institute of Agriculture and Plant Growing (KRIAPG), Almaty district, Kazakhstan. Other genotypes were received from germplasm collections of N.I.Vavilov Research Institute of Plant Industry, St.Petersburg, Russia; V.S.Pustovoit All-Russian Research Institute of Oilseeds, Krasnodar, Russia; Siberian Research Institute of Fodder Crops, Novosibirsk region, Russia; Krasnoyarsk Research Institute of Agriculture, Krasnoyarsk, Russia; Research Production Association 'Soya-Center', Minsk region, Belarus; Soybean Research Institute, Poltava region, Ukraine; US National Plant Germplasm System (USDA), Urbana, Illinois, USA; and the Research Institute of Field and Vegetable Crops, Novi Sad, Serbia. The full list of the accessions used is present in Supplementary Table S1.

Two sowing dates were used: the traditional date (late April or early May) and a delayed planting (late May or early June), to create different environmental conditions during the growing season. The field trials were conducted over two years (2023–2024) at the research fields of KRIAPG, located in South-Eastern Kazakhstan, at a latitude of 43° N. Seeds were sown to study their performance under four different environments: (1) early sowing in 2023; (2) late sowing in 2023; (3) early sowing in 2024; and (4) late sowing in 2024. The experiment was carried out using a randomized complete block design (RCBD) with three replications. The climate during the soybean sowing season of 2023 was characterized as hot with low rainfall, while 2024 was also hot but with heavy rainfall. The maximum summer air temperature reached nearly 34 °C in 2023 compared to 31 °C in 2024. In general, the climate in the region is continental with hot summers and humid winters, and estimated as 'Dfa', according to the Köppen classification [55]. The average temperature in late April is about 12–13 °C, while in early June it is 20–21 °C, and annual precipitation ranges from 100 to 60 mm (late April and early June, respectively).

2.2. Morphological, Physiological and Biochemical Characteristics of Pods in Soybean Accessions

Morphological traits related to PD such as pod length (PL), pod width (PW) and pod thickness (PT) were measured on 10 pods on the main stem in five plants using a calliper. The data received were used for calculation of ratios in pods: length/width ratio (LWR), length/thickness ratio (LTR) and width/thickness ratio (WTR).

Plant material (leaves and pods) for physiological and biochemical analyses was collected at the plant development stages R4–R5 and stored at −70 °C until used for assays. Chlorophyll and carotenoids were extracted from leaves using 80% acetone following the protocol published earlier [56]. The concentration of chlorophyll a and b, as well as carotenoids was measured at wavelengths of 646.8, 663.2 and 470 nm, respectively, using a Jenway 635031 spectrophotometer (Bibby Scientific, Staffordshire, UK).

The relative water content (RWC) was measured according to the protocol published earlier [57]. For this analysis, immature pods from at least four plants for each sample taken in the field were placed in a container with cooling elements and quickly delivered to the laboratory. After removing the seeds, pods were weighed immediately to determine fresh weight (FW) and placed in deionized distilled water for 4 h. Then, pods were weighed again to obtain turgid weight (TW), and they were then dried in an oven at 80 °C for 24 h.

Fully dried pods were reweighed to determine dry weight (DW). RWC was calculated by the following formula:

$$RWC = \frac{(FW - DW)}{(TW - DW)} \times 100\%$$

Extraction of the antioxidant enzymes superoxide dismutase (SOD) and polyphenol oxidase (PPO) was performed with a phosphate buffer containing 0.25 mM ethylenediaminetraacetic acid (EDTA), polyvinylpyrrolidone 25 (PVP-25, 2%, w/v), 1 mM ascorbic acid and glycerol (10%, w/v), according to a previously published protocol [58]. SOD activity was determined by photoreduction inhibition of nitroblue tetrazolium (NBT) according to [59] and expressed in units/g FW. This was used as substrate to determine PPO activity, which was expressed similarly as units/g FW of pyrocatechol [60].

2.3. Assessment of Pod Dehiscence

The oven-dry method was used to assess the degree of PD. All mature pods were harvested from plants at the same maturing stage (R8) and stored at room temperature for one week. After that, the PD ratio in each accession with three biological replicates was calculated as the proportion of dehiscent pods to the total number of pods after 12 h of drying at 80 °C in the oven (Drying chamber ED-115, BINDER, Tuttlingen, Germany) The percentage of PD was determined according to a 1–5 scale, where 1 represents 0% dehiscent pods; 2, 1–10%; 3, 11–25%; 4, 26–50% and 5 represents 50% dehiscent pods [61]. Based on this scale, the PD phenotypes were classified as follows: 1, very resistant; 2, resistant; 3, moderately resistant; 4, moderately sensitive; and 5, very sensitive to dehiscence.

2.4. DArTseq Genotyping and Marker Quality Control

DNA was extracted from young leaf tissue of 170 soybean accessions using a DNeasy Plant Mini kit (Qiagen, Hilden, Germany). The extracted DNA was diluted to a concentration of 100 ng/µL, and 50 µL of high-quality DNA from each genotype was submitted to Diversity Arrays Technology (DArT Pty Ltd., Canberra, Australia) (https://www.diversityarrays.com; accessed on 20 October 2024) for whole-genome analysis using DArT markers. This technology is based on sequencing of an enhanced library using a Next generation sequencing (NGS) platform with genome complexity reduction, and 60K DArT clones were applied for the study. Results of genotyping with 18,540 Silico-DArT and 41,310 DArT-SNP markers were received. Based on genome assembly, markers with unknown chromosome positions were removed from the analysis. The remaining Silico-DArT and DArT-SNP markers used for the association analysis were filtered based on a call rate of $\leq$80%, marker reproducibility at $\leq$95%, minor allele frequency (MAF) $\leq$ 5%, and missing observation fractions $\geq$ 10%.

2.5. Population Structure Analysis

STRUCTURE v2.3.3 software was applied to assess population structures utilizing a Bayesian-Markov chain-Monte Carlo (MCMC) method grounded in admixture and correlated allele frequencies [62]. The data set was run through 10,000 Markov chain-Monte Carlo iterations with an initial burn-in period of 10,000 with five replicates, considering several subgroups (K) ranging from 1 to 10. The python script of Structure Harvester 'StructureHarvester.py' [63] was used to determine the optimal k-value [64], as well as to illustrate the results obtained from STRUCTURE v2.3.3.

2.6. GWAS Analysis, QTL and Candidate Gene Prediction

To identify SNPs with significant linkage to the studied yield-related soybean traits, GWAS was conducted using the Genome association—Prediction integrated tool (GAPIT), version 3, with several models with increased power and accuracy for genome association [65]. The GAPIT models used in this study include the Bayesian-information and Linkage-disequilibrium iteratively nested keyway (BLINK) [66], the Fixed and random model circulating probability uniform (FarmCPU) [67], the Multiple loci mixed model (MLMM) [68], the Mixed linear model (MLM) [69], and the General linear model (GLM) [70]. To identify highly significant associations, in this study, rigorous logarithmic of odds (LOD $\geq$ 6.0) criteria were used, based on the Bonferroni correction test (α = 0.05) [71]. 20 kb genomic flanking regions were used to detect a candidate for each molecular marker, and annotation was based on the soybean reference genome Williams 82 (Wm82.a2.v1). The chromosome fragments for $\pm$20 kb and even less were often used in similar studies [72]. Protein database UniProt was also implemented to find protein functions of detected QTLs (https://www.uniprot.org; accessed on 10 July 2025).

2.7. Genomic Prediction (GP) for Genomic Selection of Pod Dehiscence

The fallowing six models were used to perform GP: Genomic best linear unbiased prediction (gBLUP) [73], Bayes LASSO (BL) [74], marker-assisted BLUP (maBLUP) [75], Random Forest (RF) [76], Reproducing Kernel Hilbert Space regression (RKHS) [77], and ridge regression best linear unbiased prediction (RR-BLUP) from the rrBLUP package [78]. GP was conducted for two marker sets. The first marker set included all markers from 170 soybean genotypes for predicting genomic estimated breeding values (GEBV) in GP, while the second marker set contained only 51 GWAS-derived markers. Two scenarios were used for GP using each marker set in the GAPIT package (version 3.5.0): (1) A five-fold cross-validation approach was implemented, 80% of all genotypes were used as the training set, and the remaining 20% of genotypes were employed as the testing set. (2) the population was divided into two subgroups (1 and 2) according to their origin. Subgroup1 included all genotypes except genotypes from North America and China, and this subgroup-1 was employed as the training set. Conversely, subgroup-2 containing all genotypes from both North America and China was used as the testing set. The accuracy of genomic selection was derived from the results of 100 replications and compared between GP models and illustrated by boxplots generated using the ggplot2 package in R version 4.5.1.

2.8. Statistical Analysis

All statistical analysis and data visualizations were performed in R version 4.5.1 [79]. Boxplots and histograms were illustrated using the 'ggplot2 package'. Pearson correlation analysis was conducted using the R package 'Performance Analytics', and multiple testing correction was applied using the FDR method. The R function was used for one-way ANOVA.

3. Results

3.1. Pods and Pod-Related Agronomic Traits

The PD in 170 soybean accessions was evaluated along with associated pod-related traits (PL, PW, PT, LWR, LTR and WTR), as well as photosynthetic pigments, antioxidants, and RWCP parameters (Table 1). The PD levels in the most sensitive and most resistant genotypes were 88.25% and 0%, respectively. The average PL was approximately 4.53 cm, with the shortest measuring 3.37 cm and the longest reaching up-to 6.8 cm. PW ranged from 0.7 to 1.21 cm, while the PT varied between 0.52–0.84 cm.

Table 1. Mean, range and standard deviation for observed PD-related traits of 170 soybean genotypes grown in 2023–2024. PD, pod dehiscence; RWCP, Relative water content in pods; PL, pod length; PW, pod width; PT, pod thickness; LWR, length/width ratio; LTR, length/thickness ratio; WTR, width/thickness ratio; Ca, chlorophyll a; Cb, chlorophyll b; Cx+c, total carotenoids; SOD, superoxide dismutase; PPO, polyphenol oxidase.

Traits	Mean	Min [1]	Max	SD	Mean Square
PD (%)	21.6	0	88.25	20.81	897.04 ***
RWCP (%)	92.23	85.47	95.9	1.69	8510.03 ***
PL (cm)	4.53	3.37	6.8	0.35	20.62 **
PW (cm)	0.93	0.7	1.21	0.06	0.88 *
PT (cm)	0.7	0.52	0.84	0.05	0.49 **
LWR	4.87	4.15	7.1	0.33	23.84 ***
LTR	6.54	5.17	10.11	0.62	43.16 ***
WTR	1.35	1.12	1.72	0.1	1.82 ***
Ca (mg/g)	1.88	1.34	2.85	0.27	3.61 **
Cb (mg/g)	0.63	0.41	1.12	0.12	0.41 **
Cx+c (mg/g)	0.51	0.21	0.69	0.1	0.27 ***
SOD (units/g FW)	98.63	43.5	141.4	19.44	10,103.16 ***
PPO (units/g FW)	66.83	11.25	296.88	48.59	6813.8 ***

[1] Min, minimal value; Max, maximal value; SD, standard deviation; * $p < 0.05$; ** $p < 0.01$; *** $p < 0.001$.

After treatment at 80 °C, almost all pods were open in soybean genotypes sensitive to PD, whereas no opening pods were observed in accessions highly-resistant to PD (Figure 1). According to the results of the oven-dry methods, 170 soybean samples were classified into five groups (Figure 2), including 4.1% very resistant, 32.4% resistant, 34.1% moderately resistant, 19.4% moderately sensitive, and 10% very sensitive to PD.

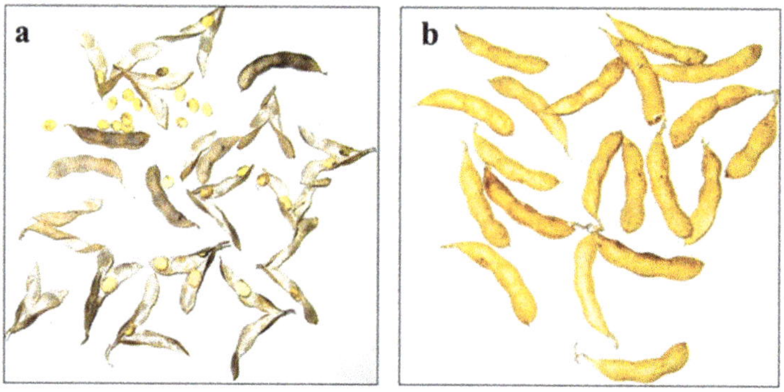

Figure 1. Pod dehiscence in soybean genotypes: (**a**) cv. Kollekcyina with very high level of pod dehiscence; (**b**) cv. Zispida with low level of pod dehiscence.

The studied agronomic traits potentially involved in PD were compared across the five groups of genotypes. As a result, three traits, pod length, pod thickness, and length/wide ratio of pods, showed highly significant differences in soybean accessions with contrasting levels of PD (Figure 3). Genotypes belonging to high-resistance Group 1 exhibited significantly lower values of PD compared to those accessions in other Groups. For instance, the average pod length was 4.1 cm in Group 1, whereas in other Groups 2–5, it ranged between 4.57 and 4.61 cm ($F = 4.021$, $p = 0.0038$). The pod thickness in high-resistance Group 1, was 0.65 cm, while it was around 0.70 cm in other Groups on average ($F = 3.261$, $p = 0.0132$). The length/width ratio of pods in Group 1 was 0.463 units, whereas in highly-sensitive Group 5, it reached up to 0.495 units ($F = 2.838$, $p = 0.0261$).

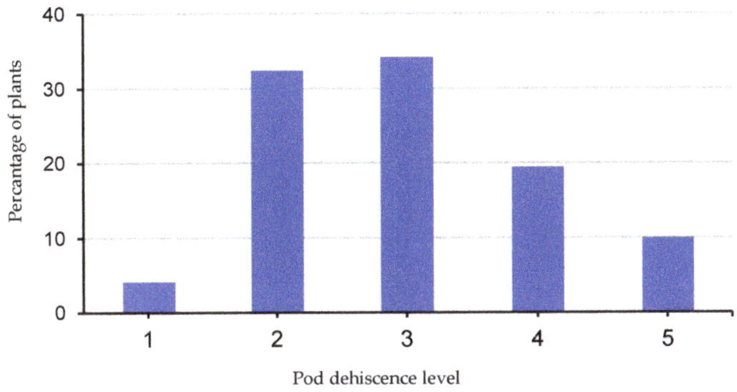

Figure 2. Distribution of 170 soybean accessions according to their pod dehiscence level: 1, very resistant; 2, resistant; 3, moderately resistant; 4, moderately sensitive; and 5, very sensitive.

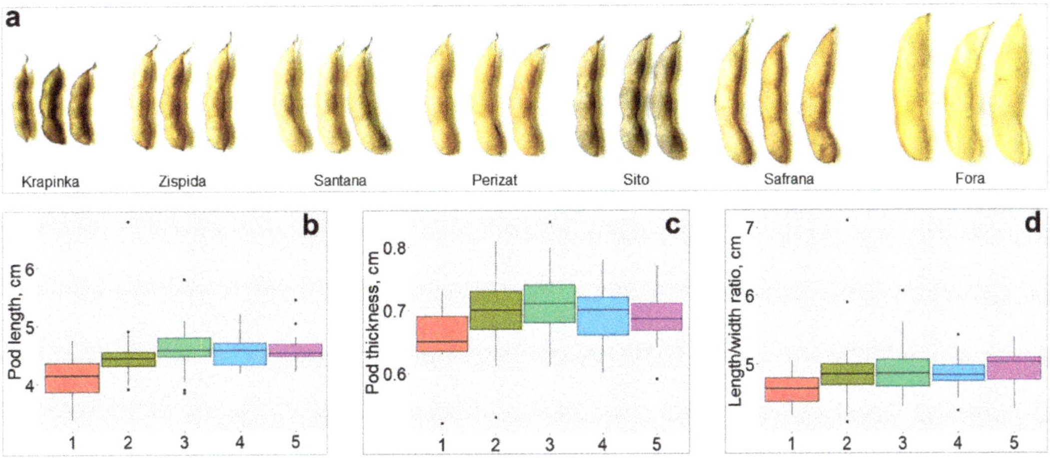

Figure 3. Pod-related traits and their distribution across various PD groups. (**a**) pod images from smallest to largest in size with corresponding PD: Krapinka, very resistant; Zispida, resistant; Santana and Perizat, moderately sensitive; Sito and Safrana, moderately resistant; Fora, very sensitive. Variability of PL (**b**): PT (**c**) and LWR (**d**) are shown in soybean accessions distributed among five PD groups: 1, very resistant; 2, resistant; 3, moderately resistant; 4, moderately sensitive; and 5, very sensitive.

The correlation analysis indicated that RWCP, PL, LTR and WTR pod-related traits showed significant positive correlations with pod dehiscence level ($r = 0.20$, 0.15, 0.15 and 0.15, respectively) (Figure 4). This may indicate that these pod-related traits have a strong effect on PD in soybean, and the lower values of these traits are related to cultivars with a high level of dehiscence. In contrast, negative association between PD and Cb was observed in the current study, suggesting that photosynthesis is more active in non-dehiscent genotypes. However, no significant correlations of PD with antioxidants were found. Nevertheless, antioxidants remain the most crucial components to prevent PD by regulating ROS levels that have a negative effect on the PD zone.

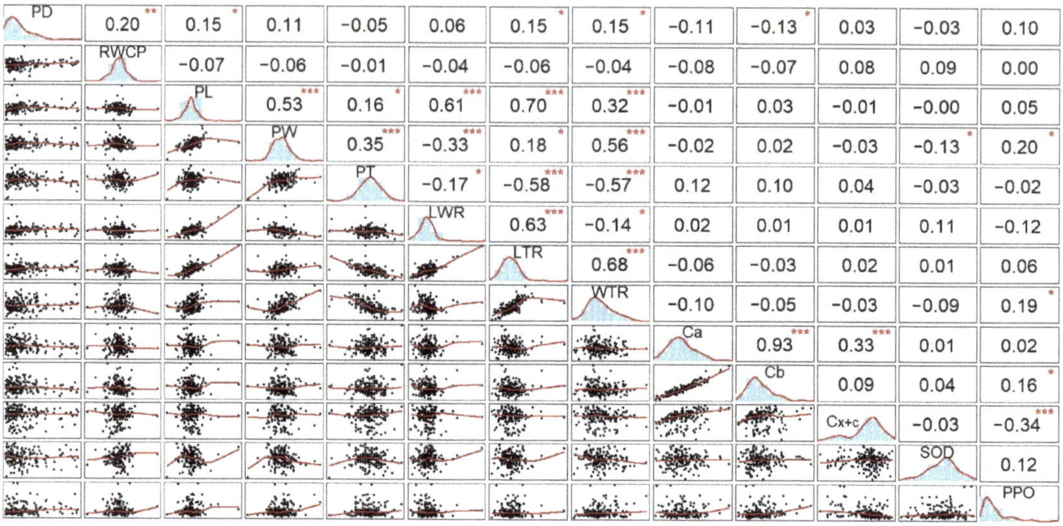

Figure 4. Pearson's correlation coefficients for 13 studied pod-related traits in 170 soybean genotypes. Abbreviations: PD, pod dehiscence; RWCP, relative water content in pod; PL, pod length; PW, pod width; PT, pod thickness; LWR, length/width ratio; LTR, length/thickness ratio; WTR, width/thickness ratio; Ca, chlorophyll a; Cb, chlorophyll b; Cx+c, total carotenoids; SOD, superoxide dismutase, and PPO, polyphenol oxidase. In this correlation matrix, the lower triangle displays scatterplots, the upper triangle reports the correlation coefficients, and the diagonal shows the density distributions. Asterisks indicate statistical significance: * $p \leq 0.05$, ** $p \leq 0.01$, *** $p \leq 0.001$.

3.2. DArTseq Genotyping and Marker Quality Control

DArTseq analysis yielded a total of 59,850 molecular markers including 18,540 Silico-DArT and 41,310 DArT-SNPs. After quality control, 13,765 Silico-DArT and 16,914 DArT-SNP markers remained for further research. The distribution of molecular markers across soybean chromosomes was illustrated in Figure 5, ranging nearly from 500 to 900 for Silico-DArT and from 700 to 1100 for SNPs. All remaining markers had an average read depth of 19, reproducibility of 0.995, and a call rate of 0.96

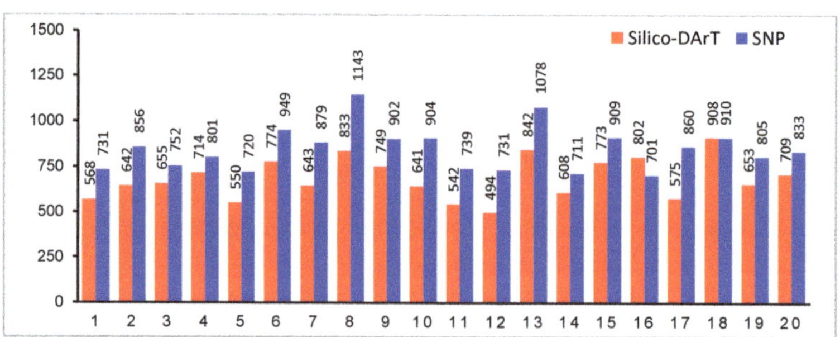

Figure 5. Distribution of filtered molecular markers across all soybean chromosomes. X-axis represents soybean chromosomes; Y-axis represents total number of markers across chromosomes.

3.3. Population Structure Analysis

In population structure analysis, 170 soybean genotypes were divided into three groups based on the optimal delta K (ΔK) values (Figure 6). Subgroup S1 contained 48 genotypes belonging to maturity group MG 00-I, and 10 of them were considered admixed. The remaining genotypes in the subgroup S1 mainly originated from North America (5), Russia (15) and other European countries (15). The largest proportion of the genotypes in the subgroup S2 belong to MG II–III, and most of these originated from Kazakhstan (15) and Europe (21), among which 13 genotypes were admixed. Subgroup S3 consisted of 62 MG 00-I genotypes including 12 admixtures, and genotypes in the subgroup S3 were mainly from countries in Asia, North America and Europe.

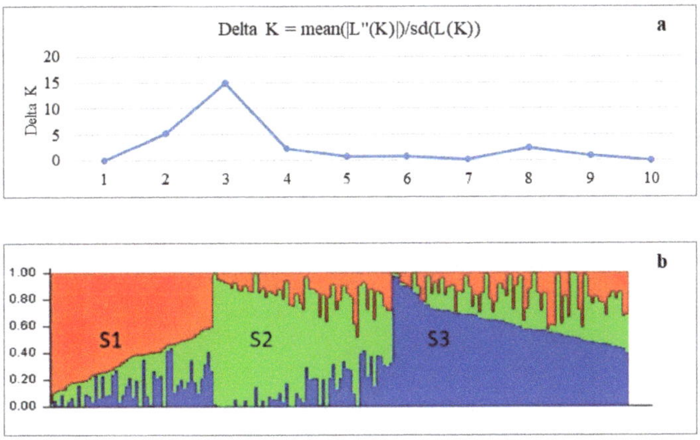

Figure 6. Population structure analysis of studied soybean accessions. (**a**) Delta K values for different numbers of populations assumed (K) in the STRUCTURE analysis. (**b**) Classification of soybean accessions into three sub-populations (K = 3) using STRUCTURE 2.3.3. The distribution of the accessions to different populations is indicated by the colour code. Numbers on the Y-axis show subgroup membership, whereas the X-axis shows the distribution of 170 studied soybean genotypes.

3.4. GWAS Analysis, QTL and Candidate Gene Prediction

GWAS was conducted utilising 170 soybean accessions evaluated across four distinct environments (Figure 7).

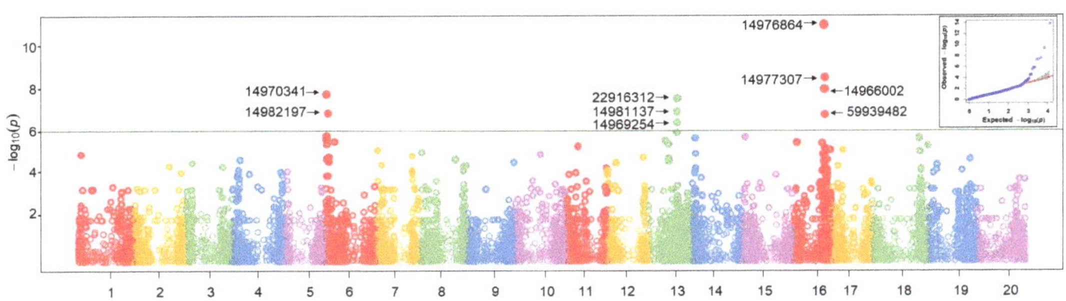

Figure 7. Manhattan plot of association mapping conducted using GLM. X-axis represents chromosome positions, Y-axis represents log10 (*p*-values). The green line in the Manhattan plot indicates the threshold (*p*-value < 1×10^{-6}). The dots with numbers located above the threshold line represents QTLs that were significantly associated with PD. The Q-Q plot for the GLM model is located in the top-right corner.

Five analytical models, including 2 single-locus and 3 multi-locus models, were used to find QTLs associated with PD and other pod-related traits. This multi-model strategy enabled a more reliable identification of marker-trait associations, elucidating the genetic background of pod dehiscence in soybean. GWAS results were illustrated using Manhattan plots, and Figure 7 shows an example of GWAS results where LOD score ≥ 6.0 used as a threshold.

According to the GWAS results, 48 QTLs were identified across 14 chromosomes (Supplementary Table S2). The majority of these identified QTLs were located on chromosome 16 (14 QTLs), followed by chromosomes 13 and 6, with 9 and 8 QTLs, respectively.

QTLs identified at least twice by different GWAS models, or detected in more than two environments, were considered as consistent and used for further investigation. As a result, 14 consistent QTLs were identified: eight QTLs for PD, two for PL, one for PW, one for WTR, one for RWCP and one for Ca+b. Subsequently, candidate genes were mined in the 20 Kb genomic flanking regions of identified QTLs. However, in most cases, candidate genes were selected if the identified SNP was located within the gene. As a result, 14 genes associated with the pod-related traits and physiological parameters were identified in studied soybean genotypes (Table 2).

Table 2. The 14 putative candidate genes identified by the stable markers associated with the studied pod dehiscence and related traits.

QTL	DArT Marker	Chr	Relevant Gene	Gene Position	Annotated Description
			Pod dehiscence, PD		
QTL_PD2	14982197	6	Glyma.06G011600	866275-867210	DUF241 domain protein; protein BPS1
QTL_PD3	14970341	6	Glyma.06G034500	2677248-2679092	RING finger protein 5
QTL_PD6	22919889	7	Glyma.07G125300	14920233-14979323	DNA-directed RNA polymerase family protein
QTL_PD8	14969254	13	Glyma.13G184600	29837154-29848121	Pollen Defective in Guidance 1 protein, isoform X1
QTL-PD11	14981137	13	Glyma.13G184500	29829471-29834027	DNA-binding bromodomain-containing protein
QTL_PD14	50681234	16	Glyma.16G141100	29893194-29896920	C2H2 zinc finger protein
QTL_PD15	14976864	16	Glyma.16G145100	30578388-30581421	Methyl esterase 1
QTL_PD16	14976250	16	Glyma.16G146900	30763467-30766039	Mitochondrial phosphate carrier protein 3
			Pod length, PL		
QTL1_PL3	22917169	13	Glyma.13G039900	12319075-12321626	Receptor-like kinase 1
QTL2_PL4	22919231	16	Glyma.16G025800	2499822-2508300	Nucleotidyltransferase
			Pod width, PW		
QTL_PW2	100064442	6	Glyma.06G013500	1009335-1010213	Hydroxyproline-rich glycoprotein family protein
			Width/thickness ratio of pods, WTR		
QTL-WTR2	50680953	19	Glyma.19G123600	38159905-38162342	Major intrinsic protein (MIP) family transporter
			Relative water content in pods, RWCP		
QTL_RWCP1	14970004	1	Glyma.01G001300	227021-237987	Cycloartenol synthase 1
			Chlorophyll a and b, Ca+b		
QTL1_Ca+b	22914385	9	Glyma.09G163700	38808617-38809561	Kunitz trypsin inhibitor 1

Particularly, three stable QTLs were revealed on chromosome 16, corresponding to genes encoding a C2H2 zinc finger protein, methyl esterase 1, and a mitochondrial phosphate carrier protein, indicating a major role of this genetic region in PD resistance. Additionally, two tightly linked PD candidate genes, coding for Pollen Defective in Guidance 1 protein, isoform X1 (Glyma.13G184600) and DNA-binding bromodomain-containing protein (Glyma.13G184500), were identified on chromosome 13. There were also two genes related to PD on chromosome 6 including genes Glyma.06G011600 (Protein BPS1) and Glyma.06G034500 (RING finger protein 5). Two putative genes, Glyma.13G039900 (Receptor-like kinase) and Glyma.16G025800 (Nucleotidyltransferase), were found for PL on chromosomes 13 and 16, respectively. One QTL linked to hydroxyproline-rich glycoprotein gene was found on chromosome 6, while the pod-related trait WTR was associated with a major intrinsic protein family transporter on chromosome 19. RWCP was linked to Cycloartenol synthase 1 on chromosome 1. Finally, a candidate gene encoding a Kunitz trypsin inhibitor for Ca+b was identified on chromosome 19 (Table 2).

3.5. Genomic Prediction (GP) for Pod Dehiscence

GP for pod dehiscence was performed across two marker sets, the whole set of markers and GWAS-derived markers, and two types of GP strategy were employed: cross-prediction and across-prediction. Cross-prediction was conducted using cross validation methods, where the population was divided into a training set (80%) and testing set (20%). While in across-prediction, models were trained in one population and tested in a different population.

In this study, GP was employed using six models including BL, gBLUP, maBLUP, RF, RKHS and RR-BLUP. GP was yielded for prediction accuracy ranging from 0.74 to 0.83, when it was conducted with cross validation methods (cross-prediction) in the whole marker set (Figure 8a). With the exception of maBLUP, all five other models showed nearly the same prediction accuracy of slightly above 0.80. The accuracy of GP decreased over a range 0.59–0.67 compared to cross-prediction, when across-prediction was performed using the whole marker set.

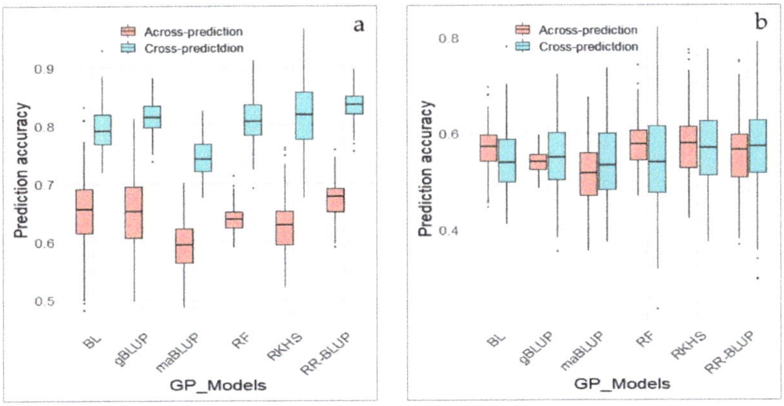

Figure 8. Genomic prediction of pod dehiscence using the whole and GWAS-derived marker sets. (**a**) GP of PD based on the whole markers; (**b**) GP of PD based on GWAS-derived markers. X-axis represents GP models, Y-axis represents prediction accuracy.

The accuracy prediction ranged from 0.544 to 0.57 in cross-prediction and from 0.543 to 0.58 in across-prediction, when the GWAS-derived marker set was implemented in GP (Figure 8b). Overall, the prediction accuracy in both cross-prediction and across-prediction remained roughly unchanged. This finding suggests that all GP models based on both whole and GWAS-derived marker sets are effective for selection of favorable genotypes resistant to PD.

Finally, GP yielded genomic breeding values (GBV) for each genotype ranging from −45 to 19.0 units. This could help to select genotypes with best resistance and sensitivity to PD. Thus, based on GP, eight PD resistant genotypes were identified with low GBV: 186.1, Jevrika, Zen, Malvina, Zakat, VNIIS1, Sponsor and 209.1. In contrast, ten genotypes with highest GBV were selected as sensitive to PD, including: Dina, Fora, Warsawska, PJeP26, Carola, Nawiko, Kollekcyina, PJeP27, Svetlaia and Jaselda.

4. Discussion

Pod dehiscence is one of the main traits that can dramatically reduce soybean yield. In Kazakhstan, soybean is cultivated mainly in the south-east of the country where the summer climate is dry and hot, and this can increase the likelihood of PD during harvest. Therefore, it is crucial to develop non-dehiscent soybean cultivars adapted to the continental climate of South-Eastern Kazakhstan.

PD is caused by many factors, including morphological and physiological characters such as PL, PW RWC in pods, and other pod-related traits [12,80]. In this study, pod-related traits that have a significant effect on PD levels in soybean were investigated. The values of PL, PW and LWR in genotypes highly resistant to PD were found to be significantly lower than in soybean genotypes sensitive to PD (Figure 3). Similar results were reported earlier confirming that PL, PW, LWR and WTR were reliable indicators for assessing the PD level in soybean, and that the longer and thicker the pods were, the more sensitive they were to PD [13,81].

Further, significant positive correlation of PD with PL, LTR, WTR and RWCP was revealed in the current study (Figure 4). This result was supported by the recent publication in common bean where positive associations of PD with PL, PW, LTR and WTR were reported [14]. Furthermore, soybean genotypes sensitive to PD were reported to have longer PL and wider PW [21]. Additionally, the water content in pod cell walls plays an important role on PD because higher moisture content in pod cell walls has been observed in soybean genotypes resistant to PD [11,12,82].

In the current study, a negative correlation between PD and Cb was found which could potentially point to an important effect of photosynthetic pigments on PD. This kind of negative association was consistent with a recent report [30] and also supported by previous studies [12,83]. According to another published paper, soybean plants with a strong level of pod dehiscence had higher photosynthetic activity during seed development stages R5 and R6. However, the reduction of photosynthesis can act as a signal for the conversion of metabolites and build-up of cellulose, lignin and total fiber in pod cells [21]. Carbohydrates produced by photosynthesis are essential for cellulose synthesis, which is considered as one of the main enzymatic pathways important for the PD trait [84].

Positive correlations of PD with both SOD and CAT were observed in a recent study, which also reported that ROS levels can increase during PD [30]. Phases R5 and R6 represent the most vulnerable pod-setting stages in the development of soybean plants in response to drought and heat stresses [85,86]. Therefore, increased activity of antioxidant enzymes in leaves can help to support resistance of plants to unfavourable conditions. In mature plants in pod maturation stage R8, all metabolic processes were completed and the level of antioxidant enzymes decreased [87]. Our assessment of SOD and PPO antioxidant enzyme activity in soybean leaves in R5 and R6 stages did not reveal a direct relationship with PD. However, it might have indirect effects on yield components in studied genotypes like the number of pods, size of pods, seed weight, etc., which ultimately can affect the PD trait.

For DArT analysis, more than 30,000 high-quality DArT molecular markers with sufficient read depth were utilized to perform GWAS in 170 soybean genotypes. Soybean accessions were split into three distinct sub-populations, S1, S2 and S3, based on population structure analysis with these DArT markers. Two sub-populations, S1 and S3, contained early-maturing soybean genotypes originating from different countries worldwide, which might indicate for local gene flow as reported earlier [88,89]. On the other hand, sub-population S2 contained genotypes from Kazakhstan and Europe with a middle-late maturity time, suggesting a more uniform background created by local breeding programs [90].

In the current study, GWAS conducted in 170 soybean genotypes revealed 48 QTLs, of which 14 were stable QTLs across different environments (Supplementary Table S2 and Table 2). Candidate genes linked to PD were located on chromosomes 6, 7, 13 and 16. Previously, the main QTL for PD had been detected on chromosome 16, namely *pdh1* (Glyma.16g141400) and *SHAT1-5* (Glyma16g019400) [9,36]. More recently, Li et al. [47] detected a new gene, *Sh1*, on chromosome 16, whereas another potential PD gene, Glyma.16g076600, was reported controlling the synthesis of a specific protein related to ABA catabolism [46]. In our study, three PD tentative candidate genes, Glyma.16G141100, Glyma.16G145100 and Glyma.16G146900, were found on chromosome 16. Among them, Glyma.16G141100 encoding the C2-H2 zinc finger protein was identified across all environments, and it was identical to the described *Sh1*. This *Sh1* gene (Glyma.16G141100) was located very close to *Pdh1* (Glyma.16G141400) and it was shown to down-regulate expression of *SHAT1-5* binding its promotor region, thickening secondary cell walls in lignified fiber cap cells [47].

Transcription factors (TF), bHLH, WRKY, NAC, MYB and MYB-related families, were shown to have a significant effect on regulating PD [91–93]. For instance, members of MYB families in *Arabidopsis* activate other TFs which can lead to PD by regulating secondary cell wall formation [94]. On chromosome 13, three DArT markers corresponding two genes (Glyma.13G184500 and Glyma.13G184600) were identified (Figure 7 and Table 2). Glyma.13G184500 encodes a MYB TF indicating that it might be a potential regulator in PD. However, this requires further functional validation to confirm its role in PD development. On chromosome 6, two stable DArT markers (Figure 7 and Table 2) and two SNPs (Supplementary Table S2) were identified. QTL_PD2 with Glyma.06G011600 was highly expressed in pods according to Libault et al. [95], while one of the SNPs in gene Glyma.06G012500 encoding 3-ketoacyl-CoA synthase 4 was tightly linked to Glyma.06G012200 from the bHLH family protein, which is an important TF for PD [92,96]. Other stable QTLs did not show any homology to known genes or TFs associated with PD. Nevertheless, two QTLs with tentative genes Glyma.03G033300 and Glyma.06G199800 were identified in only one environment. They were closely related to TFs for PD. For example, Glyma.03G033300 linked to PW was located next to the Glyma.03G033600 (C2H2 family protein), which is a crucial TF regulating PD in legumes including soybean [47,97]. Glyma.06G199800, corresponding to WTR, is in close proximity to Glyma.06G199600, representing a bHLH family protein. Such bHLH TF genes in *Arabidopsis* induce a non-lignified cell layer in the dehiscence zone, which leads to silique expansion [96,98].

Various GP models have been successfully used in evaluating complex agronomic traits and disease resistance. Genomic prediction is a very powerful approach that can dramatically accelerate the breeding process. Prediction accuracy in GP is important and it can be obtained from correlation between observed phenotypes and predicted genomic estimated breeding values (GEBVs) [99,100]. Prediction accuracy depends on model types as well as optimal marker set, which can be implemented in different random marker sets, ranging from several to thousands of markers and GWAS-derived best SNPs. For instance, salt tolerance in soybean germplasm was utilized using GP with various marker sets ranging from 10 to 10,000 [101]. In the result, the marker set with 10,000 SNPs was reported to show high prediction accuracy along with the 10 best GWAS-derived SNPs. In a similar investigation for GP of arginine content in soybean, it was concluded that marker sets with more SNPs and top SNPs derived from GWAS lead to better prediction accuracy in GP [102].

Apart from the optimal marker set, the training and testing populations also play an important role in prediction accuracy in GP. In general, models trained and tested within the same population produce higher accuracy compared to models trained and tested across different populations [102–104].

In the current study, two different marker sets were employed through cross-prediction and across prediction approaches. Genomic prediction models using different marker sets conducted in 170 soybean genotypes showed high prediction accuracies. Cross-prediction based on the whole set of markers resulted in higher prediction accuracy compared to GWAS-derived markers. However, when across-prediction was utilized on the whole marker set, prediction accuracy decreased from an average 0.80 to around 0.65. In contrast, GWAS-derived markers yielded similar prediction accuracy values (approximately 0.55) in both cross-prediction and across-prediction of PD. Across-prediction based on training models in one population and testing them in a new, unknown population, indicates that GWAS-derived marker set is more stable for selecting PD when evaluating new soybean genotypes. In addition, GP provide results for the selection of soybean genotypes most resistant as well as most sensitive to PD.

In general terms, for GP accuracy, there is no a 'Golden standard', as it varies widely depending on several factors, including the studied traits, population structure, marker density, and statistical models used. For example, in soybean, GP accuracy for the same trait can differ considerably across studies. For soybean seed weight, GP accuracies ranged between 0.75 and 0.87 [105]. It also was reported that average prediction accuracy of 0.39 for soybean yield [99], and up to 0.80 in some populations [106], for plant height 0.86 and yield per plant 0.47 [107], and for grain yield (0.58–0.60), plant height (0.43–0.45), and days to maturity (0.67–0.68) [106]. In the current study, the genomic prediction accuracy was approximately 0.60, indicating for good predictive performance and suggests that the model captured a substantial proportion of the genetic variance for the evaluated traits.

5. Conclusions

In this study, GWAS was carried out using single and multi-locus models to reveal QTLs corresponding to PD in soybean. As a result, 48 QTLs in total were identified with 14 stable QTLs, mostly located on chromosomes 6, 13 and 16, corresponding to pod dehiscence and pod-related traits. From putative candidate genes, two most stable and important genes for PD with known functions were emphasised from the QTLs: Glyma.13G184500 and Glyma.16G141100, encoding transcription factors, DNA-binding bromodomain-containing protein and C2H2 zinc finger protein, respectively. Additionally, genomic prediction (GP) was performed employing different marker sets. Cross-validation and across-prediction were used to determine the prediction accuracies of models. This demonstrated stability of GWAS-derived markers through across-prediction. Finally, GP helped to select the most PD resistant and sensitive soybean genotypes according to predicted genomic estimated breeding values. Therefore, the identified putative candidate genes will be used to improve the efficiency of marker-assisted selection for developing PD resistant varieties in soybean. Furthermore, assessed genomic prediction (GP) models for across-population prediction will be implemented to identify new PD resistant cultivars in future soybean breeding programs.

Supplementary Materials: The following supporting information can be downloaded at: https://www.mdpi.com/article/10.3390/plants14223505/s1, Table S1: Information about used soybean accessions: Names, maturity groups and origin; Table S2: Identified QTLs and putative genes associated with studied traits.

Author Contributions: Conceptualization, S.M. and K.B.; methodology, S.D., D.B. and S.K.; software, S.M.; validation, D.B. and S.K.; formal analysis investigation, A.T., A.Y. and N.S.; resources, S.D.; data curation, S.M.; writing—original draft preparation, S.M.; writing—review and editing, K.B. and Y.S.; visualization, S.M.; supervision, K.B., S.D. and S.M.; project administration, S.B.; funding acquisition, S.D. and S.B. All authors have read and agreed to the published version of the manuscript.

Funding: This research was funded by the Science Committee of the Ministry of Science and Higher Education of the Republic of Kazakhstan (Grant No. AP23487849).

Data Availability Statement: The data presented in the manuscript are available in the Supplementary Materials.

Acknowledgments: We want to thank Carly Schramm for critical comments during editing of the manuscript. Additionally, special thanks to the staff and students of our Research Institute and University for their support in this research and help with the manuscript preparation.

Conflicts of Interest: The authors declare no conflicts of interest. The funders had no role in the design of the study; in the collection, analyses, or interpretation of data; in the writing of the manuscript; or in the decision to publish the results.

References

1. Qiao, C.; Cheng, C.; Ali, T. How climate change and international trade will shape the future global soybean security pattern. *J. Clean. Prod.* **2023**, *422*, 138603. [CrossRef]
2. Hartman, G.L.; West, E.D.; Herman, T.K. Crops that feed the World 2. Soybean—Worldwide production, use, and constraints caused by pathogens and pests. *Food Secur.* **2011**, *3*, 5–17. [CrossRef]
3. Boerema, A.; Peeters, A.; Swolfs, S.; Vandevenne, F.; Jacobs, S.; Staes, J.; Meire, P. Soybean trade: Balancing environmental and socio-economic impacts of an intercontinental market. *PLoS ONE* **2016**, *11*, e0155222. [CrossRef]
4. Kumari, S.; Dambale, A.S.; Samantara, R.; Jincy, M.; Bains, G. Introduction, history, geographical distribution, importance, and uses of soybean (*Glycine max* L.). In *Soybean Production Technology: Physiology, Production and Processing*; Singh, K.P., Singh, N.K., Aravind, T., Eds.; Springer-Nature: Singapore, 2025; pp. 1–17. [CrossRef]
5. FAOSTAS. Available online: http://www.fao.org/faostat (accessed on 20 July 2025).
6. Yerzhebayeva, R.; Didorenko, S.; Amangeldiyeva, A.; Daniyarova, A.; Mazkirat, S.; Zinchenko, A.; Shavrukov, Y. Marker-assisted selection for early maturing *E* loci in soybean yielded prospective breeding lines for high latitudes of Northern Kazakhstan. *Biomolecules* **2023**, *13*, 1146. [CrossRef]
7. Kantar, M.B.; Nashoba, A.R.; Anderson, J.E.; Blackman, B.K.; Rieseberg, L.H. The genetics and genomics of plant domestication. *Bioscience* **2017**, *67*, 971–982. [CrossRef]
8. Maity, A.; Lamichaney, A.; Joshi, D.C.; Bajwa, A.; Subramanian, N.; Walsh, M.; Bagavathiannan, M. Seed shattering: A trait of evolutionary importance in plants. *Front. Plant Sci.* **2021**, *12*, 1180. [CrossRef]
9. Funatsuki, H.; Suzuki, M.; Hirose, A.; Inaba, H.; Yamada, T.; Hajika, M.; Komatsu, K.; Katayama, T.; Sayama, T.; Ishimoto, M.; et al. Molecular basis of a shattering resistance boosting global dissemination of soybean. *Proc. Natl. Acad. Sci. USA* **2014**, *111*, 17797–17802. [CrossRef]
10. Liu, J.; Zhang, Y.; Jiang, Y.; Sun, H.; Duan, R.; Qu, J.; Yao, D.; Liu, S.; Guan, S. Formation mechanism and occurrence law of pod shattering in soybean: A review. *Phyton* **2022**, *91*, 1327–1340. [CrossRef]
11. Bara, N.; Khare, D.; Shrivastava, A.N. Studies on the factors affecting pod shattering in soybean. *Indian J. Genet. Plant Breed.* **2013**, *73*, 270–277. [CrossRef]
12. Kuai, J.; Sun, Y.; Liu, T.; Zhang, P.; Zhou, M.; Wu, J.; Zhou, G. Physiological mechanisms behind differences in pod shattering resistance in rapeseed (*Brassica napus* L.) varieties. *PLoS ONE* **2016**, *11*, e0157341. [CrossRef] [PubMed]
13. Krisnawati, A.; Adie, M.M. Identification of soybean genotypes for pod shattering resistance associated with agronomical and morphological characters. *Biosaintifika J. Biol. Biol. Educ.* **2017**, *9*, 193–200. [CrossRef]
14. Fatima, S.; Rani, S.; Shafi, S.; Zaffar, A.; Riyaz, I.; Wani, M.A.; Zargar, S.M.; Vara Prasad, P.V.; Sofi, P.A. Pod physical traits significantly implicate shattering response of pods in beans (*Phaseolus vulgaris* L.). *Biol. Plant.* **2024**, *68*, 107–116. [CrossRef]
15. Evans, L.S. Comparative anatomy of pod dehiscence of legumes1. *J. Torrey Bot. Soc.* **2024**, *151*, 116–123. [CrossRef]
16. Moura, L.O.; da Silva, M.F.; da Cunha, F.F.; Picoli, E.A.T.; Silva, F.C.S.; da Silva, F.L. Water deficit as a trigger to immature soybean pod opening. *J. Agron. Crop Sci.* **2023**, *209*, 390–401. [CrossRef]
17. Buttery, B.R.; Buzzell, R.I. The relationship between chlorophyll content and rate of photosynthesis in soybeans. *Can. J. Plant Sci.* **1977**, *57*, 1–5. [CrossRef]

18. Shi, H.; Guo, J.; An, J.; Tang, Z.; Wang, X.; Li, W.; Zhao, X.; Jin, L.; Xiang, Y.; Li, Z.; et al. Estimation of chlorophyll content in soybean crop at different growth stages based on optimal spectral index. *Agronomy* **2023**, *13*, 663. [CrossRef]
19. Thompson, J.A.; Schweitzer, L.E.; Nelson, R.L. Association of specific leaf weight, an estimate of chlorophyll, and chlorophyll concentration with apparent photosynthesis in soybean. *Photosynth. Res.* **1996**, *49*, 1–10. [CrossRef] [PubMed]
20. Tu, B.; Liu, C.; Zhang, Q.; Liu, X. Quantitative differences in pod valve composition affect shattering in vegetable and grain soybean. *Ital. J. Agron.* **2025**, *20*, 100050. [CrossRef]
21. Kataliko, R.K.; Kimani, P.M.; Muthomi, J.W.; Wanderi, W.S.; Olubayo, F.M.; Nzuve, F.M. Resistance and correlation of pod shattering and selected agronomic traits in soybeans. *J. Plant Stud.* **2019**, *8*, 1–39. [CrossRef]
22. Summers, J.E.; Bruce, D.M.; Vancanneyt, G.; Redig, P.; Werner, C.P.; Morgan, C.; Child, R.D. Pod shatter resistance in the resynthesized *Brassica napus* line DK142. *J. Agric. Sci.* **2023**, *140*, 43–52. [CrossRef]
23. Ogutcen, E.; Pandey, A.; Khan, M.K.; Marques, E.; Penmetsa, R.V.; Kahraman, A.; Von Wettberg, E.J. Pod shattering: A homologous series of variation underlying domestication and an avenue for crop improvement. *Agronomy* **2018**, *8*, 137. [CrossRef]
24. Tsuchiya, T. Physiological and genetic analysis of pod shattering in soybean. *Jpn. Agric. Res. Q.* **1987**, *21*, 166–175.
25. Sachdev, S.; Ansari, S.A.; Ansari, M.I.; Fujita, M.; Hasanuzzaman, M. Abiotic stress and reactive oxygen species: Generation, signaling, and defense mechanisms. *Antioxidants* **2021**, *10*, 277. [CrossRef] [PubMed]
26. Haider, S.; Raza, A.; Iqbal, J.; Shaukat, M.; Mahmood, T. Analyzing the regulatory role of heat shock transcription factors in plant heat stress resistance: A brief appraisal. *Mol. Biol. Rep.* **2022**, *49*, 5771–5785. [CrossRef]
27. Kanase, T.; Guhey, A.; Gawas, D. Activity of antioxidant enzymes in soybean genotypes under drought stress. *Int. J. Curr. Microbiol. App. Sci.* **2019**, *8*, 2323–2330. [CrossRef]
28. Ebone, L.A.; Caverzan, A.; Silveira, D.C.; Siqueira, L.D.O.; Lângaro, N.C.; Chiomento, J.L.T.; Chavarria, G. Biochemical profile of the soybean seed embryonic axis and its changes during accelerated aging. *Biology* **2020**, *9*, 186. [CrossRef]
29. Kahrizi, D.; Allahvarand, T. Estimation and interrelationships of genetic variability parameters of some morpho-phenological traits in spring rapeseed (*Brassica napus* L.). *Asian J. Biol. Sci.* **2012**, *5*, 358–364. [CrossRef]
30. Ali, M.F.; Mahmood, I.; Shamsi, I.H.; Asad, S.A.; Ali, H.; Shah, G.A.; Tung, S.A.; Raja, N.I.; Demirkiran, A.R.; Hussain, M.I. Integrated application of nitrogen and mepiquat chloride improves yield of *Brassica napus* L. by modulating shattering, lodging and nitrogen use efficiency under rainfed conditions. *Plant Growth Regul.* **2025**, *105*, 1657–1677. [CrossRef]
31. Ding, C.; Alghabari, F.; Rauf, M.; Zhao, T.; Javed, M.M.; Alshamrani, R.; Ghazy, A.-H.; Al-Doss, A.A.; Khalid, T.; Yang, S.H.; et al. Optimization of soybean physiochemical, agronomic, and genetic responses under varying regimes of day and night temperatures. *Front. Plant Sci.* **2024**, *14*, 1332414. [CrossRef]
32. Malenčić, D.; Popović, M.; Miladinović, J. Stress tolerance parameters in different genotypes of soybean. *Biol. Plant.* **2003**, *46*, 141–143. [CrossRef]
33. Baisak, R.; Rana, D.; Acharya, P.B.; Kar, M. Alterations in the activities of active oxygen scavenging enzymes of wheat leaves subjected to water stress. *Plant Cell Physiol.* **1994**, *35*, 489–495. [CrossRef]
34. Collard, B.C.Y.; Jahufer, M.Z.Z.; Brouwer, J.B.; Pang, E.C.K. An introduction to markers, quantitative trait loci (QTL) mapping and marker-assisted selection for crop improvement: The basic concepts. *Euphytica* **2005**, *142*, 169–196. [CrossRef]
35. Semagn, K.; Bjornstad, A.; Xu, Y. The genetic dissection of quantitative traits in crops. *Electron. J. Biotechnol.* **2010**, *13*, ej10060. [CrossRef]
36. Bailey, M.A.; Mian, M.A.R.; Carter, T.E.; Ashley, D.A.; Boerma, H.R. Pod dehiscence of soybean: Identification of quantitative trait loci. *J. Hered.* **1997**, *88*, 152–154. [CrossRef]
37. Fourquin, C.; del Cerro, C.; Victoria, F.C.; Vialette-Guiraud, A.; de Oliveira, A.C.; Ferrándiz, C. A change in shatterproof protein lies at the origin of a fruit morphological novelty and a new strategy for seed dispersal in *Medicago* genus. *Plant Physiol.* **2013**, *162*, 907–917. [CrossRef] [PubMed]
38. Kang, S.T.; Kwak, M.; Kim, H.K.; Choung, M.G.; Han, W.Y.; Baek, I.Y.; Lee, S.H. Population-specific QTLs and their different epistatic interactions for pod dehiscence in soybean [*Glycine max* (L.) Merr.]. *Euphytica* **2009**, *166*, 15–24. [CrossRef]
39. Dong, Y.; Yang, X.; Liu, J.; Wang, B.H.; Liu, B.L.; Wang, Y.Z. Pod shattering resistance associated with domestication is mediated by a NAC gene in soybean. *Nat. Commun.* **2014**, *5*, 3352. [CrossRef]
40. Han, J.; Han, D.; Guo, Y.; Yan, H.; Wei, Z.; Tian, Y.; Qiu, L. QTL mapping pod dehiscence resistance in soybean (*Glycine max* L.) using specific-locus amplified fragment sequencing. *Theor. Appl. Genet.* **2019**, *132*, 2253–2272. [CrossRef]
41. Kang, X.; Cai, J.; Chen, Y.; Yan, Y.; Yang, S.; He, R.; Zhu, Y. Pod-shattering characteristics differences between two groups of soybeans are associated with specific changes in gene expression. *Funct. Integr. Genom.* **2020**, *20*, 201–210. [CrossRef]
42. He, J.; Gai, J. Genome-wide association studies (GWAS). In *Plant Genotyping: Methods and Protocols*; Shavrukov, Y., Ed.; Methods in Molecular Biology; Springer-Nature: New York, NY, USA, 2023; Volume 2638, pp. 123–146. [CrossRef]

43. Supritha, D.S.R.; Swaroop, K.O.; Shreeraksha, R.J.; Nandigavi, V.S.; Akshitha, N.Y.; Kiranakumara, D.M.; Gangashetty, P.I.; Aski, M.S.; Thudi, M.; Patil, B.S.; et al. Global status of genetic, genomic, and bioinformatics resources for pulse crop improvement. In *Breeding Climate Resilient and Future Ready Pulse Crops*; Pandey, M.K., Mallikarjuna, M.G., Lohithaswa, H.C., Aski, M.S., Gupta, S., Eds.; Springer-Nature: Singapore, 2025; pp. 71–129. [CrossRef]
44. Hu, D.; Kan, G.; Hu, W.; Li, Y.; Hao, D.; Li, X.; Yu, D. Identification of loci and candidate genes responsible for pod dehiscence in soybean via genome-wide association analysis across multiple environments. *Front. Plant Sci.* **2019**, *10*, 811. [CrossRef]
45. Chun, J.; Hwang, S. Genome-wide association and selection studies for pod dehiscence resistance in the USDA soybean germplasm collection. *PLoS ONE* **2025**, *20*, e0318815. [CrossRef]
46. Seo, J.H.; Kang, B.K.; Dhungana, S.K.; Oh, J.H.; Choi, M.S.; Park, J.H.; Shin, J.H.; Kim, H.S.; Baek, I.Y.; Sung, J.S.; et al. QTL mapping and candidate gene analysis for pod shattering resistance in soybean (*Glycine max* L.). *Plants* **2020**, *9*, 1163. [CrossRef]
47. Li, S.; Wang, W.; Sun, L.; Zhu, H.; Hou, R.; Zhang, H.; Tang, X.; Clark, C.B.; Swarm, S.A.; Nelson, R.L.; et al. Artificial selection of mutations in two nearby genes gave rise to shattering resistance in soybean. *Nat. Commun.* **2024**, *15*, 7588. [CrossRef] [PubMed]
48. Hickey, J.M.; Chiurugwi, T.; Mackay, I.; Powell, W. Genomic prediction unifies animal and plant breeding programs to form platforms for biological discovery. *Nat. Genet.* **2017**, *49*, 1297–1303. [CrossRef] [PubMed]
49. Zhao, T.; Wang, F.; Qi, J.; Chen, Q.; Zhu, L.; Liu, L.; Yan, L.; Chen, Y.; Yang, C.; Qin, J. Genome-wide association analysis study and genomic prediction for resistance to soybean mosaic virus in soybean population. *BMC Plant Biol.* **2025**, *25*, 837. [CrossRef] [PubMed]
50. Keller, B.; Ariza-Suarez, D.; de la Hoz, J.; Aparicio, J.S.; Portilla-Benavides, A.E.; Buendia, H.F.; Mayor, V.M.; Studer, B.; Raatz, B. Genomic prediction of agronomic traits in common bean (*Phaseolus vulgaris* L.) under environmental stress. *Front. Plant Sci.* **2020**, *11*, 1001. [CrossRef]
51. Jean, M.; Cober, E.; O'Donoughue, L.; Rajcan, I.; Belzile, F. Improvement of key agronomical traits in soybean through genomic prediction of superior crosses. *Crop Sci.* **2021**, *61*, 3908–3918. [CrossRef]
52. Bhat, J.A.; Adeboye, K.A.; Ganie, S.A.; Barmukh, R.; Hu, D.; Varshney, R.K.; Yu, D. Genome-wide association study, haplotype analysis, and genomic prediction reveal the genetic basis of yield-related traits in soybean (*Glycine max* L.). *Front. Genet.* **2022**, *13*, 953833. [CrossRef]
53. Xiong, H.; Chen, Y.; Pan, Y.B.; Wang, J.; Lu, W.; Shi, A. A genome-wide association study and genomic prediction for *Phakopsora pachyrhizi* resistance in soybean. *Front. Plant Sci.* **2023**, *14*, 1179357. [CrossRef]
54. Budhlakoti, N.; Kushwaha, A.K.; Rai, A.; Chaturvedi, K.K.; Kumar, A.; Pradhan, A.K.; Kunmar, U.; Kumar, R.R.; Juliana, P.; Mishra, D.C.; et al. Genomic selection: A tool for accelerating the efficiency of molecular breeding for development of climate-resilient crops. *Front. Genet.* **2022**, *13*, 832153. [CrossRef]
55. Peel, M.C.; Finlayson, B.L.; McMahon, T.A. Updated world map of the Köppen–Geiger climate classification. *Hydrol. Earth Syst. Sci.* **2007**, *11*, 1633–1644. [CrossRef]
56. Wellburn, A.R. The spectral determination of chlorophylls *a* and *b*, as well as total carotenoids, using various solvents with spectrophotometers of different resolution. *J. Plant Physiol.* **1994**, *144*, 307–313. [CrossRef]
57. Barrs, H.D.; Weatherley, P.E. A re-examination of the relative turgidity technique for estimating water deficits in leaves. *Aust. J. Biol. Sci.* **1962**, *15*, 413–428. [CrossRef]
58. Khan, M.H.; Panda, S.K. Alterations in root lipid peroxidation and antioxidative responses in two rice cultivars under NaCl-salinity stress. *Acta Physiol. Plan.* **2008**, *30*, 81–89. [CrossRef]
59. Beauchamp, C.; Fridovich, I. Superoxide dismutase: Improved assays and an assay applicable to acrylamide gels. *Anal Biochem.* **1971**, *44*, 276–287. [CrossRef]
60. Liu, H.; Jiang, W.; Bi, Y.; Luo, Y. Postharvest BTH treatment induces resistance of peach (*Prunus persica* L. cv. Jiubao) fruit to infection by *Penicillium expansum* and enhances activity of fruit defense mechanisms. *Postharvest Biol. Technol.* **2005**, *35*, 263–269. [CrossRef]
61. Tukamuhabwa, P.; Rubaihayo, P.; Dashiell, K.E. Genetic components of pod shattering in soybean. *Euphytica* **2002**, *125*, 29–34. [CrossRef]
62. Evanno, G.; Regnaut, S.; Goudet, J. Detecting the number of clusters of individuals using the software STRUCTURE: A simulation study. *Mol. Ecol.* **2005**, *14*, 2611–2620. [CrossRef]
63. Earl, D.A.; von Holdt, B.M. STRUCTURE HARVESTER: A website and program for visualizing STRUCTURE output and implementing the Evanno method. *Conserv. Genet. Resour.* **2012**, *4*, 359–361. [CrossRef]
64. Structure Harvester. 2022. Available online: https://github.com/dentearl/structureHarvester (accessed on 15 July 2025).
65. Wang, J.; Zhang, Z. GAPIT Version 3: Boosting power and accuracy for genomic association and prediction. *Genom. Proteom. Bioinform.* **2021**, *19*, 629–640. [CrossRef]
66. Huang, M.; Liu, X.; Zhou, Y.; Summers, R.M.; Zhang, Z. BLINK: A package for the next level of genome-wide association studies with both individuals and markers in the millions. *Gigascience* **2019**, *8*, giy154. [CrossRef]

67. Liu, X.; Huang, M.; Fan, B.; Buckler, E.S.; Zhang, Z. Iterative usage of fixed and random effect models for powerful and efficient genome-wide association studies. *PLoS Genet.* **2016**, *12*, e1005767. [CrossRef]
68. Segura, V.; Vilhjálmsson, B.J.; Platt, A.; Korte, A.; Seren, Ü.; Long, Q.; Nordborg, M. An efficient multi-locus mixed-model approach for genome-wide association studies in structured populations. *Nat. Genet.* **2012**, *44*, 825–830. [CrossRef] [PubMed]
69. Zhang, Z.; Ersoz, E.; Lai, C.Q.; Todhunter, R.J.; Tiwari, H.K.; Gore, M.A.; Bradbury, P.J.; Yu, J.; Arnett, D.K.; Ordovas, J.M.; et al. Mixed linear model approach adapted for genome-wide association studies. *Nat. Genet.* **2010**, *42*, 355–360. [CrossRef] [PubMed]
70. Nelder, J.A.; Wedderburn, R.W.M. Generalized linear models. *J. R. Stat. Soc. Ser. A (Stat. Soc.)* **1972**, *135*, 370–384. [CrossRef]
71. Bland, J.M.; Altman, D.G. Multiple significance tests: The Bonferroni method. *BMJ* **1995**, *310*, 170. [CrossRef]
72. Qin, J.; Shi, A.; Song, Q.; Li, S.; Wang, F.; Cao, Y.; Ravelombola, W.; Song, Q.; Yang, C.; Zhang, M. Genome Wide Association Study and Genomic Selection of Amino Acid Concentrations in Soybean Seeds. *Front. Plant Sci.* **2019**, *10*, 1445. [CrossRef]
73. VanRaden, P.M. Efficient methods to compute genomic predictions. *J. Dairy Sci.* **2008**, *91*, 4414–4423. [CrossRef]
74. Park, T.; Casella, G. The Bayesian Lasso. *J. Am. Stat. Assoc.* **2008**, *103*, 681–686. [CrossRef]
75. Fernando, R.L.; Grossman, M. Marker assisted selection using best linear unbiased prediction. *Genet. Sel. Evol.* **1989**, *21*, 467–477. [CrossRef]
76. Heslot, N.; Yang, H.P.; Sorrells, M.E.; Jannink, J.L. Genomic selection in plant breeding: A comparison of models. *Crop Sci.* **2012**, *52*, 146–160. [CrossRef]
77. Gianola, D.; Van Kaam, J.B. Reproducing kernel Hilbert spaces regression methods for genomic assisted prediction of quantitative traits. *Genetics* **2008**, *178*, 2289–2303. [CrossRef] [PubMed]
78. Endelman, J.B. Ridge regression and other kernels for genomic selection with R package rrBLUP. *Plant Genome* **2011**, *4*, 250–255. [CrossRef]
79. R Core Team. *R: A Language and Environment for Statistical Computing*; R Foundation for Statistical Computing: Vienna, Austria, 2025; Available online: https://www.R-project.org (accessed on 20 July 2025).
80. Zhang, Q.; Tu, B.; Liu, C.; Liu, X. Pod anatomy, morphology and dehiscing forces in pod dehiscence of soybean (*Glycine max* L.). *Flora* **2018**, *248*, 48–53. [CrossRef]
81. Adeyeye, A.S.; Togun, A.O.; Akanbi, W.B.; Adepoju, I.O.; Ibirinde, D.O. Pod shattering of different soybean varieties, *Glycine max* (L) Merrill, as affected by some growth and yield parameters. *Int. J. Agric. Pol. Res.* **2014**, *2*, 010–015.
82. Romkaew, J.; Umezaki, T.; Suzuki, K.; Nagaya, Y. Pod dehiscence in relation to pod position and moisture content in soybean. *Plant Prod. Sci.* **2007**, *10*, 292–296. [CrossRef]
83. Neubohn, B.; Gubatz, S.; Wobus, U.; Weber, H. Sugar levels altered by ectopic expression of a yeast-derived invertase affect cellular differentiation of developing cotyledons of *Vicia narbonensis* L. *Planta* **2000**, *211*, 325–334. [CrossRef]
84. de Carvalho, R.C.; da Silva, A.B.; Branquinho, C.; da Silva, J.M. Influence of dehydration rate on cell sucrose and water relations parameters in an inducible desiccation resistant aquatic bryophyte. *Environ. Exper. Bot.* **2015**, *120*, 18–22. [CrossRef]
85. Wang, X.; Wu, Z.; Zhou, Q.; Wang, X.; Song, S.; Dong, S. Physiological response of soybean plants to water deficit. *Front. Plant Sci.* **2022**, *12*, 809692. [CrossRef]
86. Amjid, M.; Üstün, R. Selection of soybean genotypes exhibiting drought resistance by assessing morphological and yield traits. *Euphytica* **2025**, *221*, 44. [CrossRef]
87. Prochazkova, D.; Wilhelmova, N. Leaf senescence and activities of the antioxidant enzymes. *Biol. Plant.* **2007**, *51*, 401–406. [CrossRef]
88. Zhou, Z.; Jiang, Y.; Wang, Z.; Gou, Z.; Lyu, J.; Li, W.; Yu, Y.; Shu, L.; Zhao, Y.; Ma, Y.; et al. Resequencing 302 wild and cultivated accessions identifies genes related to domestication and improvement in soybean. *Nat. Biotechnol.* **2015**, *33*, 408–414. [CrossRef] [PubMed]
89. Li, Y.H.; Smulders, M.J.; Chang, R.Z.; Qiu, L.J. Genetic diversity and association mapping in a collection of selected Chinese soybean accessions based on SSR marker analysis. *Conserv. Genet.* **2011**, *12*, 1145–1157. [CrossRef]
90. Bandillo, N.; Jarquin, D.; Song, Q.; Nelson, R.; Cregan, P.; Specht, J.; Lorenz, A. A population structure and genome-wide association analysis on the USDA soybean germplasm collection. *Plant Genome* **2015**, *8*, 1–14. [CrossRef]
91. Parker, T.A.; Lo, S.; Gepts, P. Pod shattering in grain legumes: Emerging genetic and environment-related patterns. *Plant Cell* **2021**, *33*, 179–199. [CrossRef] [PubMed]
92. Mahmood, U.; Li, X.; Qian, M.; Fan, Y.; Yu, M.; Li, S.; Shahzad, A.; Qu, C.; Li, J.; Liu, L.; et al. Comparative transcriptome and co-expression network analysis revealed the genes associated with senescence and polygalacturonase activity involved in pod shattering of rapeseed. *Biotechnol. Biofuels Bioprod.* **2023**, *16*, 20. [CrossRef]
93. Liu, L.; Javed, H.H.; Hu, Y.; Luo, Y.Q.; Peng, X.; Wu, Y.C. Research progress and mitigation strategies for pod shattering resistance in rapeseed. *Peer J* **2024**, *12*, e18105. [CrossRef]
94. Li, C.; Ma, X.; Yu, H.; Fu, Y.; Luo, K. Ectopic expression of *PtoMYB74* in poplar and Arabidopsis promotes secondary cell wall formation. *Front. Plant Sci.* **2018**, *9*, 1262. [CrossRef]

95. Libault, M.; Farmer, A.; Joshi, T.; Takahashi, K.; Langley, R.J.; Franklin, L.D.; He, J.; Xu, D.; May, G.; Stacey, G. An integrated transcriptome atlas of the crop model *Glycine max*, and its use in comparative analyses in plants. *Plant J.* **2010**, *63*, 86–99. [CrossRef]
96. Moya-Cuevas, J.; Ortiz-Gutiérrez, E.; López-Sánchez, P.; Simón-Moya, M.; Ballester, P.; Álvarez-Buylla, E.R.; Ferrándiz, C. A Model for the gene regulatory network along the Arabidopsis fruit medio-lateral axis: Rewiring the pod shatter process. *Plants* **2024**, *13*, 2927. [CrossRef]
97. Marsh, J.I.; Nestor, B.J.; Petereit, J.; Fernandez, T.C.G.; Bayer, P.E.; Batley, J.; Edwards, D. Legume-wide comparative analysis of pod shatter locus *PDH1* reveals phaseoloid specificity, high cowpea expression, and stress responsive genomic context. *Plant J.* **2023**, *115*, 68–80. [CrossRef]
98. Rajani, S.; Sundaresan, V. The *Arabidopsis* myc/bHLH gene *ALCATRAZ* enables cell separation in fruit dehiscence. *Curr. Biol.* **2001**, *11*, 1914–1922. [CrossRef]
99. Duhnen, A.; Gras, A.; Teyssèdre, S.; Romestant, M.; Claustres, B.; Daydé, J.; Mangin, B. Genomic selection for yield and seed protein content in soybean: A study of breeding program data and assessment of prediction accuracy. *Crop Sci.* **2017**, *57*, 1325–1337. [CrossRef]
100. Alemu, A.; Astrand, J.; Montesinos-Lopez, O.A.; y Sanchez, J.I.; Fernandez-Gonzalez, J.; Tadesse, W.; Vetukuri, R.R.; Carlsson, A.S.; Ceplitis, A.; Crossa, J.; et al. Genomic selection in plant breeding: Key factors shaping two decades of progress. *Mol. Plant* **2024**, *17*, 552–578. [CrossRef]
101. Xu, R.; Yang, Q.; Liu, Z.; Shi, X.; Wu, X.; Chen, Y.; Du, X.; Gao, Q.; He, D.; Shi, A.; et al. Genome-wide association analysis and genomic prediction of salt resistance trait in soybean germplasm. *Front. Plant Sci.* **2024**, *15*, 1494551. [CrossRef]
102. Ma, J.; Yang, Q.; Yu, C.; Liu, Z.; Shi, X.; Wu, X.; Xu, R.; Shen, P.; Zhang, Y.; Shi, F.; et al. Identification of loci and candidate genes associated with arginine content in soybean. *Agronomy* **2025**, *15*, 1339. [CrossRef]
103. Spindel, J.; Begum, H.; Akdemir, D.; Virk, P.; Collard, B.; Redona, E.; Altin, G.; Jannink, J.; McCouch, S.R. Genomic selection and association mapping in rice (*Oryza sativa*): Effect of trait genetic architecture, training population composition, marker number and statistical model on accuracy of rice genomic selection in elite, tropical rice breeding lines. *PLoS Genet.* **2015**, *11*, e1004982. [CrossRef]
104. Sun, J.; Poland, J.A.; Mondal, S.; Crossa, J.; Juliana, P.; Singh, R.P.; Rutkoski, J.E.; Jannink, J.L.; Crespo-Herrera, L.; Velu, G.; et al. High-throughput phenotyping platforms enhance genomic selection for wheat grain yield across populations and cycles in early stage. *Theor. Appl. Genet.* **2019**, *132*, 1705–1720. [CrossRef]
105. Zhang, J.; Song, Q.; Cregan, P.B.; Jiang, G. Genome-wide association study, genomic prediction and marker-assisted selection for seed weight in soybean (*Glycine max*). *Theor. Appl. Genet.* **2016**, *129*, 117–130. [CrossRef]
106. Jarquín, D.; Kocak, K.; Posadas, L.; Hyma, K.; Jedlicka, J.; Graef, G.; Lorenz, A. Genotyping by sequencing for genomic prediction in a soybean breeding population. *BMC Genom.* **2014**, *15*, 740. [CrossRef]
107. Ma, Y.; Reif, J.C.; Jiang, Y.; Wen, Z.; Wang, D.; Liu, Z.; Guo, Y.; Wei, S.; Wang, S.; Yang, C.; et al. Potential of marker selection to increase prediction accuracy of genomic selection in soybean (*Glycine max* L.). *Mol. Breed.* **2016**, *36*, 113. [CrossRef]

plants

MDPI

Article

Genus-Wide Pan-Genome Analysis of *Oryza* Calcium-Dependent Protein Kinase Genes and Their Related Kinases Highlights the Complexity of Protein Domain Architectures and Expression Dynamics

Fu Shi [1,†], Li Li [1,†], Mingjie Chen [1], Junli Chang [1], Min Tu [2,*], Guangyuan He [1,*], Yin Li [1,*] and Guangxiao Yang [1,*]

1 The Genetic Engineering International Cooperation Base of Chinese Ministry of Science and Technology, The Key Laboratory of Molecular Biophysics of Chinese Ministry of Education, College of Life Science and Technology, Huazhong University of Science & Technology, Wuhan 430074, China; 2024511001@hust.edu.cn (L.L.); cmj@hust.edu.cn (M.C.); cjl@hust.edu.cn (J.C.)

2 Hubei Province Key Laboratory of Agricultural Waste Resource Utilization, Hubei Technical Engineering Research Center for Chemical Utilization and Engineering Development of Agricultural and Byproduct Resources, School of Chemical and Environmental Engineering, Wuhan Polytechnic University, Wuhan 430023, China

* Correspondence: 12739@whpu.edu.cn (M.T.); hegy@hust.edu.cn (G.H.); yinli2021@hust.edu.cn (Y.L.); ygx@hust.edu.cn (G.Y.)

† These authors contributed equally to this work.

Abstract: The *Oryza* genus serves not only as a gene pool for rice improvement but also as a model system for plant evolutionary research. Calcium-dependent protein kinases (CPKs) function as both effectors and sensors in calcium signaling and play versatile roles in plant development and stress responses. Four kinase families, namely CPK-related kinases (CRKs), phosphoenolpyruvate carboxylase kinases (PPCKs), PPCK-related kinases (PEPRKs), and calcium- and calmodulin-dependent kinases (CCaMKs), are frequently called CPK-related kinases. This study utilized evolutionary genomics approaches and provided the pan-genome repertoires of *CPKs* and their related kinases in 34 *Oryza* genomes by leveraging the rich genomics resources of the *Orzya* genus. Gene duplication analysis revealed that distinct duplication types contributed to expanding *CPKs* and their related kinases in wild rice. We depicted the protein domain architectures of CPKs and their related kinases, highlighting the complexity of EF-hand motifs in CPKs and CCaMKs. Transcriptome analysis determined that alternative splicing was a mechanism contributing to the diversity in the domain architectures of CPKs and CCaMKs. We also generated the expression atlas of *CPKs* and their related kinases in multiple species of *Oryza* genus, emphasizing divergent homoeolog expression patterns across tissues and species in allotetraploid wild rice. Collectively, our *Oryza*-wide analysis of *CPKs* and their related kinases revealed their evolutionary trajectories and highlighted their diversified domain architectures and expression dynamics, providing gene resources of wild relatives for rice improvement.

Keywords: calcium-dependent protein kinases; CPKs and their related kinases; *Oryza*; duplicated gene; EF-hand motifs; transcriptome analysis

1. Introduction

Rice is the most widely consumed staple crop, feeding over half of the world's population [1]. However, rice production is facing enormous challenges: the increasing demands arising from a growing global population and the escalating threats to rice yield posed by climate changes. To address these challenges, increasing yield and environmental adaptability is urgent for rice improvement. The species of *Oryza* genus has been considered an important gene pool for cultivated rice and contains a vast array of genetic diversity that can be harnessed to enhance various traits in cultivated rice, such as disease resistance, stress tolerance, and yield potential [2].

The *Oryza* genus belongs to the grass family (*Poaceae*) and encompasses 2 cultivated species (Asian cultivated rice (*Oryza sativa*) and African cultivated rice (*Oryza glaberrima*)) and 25 wild relatives, including 11 distinct genome types (AA, BB, CC, EE, FF, GG, BBCC, CCDD, HHJJ, HHKK, and KKLL) [3]. Additionally, these wild relatives exhibit significant phenotypic diversity. On one hand, wild relatives have numerous desirable traits enabling their great adaptability to different ecosystems, owing to their diverse genotypes and extensive geographical distribution. For example, *Oryza longistaminata* possesses special traits that give it strong resistance to biotic and abiotic stresses and rhizomatousness [4]. On the other hand, some favorable genes were lost during rice domestication and/or the improvement of rice, and the genetic diversity of modern rice varieties has been reduced by human selection during rice improvement and breeding processes. Thus, mining favorable genes from wild relatives and re-identifying lost genes or alleles from wild relatives serve as promising strategies for rice improvement.

Recently, high-quality genome assemblies of the species of *Oryza* genus have been available for comparative and functional genomics studies due to the advances in genomics technologies. To date, more than 68 genome assemblies of wild rice species have been published, covering almost all genome types in *Oryza*, except for HHJJ and HHKK [4–10]. For cultivated rice, based on the 3000 Rice Genomes Project [11], Asian cultivated rice has been classified into nine subpopulations. Subsequently, more than 40 high-quality genomes [12–14], representing subpopulations of Asian cultivated rice, have been de novo assembled. These genomic resources allow us to mine favorable genes and investigate the evolutionary relationships of gene families in *Oryza*. Importantly, the *Oryza* genus has diversified genotypes and phenotypes within a narrow evolutionary time scale (~15 million years) with several closely spaced speciation events, recording a nearly stepwise evolutionary history [15]. *O. sativa* and *O. glaberrima* were independently domesticated from *Oryza rufipogon* and *Oryza barthii*, respectively [16]. Recently, de-domestication has been revealed in weedy rice (*Oryza sativa f. spontanea*), which are mostly derived from local cultivated rice [17]. The *Oryza* genus has roles in evolutionary dynamics, domestication, de-domestication, and speciation, making it a suitable genus model for evolutionary and comparative genomics studies.

In parallel to the advances in monocot genome assemblies (such as those in the *Oryza* genus), evolutionary genomics has unraveled several ancestral whole genome duplication (WGD) events, important for genetic innovation and expanding genetic contents [18–21]. In particular, lineage-specific features of gene duplication have been identified with potential contributions to the speciation and improvement of crop species at both the whole genome and gene family levels, such as in the Triticeae tribe [22,23]. In line with this concept, a new comparative genomics strategy has been proposed. This strategy emphasizes integrating gene duplication and genome-wide analyses (iGG) to facilitate cross-species gene identification and the reverse-genetics discovery of functional genes [23]. Although these new resources and strategies have been established for crop genomics, their utilization has been scarce in *Oryza*, an agriculturally important genus with probably the richest genomics resources.

This study aimed to leverage the rich genomics resources and to exemplify the utilization of the iGG strategy in the *Oryza* genus. As such, we chose calcium-dependent protein kinases (CPKs) and their related kinases to provide comprehensive bioinformatic characterization, because they represent a set of protein kinase families crucial for plant development and environmental adaptability. We believe that this study will serve as a starting point for utilizing evolutionary genomics-based strategies to identify gene members valuable for rice improvement. To rapidly respond to environment stimulus and coordinate environmental conditions and growth, plants have evolved several protein kinase families to facilitate calcium signaling and cellular signaling transduction. Among these kinases, CPKs and the four families, namely CPK-related kinases (CRKs), phosphoenolpyruvate carboxylase kinases (PPCKs), PPCK-related kinases (PEPRKs), and calcium- and calmodulin-dependent kinases (CCaMKs), are functionally important. CPKs serve as both sensors and effectors of Ca^{2+} signaling, while the other four related families are evolutionarily close to CPKs with differentiated structures, thus being frequently referred as CPK-related kinases [24–27]. Hereafter, we refer to CPKs and these above-mentioned kinases as CPKs and their related kinases.

In terms of distribution across taxa, *CPKs* are found in plants, as well as in green algae and protozoa (ciliates and apicomplexans), whereas *CRKs*, *CCaMKs*, *PPCKs*, and *PEPRKs* are plant-specific, despite the absence of *CCaMK* in *Arabidopsis* [24]. In rice, at least twenty-nine *OsCPKs*, five *OsCRKs*, two *OsPEPRKs*, three *OsPPCKs*, and one *OsCCaMK* have been identified [28–30]. Functionally, CPKs play versatile roles in plant development, stress tolerance, and plant immunity [31–33]. Unlike CPKs, relatively fewer functions have been reported for CRKs in plants, revealing their roles in root growth and hormonal responses [34,35]. CCaMKs function as regulators in root nodules and Arbuscular mycorrhizal symbioses in legumes [36]. OsCCaMK/OsDMI3 is reported to be involved in the response to abiotic and biotic stresses [37]. PPCKs phosphorylate PEP carboxylase, a key enzyme in C4 and CAM pathways. *Arabidopsis AtPPCKs* are reported to respond to phosphate and salt stresses [38]. Regarding PEPRKs, expression analyses suggest that they might be transcriptionally regulated in response to abiotic stresses and phytohormones.

Structurally, CPKs and their related kinases contain distinct regulatory domains. CPKs consist of four domains, namely a variable N-terminal domain (VNTD), a kinase domain, an autoinhibitory junction domain (JD), and a C-terminal calmodulin (CaM)-like domain (CaMLD) with four EF-hand motifs. Most CPKs contain potential myristoylation and palmitoylation sites in the VNTD, and these sites are associated with subcellular localization. Ca^{2+} could bind EF-hand motifs and regulate the activity of CPKs. CRKs share similar structures with CPKs but lack functional EF-hand motifs in their CaMLDs, making them unable to bind to Ca^{2+}. Some CRKs are reported to interact with CaM for regulation. PPCKs are characterized by a single kinase domain, while regulatory domains are absent in the C-terminals. PEPRKs possess a kinase domain and a distinct C-terminal domain. Their kinase domains are closely related to PPCKs, while their C-terminal domains differ from the remaining CPKs and their related kinases. In contrast, CCaMKs contain a kinase domain, an overlapping autoinhibitory region/CaM-binding domain, and a visinin-like Ca^{2+}-binding domain. Due to the unique structures, CCaMKs are regulated through both Ca^{2+} and CaM [24,26]. Because of the significance of Ca^{2+} signaling in the plant response to environmental cues and the critical roles of CPKs and their related kinases in environmental adaptability and coordination between growth and environmental conditions, we speculated that the gene copy numbers and/or protein structures of CPKs and their related kinases might exhibit dynamic variation within the *Oryza* genus, especially between the cultivated rice species and their wild relatives.

Here, we provided the first pan-genome repertoire of *CPKs* and their related kinases in the 34 *Oryza* genomes. Gene duplication analysis revealed that different types of duplication events contributed to the expansion of *CPKs* and their related kinases in wild rice. We depicted protein domain architectures of CPKs and their related kinases, highlighting the complexity of EF-hand motifs in *Oryza* CPKs and CCaMKs. Transcriptome analysis found that alternative splicing contributed to the diversity in the protein domain architectures of CPKs and CCaMKs. We also generated the expression atlas of *CPKs* and their related kinases in multiple species of *Oryza* genus. Our homoeolog expression analysis uncovered divergent homoeolog expression bias patterns between species and tissues in allotetraploid wild rice. This updated knowledge of *CPKs* and their related kinases deepens our understanding of the regulation mechanisms of CPKs and their related kinases and provides gene resources from wild relatives for rice improvement.

2. Results

2.1. The Pan-Genome Analyses Enabled Understanding the Repertoire of CPKs and Their Related Kinases in Oryza

To obtain a comprehensive *Oryza* catalog of the genes encoding CPKs and their related kinases, we selected 34 representative genome assemblies in *Oryza* covering nine genome types (AA, BB, CC, EE, FF, GG, BBCC, CCDD, and KKLL), including 20 wild rice, 12 cultivated rice, and 2 weedy rice accessions, with *Leersia perrieri* being the outgroup (Figure 1 and Table S1). Firstly, we systematically re-identified these gene families in three publicly available Nipponbare reference annotations (MUS [39], AGIS-1.0 [40], and RAP-DB [41]) via Bitacora [42], an innovative pipeline integrating sequence similarity-based search algorithms (BLAST and HMMER) and a homology-based gene prediction program GeMoMa v1.9 [43], allowing us to scan genomic sequences to precisely identify gene family members not curated in genome annotation. A total of 40 genes, comprising 1 *CCaMK*, 29 *CPKs*, 5 *CRKs*, 2 *PEPRKs*, and 3 *PPCKs*, were identified in each annotation without newly identified genes (Table S2). We further characterized the gene structures of the encoding genes and protein domain architectures of OsCPKs and their related kinases (Figure S1). Considering the accuracy and completeness of annotations, our identified CPKs and their related kinases from the Nipponbare MUS annotation and *Arabidopsis thaliana* Col-0 (TAIR10) were used as the seeds for the gene identification of CPKs and their related kinases in the *Oryza* genus.

We systematically identified 1676 genes encoding CPKs and their related kinases, including 41 *CCaMKs*, 1217 *CPKs*, 212 *CRKs*, 83 *PEPPKs*, and 123 *PPCKs*, in *Oryza* (Figure 1 and Table S3). Notably, the cultivated rice and weedy rice accessions shared the same gene numbers of *CPKs* and their related kinases, while some wild rice accessions had more kinase-encoding genes than those in the cultivated rice (e.g., *PPCKs* in *Oryza australiensis*, *CRKs* in *Oryza alta* and *Oryza punctata* (BB), and *CPKs* in *Oryza malampuzhaensis*), underlining species-specific expansions of certain kinase families in the *Oryza* genus (Figure 1). Our pan-genome catalogs of *CPKs* and their related kinases in *Oryza* laid the foundation for their evolutionary relationships, protein domain architectures, and expression analysis.

To understand the evolutionary relationships of CPKs and their related kinases in *Oryza*, we performed phylogenetic analysis using the protein sequences of the identified CPKs and their related kinases, together with thirty-four AtCPKs, eight AtCRKs, two AtPPCKs, and two AtPEPRKs [24]. CPKs were clustered into four subgroups, designated as CPK subgroups I-IV, while CRKs, PPCKs, PEPPKs, and CCaMKs were separately clustered, designated as CRKs, PPCKs, PEPPKs, and CCaMKs, respectively. Furthermore, CPKs from subgroup IV were closer to CRKs and CCaMKs than to CPKs from other subgroups, suggesting the more recent divergence of CRKs, at least in *Oryza* (Figure 2). In our analysis, we identified 11 duplicated gene pairs that commonly occurred in these species of *Oryza* genus (e.g., *CPK11*/*CPK17* in CPK subgroup I, *CPK1*/*CPK15* in CPK subgroup II, *CPK3*/*CPK16* in CPK subgroup III, *PPCK1*/*PPCK3*, and *CRK1*/*CRK4*), implying that these gene pairs had been duplicated before *Oryza* speciation, while no gene pair was found in CPK subgroup IV (Figure 2). Our phylogenetic analysis established the orthologous relationships of CPKs and their related kinases among these species of *Oryza* genus, facilitating a detailed analysis of gene duplication and copy number variations for genes encoding these kinases.

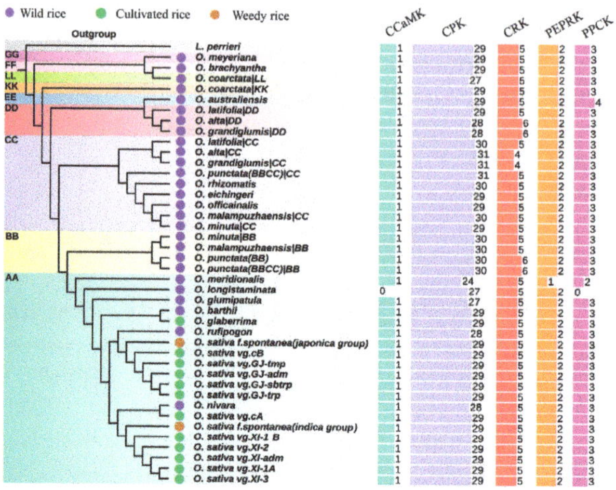

Figure 1. Gene numbers of *CPKs* and their related kinases in *Oryza*. The identified gene numbers of *CPKs* and their related kinases in 34 *Oryza* genomes and the *L. perrieri* outgroup. Please note that the subgenomes of allotetraploid rice have been separated and treated as a diploid genome in this figure. The colored bars represent the gene numbers of the *CCaMK*, *CPK*, *CRK*, *PEPRK*, and *PPCK* gene families. The evolutionary tree was constructed based on 2297 orthologous genes, with *L. perrieri* being the outgroup. Wild, cultivated, and weedy rice are indicated in the tree with purple, green, and brown circles, respectively. Corresponding genome/subgenome types (AA, BB, CC, DD, EE, KK, LL, FF, and GG) are labeled in the evolutionary tree.

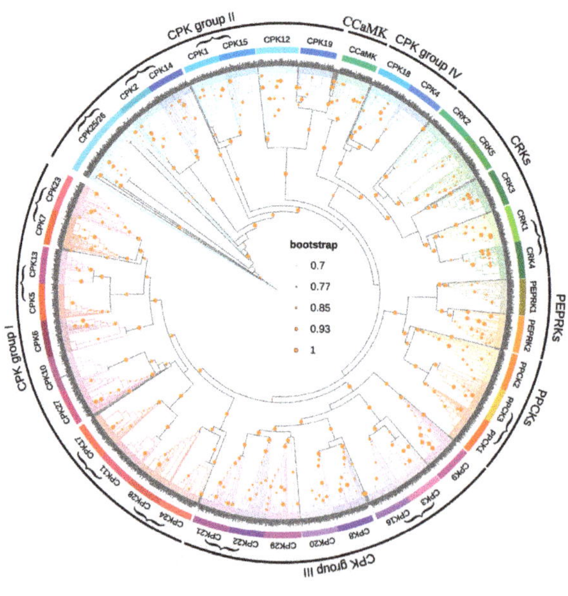

Figure 2. A phylogenetic analysis of the CPKs and their related kinases. The identified CPKs and their related kinases from wild, cultivated, and weedy rice, along with their members in *Arabidopsis*, were used in the phylogenetic analysis. The 1676 identified CPKs and their related kinases were cladded into eight groups, namely CPK subgroups I-IV, CRKs, PPCKs, PEPPKs, and CCaMK. Bootstrap values > 0.7 are indicated as orange circles. Duplicated gene pairs shared in *Oryza* genomes are labeled with brackets.

2.2. Several Types of Gene Duplication Contribute to the Expansion of CPKs and Their Related Kinases

To establish the syntenic orthologous/paralogous relationships of the *Oryza* CPKs and their related kinases, we constructed collinear graphs at the inter- and intra-genome/subgenome levels. Regarding the copy number variation (CNV) in *CPKs* and their relatives in genomes/subgenomes (i.e., AA, BB, CC, DD, EE, FF, GG, KK, and LL), cultivated rice and weedy rice accessions shared the same gene numbers for each of the kinase families (one *CCaMK*, twenty-nine *CPKs*, five *CRKs*, two *PEPPKs*, and three *PPCKs*), implying that these kinase families might be fixed in cultivated rice and weedy rice (Figure 3 and Table S3). In contrast, wild rice exhibited lineage- or species-specific expansion patterns. *CCaMK* was present in single-copy form in all genomes/subgenomes, excluding its absence in *O. longistaminata* (Figure S2), which may have been due to the genome assembly quality. *CPK12s* were expanded in subgenomes BB and LL (Figure S3), and *PPCK2* was expanded in *Oryza australiensis* (EE), showing lineage-specific patterns. *CPK15* and *CPK21* were uniquely expanded in *O. malampuzhaensis* and *Oryza officainalis*, while *PEPPK1* was exclusively absent in *Oryza meridionalis*. Notably, *CPK7* was absent in the subgenome CC of *O. alta* and *Oryza grandiglumis*, while it was preserved as two copies in the subgenome DD of *O. alta* and *O. grandiglumis*, resulting in unaltered total gene numbers. A similar phenomenon was also found in *CRK2* (Figure 3 and Table S3).

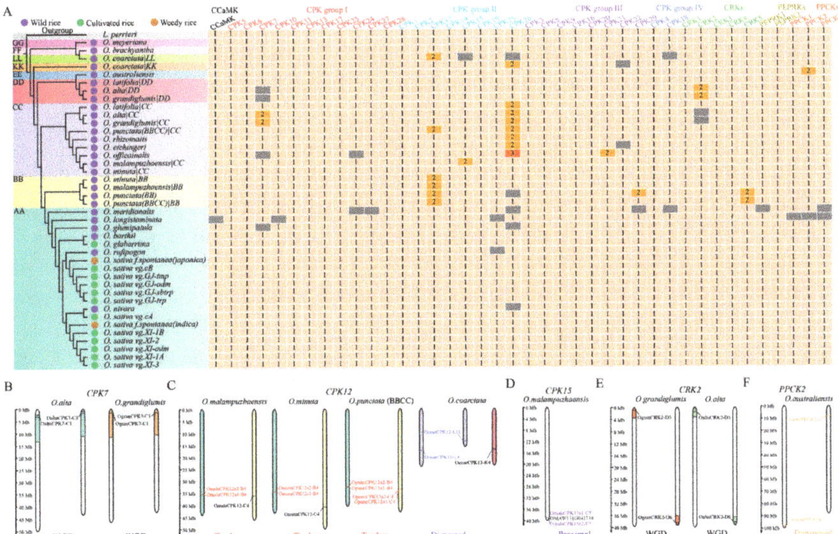

Figure 3. Gene duplication contributes to the expansion of *CPKs* and their related kinases in *Oryza*. (**A**) The copy numbers of genes encoding CPKs and their related kinases in genomes/subgenomes are indicated in the boxes. Gray indicates that genes are absent in corresponding genomes/subgenomes. A coalescent-based evolutionary tree was constructed based on 2297 single-copy orthologous genes with *L. perrieri* as the outgroup. Wild, cultivated, and weedy rice are indicated in tree with purple, green, and brown circles, respectively. Corresponding genome/subgenome types (AA, BB, CC, DD, EE, KK, LL, FF, and GG) are labeled in the evolutionary tree. Gene names in different colors are ordered by their evolutionary groups, namely CCaMK, CPK group I-IV, CRKs, PEPRKs, and PPCKs. (**B–F**) The visualization of representative gene duplication examples that contributed to the expansion of *CPK7* (**B**), *CPK12* (**C**), *CPK15* (**D**), *CRK2* (**E**), and *PPCK2* (**F**). WGD: whole genome duplication; tandem, tandem duplication; dispersed, dispersed duplication; proximal, proximal duplication; transposed, transposed duplication.

Whole genome duplication (WGD), tandem duplication (TD), proximal duplication (PD), transposed duplication (TRD), and dispersed duplication (DSD) drive the expansion of gene families in plant genomes, contributing to the diversification of gene contents, structures, and functions; complex molecular network constructions; and phenotypic innovations [39]. To explore the drivers behind the expansion of *CPKs* and their related kinases in the wild rice species, we analyzed the gene duplication modes of duplicated genes, and these above-mentioned drivers contributed jointly to expanding *CPKs* and their related kinases. For instance, WGD-derived *CRK2, CPK7,* and *CPK21* generated two copies in *O. alta, O. grandiglumis,* and *O. officainalis,* respectively (Figure 3). In contrast, TRD contributed to expanding *CPK29* in *O. punctata* (BB) and *PPCK2* in *O. australiensis,* while PD drove *CPK15* expansion in *O. malampuzhaensis* (Figure 3). We also observed that the expansions of *CRK5s* in *O. punctata* (BB) and *O. punctata* (BBCC) were influenced by both WGD and TRD.

Furthermore, we identified gene duplication modes that were conservative or divergent in a genome- or subgenome-specific way. TD contributed to the expansion of *CPK12s* in the subgenome BB, whereas DSD drove *CPK12* expansion in the subgenome LL, suggesting subgenome-specific evolutionary trajectories with independent contributions to gene expansion (Figure 3). *CPK26s* gave rise to two copies with genome-specific mechanisms. TRD was the dominant driver in the subgenome CC, except for *Oryza rhizomatis* and *O. officainalis.* DSD mediated *CPK26* expansion in *O. rhizomatis,* while both WGD and PD contributed to *CPK26* expansion in *O. officainalis.* Taken together, our analyses revealed conservative and divergent gene duplication mechanisms in *Oryza*, establishing a framework for exploring how duplication modes influence the protein domain architectures and expression patterns of CPKs and their related kinases.

2.3. Myristoylation and Palmitoylation of CPKs and Their Related Kinases

Myristoylation and palmitoylation sites in the N-termini of CPKs and CRKs play critical roles in membrane association and biological functions, and these sites could be predicted with established bioinformatic tools [44,45]. Our analysis results showed that neither myristoylation nor palmitoylation sites were detected in all CCaMKs. Moreover, no myristoylation or palmitoylation sites were predicted in most PPCKs and PEPRKs. In contrast, myristoylation and/or palmitoylation sites were predicted in most CRKs and CPKs, except CPK24 and CPK28 (Table S3). The predicted subcellular localizations of CPKs and their related kinases also supported their myristoylation and palmitoylation sites. CPK24, CPK28, and CCaMKs without myristoylation or palmitoylation sites were predicted to be localized in cytoplasm or nucleus, while those proteins containing myristoylation and/or palmitoylation sites were predicted to target the cell membrane. The information for myristoylation and palmitoylation and predicted subcellular localization suggest differences in membrane targeting and potential functional divergence among CPKs and their related kinases.

2.4. Molecular Properties of CPKs and Their Related Kinases

We predicted the molecular properties (i.e., protein length, molecular weight, and isoelectric point (pI)) of 1676 identified CPKs and related kinases (Table S3). These proteins varied in length, ranging from 119 amino acids (AAs) to 1246 AAs, with their molecular weights ranging from 12.6 kDa to 140.3 kDa. These proteins also had a wide range of pI, varying from 4.39 to 9.79. CPKs from group IV displayed the narrowest length distribution (428–609 AAs), recording molecular weights between 48.5 and 68.2 kDa. In contrast, PEPRKs had the largest variations in protein length (119–1172 AAs) and molecular weight

(12.6–127.3 kDa). These data demonstrated the diversity among subgroups of CPKs and their related kinases.

Gene duplication might lead to functional divergence by shaping protein domain architectures, which could be revealed via molecular characteristics. To explore the effects of gene duplication on these proteins, we compared molecular properties between duplicated gene pairs. Most duplicated gene pairs shared similar protein characteristics. For example, WGD-driven CPK7 in *O. alta* had the same protein length (547 AAs), a similar molecular weight (62.0 kDa vs. 61.9 kDa), and the same pI (5.5). However, we found changed properties of duplicated gene pairs in some cases, for example, TRD-derived *CPK26* in *Oryza eichingeri*.

2.5. Landscapes of Domain Architectures Reveal the Diversity of EF-Hand Motifs in CPKs and CCaMKs

We analyzed the protein domains in the CPKs and related kinases, depicted the landscape of domain architectures, and revealed complexity in domain organization. CPKs and CCaMKs consisted of a kinase domain and variable EF-hand motifs, while PEPRKs and PPCKs only possessed a kinase domain. Compared with three EF-hand motifs in all CCaMKs, the number of EF-hand motifs in *Oryza* CPKs varied from three to four. Notably, CRK4 harbored a kinase domain and a single degenerated EF-hand motif that could not bind Ca^{2+}, which is consistent with AtCRK3 lacking Ca^{2+}-binding capacity despite containing two analogous EF-hand motifs [46]. The remaining CRKs only contained a kinase domain, resembling PEPRKs and PPCKs (Figures 4 and S4).

According to the previous studies and our analysis, the major differences in domain architectures of CPKs and their related kinases lie in their C-terminal domains, especially in the numbers and variants of EF-hand motifs [24,26]. To address whether EF-hand motifs and Ca^{2+} binding sites of CPKs and CCaMKs had become divergent in *Oryza*, we compared the organizations of EF-hand motifs with a focus on Ca^{2+} binding sites based on identified CPKs and CCaMKs. We excluded the solo types per CPK or CCaMK and clustered CPKs and CCaMKs, respectively, according to the organization of EF-hand motifs and Ca^{2+} binding sites. Protein domain architectures supported by most CPKs or CCaMKs were considered the representative types, while those supported by minor CPKs or CCaMKs were regarded as the variants. By comparing the protein domain architectures, we found that most canonical CPKs generally contained four EF-hand motifs, while the duplicated gene pair CPK7/CPK23 contained three EF-hand motifs (Figure 4). In some canonical CPKs, some EF-hand motifs lacked functional Ca^{2+}-binding sites. Notably, CPK7/CPK23 (the gene pair in group I) held three EF-hand motifs, but only with a single Ca^{2+}-binding site in the C-lobe of the CaMLD domain. CPK3/CPK16 (the gene pair in group III) and CPK29 possessed four EF-hand motifs, with a degenerated Ca^{2+}-binding site in the N-lobe of the CaMLD domain. Owing to the differences in Ca^{2+} affinity between the N-lobe and C-lobe EF-hand motifs, the altered number of functional EF-hand motifs could be associated with distinct kinase activity levels in response to basal and elevated calcium concentrations, contributing to decode calcium signals with distinct dynamics.

Comparing EF-hand motifs between variants and the corresponding representative types, some CPKs within group I (CPK5, CPK6, CPK10, CPK17, and CPK28), group II (CPK15 and CPK19), and group III (CPK29), and all group IV CPK members (CPK4 and CPK18), exhibited conservative organization of EF-hand motifs (Figure 4). Compared with the corresponding representative domain architectures, CPK12, CPK21, and CPK26 contained fewer EF-hand motifs and Ca^{2+}-binding sites, while CPK8, CPK9, CPK22, and CPK27 had more Ca^{2+}-binding sites, with their numbers of EF-hand motifs being unaltered. In some cases, CPKs within group I (CPK11 and CPK24), group II (CPK2), and group III (CPK20) had longer amino acid sequences after the EF-hand motifs in the C-terminus. Additionally, CPK3, CPK13,

CPK14, and CPK21 contained longer interspaced sequences between the first and second EF-hand motifs (Figure 4). The divergence in EF-hand motifs and the proximal sequences could contribute to the distinct CPK-involved regulatory mechanisms.

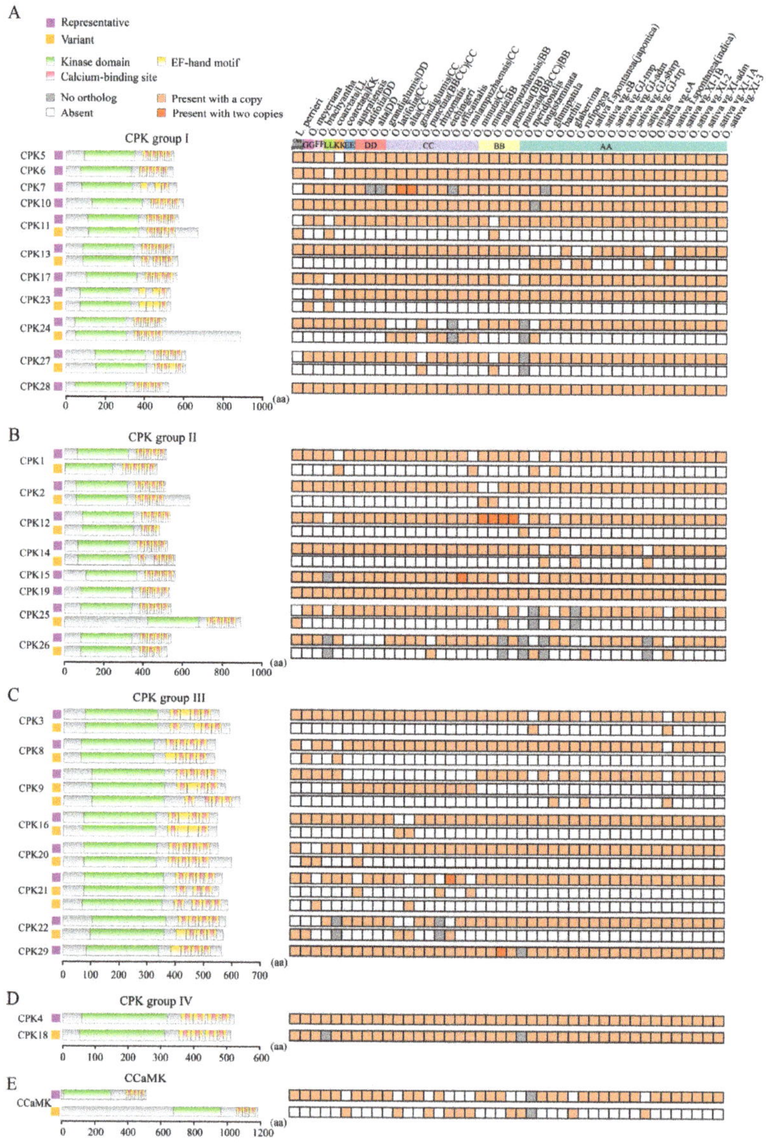

Figure 4. Comparative analysis of EF-hand motifs of CPKs and CCaMKs in *Oryza*. (**A–E**) Predicted kinase domains, EF-hand motifs, and calcium-binding sites of CPKs and CCaMKs are represented with green, yellow, and red rectangles, respectively. CPKs and CCaMKs are ordered based on evolutionary subgroups, including CPK groups I-IV (**A–D**) and CCaMKs (**E**). Protein domain architectures are clustered into two types: those supported by the majority of CPKs or CCaMKs were considered as the representative types (purple squares), and those supported by the minor members were regarded as variants (orange squares). The presence and absence of protein domain architecture types in corresponding genomes/subgenomes are indicated on the right squares in orange and white, respectively.

We further explored the impacts of gene duplication on the protein domain architectures. *CPK12* experienced diverse gene duplication events, and the domain architectures of the proteins encoded by the *CPK12* duplicates diverged. The duplicates *OcoarCPK12-L4/OcoarCPK12-L11* were generated from dispersed duplication, leading to alterations in the protein domain architectures. *CPK12* was tandemly duplicated into two copies in *O. malampuzhaensis*, *Oryza minuta*, *O. punctata* (BB), and *O. punctata* (BBCC), respectively, while the proteins encoded by these duplicated genes exhibited divergent domain architectures (Figure S4). Similarly, the proteins encoded by the transposon-mediated duplicates *OaustPPCK2a1/OaustPPCK2a2* diverged in the domain architecture, with a C-terminal truncation detected in OaustPPCK2a2 (Figure S5). Conversely, the two copies of WGD-derived *CPK7* in *O. alta* and *O. grandiglumis*, respectively, both retained identical protein domain architectures (Figure S6). Like *CPK7*, *CPK15* in *O. malampuzhaensis* underwent proximal duplication, yet it also maintained the same domain structure (Figure S7). Collectively, different types of gene duplication in *Oryza* CPKs could be associated with similar or diversified domain architectures of the corresponding encoding proteins. More importantly, we depicted the landscapes of domain architectures of CPKs and CCaMKs, highlighted the complexity of EF-hand motifs and Ca^{2+}-binding sites, and documented gene duplication events and their associated diversification of CPK domain architectures.

2.6. Alternative Splicing Contributes to the Diversity of Domain Architectures of CPKs and CCaMKs

Alternative splicing (AS) is a pivotal mechanism in transcriptional regulation and allows for producing different protein variants from the same gene loci, substantially expanding protein diversity. To explore the impact of AS in shaping protein domain architectures, we collected and processed 427 RNA-seq libraries to systematically identify the transcripts of CPKs and CCaMKs (Table S4). Protein domain architectures encoded by identified transcripts were predicted via the homology-based annotation method. With the RNA-seq-based transcript assemblies, we corrected the annotations of some genes, e.g., *OaltaCPK26a1-C12* (Figure S8). Our assembled transcript of *OaltaCPK26a1-C12.2* had a more similar gene model and protein length to the orthologous gene *OsCPK26* than the original gene (*OaltaCPK26a1-C12.1*).

Compared with representative domain architectures of CPKs or CCaMKs, the potential consequences of AS on domain architectures were categorized into nine types, namely truncated EF-hand motifs (Types 1–4), truncated kinase domain (Type 5), and remaining EF-hand motifs (Types 6–9) (Figure 5A). For CCaMKs, the representative domain architectures consisted of a kinase domain and three EF-hand motifs. We detected truncated protein variants of CCaMKs with an incomplete kinase domain (Type 5), as well as other CCaMK variants with only two or three EF-hand motifs but lacking the kinase domain (Types 7–8) (Figure 5B). CPK variants exhibited more diverse domain architecture types than CCaMK variants. Akin to CCaMK variants, we found similar architectures in CPK variants, i.e., variants with only EF-hand motifs (Types 6–7) and incomplete kinase domains (Type 5). Moreover, we identified variants containing a complete kinase domain with fewer EF-hand motifs (Types 1–3), as well as variants made up of only a complete kinase domain (Type 4) (Figure 5C). Although the functions of the truncated protein variants still need to be investigated, our analysis demonstrated that AS significantly contributes to the diversified domain architectures of CPKs and CCaMKs, enhancing the knowledge of AS-mediated Ca^{2+}-dependent regulatory mechanisms of CPKs and CCaMKs.

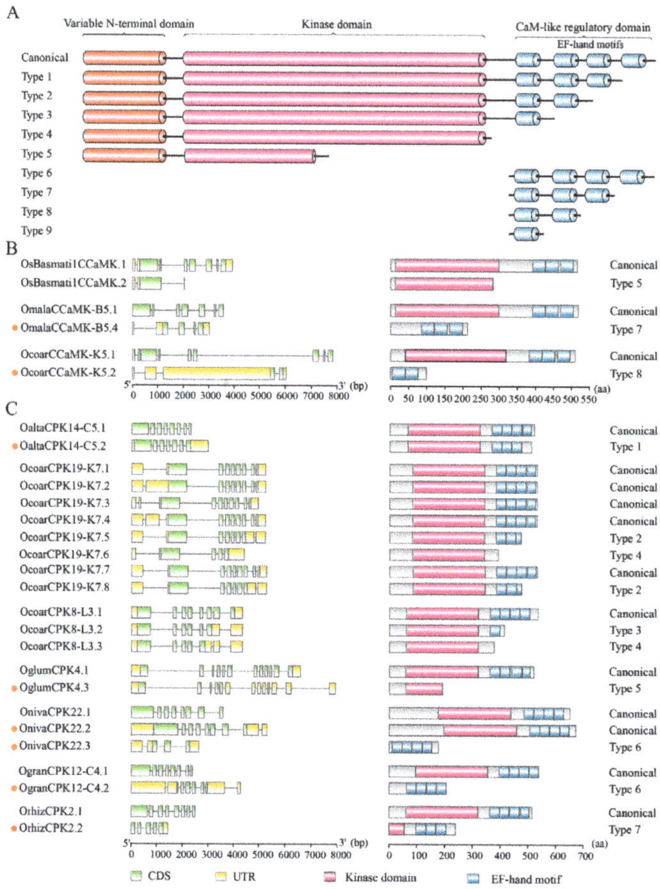

Figure 5. Alternative splicing contributes to the protein domain architectures of CPKs and CCaMKs. (**A**) A graphic illustration representing the nine potential domain architecture types of canonical and truncated CPKs, namely truncated EF-hand motifs (Types 1–4), truncated kinase domain (Type 5), and the remaining EF-hand motifs (Types 6–9). Variable N-terminal domains, the kinase domains, and the EF-hand motifs are shown with range, red, and blue cylinders, respectively. (**B**,**C**) Gene structures of representative transcripts of *CCaMKs* (**B**) and *CPKs* (**C**) and their encoded protein domain architectures. The reconstructed transcripts are highlighted with orange circles. Coding sequences (CDSs) and untranslated regions (UTRs) are shown with green and yellow rectangles, respectively. Red and blue rectangles represent the kinase domains and the EF-hand motifs, respectively. The scale bars indicate gene and protein lengths.

2.7. Homoeolog Expression Patterns of CPKs and Related Kinases Across Tissues and Species

In allopolyploid species, one homoeolog frequently has a dominant expression level over those of the others, and this is known as homoeolog expression bias (HEB). Understanding HEB will facilitate crop improvement and breeding programs by precisely modulating individual or multiple homoeologs [47]. Therefore, HEB patterns of the genes encoding CPKs and the related kinases were investigated in six allotetraploid species of wild rice accessions, comprising two genotypes: BBCC (*O. malampuzhaensis*, *O. minuta*, and *O. punctate* (BBCC)) and CCDD (*O. alta*, *O. grandiglumis*, and *Oryza latifolia*). Those that expressed genes with the 1:1 homoeologous relationship were included in the analysis. Most of the kinase-encoding genes, including *CPK3*, *PEPRK1*, and *PPCK1*, had balanced expression between subgenomes across tissues, demonstrating the stable HEB patterns of these genes. In contrast, we found dynamic HEB patterns between species and tissues. In the BBCC genome type, *CCaMK* copies from the subgenome CC were preferentially expressed in *O. malapuzhaensis* and *O. minuta*, while the BB subgenome copies were predominantly expressed in *O. punctate*. Similarly, in the CCDD genome type, *CCaMKs* in *O. alta* exhibited CC subgenome-biased expression in leaves and panicles, switching to balanced expression in roots and stems. The HEB patterns shifted in *O. grandiglumis* and *O. latifolia*, showing the DD subgenome-biased expressions in stems and panicles turned to balanced expression patterns in roots and leaves. *PEPRK2* exhibited an *O. minuta*-specific, biased expression pattern towards the CC subgenome copy across the investigated tissues, whereas the *PEPRK2* from other species of *Oryza* genus had balanced expression patterns, implying that HEB patterns are likely determined by the species rather than the subgenomes and are tissue-specific (Figure 6).

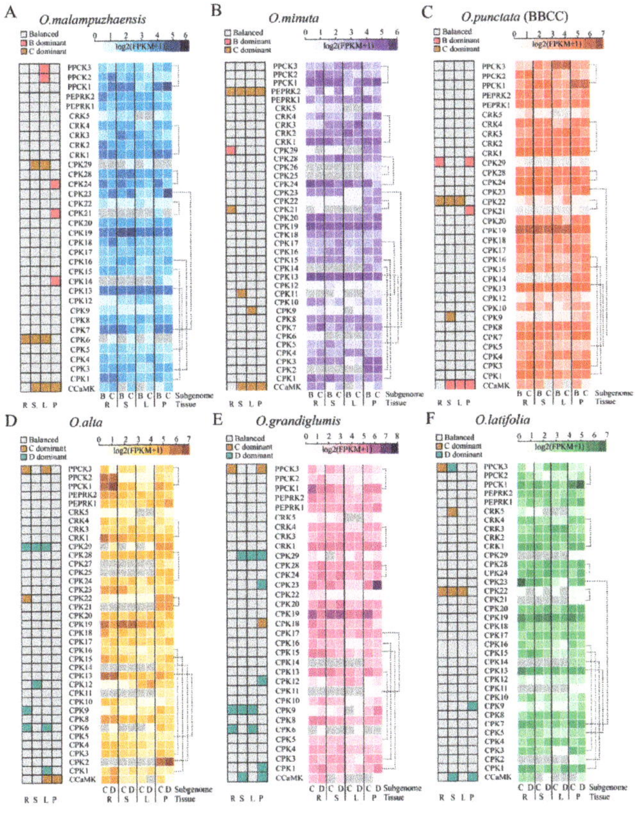

Figure 6. Homoeolog expression patterns of *CPKs* and related kinases across tissues and species. Homoeolog expression patterns were analyzed in two *Oryza* genome types, BBCC (**A–C**) and CCDD (**D–F**), in six allotetraploid species of wild rice: *O. malapuzhaensis* (**A**), *O. minuta* (**B**), *O. punctata* (**C**), *O. alta* (**D**), *O. grandiglumis* (**E**), and *O. latifolia* (**F**). Heatmaps on the left indicate the balanced expression pattern or a dominant expression towards a subgenome across roots (R), stems (S), leaves (L), and panicles (P). Gray indicates balanced expression between homoeologs. Red, brown, and cyan colors indicate preferential expression toward subgenomes BB, CC, and DD, respectively. The heatmaps on the right illustrate Log_2 (FPKM+1)-normalized expression levels per homoeolog in root (R), stem (S), leaf (L), and panicle (P). Gray indicates that the gene was not expressed. The dashed lines link gene pairs.

We observed similar HEB patterns between the duplicated gene pairs. For example, *CPK3/CPK16* and *CRK1/CRK4* in the selected genomes both showed balanced expression between the homoeologs across tissues, suggesting still-conserved expression patterns after gene duplication events. However, in some cases, the duplicated pairs exhibited diverse HEB patterns. *PPCK1/PPCK3* in *O. minuta* and *O. punctate* showed balanced expression across tissues, while distinct expression patterns were observed in the other species of *Oryza* genus. *CPK24/CPK28* in *O. malampuzhaensis* had disparate expression patterns in the panicle but balanced patterns in the root, stem, and leaf, confirming that HEB patterns of duplicated gene pairs shifted between genomes and tissues (Figure 6). Collectively, our HEB analysis provided insights into homoeolog expression patterns in the allotetraploid wild rice species, enhanced our knowledge of the allotetraploidization processes, and contributed to rice improvement.

2.8. *Transcriptomal Atlas of CPKs and Their Related Kinases Underlines the Expression Dynamics and Indicates Potential Functions*

To explore the expression dynamics and potential functions of genes encoding CPKs and related kinases, we complied multiple transcriptome datasets to form a comprehensive expression atlas in both wild and cultivated rice accessions. Firstly, we detected the expression patterns between the phylogenetic clades. CCaMKs were predominantly expressed in the roots of both wild and cultivated rice accessions, implying their conserved roles in root development or stress responses (Figure 7). However, there was no significant expression specificity among members within CPK subgroups, PPCKs, and PEPRKs, revealing the multi-faceted roles in different aspects of plant development.

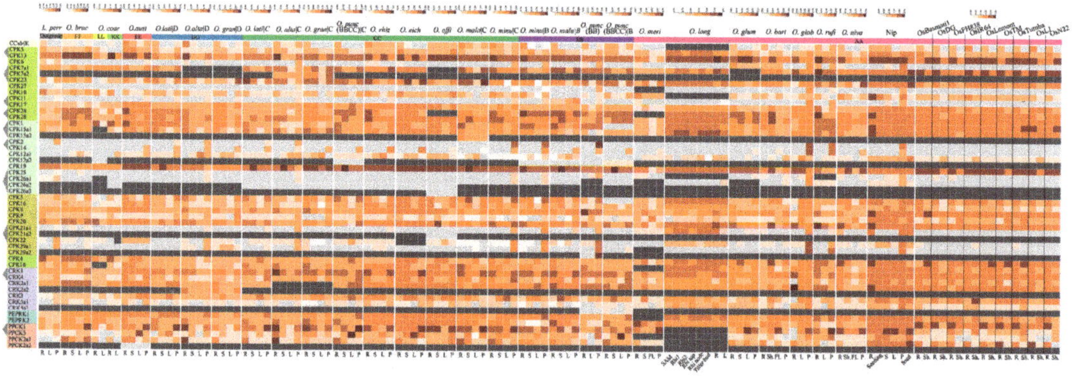

Figure 7. The expression patterns of the genes encoding CPKs and their related kinases in wild and cultivated rice. The expression levels of *CPKs* and their related kinases are visualized in the heatmap in $\log_{10}$ (FPKM). On the left of the heatmap, different colors highlight evolutionary groups of CPKs and their related kinases (from top to bottom, CCaMK, CPK subgroups I–IV, CRKs, PEPRKs, and PPCKs, respectively), while gray triangles connect duplicated gene pairs. Light gray indicates genes that were not expressed in tissues or species. Dark gray indicates orthologous genes absent in genomes/subgenomes. R, roots; L, leaves; S, stems; P, panicles; SAM, shoot apical meristem; Rhi, rhizomes; Sh, shoots.

Secondly, we compared the expression patterns of the duplicated gene pairs in *Oryza*. Our analysis demonstrated that the duplicated gene pairs exhibited different expression patterns. For example, for the gene pair *CPK5/CPK13*, expression levels of *CPK13* were generally higher than those of *CPK5* in all tested tissues and species, implying that *CPK5* and *CPK13* experienced a conservative evolutionary trajectory but diverse functions in *Oryza* (Figure 7). We observed asymmetric expression patterns in the *CPK11/CPK17* gene pair: *CPK17* exhibited dominant expression in most tested species, whereas *CPK11* was preferentially expressed in the panicles of *O. malampuzhaensis*, implying the subfunctionalization of *CPK11* in *O. malampuzhaensis*. For *CPK25/CPK26*, similar expression patterns for *CPK25/CPK26* were observed in panicles across species, indicating that they might function redundantly in panicle development.

Thirdly, we analyzed the expression patterns of species-specific duplicated gene pairs. *PPCK2a2* in *O. australiensis* and *CPK12a2* in *O. minuta* and *O. malampuzhaensis* were not expressed in the examined tissues, in contrast to their expressed homologs (*PPCK2a1* and *CPK12a1*), suggesting that they were undergoing pseudogenization (Figure 7). There was no significant expression difference between *CPK15a1* and *CPK15a2* in *O. malampuzhaensis*, indicating that they might function redundantly. *CPK29a2* in *O. punctata* (BB) had a higher expression level in the panicles than that of *CPK29a1*, implying neofunctionalization in *CPK29a2*. *CPK7a1* and *CPK7a2* in *O. alta* and *O. grandiglumis* were similarly expressed in the panicles, whereas biased expression was identified in the stem, showing that these elements could be functionally redundant in the panicles while playing divergent roles during stem development. Our analysis established a comprehensive expression atlas, exhibited the expression dynamics of CPKs and their related kinases across tissues and species, and characterized their potential functional roles, providing favorable genes encoding CPKs and their related kinases for rice improvement.

3. Discussion

The *Oryza* genus serves as a crucial reservoir of beneficial genes for rice improvement [10]. Several genes, such as *PROG1*, *Bph14*, and *SUB1A-1* from wild rice, have been identified and applied in rice breeding [48–50]. These examples highlight the value of species of *Oryza* genus, especially wild relatives, in addressing agricultural challenges. Here, with the rich genomic resources in *Oryza*, we selected the five evolutionarily close and functionally important kinases, CPKs and their related kinases, to apply an iGG strategy for genus-wide evolutionary analysis and favorable candidate gene mining. Previous studies in rice about *CPKs* and their related kinase gene families are limited in cultivated rice; however, *Oryza* species represent abundant favorable gene resources [28–30]. In this study, we identified 1217 *CPKs*, 212 *CRKs*, 123 *PPCKs*, 83 *PEPRKs*, and 41 *CCaMKs* in 34 *Oryza* genomes, representing almost all genome types and all three ecotypes in *Oryza* (Figures 1 and 2 and Table S3). Our genus-wide analyses provided repertoires of *CPKs* and their related kinase gene families in *Oryza*, allowing us to investigate their evolutionary relationships, protein domain architectures, and expression dynamics, serving as a useful starting point for identifying *CPKs* and their related kinase gene families for rice improvement. It is greatly significant to investigate the five evolutionarily related and functionally important kinase family from a genus. Currently, population genomic data have been focused on two cultivated rice accessions and their ancestors (*O. rufipogon* and *O. barthii*), while there is a lack of data on the populations of the remaining wild relatives [11,51,52]. In the future, it will be worth exploring structural variations and single-nucleotide variations in gene family members considering population genomics.

The canonical CPKs and their related kinases were featured with distinct domain architectures. CCaMKs possessed a kinase domain, autoinhibitory domain, and a visinin-like domain containing three functional EF-hand motifs. The CPKs contained a variable N-terminal domain with the myristoylation and/or palmitoylation sites, followed by a serine/threonine kinase domain, an auto-inhibitory junction domain, and a CaMLD domain with four functional EF-hand motifs. CRKs shared similar architectures but with degenerated EF hands, which could not bind Ca^{2+}. PPCKs and PEPRKs consisted only of a kinase domain, with differing C-termini [24,26]. In our analysis, most CPKs and their related kinases were in line with their canonical domain architectures. However, reduced numbers of functional EF-hand motifs, extended sequences in the C-terminus, and longer interspaced sequences between EF-hand motifs were observed in some CPKs (Figure 4). CPKs and CRKs have a common ancestor, originating from the fusion of the CaM-dependent protein kinase and CaM [25,27]. These variations in CPKs were possibly caused by fusion processes, recording *Oryza* evolution. The functions and Ca^{2+} affinities of these variations still need to be explored. According to our analysis, the domain architectures of duplicated gene pairs in CPKs and their related kinases were similar, suggesting the conservation of protein domain architectures in *Oryza* evolution. However, in the canonical CPKs, we identified the distinct arrangements of functional EF-hand motifs. CPKs in group I (CPK7 and CPK23) lacked EF-hand motifs in the N-lobe and C-lobe, while CPKs in group III (CPK3, CPK16, and CPK29) lost functional EF-hand motifs in the N-lobe of CaMLD (Figure 3). Considering the differential affinities in the N- and C-lobes of the CaMLD domain, these distinct arrangements of functional EF-hand motifs might have different consequences with two patterns toward calcium sensitivity. We speculated that the kinase activation levels of CPK7 and CPK23 might be affected at basal and elevated calcium concentrations, while the kinase activation levels of CPK3, CPK16, and CPK29 might not be affected at basal calcium concentrations but at elevated calcium concentrations. Our domain analysis revealed characteristic protein domain architectures in their related kinases and highlighted the diverse EF-hand motifs in CPKs and CCaMKs, contributing to heterogeneity in calcium sensitivity among CPKs and CCaMKs in *Oryza*.

Up to now, only limited articles have reported that the AS of CPKs and CCaMKs is involved in regulating protein domain architectures. No study has reported that CCaMKs are involved in AS-regulated protein domain architectures. The intron-retained *AtCPK28* spliced variant encodes a truncated protein lacking two high-affinity EF hands, and functions as a negative regulator in immune responses [53]. The AS-regulated truncated OsCPK17, lacking both the JD and CaM-like regulatory domain (similar to Type 4), loses Ca^{2+}-binding capacity and in vitro kinase activity [54]. AtCPK25 is characterized by degenerated EF-hand motifs when compared with other AtCPKs (similar to Type 3), and displays Ca^{2+}-independent kinase activity [55]. Collectively, these observations suggest that truncated CPKs with reduced functional EF-hand motifs in the CaM-like regulatory domain (Types 1–4) may partially or fully lose their ability to bind Ca^{2+}. The variants of *OsCPK17* have been experimentally validated; however, so far, these cloned transcripts have not been completely curated in three Nipponbare reference genome annotations [56]. Therefore, it is possible that AS variants of *CPKs* and *CCaMKs* are not adequately curated. Thus, systematic efforts to improve transcript annotations are urgently required to explore the AS variants of *CPKs* and *CCaMKs*. Here, we utilized 427 RNA-seq libraries to analyze the effects of AS on protein domain architectures in *Oryza* and clarified the protein domain architectures of identified truncated protein variants, contributing to expanding our knowledge of the regulatory mechanisms of CPKs and CCaMKs (Figure 5). However, there are still some challenges to address: the large-scale identification of AS events and novel transcripts, the in vivo evidence for AS transcripts of *CPKs* and *CCaMKs*, and the modified functions of resulting truncated protein variants. Large-scale and high-throughput sequencing technologies, such as ISO-seq and Direct RNA sequencing, might be effective methods for identifying AS variants. Proteomics data mining and function studies might be the solutions for the veracity and modified functions of AS variants.

Previous studies have reported expression patterns of *OsCPKs* and *OsCRKs* in cultivated rice, and their results showed that *OsCPKs* and *OsCRKs* are expressed broadly but in a tissue-specific manner [28,30,57–59]. In our analysis, *CCaMK* displayed dominant expression in roots among wild rice and cultivated rice, while no apparent tissue-specific expression patterns among other groups were observed. Regarding the expression of duplicated gene pairs, our analysis results are supported by previous reports [28,59], which also found similar expression levels between gene pairs *CPK25/CPK26* and preferential expression towards *CPK13* compared with *CPK5*. We further compared the expression patterns of species-specific duplicated gene pairs to infer their potential functional conservation and divergence, e.g., *CPK15a1/CPK15a2* were similarly expressed in *O. malampuzhaensis*, indicating their redundant functions. The altered expression patterns of *CPK29a2* in *O. punctata* (BB) suggested their functional divergence (Figure 7). As reported, *CPKs* and their related kinase genes respond to biotic and abiotic stresses and phytohormones, highlighting the shortcomings of our study. Due to limited transcriptome data on wild rice and weedy rice, no public stress- and phytohormone-treated transcriptome data were included in our analysis. Our expression profile analysis exhibited the transcriptome landscapes of genes encoding CPKs and their related kinases in *Oryza*, emphasizing the expression patterns of duplicated gene pairs from the perspective of the genus.

4. Materials and Methods

4.1. Gene Identification of CPKs and Their Related Kinases

We conducted this study to comprehensively identify genes encoding CPKs and their related kinases, CRKs, PEPRKs, PPCKs, and CCaMKs, in *Oryza*. The 34 *Oryz* genome assemblies, including 20 wild rice, 12 cultivated rice, and 2 weedy rice accessions, along with *L. perrieri* as an outgroup, were selected for further analysis. The detailed information of the above-mentioned genome assemblies is provided in Table S1. To identify genes encoding CPKs and their related kinases in *Oryza*, we initially re-searched these gene families via Bitacora v1.4 [42] using the full mode against three Nipponbare reference annotations, including MUS [39], AGIS-1.0 [40], and RAP-DB [41]. Then, the protein sequences of CPKs and their related kinases from the MUS annotation and *Arabidopsis* were used as seeds for BLASTP v2.16.0+ and HMMER v3.4 searches with the protein kinase domain (PF00069) and Ca^{2+}-binding domain (PF13499 and PF13833). The proteins searched via BLASTP and HMMER, together with orthologous and paralogous proteins identified via MCScanX [60] and GeneTribe v1.2.1 [61], were used as putative candidates. The protein domains of putative candidates were validated via local InterProScan [62] (version: 5.71–102.0) and ScanProsite [63] (Release version: 5-February-2025). The confirmed proteins were categorized into *CCaMK*, *CPK*, *CRK*, *PEPRK*, and *PPCK* gene families based on protein similarity matrix and their contained domains.

4.2. Construction of Phylogenetic Tree

For the construction of a phylogenetic tree, the protein sequences of CPKs and their related kinases in *Oryza* and *Arabidopsis* were aligned using MUSCLE v3.8.31 [64] with default parameters. After alignment trimming via trimAl v1.5 [65], the phylogenetic tree was inferred via Maximum Likelihood implemented in RaxML v8.2.12 [66] using the PROTGAMMAJTT algorithms with 1000 bootstrap replicates.

For constructing the species tree, a coalescent-based species tree approach was employed to infer species phylogeny. Allotetraploid wild rice accessions were divided based on their genomic compositions. Regarding *O. coarctata*, its genome was aligned using nucmer v4.0.0beta2 [67] against OcoaRS1 (GCA_036417745.1), a non-public genome assembly version. Meanwhile, subgenomes of *O. coarctata* were phased via SubPhaser v1.2.6 [68]. Orthofinder v2.5.4 [69] was employed to infer orthologous genes based on the longest transcripts. The 2297 single-copy orthologous genes were aligned using MUSCLE v3.8.31 [64]. Subsequently, the gene trees were constructed using RaxML v8.2.12 [66] with 100 bootstrap replicates using the PROTGAMMAJTT model. The coalescent-based species tree was estimated via ASTRAL v5.7.8 [70] based on constructed gene trees, with *L. perrieri* serving as the outgroup. The phylogenetic trees were visualized and annotated using Interactive Tree of Life (https://itol.embl.de/ (accessed on 21 February 2025)) [71] for graphical optimization and topological annotation.

4.3. Gene Structure and Protein Domain Analysis

For gene structure analysis, gene features were extracted from genomics data and reconstructed transcripts and visualized using TBtools-II v2.225 [72]. The protein domain architectures of identified CPKs and their related kinases were scanned with ScanProsite. Profiles PS50011 and PS50222 were regarded as a protein kinase domain and an EF-hand motif, respectively. Pattern PS00018 was considered a Ca^{2+}-binding site, namely the functional EF-hand motifs.

Myristoylation and palmitoylation sites were predicted using GPS-Lipid [73]. Protein parameters were computed with ProtParam in Expasy (https://web.expasy.org/protparam/ (accessed on 25 February 2025)). Subcellular localizations of CPKs and their related kinases were predicted using DeepLoc (https://services.healthtech.dtu.dk/services/DeepLoc-2.0/ (accessed on 23 March 2025)) [74].

4.4. Gene Duplication and Synteny Analysis

All-versus-all BLASTP (the E-value cutoff of 1×10^{-10}) between genomes and subgenomes were first performed. Subsequently, the collinearity and homology between inter-genomes were inferred with GeneTribe v1.2.1 [61]. The collinearity blocks between intra-genomes were scanned via MCScanX [60]. Gene duplication types were categorized into WGD, TD, PD, TRD, and DSD using the DupGen finder pipeline [75].

4.5. Reference-Based Transcriptome Assembly and Expression Quantification

First, Raw RNA-seq reads were processed using fastp v0.23.4 [76] to trim low-quality bases and remove adapters. Then, the clean reads were mapped to the corresponding genomes with HISAT2 v2.1.0 [77], with the intron length ranging from 50 to 5000 bp. Followed by transcript assembly for each library performed with StringTie v2.2.3 [78] and Scallop v0.10.5 [79], the merge mode implemented in StringTie v2.2.3 was used to integrate reconstructed transcripts in all samples and generate nonredundant transcripts. Subsequently, novel transcripts were filtered using GffCompare v0.12.6 [80], retaining those with class codes (k, m, n, x, i, and o). The protein-coding potentials of the novel transcripts were predicted via CPC2 v0.1 [81]. Finally, TransDecoder v5.7.1 (Haas, BJ. https://github.com/TransDecoder/TransDecoder (accessed on 25 November 2024)) was used to identify the candidate coding regions of novel coding transcripts via combination with the search for Pfam using hmmscan and BLASTP results against the local protein database, including SwissProt, Uniprot, and protein sequences from reference genomes and pan-genomes of rice [12,13,17,52], *Arabidopsis* [82], barley [83], and wheat [84]. To analyze the effects of AS events on protein domain architectures, the reconstructed transcripts of the identified genes encoding CPKs and CCaMKs were selected, and combined with transcripts in genome annotations. Subsequently, their corresponding protein sequences were used for protein domain analysis.

For gene expression quantification, StringTie v2.2.3 [78] was utilized to estimate the fragments per kilobase of transcript per million mapped reads (FPKMs) as the gene expression levels. Expression profiles of Nipponbare and *O. glaberrima* were obtained from PPRD (https://plantrnadb.com/ricerna/ (accessed on 28 February 2025)) with accessions (PRJNA482217 and PRJNA13765) [85]. Homoeolog expression bias (HEB) was performed according to the previous method [47]. We included expressed homologs (FPKM > 0.5) with the 1:1 homoeologous relationship between subgenomes, calculated the relative contributions of each subgenome per homeolog, and assigned the homoeolog expression bias category based on their Euclidean distances from observed normalized expression to three ideal categories. The ideal normalized expression bias patterns for the three categories are provided in Table S6.

5. Conclusions

In summary, we applied an iGG strategy for genus-wide evolutionary analysis of the five evolutionarily close and functionally important kinases, namely *CPKs* and their related kinases, and provided several valuable findings regarding CPKs and their related kinases. (1) Through *Oryza*-wide identification and analysis, we provide comprehensive repertoires of *CPKs* and their related kinases in 34 *Oryza* genomes for evolutionary relationships, protein domain architectures, and expression analysis. (2) We elucidate distinct gene duplication types contributing to the expansion of genes encoding *CPKs* and their related kinases in wild rice by establishing a colinear map in *Oryza*. (3) Protein domain architecture analysis reveals the divergent EF-hand motifs of CPKs and CCaMKs in *Oryza*, and we find that alternative splicing regulates the domain architecture rearrangement of CPKs and CCaMKs, deepening our knowledge of their regulation mechanisms. (4) Transcriptome landscapes of genes encoding CPKs and their related kinases exhibit their expression dynamics and potential functions in *Oryza*, emphasizing divergent homoeolog expression patterns between species and tissues. In future studies, our findings from analyzing protein domain architectures will undergo experimental investigations to elucidate regulation mechanisms, and mining gene resources from wild relatives will be applied to other grass or plant species.

Supplementary Materials: The following supporting information can be downloaded at: https://www.mdpi.com/article/10.3390/plants14101542/s1, Figure S1: Re-identification of genes encoding CPKs and their related kinases in three Nipponbare reference genomes; Figure S2: Conservative *CCaMK* locus in *Oryza*; Figure S3: Expansion of *CPK12* in *Oryza*; Figure S4: Divergent protein domain architectures encoded by duplicates of *CPK12* gene pairs; Figure S5: Conserved domain architectures between PPCK2 copies mediated by transposed duplication; Figure S6: Conserved domain architectures encoded by WGD-driven *CPK7* gene pairs; Figure S7: Conserved domain architectures encoded by *CPK15* gene pairs mediated by proximal duplication; Figure S8: Homoeolog expression patterns of *CPKs* and their related kinases across species and tissues; Figure S9: Revised annotation of *OaltaCPK26a1*; Table S1: Meta information of 35 genome assemblies used in this study. Table S2: Information of the identified CPKs and their related kinases in three Nipponbare reference genomes. Table S3: Information of the identified CPKs and their related kinases in the present study. Table S4: RNA-seq datasets used in this study. Table S5: Information of reconstructed transcripts of genes encoding CPKs and their related kinases. Table S6: Definition of homoeolog expression bias categories.

Author Contributions: Conceptualization, project administration, resources, supervision, funding acquisition, M.T., G.H., Y.L. and G.Y.; methodology, investigation, writing—original draft preparation, F.S. and L.L.; Data analysis, F.S., L.L., M.T. and Y.L.; writing, review and editing, M.C., J.C., M.T., G.H., Y.L. and G.Y. All authors have read and agreed to the published version of the manuscript.

Funding: This research was funded by the National Natural Science Foundation of China (31570261), the National Natural Science Foundation of Hubei China (2016CFB549), the Research Support Programs for the HUST Research Facilities and Bases (5001170128), the International Science and Technology Collaboration Project of Hubei Province (2024EHA056) and the Fundamental Research Funds for Central Universities, HUST to Yin Li (2021XXJS070 and 3004170157). The research was also supported by the Natural Science Foundation of Hubei Province of China (2024AFB955), the Start-Up Research Funding of Wuhan Polytechnic University (2021RZ100, 53210052172).

Data Availability Statement: The data presented in this study are available in the article and the Supplementary Materials.

Acknowledgments: We acknowledge the Core Facility of Life Sciences, HUST, for providing instrumental service.

Conflicts of Interest: The authors declare no conflicts of interest.

Abbreviations

The following abbreviations are used in this manuscript:

CPKs	Calcium-dependent protein kinases
CRKs	CPK-related kinases
PPCKs	Phosphoenolpyruvate carboxylase kinases
PEPRKs	PPCK-related kinases
CCaMKs	Calcium and calmodulin-dependent kinases
VNTD	Variable N-terminal domain
JD	Junction domain
CaM	Calmodulin
CaMLD	Calmodulin (CaM)-like domain
CNV	Copy number variation
WGD	Whole-genome duplication
TD	Tandem duplication
PD	Proximal duplication
TRD	Transposed duplication
DSD	Dispersed duplication
AS	Alternative splicing
HEB	Homoeolog expression bias

References

1. Wing, R.A.; Purugganan, M.D.; Zhang, Q. The rice genome revolution: From an ancient grain to Green Super Rice. *Nat. Rev. Genet.* **2018**, *19*, 505–517. [CrossRef] [PubMed]
2. Zheng, X.; Peng, Y.; Qiao, J.; Henry, R.; Qian, Q. Wild rice: Unlocking the future of rice breeding. *Plant Biotechnol. J.* **2024**, *22*, 3218–3226. [CrossRef] [PubMed]
3. Huang, C.; Chen, Z.; Liang, C. *Oryza* pan-genomics: A new foundation for future rice research and improvement. *Crop J.* **2021**, *9*, 622–632. [CrossRef]
4. Zhang, Y.; Zhang, S.; Liu, H.; Fu, B.; Li, L.; Xie, M.; Song, Y.; Li, X.; Cai, J.; Wan, W.; et al. Genome and comparative transcriptomics of african wild rice *Oryza longistaminata* provide insights into molecular mechanism of rhizomatousness and self-incompatibility. *Mol. Plant* **2015**, *8*, 1683–1686. [CrossRef]
5. Stein, J.C.; Yu, Y.; Copetti, D.; Zwickl, D.J.; Zhang, L.; Zhang, C.; Chougule, K.; Gao, D.; Iwata, A.; Goicoechea, J.L.; et al. Genomes of 13 domesticated and wild rice relatives highlight genetic conservation, turnover and innovation across the genus *Oryza*. *Nat. Genet.* **2018**, *50*, 285–296. [CrossRef]
6. Wang, M.; Yu, Y.; Haberer, G.; Marri, P.R.; Fan, C.; Goicoechea, J.L.; Zuccolo, A.; Song, X.; Kudrna, D.; Ammiraju, J.S.; et al. The genome sequence of African rice (*Oryza glaberrima*) and evidence for independent domestication. *Nat. Genet.* **2014**, *46*, 982–988. [CrossRef]
7. Zhao, H.; Wang, W.; Yang, Y.; Wang, Z.; Sun, J.; Yuan, K.; Rabbi, S.; Khanam, M.; Kabir, M.S.; Seraj, Z.I.; et al. A high-quality chromosome-level wild rice genome of *Oryza coarctata*. *Sci. Data* **2023**, *10*, 701. [CrossRef]
8. Long, W.; He, Q.; Wang, Y.; Wang, Y.; Wang, J.; Yuan, Z.; Wang, M.; Chen, W.; Luo, L.; Luo, L.; et al. Genome evolution and diversity of wild and cultivated rice species. *Nat. Commun.* **2024**, *15*, 9994. [CrossRef]
9. Xie, X.; Du, H.; Tang, H.; Tang, J.; Tan, X.; Liu, W.; Li, T.; Lin, Z.; Liang, C.; Liu, Y.G. A chromosome-level genome assembly of the wild rice Oryza rufipogon facilitates tracing the origins of Asian cultivated rice. *Sci. China Life Sci.* **2021**, *64*, 282–293. [CrossRef]
10. Cai, X.; He, W.; Qian, Q.; Shang, L. Genetic resource utilization in wild rice species: Genomes and gene bank. *New Crops* **2025**, *2*, 100065. [CrossRef]
11. Wang, W.; Mauleon, R.; Hu, Z.; Chebotarov, D.; Tai, S.; Wu, Z.; Li, M.; Zheng, T.; Fuentes, R.R.; Zhang, F.; et al. Genomic variation in 3,010 diverse accessions of Asian cultivated rice. *Nature* **2018**, *557*, 43–49. [CrossRef] [PubMed]

12. Qin, P.; Lu, H.; Du, H.; Wang, H.; Chen, W.; Chen, Z.; He, Q.; Ou, S.; Zhang, H.; Li, X.; et al. Pan-genome analysis of 33 genetically diverse rice accessions reveals hidden genomic variations. *Cell* **2021**, *184*, 3542–3558. [CrossRef] [PubMed]
13. Yu, Z.; Chen, Y.; Zhou, Y.; Zhang, Y.; Li, M.; Ouyang, Y.; Chebotarov, D.; Mauleon, R.; Zhao, H.; Xie, W.; et al. Rice Gene Index: A comprehensive pan-genome database for comparative and functional genomics of Asian rice. *Mol. Plant* **2023**, *16*, 798–801. [CrossRef]
14. Song, J.M.; Xie, W.Z.; Wang, S.; Guo, Y.X.; Koo, D.H.; Kudrna, D.; Gong, C.; Huang, Y.; Feng, J.W.; Zhang, W.; et al. Two gap-free reference genomes and a global view of the centromere architecture in rice. *Mol. Plant* **2021**, *14*, 1757–1767. [CrossRef]
15. Ammiraju, J.S.; Lu, F.; Sanyal, A.; Yu, Y.; Song, X.; Jiang, N.; Pontaroli, A.C.; Rambo, T.; Currie, J.; Collura, K.; et al. Dynamic evolution of *oryza* genomes is revealed by comparative genomic analysis of a genus-wide vertical data set. *Plant Cell* **2008**, *20*, 3191–3209. [CrossRef]
16. Chen, E.; Huang, X.; Tian, Z.; Wing, R.A.; Han, B. The genomics of *oryza* species provides insights into rice domestication and heterosis. *Annu. Rev. Plant Biol.* **2019**, *70*, 639–665. [CrossRef]
17. Wu, D.; Xie, L.; Sun, Y.; Huang, Y.; Jia, L.; Dong, C.; Shen, E.; Ye, C.Y.; Qian, Q.; Fan, L. A syntelog-based pan-genome provides insights into rice domestication and de-domestication. *Genome Biol.* **2023**, *24*, 179. [CrossRef]
18. Bowers, J.E.; Chapman, B.A.; Rong, J.; Paterson, A.H. Unravelling angiosperm genome evolution by phylogenetic analysis of chromosomal duplication events. *Nature* **2003**, *422*, 433–438. [CrossRef]
19. Paterson, A.H.; Bowers, J.E.; Chapman, B.A. Ancient polyploidization predating divergence of the cereals, and its consequences for comparative genomics. *Proc. Natl. Acad. Sci. USA* **2004**, *101*, 9903–9908. [CrossRef]
20. Jiao, Y.; Li, J.; Tang, H.; Paterson, A.H. Integrated syntenic and phylogenomic analyses reveal an ancient genome duplication in monocots. *Plant Cell* **2014**, *26*, 2792–2802. [CrossRef]
21. Tang, H.; Bowers, J.E.; Wang, X.; Paterson, A.H. Angiosperm genome comparisons reveal early polyploidy in the monocot lineage. *Proc. Natl. Acad. Sci. USA* **2010**, *107*, 472–477. [CrossRef] [PubMed]
22. Wang, X.; Yan, X.; Hu, Y.; Qin, L.; Wang, D.; Jia, J.; Jiao, Y. A recent burst of gene duplications in Triticeae. *Plant Commun.* **2022**, *3*, 100268. [CrossRef] [PubMed]
23. Wu, Y.; Feng, J.; Zhang, Q.; Wang, Y.; Guan, Y.; Wang, R.; Shi, F.; Zeng, F.; Wang, Y.; Chen, M.; et al. Integrative gene duplication and genome-wide analysis as an approach to facilitate wheat reverse genetics: An example in the *TaCIPK* family. *J. Adv. Res.* **2024**, *61*, 19–33. [CrossRef]
24. Hrabak, E.M.; Chan, C.W.; Gribskov, M.; Harper, J.F.; Choi, J.H.; Halford, N.; Kudla, J.; Luan, S.; Nimmo, H.G.; Sussman, M.R.; et al. The *Arabidopsis* CDPK-SnRK superfamily of protein kinases. *Plant Physiol.* **2003**, *132*, 666–680. [CrossRef]
25. Chen, F.; Zhang, L.; Cheng, Z.M. The calmodulin fused kinase novel gene family is the major system in plants converting Ca^{2+} signals to protein phosphorylation responses. *Sci. Rep.* **2017**, *7*, 4127. [CrossRef]
26. Harmon, A.C.; Gribskov, M.; Gubrium, E.; Harper, J.F. The CDPK superfamily of protein kinases. *New Phytol.* **2001**, *151*, 175–183. [CrossRef]
27. Zhang, X.S.; Choi, J.H. Molecular evolution of calmodulin-like domain protein kinases (CDPKs) in plants and protists. *J. Mol. Evol.* **2001**, *53*, 214–224. [CrossRef]
28. Ye, S.; Wang, L.; Xie, W.; Wan, B.; Li, X.; Lin, Y. Expression profile of calcium-dependent protein kinase (CDPKs) genes during the whole lifespan and under phytohormone treatment conditions in rice (*Oryza sativa* L. ssp. indica). *Plant Mol. Biol.* **2009**, *70*, 311–325. [CrossRef]
29. Asano, T.; Tanaka, N.; Yang, G.; Hayashi, N.; Komatsu, S. Genome-wide identification of the rice calcium-dependent protein kinase and its closely related kinase gene families: Comprehensive analysis of the *CDPKs* gene family in rice. *Plant Cell Physiol.* **2005**, *46*, 356–366. [CrossRef]
30. Ray, S.; Agarwal, P.; Arora, R.; Kapoor, S.; Tyagi, A.K. Expression analysis of calcium-dependent protein kinase gene family during reproductive development and abiotic stress conditions in rice (*Oryza sativa* L. ssp. indica). *Mol. Genet. Genom.* **2007**, *278*, 493–505. [CrossRef]
31. Zhang, J.; Lyu, H.; Chen, J.; Cao, X.; Du, R.; Ma, L.; Wang, N.; Zhu, Z.; Rao, J.; Wang, J.; et al. Releasing a sugar brake generates sweeter tomato without yield penalty. *Nature* **2024**, *635*, 647–656. [CrossRef] [PubMed]
32. Zhang, Z.; Wang, Q.; Yan, H.; Cang, X.; Li, W.; He, J.; Zhang, M.; Lou, L.; Wang, R.; Chang, M. Lighting-up wars: Stories of Ca^{2+} signaling in plant immunity. *New Crops* **2024**, *1*, 100027. [CrossRef]
33. Yip Delormel, T.; Boudsocq, M. Properties and functions of calcium-dependent protein kinases and their relatives in *Arabidopsis thaliana*. *New Phytol.* **2019**, *224*, 585–604. [CrossRef]
34. Baba, A.I.; Rigo, G.; Ayaydin, F.; Rehman, A.U.; Andrasi, N.; Zsigmond, L.; Valkai, I.; Urbancsok, J.; Vass, I.; Pasternak, T.; et al. Functional analysis of the *Arabidopsis thaliana* CDPK-related kinase family: AtCRK1 regulates responses to continuous light. *Int. J. Mol. Sci.* **2018**, *19*, 1282. [CrossRef]

35. Rigo, G.; Ayaydin, F.; Tietz, O.; Zsigmond, L.; Kovacs, H.; Pay, A.; Salchert, K.; Darula, Z.; Medzihradszky, K.F.; Szabados, L. et al. Inactivation of plasma membrane-localized CDPK-RELATED KINASE5 decelerates PIN2 exocytosis and root gravitropic response in *Arabidopsis*. *Plant Cell* **2013**, *25*, 1592–1608. [CrossRef]
36. Wang, T.; Guo, J.; Peng, Y.; Lyu, X.; Liu, B.; Sun, S.; Wang, X. Light-induced mobile factors from shoots regulate rhizobium triggered soybean root nodulation. *Science* **2021**, *374*, 65–71. [CrossRef]
37. Wang, Q.; Shen, T.; Ni, L.; Chen, C.; Jiang, J.; Cui, Z.; Wang, S.; Xu, F.; Yan, R.; Jiang, M. Phosphorylation of OsRbohB by the protein kinase OsDMI3 promotes H_2O_2 production to potentiate ABA responses in rice. *Mol. Plant* **2023**, *16*, 882–902. [CrossRef]
38. Feria, A.B.; Bosch, N.; Sanchez, A.; Nieto-Ingelmo, A.I.; de la Osa, C.; Echevarria, C.; Garcia-Maurino, S.; Monreal, J.A. Phosphoenolpyruvate carboxylase (PEPC) and PEPC-kinase (PEPC-k) isoenzymes in *Arabidopsis thaliana*: Role in control and abiotic stress conditions. *Planta* **2016**, *244*, 901–913. [CrossRef]
39. Hamilton, J.P.; Li, C.; Buell, C.R. The rice genome annotation project: An updated database for mining the rice genome. *Nucleic Acids Res.* **2025**, *53*, D1614–D1622. [CrossRef]
40. Shang, L.; He, W.; Wang, T.; Yang, Y.; Xu, Q.; Zhao, X.; Yang, L.; Zhang, H.; Li, X.; Lv, Y.; et al. A complete assembly of the rice Nipponbare reference genome. *Mol. Plant* **2023**, *16*, 1232–1236. [CrossRef]
41. Sakai, H.; Lee, S.S.; Tanaka, T.; Numa, H.; Kim, J.; Kawahara, Y.; Wakimoto, H.; Yang, C.C.; Iwamoto, M.; Abe, T.; et al. Rice Annotation Project Database (RAP-DB): An Integrative and Interactive Database for Rice Genomics. *Plant Cell Physiol.* **2013**, *54*, e6. [CrossRef] [PubMed]
42. Vizueta, J.; Sanchez-Gracia, A.; Rozas, J. Bitacora: A comprehensive tool for the identification and annotation of gene families in genome assemblies. *Mol. Ecol. Resour.* **2020**, *20*, 1445–1452. [CrossRef] [PubMed]
43. Keilwagen, J.; Wenk, M.; Erickson, J.L.; Schattat, M.H.; Grau, J.; Hartung, F. Using intron position conservation for homology-based gene prediction. *Nucleic Acids Res.* **2016**, *44*, e89. [CrossRef]
44. Saito, S.; Hamamoto, S.; Moriya, K.; Matsuura, A.; Sato, Y.; Muto, J.; Noguchi, H.; Yamauchi, S.; Tozawa, Y.; Ueda, M.; et al. N-myristoylation and S-acylation are common modifications of Ca^{2+}-regulated *Arabidopsis* kinases and are required for activation of the SLAC1 anion channel. *New Phytol.* **2018**, *218*, 1504–1521. [CrossRef]
45. Martin, M.L.; Busconi, L. Membrane localization of a rice calcium-dependent protein kinase (CDPK) is mediated by myristoylation and palmitoylation. *Plant J.* **2000**, *24*, 429–435. [CrossRef]
46. Du, W.; Wang, Y.; Liang, S.; Lu, Y. Biochemical and expression analysis of an *Arabidopsis* calcium-dependent protein kinase-related kinase. *Plant Sci.* **2005**, *168*, 1181–1192. [CrossRef]
47. Ramirez-Gonzalez, R.H.; Borrill, P.; Lang, D.; Harrington, S.A.; Brinton, J.; Venturini, L.; Davey, M.; Jacobs, J.; van Ex, F.; Pasha, A.; et al. The transcriptional landscape of polyploid wheat. *Science* **2018**, *361*, eaar6089. [CrossRef]
48. Tan, L.; Li, X.; Liu, F.; Sun, X.; Li, C.; Zhu, Z.; Fu, Y.; Cai, H.; Wang, X.; Xie, D.; et al. Control of a key transition from prostrate to erect growth in rice domestication. *Nat. Genet.* **2008**, *40*, 1360–1364. [CrossRef]
49. Hu, L.; Wu, Y.; Wu, D.; Rao, W.; Guo, J.; Ma, Y.; Wang, Z.; Shangguan, X.; Wang, H.; Xu, C.; et al. The coiled-coil and nucleotide binding domains of BROWN PLANTHOPPER RESISTANCE14 function in signaling and resistance against planthopper in rice. *Plant Cell* **2017**, *29*, 3157–3185. [CrossRef]
50. Lin, C.C.; Lee, W.J.; Zeng, C.Y.; Chou, M.Y.; Lin, T.J.; Lin, C.S.; Ho, M.C.; Shih, M.C. SUB1A-1 anchors a regulatory cascade for epigenetic and transcriptional controls of submergence tolerance in rice. *PNAS Nexus* **2023**, *2*, pgad229. [CrossRef]
51. Shang, L.; Li, X.; He, H.; Yuan, Q.; Song, Y.; Wei, Z.; Lin, H.; Hu, M.; Zhao, F.; Zhang, C.; et al. A super pan-genomic landscape of rice. *Cell Res.* **2022**, *32*, 878–896. [CrossRef] [PubMed]
52. Zhao, Q.; Feng, Q.; Lu, H.; Li, Y.; Wang, A.; Tian, Q.; Zhan, Q.; Lu, Y.; Zhang, L.; Huang, T.; et al. Pan-genome analysis highlights the extent of genomic variation in cultivated and wild rice. *Nat. Genet.* **2018**, *50*, 278–284. [CrossRef]
53. Dressano, K.; Weckwerth, P.R.; Poretsky, E.; Takahashi, Y.; Villarreal, C.; Shen, Z.; Schroeder, J.I.; Briggs, S.P.; Huffaker, A. Dynamic regulation of Pep-induced immunity through post-translational control of defence transcript splicing. *Nat. Plants* **2020**, *6*, 1008–1019. [CrossRef]
54. Almadanim, M.C.; Goncalves, N.M.; Rosa, M.T.G.; Alexandre, B.M.; Cordeiro, A.M.; Rodrigues, M.; Saibo, N.J.M.; Soares, C.M.; Romao, C.V.; Oliveira, M.M.; et al. The rice cold-responsive calcium-dependent protein kinase OsCPK17 is regulated by alternative splicing and post-translational modifications. *Biochim. Biophys. Acta Mol. Cell Res.* **2018**, *1865*, 231–246. [CrossRef]
55. Boudsocq, M.; Droillard, M.J.; Regad, L.; Lauriere, C. Characterization of *Arabidopsis* calcium-dependent protein kinases: Activated or not by calcium? *Biochem. J.* **2012**, *447*, 291–299. [CrossRef]
56. Loranger, M.E.W.; Huffaker, A.; Monaghan, J. Truncated variants of Ca^{2+}-dependent protein kinases: A conserved regulatory mechanism? *Trends Plant Sci.* **2021**, *26*, 1002–1005. [CrossRef]
57. Wan, B.; Lin, Y.; Mou, T. Expression of rice Ca^{2+}-dependent protein kinases (CDPKs) genes under different environmental stresses. *FEBS Lett.* **2007**, *581*, 1179–1189. [CrossRef]

58. Valmonte, G.R.; Arthur, K.; Higgins, C.M.; MacDiarmid, R.M. Calcium-dependent protein kinases in plants: Evolution, expression and function. *Plant Cell Physiol.* **2014**, *55*, 551–569. [CrossRef]
59. Yadav, A.; Garg, T.; Singh, H.; Yadav, S.R. Tissue-specific expression pattern of calcium-dependent protein kinases-related kinases (CRKs) in rice. *Plant Signal. Behav.* **2020**, *15*, 1809846. [CrossRef]
60. Wang, Y.; Tang, H.; Debarry, J.D.; Tan, X.; Li, J.; Wang, X.; Lee, T.H.; Jin, H.; Marler, B.; Guo, H.; et al. MCScanX: A toolkit for detection and evolutionary analysis of gene synteny and collinearity. *Nucleic Acids Res.* **2012**, *40*, e49. [CrossRef]
61. Chen, Y.; Song, W.; Xie, X.; Wang, Z.; Guan, P.; Peng, H.; Jiao, Y.; Ni, Z.; Sun, Q.; Guo, W. A collinearity-incorporating homology inference strategy for connecting emerging assemblies in the triticeae tribe as a pilot practice in the plant pangenomic era. *Mol. Plant* **2020**, *13*, 1694–1708. [CrossRef] [PubMed]
62. Blum, M.; Andreeva, A.; Florentino, L.C.; Chuguransky, S.R.; Grego, T.; Hobbs, E.; Pinto, B.L.; Orr, A.; Paysan-Lafosse, T.; Ponamareva, I.; et al. InterPro: The protein sequence classification resource in 2025. *Nucleic Acids Res.* **2025**, *53*, D444–D456. [CrossRef]
63. de Castro, E.; Sigrist, C.J.; Gattiker, A.; Bulliard, V.; Langendijk-Genevaux, P.S.; Gasteiger, E.; Bairoch, A.; Hulo, N. ScanProsite: Detection of PROSITE signature matches and ProRule-associated functional and structural residues in proteins. *Nucleic Acids Res.* **2006**, *34*, W362–W365. [CrossRef]
64. Edgar, R.C. MUSCLE: A multiple sequence alignment method with reduced time and space complexity. *BMC Bioinform.* **2004**, *5*, 113. [CrossRef]
65. Capella-Gutierrez, S.; Silla-Martinez, J.M.; Gabaldon, T. trimAl: A tool for automated alignment trimming in large-scale phylogenetic analyses. *Bioinformatics* **2009**, *25*, 1972–1973. [CrossRef]
66. Stamatakis, A. RAxML version 8: A tool for phylogenetic analysis and post-analysis of large phylogenies. *Bioinformatics* **2014**, *30*, 1312–1313. [CrossRef]
67. Marcais, G.; Delcher, A.L.; Phillippy, A.M.; Coston, R.; Salzberg, S.L.; Zimin, A. MUMmer4: A fast and versatile genome alignment system. *PLoS Comput. Biol.* **2018**, *14*, e1005944. [CrossRef]
68. Jia, K.H.; Wang, Z.X.; Wang, L.; Li, G.Y.; Zhang, W.; Wang, X.L.; Xu, F.J.; Jiao, S.Q.; Zhou, S.S.; Liu, H.; et al. SubPhaser: A robust allopolyploid subgenome phasing method based on subgenome-specific k-mers. *New Phytol.* **2022**, *235*, 801–809. [CrossRef]
69. Emms, D.M.; Kelly, S. OrthoFinder: Phylogenetic orthology inference for comparative genomics. *Genome Biol.* **2019**, *20*, 238. [CrossRef]
70. Zhang, C.; Rabiee, M.; Sayyari, E.; Mirarab, S. ASTRAL-III: Polynomial time species tree reconstruction from partially resolved gene trees. *BMC Bioinform.* **2018**, *19*, 153. [CrossRef]
71. Letunic, I.; Bork, P. Interactive Tree of Life (iTOL) v6: Recent updates to the phylogenetic tree display and annotation tool. *Nucleic Acids Res.* **2024**, *52*, W78–W82. [CrossRef] [PubMed]
72. Chen, C.; Wu, Y.; Li, J.; Wang, X.; Zeng, Z.; Xu, J.; Liu, Y.; Feng, J.; Chen, H.; He, Y.; et al. TBtools-II: A "one for all, all for one" bioinformatics platform for biological big-data mining. *Mol. Plant* **2023**, *16*, 1733–1742. [CrossRef] [PubMed]
73. Xie, Y.; Zheng, Y.; Li, H.; Luo, X.; He, Z.; Cao, S.; Shi, Y.; Zhao, Q.; Xue, Y.; Zuo, Z.; et al. GPS-Lipid: A robust tool for the prediction of multiple lipid modification sites. *Sci. Rep.* **2016**, *6*, 28249. [CrossRef]
74. Thumuluri, V.; Almagro Armenteros, J.J.; Johansen, A.R.; Nielsen, H.; Winther, O. DeepLoc 2.0: Multi-label subcellular localization prediction using protein language models. *Nucleic Acids Res.* **2022**, *50*, W228–W234. [CrossRef]
75. Qiao, X.; Li, Q.; Yin, H.; Qi, K.; Li, L.; Wang, R.; Zhang, S.; Paterson, A.H. Gene duplication and evolution in recurring polyploidization-diploidization cycles in plants. *Genome Biol.* **2019**, *20*, 38. [CrossRef]
76. Chen, S. Ultrafast one-pass FASTQ data preprocessing, quality control, and deduplication using fastp. *Imeta* **2023**, *2*, e107. [CrossRef]
77. Kim, D.; Langmead, B.; Salzberg, S.L. HISAT: A fast spliced aligner with low memory requirements. *Nat. Methods* **2015**, *12*, 357–360. [CrossRef]
78. Pertea, M.; Pertea, G.M.; Antonescu, C.M.; Chang, T.C.; Mendell, J.T.; Salzberg, S.L. StringTie enables improved reconstruction of a transcriptome from RNA-seq reads. *Nat. Biotechnol.* **2015**, *33*, 290–295. [CrossRef]
79. Shao, M.; Kingsford, C. Accurate assembly of transcripts through phase-preserving graph decomposition. *Nat. Biotechnol.* **2017**, *35*, 1167–1169. [CrossRef]
80. Pertea, G.; Pertea, M. GFF Utilities: GffRead and GffCompare. *F1000Research* **2020**, *9*, ISCB. [CrossRef]
81. Kang, Y.J.; Yang, D.C.; Kong, L.; Hou, M.; Meng, Y.Q.; Wei, L.; Gao, G. CPC2: A fast and accurate coding potential calculator based on sequence intrinsic features. *Nucleic Acids Res.* **2017**, *45*, W12–W16. [CrossRef] [PubMed]
82. Lian, Q.; Huettel, B.; Walkemeier, B.; Mayjonade, B.; Lopez-Roques, C.; Gil, L.; Roux, F.; Schneeberger, K.; Mercier, R. A pangenome of 69 *Arabidopsis thaliana* accessions reveals a conserved genome structure throughout the global species range. *Nat. Genet.* **2024**, *56*, 982–991. [CrossRef]
83. Jayakodi, M.; Lu, Q.; Pidon, H.; Rabanus-Wallace, M.T.; Bayer, M.; Lux, T.; Guo, Y.; Jaegle, B.; Badea, A.; Bekele, W.; et al. Structural variation in the pangenome of wild and domesticated barley. *Nature* **2024**, *636*, 654–662. [CrossRef]

84. Jiao, C.; Xie, X.; Hao, C.; Chen, L.; Xie, Y.; Garg, V.; Zhao, L.; Wang, Z.; Zhang, Y.; Li, T.; et al. Pan-genome bridges wheat structural variations with habitat and breeding. *Nature* **2025**, *637*, 384–393. [CrossRef]
85. Yu, Y.; Zhang, H.; Long, Y.; Shu, Y.; Zhai, J. Plant Public RNA-seq Database: A comprehensive online database for expression analysis of ~45,000 plant public RNA-Seq libraries. *Plant Biotechnol. J.* **2022**, *20*, 806–808. [CrossRef]

plants

MDPI

Article

Integrative Transcriptomic and Evolutionary Analysis of Drought and Heat Stress Responses in *Solanum tuberosum* and *Solanum lycopersicum*

Eugeniya I. Bondar [1,2,*], Ulyana S. Zubairova [3,4], Aleksandr V. Bobrovskikh [3,5] and Alexey V. Doroshkov [1,3,*]

1 Institute of Fundamental Biology and Biotechnology, Siberian Federal University, 660041 Krasnoyarsk, Russia
2 Federal Research Center "Krasnoyarsk Science Center of the Siberian Branch of the Russian Academy of Sciences", 660036 Krasnoyarsk, Russia
3 The Federal Research Center Institute of Cytology and Genetics, Siberian Branch of the Russian Academy of Sciences, 630090 Novosibirsk, Russia; ulyanochka@bionet.nsc.ru (U.S.Z.); avb@bionet.nsc.ru (A.V.B.)
4 Department of Information Technologies, Novosibirsk State University, 630090 Novosibirsk, Russia
5 Department of Natural Sciences, Novosibirsk State University, 630090 Novosibirsk, Russia
* Correspondence: ebondar@sfu-kras.ru (E.I.B.); ad@bionet.nsc.ru (A.V.D.)

Abstract: Abiotic stresses such as drought and heat severely constrain the growth and productivity of Solanaceae crops, including potato (*Solanum tuberosum* L.) and tomato (*Solanum lycopersicum* L.), yet the conserved regulatory mechanisms underlying their stress adaptation remain incompletely understood. Here, we performed an integrative meta-analysis of publicly available transcriptomic datasets, complemented by comparative and evolutionary analyses across the *Solanum* genus. Functional annotation revealed coordinated transcriptional reprogramming characterized by induction of protective processes, including molecular chaperone activity, oxidative stress responses, and immune signaling, accompanied by repression of photosynthetic and primary metabolic pathways, reflecting energy reallocation under stress conditions. Promoter motif and transcription factor enrichment analyses implicated the bZIP, bHLH, DOF, and BBR/BPC families as central regulators of drought- and heat-induced transcriptional programs. Orthogroup inference and K_a/K_s analysis across representative *Solanum* species demonstrated a predominance of purifying selection, indicating evolutionary conservation of regulatory network architecture. Integration of motif occurrence, co-expression profiles, and protein–protein interaction data enabled reconstruction of regulatory networks and identification of conserved hub transcription factors coordinating stress responses. Comparative analysis revealed distinct but conserved transcriptional signatures for heat and drought shared between potato and tomato, indicative of conserved abiotic stress strategies across Solanaceae.

Keywords: potato; tomato; gene regulatory networks; transcriptomic meta-analysis; orthologous genes; differential gene expression; transcription factors; co-expression analysis; stress adaptation; systems biology

1. Introduction

Solanaceous crops, particularly *Solanum tuberosum* (potato) and *Solanum lycopersicum* (tomato), represent cornerstone species in global agriculture, ranking among the ten most economically and nutritionally significant food crops. Beyond their economic relevance, these species are key models for dissecting the genetic and physiological mechanisms underlying plant responses to environmental constraints due to their well-characterized genomes and extensive agronomic data.

However, global climate change represents an escalating challenge to the stability of *Solanum* crop yields. Projections suggest that an increase in mean global temperatures by 1–7 °C by the end of the century will profoundly alter precipitation regimes and regional weather patterns [1,2]. In drought-prone regions, this shift is expected to exacerbate water scarcity and intensify abiotic stress, particularly in drought-sensitive crops such as potato [3,4]. These environmental perturbations are predicted to diminish tuber yield and quality, reducing dry matter accumulation while increasing the content of reducing sugars [5]. Concurrently, increased evapotranspiration and irregular rainfall will heighten crop water demand, collectively imposing prolonged and more frequent drought episodes that threaten agricultural productivity and food security.

Potato plants (*Solanum tuberosum*) are particularly susceptible to both drought and heat stresses [6,7], mainly due to its shallow, adventitious root system [8] with low water absorption ability, and adaptation to moderate temperatures [9–11]. The drought itself inhibits plant growth [12], shortens growing period [13], and reduces amount [14] and size [15] of the tubers. Heat stress that often accompanies water deficit also has critical influence on potato yield [16–18]. The optimum temperature for the initiation and bulking of the underground tubers are known to be 15–20 °C [11]. Chronic increase in temperature to 13.8–33 °C leads to decrease in ABA and auxin levels in leaves and tuber tissues, and causes reduction in the average number of tubers per plant [19], greatly affecting the productivity [20]. Heat stress delays tuber set and growth and causes tuber quality problems such as deformation and necrosis [7,21,22].

Tomato plants (*Solanum lycopersicum*) are highly susceptible to abiotic stresses such as high temperature and drought, which severely impair vegetative growth, reproductive development, and overall yield [23]. Heat stress negatively impacts tomato growth, limits nutrient availability, inhibits photosynthesis, impairs reproduction, denatures proteins, disrupts signaling pathways, damages cell membranes, and causes excessive production of reactive oxygen species [23]. All of these factors are toxic to tomato plants. High-temperature stress disrupts pollen viability, fruit set, and photosynthetic performance, leading to substantial yield losses [24,25]. One of the problems facing tomatoes under heat stress is oxidative stress and fruit set. Currently, this problem has been proposed to be addressed by using mutant parthenocarpic tomato varieties [26]. Under heat stress, the parthenocarpic genotypes studied exhibited lower electrolyte loss characteristics and lower H_2O_2 concentrations, as well as higher antioxidant activity compared to wild tomatoes under heat stress. Although thermotolerance is present among tomato varieties, the mechanisms underlying this tolerance require further study. The precise roles and interactions of signaling pathways in response to heat stress remain unclear and require systematic integration of diverse data [23]. The authors note that significant progress remains to be made in identifying mechanisms that can be modified to create plants that are resilient to current heat stress and resilient to the prospect of increased global warming. The known part of molecular basis of thermotolerance in tomato involves a hierarchical network of heat shock transcription factors (HSFs): HsfA1a acts as the master regulator initiating the early heat response, HsfA2 mediates acquired thermotolerance during prolonged exposure, while HsfB1 functions as both co-activator and repressor to fine-tune the balance between growth and protection [27,28]. Additional regulatory layers include C2H2-type zinc finger proteins and microRNAs that orchestrate transcriptional reprogramming under both individual and combined heat and drought stress [29,30]. Exogenous application of ascorbic acid and inoculation with heat-tolerant plant-growth-promoting rhizobacteria further mitigate stress damage by enhancing antioxidant defenses and stabilizing cellular homeostasis [31,32]. Together, these findings establish tomato as an important model for studying molecular and physiological adaptations to environmental stress.

Unlike visible phenotypic traits or metabolite accumulation that often can be explained by a few genetic variants, a plant's response to abiotic stress is a complex, multilayered process largely governed by transcriptional reprogramming. Although extensive progress has been made in elucidating the molecular mechanisms of stress tolerance in model plants such as *Arabidopsis thaliana*, the specific features of stress responses in *Solanum* species remain insufficiently characterized. Recent genome-wide transcriptomic studies have identified numerous candidate genes involved in responses to heat stress and water deficit; however, most individual experiments provide limited mechanistic insight, are restricted to a narrow set of genotypes, and often focus on a single environmental condition. Consequently, the broader architecture of transcriptional regulation, including gene co-expression networks and major transcription factors with their downstream targets, remains poorly understood in potato and tomato. Addressing these gaps requires an integrative analytical framework capable of comparing heterogeneous transcriptomic datasets to reveal conserved regulatory patterns and clarify shared molecular pathways underlying stress adaptation. To date, a number of transcriptomic studies have investigated changes in gene expression profiles under heat and water deficit stress conditions in both *S. tuberosum* [33–39] and *S. lycopersicum* [27,40–44]. However, comparative analyses examining conserved and species-specific transcriptional responses in these crops remain limited [45]. A systematic meta-analysis of stress-induced transcriptomes can reveal conserved regulatory patterns and clarify shared molecular pathways underlying adaptation. Previously, we demonstrated the effectiveness of this bioinformatic strategy by integrating multiple transcriptomic datasets on high-light stress in *A. thaliana*, first across five independent experiments [46] and subsequently across twenty-one datasets encompassing 58 distinct light-stress conditions [47]. These analyses collectively identified 218,000 instances of differentially expressed genes corresponding to 19,000 unique loci, delineating distinct transcriptional programs between short- and long-term light stress responses and refining the understanding of core regulatory modules. Drawing on the demonstrated effectiveness of this approach in *A. thaliana*, we reasoned that this integrative framework may provide valuable insights into conserved and species-specific transcriptional programs underlying abiotic stress adaptation in *Solanum* species.

In today's world of widespread access to diverse data, studies involving large-scale pan-omic analyses are essential for understanding the variability arising from complex interactions between genes, proteins, metabolites, and regulatory networks within species [48]. Comparative studies of the composition of essential, variable, and rare or unique components in different samples help identify new genes, proteins, and metabolites responsible for stress resistance. For example, in tomato, integrating omics data into a single, unified picture helped uncover mechanisms of resistance to biotic stress [49] and identify protein-coding genes and non-coding RNAs that could be targets for genetic editing to create pathogen-resistant varieties. More powerful methods include approaches that integrate omics data across a sample of related species [50,51]. Matthias Benoit and co-authors [52] conducted extensive work structuring pangenome data for the genus Solanum, integrating functional genomics and pangenetics. The authors demonstrated that at high levels of chromosomal synteny, gene duplication and subsequent paralogous diversification are the main obstacles to the predictability of genotype-phenotype relationships. For example, they found that the loss of the redundant paralog of the *CLAVATA3 (CLV3)* gene can be compensated for by tandem duplication followed by gene fusion, creating a single fused *CLV3* allele capable of regulating fetal organ number in conjunction with the enzymatic gene controlling the same trait [52]. Thus, when making interspecific comparisons, it is necessary to take into account both the composition of orthologs of each species and the characteristics of substitution accumulation.

In this study, we perform a meta-analysis of differential gene expression under drought and heat stress in *Solanum tuberosum* and *Solanum lycopersicum*, integrating data from several transcriptomic experiments. Furthermore, we analyze orthogroups among multiple *Solanum* species to identify conserved and divergent responses to abiotic stress, aiming to uncover core stress response modules and species-specific adaptations.

2. Results

2.1. Search and Selection of Relevant Transcriptomic Experiments for Analysis

To comprehensively investigate the transcriptomic responses of *Solanum tuberosum* and *Solanum lycopersicum* to drought and heat stress, we collected and processed all publicly available RNA-seq datasets for these species. Transcriptomic experiments were retrieved from the NCBI Sequence Read Archive (https://www.ncbi.nlm.nih.gov/sra, (accessed on 29 October 2025)) and BioProject (https://www.ncbi.nlm.nih.gov/bioproject/, (accessed on 29 October 2025)) databases using targeted searches with the keywords "drought," "water deficit," "dehydration," "heat," and "high temperature," in combination with the species taxonomy identifiers `txid4081` and `txid4113`. The resulting dataset list was manually curated to remove non-relevant, redundant, or low-quality experiments.

After the initial screening and quality control steps, a total of 450 RNA-seq samples were selected for downstream analysis. Specifically, the dataset included 48 transcriptomic samples corresponding to three independent experiments on heat stress in potato, 184 samples from six drought stress experiments in potato, 143 samples from six heat stress experiments in tomato, and 75 samples from six drought stress experiments in tomato (listed in Table 1). Detailed information on all analyzed experiments, including accession numbers, sample descriptions, and stress treatment conditions, is provided in Supplementary File S1.

The overall summary of sequencing depth, mapping statistics, and sample variance for both species under drought and heat stress is presented in Figure 1. Across all samples, the average alignment rate obtained with `Hisat2` was 92% for tomato and 77% for potato. The mean proportion of successfully assigned reads, as calculated by `featureCounts`, reached 85% for tomato and 70% for potato (Figure 1A; Supplementary Table S2). Principal component analysis (PCA) of normalized expression profiles, performed separately for each species and stress condition after batch-effect correction using the `removeBatchEffect` function from the `limma` package, revealed clear clustering of biological replicates within each dataset (Figure 1B).

2.2. Differential Gene Expression Analysis

Differential expression analysis of the combined RNA-seq datasets using `DESeq2` identified a total of 4536 significant genes in *S. tuberosum* under heat stress and 5484 under drought stress. In *S. lycopersicum*, 5527 genes were found to be significantly differentially expressed under heat stress, whereas 440 DEGs were detected under drought stress (Supplementary Table S2).

In parallel, differential expression was analyzed within each individual experiment to assess the reproducibility of transcriptional responses under comparable stress conditions (Supplementary Table S2; Figure 2A). Genes were ranked according to the frequency with which they were detected as DEGs across independent pairwise comparisons. This approach allowed us to identify groups of genes that were consistently differentially expressed across multiple experiments, reflecting stable components of the stress response network. Specifically, in *S. lycopersicum*, DEGs detected in at least 10 out of 17 pairwise comparisons under heat stress were retained. In *S. tuberosum*, DEGs identified in at least 6 out of 8 pairwise comparisons under heat stress and in at least 13 out of 18 comparisons under drought stress were selected.

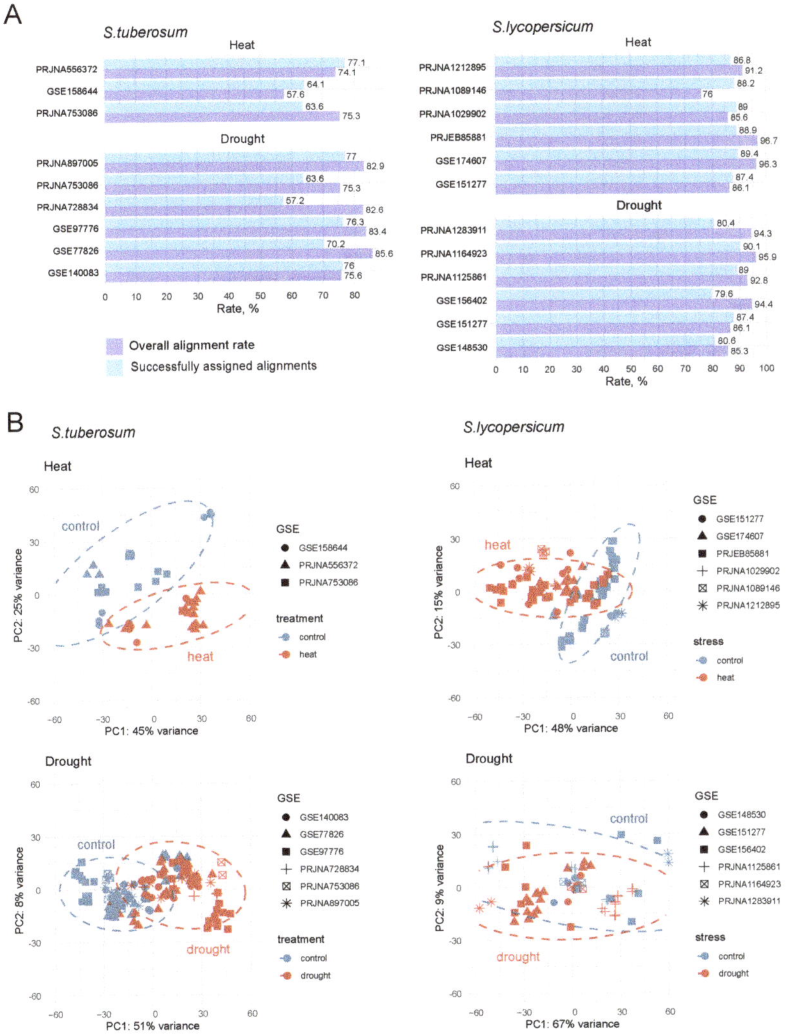

Figure 1. Summary of RNA-seq data quality and sample variance for *S. tuberosum* and *S. lycopersicum* under drought and heat stress. (**A**) Overall alignment rates obtained with `Hisat2` (pale purple) and percentages of successfully assigned reads calculated with `featureCounts` (pale blue) across all analyzed experiments. (**B**) Principal component analysis (PCA) of normalized expression data after batch effect removal using the `removeBatchEffect` function from the `limma` package. The plot displays the first and second principal components (PC1 and PC2), with the proportion of explained variance indicated on each axis. Stress-treated samples are shown in red, and control samples in pale blue.

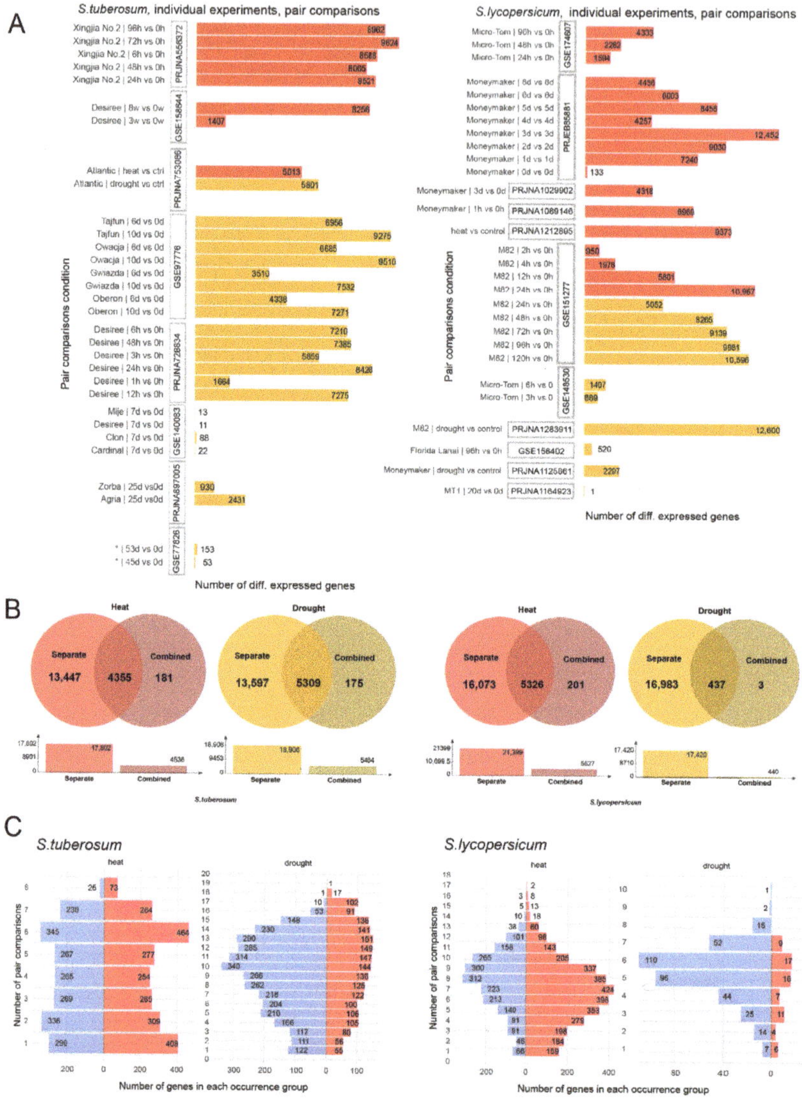

Figure 2. Differential gene expression analysis for *S. tuberosum* and *S. lycopersicum* under drought and heat stress. (**A**) Numbers of differentially expressed genes (DEGs) identified in each pairwise comparison from individual experiments. Experiments involving heat stress are shown in red, and those involving drought stress in yellow. The cultivar and exposure time are indicated on the Y-axis, and the total number of DEGs is shown at the right end of each bar. * Milva, Alegria, Desiree and Saturna cultivars were used for GSE77826 experiment. (**B**) Overlap between DEGs identified using two analytical approaches: individual-experiment analysis and combined-experiment analysis. Each Venn diagram represents one species–stress combination (*S. tuberosum* or *S. lycopersicum* under drought or heat stress). Bars below the Venn diagrams indicate the size of each DEG set (i.e., the number of genes). (**C**) Distribution of DEGs across occurrence groups. DEGs were obtained from the intersection of the two analytical approaches, and groups were defined by the number of pairwise comparisons in which a gene was detected as differentially expressed. Each group includes both upregulated (red) and downregulated (blue) genes, displayed separately in the plot. The number of pairwise comparisons is shown on the Y-axis, and the counts of up- and downregulated genes within each group are displayed at the end of the bars.

The overlap between these consistently identified DEGs from individual experiments and the DEGs obtained from the combined-analysis approach is shown in Figure 2B. This intersection was considered a validated set of differentially expressed genes, representing high-confidence stress-responsive candidates. From this validated DEG set, the top 1000 genes (500 upregulated and 500 downregulated) were selected for each species–stress combination for further analysis. Within each occurrence group, genes were ordered by the absolute value of their $\log_2$ fold change ($|LFC|$), ensuring that the most strongly regulated genes were prioritized within each group. The distribution of DEGs across occurrence groups, separated into up- and downregulated subsets, is illustrated in Figure 2C.

For *S. lycopersicum* under drought stress, only 440 DEGs were identified in the combined analysis; therefore, all of them were included in downstream analyses. Overall, integrating the two analytical strategies (Figure 2) enabled us to identify a robust and reproducible core set of stress-responsive genes while minimizing experiment-specific variability.

2.3. *Functional and Regulatory Analysis of Differentially Expressed Genes Under Drought and Heat Stress*

2.3.1. Functional Annotation of Differentially Expressed Genes

Functional enrichment analysis was performed using the `clusterProfiler` package in R (v. 4.4.1) to identify overrepresented Gene Ontology (GO) terms among the top differentially expressed genes (DEGs) in each species–stress combination. The analysis revealed distinct but partially overlapping sets of biological processes associated with upregulated and downregulated genes (Figure 3).

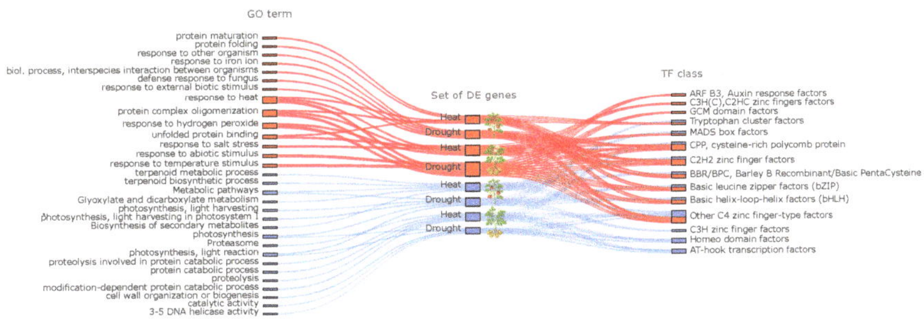

Figure 3. Sankey diagram illustrating the relationships between the top differentially expressed genes (DEGs) in each species–stress combination, their enriched Gene Ontology (GO) terms, and transcription factor (TF) classes predicted in the promoters of these DEGs. The central nodes represent eight DEG groups, defined by the combination of species (*S. tuberosum* or *S. lycopersicum*), stress type (drought or heat), and regulation direction (up- or downregulation, shown in red and blue, respectively). Links on the left connect each DEG group to significantly enriched GO terms, whereas links on the right connect DEG groups to associated TF classes.

Upregulated DEGs in both *S. tuberosum* and *S. lycopersicum* were predominantly enriched in GO terms related to cellular protection, protein quality control, and stress-responsive signaling pathways. These enriched categories can be grouped into three major functional modules:

(1) Protein quality control and protection against denaturation. Enriched terms included "protein folding," "protein maturation," "unfolded protein binding," and "protein complex oligomerization." The strong induction of chaperones (e.g., *HSP70*, *HSP90*) is a cornerstone of thermotolerance in crop plants. For *S. lycopersicum* (tomato), this response is vital for protecting the photosynthetic apparatus and reproductive development during heat waves, which are major causes of fruit set failure and yield loss [26,53]. In *S. tuberosum*

(potato), a crop highly sensitive to soil temperature fluctuations, robust protein quality control is essential for protecting the enzymatic pathways involved in starch biosynthesis and tuberization, processes that are critically disrupted by heat stress.

(2) Protection against oxidative damage and activation of stress-associated pathways. Key enriched terms included "response to hydrogen peroxide," "response to iron ion," "defense response to fungus," and "response to external biotic stimulus." This indicates a coordinated activation of antioxidant and immune defense mechanisms. The cross-talk between abiotic and biotic stress responses, as seen in the enrichment of defense-related terms, is particularly relevant for these species. In tomato, drought and heat stress can predispose plants to fungal pathogens like *Botrytis cinerea* [54–56]; the concurrent activation of these pathways may represent a pre-emptive defense priming. Furthermore, managing ROS signaling is crucial for fruit development in tomatoes [23,57], highlighting the central role of these antioxidant systems in agronomically important traits.

(3) Stress-specific adaptive responses. Processes such as "response to heat," "response to temperature stimulus," "response to salt stress," and "response to abiotic stimulus" confirm that both species deploy broad-spectrum stress tolerance mechanisms. The co-enrichment of "response to salt stress" is noteworthy, as it suggests the activation of shared osmotic adjustment pathways, which could be leveraged to improve tolerance to multiple abiotic constraints in these species through breeding.

Conversely, downregulated DEGs were enriched in categories related to primary metabolism, photosynthesis, and growth, reflecting an adaptive metabolic shift towards energy conservation under adverse conditions. Together, these results indicate that Solanaceae plants under heat and drought stress activate a coordinated transcriptional program that strengthens proteostasis, enhances redox balance, and suppresses growth-related processes in favor of survival.

2.3.2. Identification of Regulatory Motifs and Transcription Factor Classes

To elucidate potential upstream regulatory mechanisms controlling the expression of stress-responsive genes, promoter regions of the top 1000 DEGs (500 up- and 500 downregulated genes per species–stress combination) were extracted using `bedtools` [58]. *De novo* motif discovery was performed with the `XSTREME` tool from the MEME Suite, followed by functional annotation of motifs using `TOMTOM`. This approach enabled the identification of sequence motifs corresponding to known transcription factor (TF) binding sites and their classification according to TF families.

After statistical filtering based on the criterion q-value < 0.05, 305 significant motif–set associations were retained, including 119 unique motifs and 281 unique motif–set combinations. The annotated motifs were classified into 14 TF families, many of which are known regulators of stress, photomorphogenesis, and metabolic adaptation. Among the most represented families were basic helix–loop–helix (bHLH) and basic leucine zipper (bZIP) factors, consistent with their well-established roles in abiotic and biotic stress responses (Figure 3).

Several motif–TF associations were particularly notable: the *BPC6*-like motif (BBR/BPC family) was enriched in the heat-associated set of downregulated genes in *S. lycopersicum*, suggesting repression of heat-response genes; the *DOF3.4*-like motif (DOF family) appeared in the drought-associated set of upregulated genes in *S. lycopersicum*, reflecting its known involvement in drought response; a G-box–like motif (CACGTG), characteristic of bZIP, bHLH, and GBF factors, was found in the drought-associated set of downregulated group of genes; and A/T-rich motifs (e.g., TTAAAAA, TATTTTAAAT) were prevalent in the heat-associated sets of up- and downregulated genes in *S. tuberosum*, possibly indicating bZIP/ERF-like binding preferences or promoter composition bias.

Overall, the comparative analysis of promoter motifs and TF class distribution highlights a complex and multilayered regulatory landscape underlying transcriptional stress responses in Solanaceae. The Sankey diagram (Figure 3) integrates the relationships between DEG groups, enriched GO terms, and transcription factor classes, providing a unified overview of functional and regulatory modules involved in plant adaptation to drought and heat stress.

2.4. Comparative Genomic and Orthogroup Analysis of Solanaceae Species

2.4.1. Identification of Orthologous Groups

To place transcriptomic results in an evolutionary context and assess the conservation of stress-responsive genes across related taxa, a comparative genomic analysis was conducted using representative Solanaceae species and several wild potato genomes. The analysis aimed to identify orthologous gene groups, detect lineage-specific gene expansions, and characterize selective constraints influencing the evolution of stress-associated genes in *S. tuberosum* and *S. lycopersicum*.

Phylogenetic orthology was inferred using OrthoFinder [59] based on predicted proteomes from 11 plant genomes. The dataset included eight reference Solanaceae genomes retrieved from the PLAZA database (https://bioinformatics.psb.ugent.be/plaza.dev) and three wild potato genomes (*Solanum neorossii* PG6243, *S. pinnatisectum* PG1013, and *S. burkartii* PG4005) reported in [60]. Protein sequences were extracted from genome FASTA and GFF annotation files, filtered to remove incomplete or truncated transcripts, and merged into nonredundant loci. Orthogroups were computed for both the full set of *Solanum* species and a representative subset of model Solanaceae genomes (*Nicotiana tabacum*, *Petunia axillaris*, *Capsicum annuum*, *S. lycopersicum*, *S. pennellii*, and *S. tuberosum*). The complete list of analyzed species and genome sources is provided in Supplementary File S3.

The resulting orthogroup statistics and phylogenetic relationships among the 11 analyzed genomes are summarized in Figure 4, and detailed orthogroup composition data are presented in Supplementary File S4. Across all species, more than 90% of protein-coding genes were assigned to orthogroups, confirming the robustness of clustering. The total number of predicted genes varied substantially among species, from 27,183 in *P. axillaris* to 65,503 in *S. pinnatisectum*. A generally higher gene count was characteristic of *Solanum* species, as well as *N. tabacum* and *Olea europaea*, consistent with previously reported whole-genome and segmental duplications within these evolutionary lineages.

Among *Solanum* species, both cultivated and wild potatoes exhibited higher total gene counts compared with other Solanaceae members. In contrast, *S. lycopersicum* contained approximately half as many genes as *S. tuberosum* (28,988 vs. 55,905). Accordingly, the proportion of species-specific orthogroups in *S. lycopersicum* was markedly lower (0.6%) than in other *Solanum* species (2.8–3.0%). These unique orthogroups likely originated through genome rearrangements following duplication events, activation of previously non-coding regions, or gene fusion and fission processes.

This pattern suggests that homology-based inference alone may underestimate species-specific adaptations within stress-response networks, underscoring the importance of direct transcriptomic analysis in *S. tuberosum*. To further examine the representation of stress-associated genes, orthogroups containing at least one stress-related gene in any species were identified, and their distribution across taxa was analyzed (Figure 4A). The number of genes in stress-associated orthogroups correlated with total gene counts across species (Pearson's linear correlation, $p < 0.005$). Approximately one-third of all genes belonged to orthogroups that included genes differentially expressed under drought conditions, with a similar trend observed for heat stress.

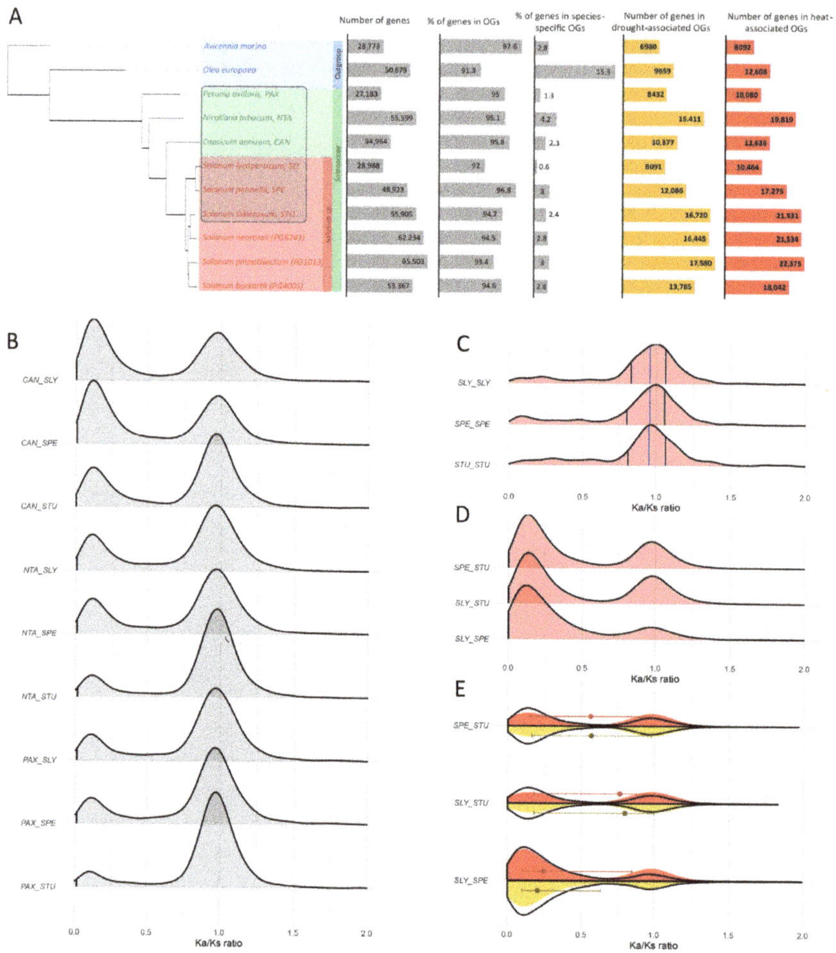

Figure 4. Phylogenetic relationships, orthogroup statistics, and K_a/K_s ratio distributions for selected plant genomes. (**A**) Phylogenetic relationships and orthogroup statistics for 11 plant genomes analyzed using OrthoFinder. Leaves are highlighted with colored rectangles indicating taxonomic groups (red, Solanum spp.; green, Solanaceae; blue, outgroup). On the right, five horizontal bar charts summarize gene content and orthogroup information for each species: (1) total number of genes, (2) percentage of genes assigned to orthogroups, (3) percentage of genes in species-specific orthogroups, (4) number of genes in drought-associated orthogroups, and (5) number of genes in heat-associated orthogroups. Each bar chart includes 11 bars corresponding to the species in the phylogenetic tree, aligned horizontally for direct comparison. (**B**) K_a/K_s ratio distributions for six species highlighted in panel A. (**C**) K_a/K_s ratio distributions for paralogous genes in *S. lycopersicum*, *S. pennellii*, and *S. tuberosum*. (**D**) K_a/K_s ratio distributions for orthologous genes in *S. lycopersicum*, *S. pennellii*, and *S. tuberosum*. (**E**) Split violin plot showing K_a/K_s ratios for orthologous genes based on DEGs derived from stress-associated HOGs. The red upper half of each split violin represents heat-associated genes, and the yellow lower half represents drought-associated genes. Solid black lines in each half indicate the K_a/K_s distributions for genes not belonging to stress-associated HOGs.

2.4.2. Evolutionary Dynamics and Selective Constraints in Orthologous Genes

To explore evolutionary patterns within the Solanaceae family, we compared nonsynonymous-to-synonymous substitution rate ratios (K_a/K_s) among orthologous and paralogous gene pairs across six representative species: *Capsicum annuum*, *Nicotiana tabacum*,

Petunia axillaris, *Solanum lycopersicum*, *S. pennellii*, and *S. tuberosum*. Only hierarchical orthogroups containing genes from all species were retained to ensure comparable evolutionary depth. The full list of orthogroup assignments and K_a/K_s estimates is provided in Supplementary Files S4 and S5.

The K_a/K_s distributions obtained by comparing species of the genus *Solanum* (*S. lycopersicum*, SLY; *S. pennellii*, SPE; *S. tuberosum*, STU) with more distantly related Solanaceae taxa (*C. annuum*, CAN; *N. tabacum*, NTA; *P. axillaris*, PAX) are shown in Figure 4B. Across these comparisons, the genes displayed a broad range of substitution rate ratios from values close to zero to slightly above 1.0, indicating the coexistence of strong purifying and relaxed selection regimes. However, no excess of genes with elevated nonsynonymous substitution rates ($K_a/K_s \gg 1$) was observed, suggesting the absence of widespread positive selection and supporting the predominance of conservative evolutionary dynamics across the family.

Most orthologous gene pairs showed K_a/K_s values centered around 1.0, consistent with neutral evolution, whereas a smaller subset fell within the 0.1–0.3 range, corresponding to negative selection. Within-species paralog comparisons in *S. lycopersicum* (SLY), *S. pennellii* (SPE), and *S. tuberosum* (STU) yielded unimodal distributions centered near 1.0, indicative of neutral divergence following gene duplication (Figure 4C). The distributions did not differ significantly among the three species (Wilcoxon rank-sum test, $p > 0.05$).

Pairwise comparisons between the three *Solanum* species (SLY vs SPE, STU vs SLY, SPE vs STU) revealed a distinct bimodal K_a/K_s distribution (Figure 4D), significantly different from all previous datasets (Wilcoxon test, $p < 0.05$). The first peak (0.1–0.3) corresponded to strong purifying selection, while the second peak near 1.0 reflected neutral evolution. This pattern suggests the coexistence of conserved loci essential for fundamental cellular functions and more variable genes that may contribute to adaptive diversification. Among these, the most pronounced conservation was observed for the *S. lycopersicum*–*S. pennellii* pair, which represents the shortest evolutionary distance ($p < 0.05$ compared to other species pairs).

Stress-associated orthogroups exhibited distributions similar between drought and heat stress (Figure 4E), but when compared to all other genes, showed a moderate shift toward higher K_a/K_s values in the SLY–STU pair (Wilcoxon test, $p < 0.05$). Although purifying selection remained the dominant trend, this upward shift indicates a slight relaxation of evolutionary constraints and possibly faster sequence evolution in stress-responsive genes.

In comparisons among more distant Solanaceae taxa, the entire K_a/K_s distribution shifted toward higher values (Figure 4B), reflecting the gradual weakening of selective pressure with increasing phylogenetic distance. Nevertheless, all *Solanum*-specific comparisons consistently showed dominant low K_a/K_s values, confirming that purifying selection remains the prevailing evolutionary mode within this genus.

Overall, these results demonstrate that most genes in Solanaceae evolve under purifying selection, which persists across evolutionary distances of up to approximately 7 million years, consistent with the divergence time between *S. lycopersicum* and *S. tuberosum* (TimeTree median 8.0 MYA; adjusted 7.1 MYA) [61]. Paralogous genes evolve primarily under neutral drift, whereas stress-associated genes show mild relaxation of selective pressure. This long-term evolutionary conservatism of core stress-response components provides a robust framework for cross-species inference of regulatory mechanisms and supports the stability of transcriptional and signaling pathways underlying environmental adaptation.

2.5. Gene Network Analysis of Stress Response

Integration of motif-based, co-expression, and protein–protein interaction (PPI) data resulted in four comprehensive gene regulatory networks (GRNs) representing *S. tuberosum* and *S. lycopersicum* responses to drought and heat stress (Figures 5 and 6). Functional annotations for all clusters in the reconstructed GRNs are provided in Supplementary Table S6. Below, we describe the largest clusters (containing 100 or more genes).

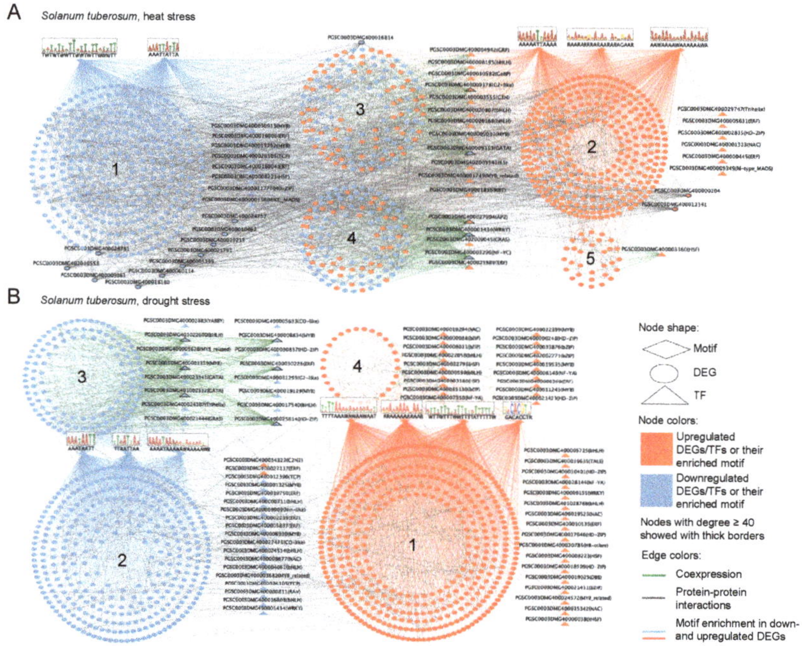

Figure 5. Integrated GRNs reconstructed for *Solanum tuberosum* under (**A**) heat and (**B**) drought stress conditions. Nodes are colored according to expression pattern: red for upregulated and blue for downregulated genes. Node shapes indicate element type: circles represent differentially expressed genes (DEGs), triangles denote transcription factors (TFs), and rhombi correspond to TF binding site motifs. Edge colors reflect the source of evidence: green for co-expression, gray for protein–protein interactions (PPIs), and red or blue for motif enrichment in up- or downregulated DEGs, respectively.

2.5.1. *Solanum tuberosum*

The reconstructed GRN of *S. tuberosum* under heat stress (Figure 5A) comprised 961 genes and 3856 interactions, including 587 co-expression links, 1664 motif-based regulatory associations, and 1605 protein–protein interactions. Four major clusters were identified; their functional annotations are summarized in Supplementary Table S6.1.

The first cluster consists of 336 genes that were predominantly downregulated in response to heat. This cluster is enriched in processes related to catabolism (GO:0009056), proteolysis involved in protein catabolism (GO:0051603), and intracellular transport (GO:0046907). Two motifs were significantly enriched in the promoter regions of genes in this cluster: `AAATTATTA` and `TWTWTWWTTWWTWTTWWWTT`.

The second cluster includes 301 genes and six transcription factors that were coordinately upregulated under heat stress: *PGSC0003DMG400000445* (AP2/ERF), *PGSC0003DMG400009349* (MADS-box protein), *PGSC0003DMG400002835* (Homeobox protein), *PGSC0003DMG400005631* (AP2/ERF), *PGSC0003DMG400029747* (Trihelix), and *PGSC0003DMG400001323* (NAC). Three motifs were significantly enriched in the promoter regions of genes in this cluster: `AAAAATTAAAA`, `RAARARRRARAARARAGAAR`, and `RAARARRRARAARARAGAAR`. Functionally, this cluster is associated with the *response to heat*

(GO:0009408), *response to oxidative stress* (GO:0006979), *macromolecule biosynthetic process* (GO:0009059), and *RNA metabolic process* (GO:0016070).

The third cluster is composed mainly of co-expression interactions linking 12 transcription factors to 140 target genes. It is enriched in the *ubiquitin-dependent protein catabolic process* (GO:0006511) and *vesicle-mediated transport* (GO:0016192). Remarkably, only one transcription factor, *PGSC0003DMG400009353* (GATA), was downregulated; this gene functions as the central hub of the cluster with 76 edges. The remaining transcription factors were upregulated, including three bHLH, two B3, and several others from different families. Overall, the cluster exhibits mixed regulation, with 67 downregulated and 73 upregulated genes.

The fourth cluster comprises five transcription factors and 124 associated genes, connected mainly through co-expression links. This cluster also displays mixed regulation (78 downregulated, 46 upregulated genes) and is functionally linked to transport processes (GO:0006810) and proteolysis (GO:0006508). Three transcription factors show high connectivity (greater than 40 edges): *PGSC0003DMG400027904* (AP2), *PGSC0003DMG400001434* (WRKY), and *PGSC0003DMG400003296* (NF-YC).

The reconstructed GRN of *S. tuberosum* under drought stress (Figure 5B) comprised 993 genes and 3498 interactions, including 450 co-expression links, 2668 motif-associated connections, and 380 protein-protein interactions. Six major clusters were identified, and their functional annotations are provided in Supplementary Table S6.2.

The first cluster is upregulated and includes 430 genes together with 17 transcription factors from diverse families. Most interactions in this cluster are motif-enriched, with four significantly overrepresented promoter motifs: `GACACGTR`, `RRAAAAAARAAR`, `TTTTAAAWAWAAWAAT`, and `WTTWTTTWWTTTTATTTTTW`. Functionally, the cluster is associated with protein metabolism (GO:0019538), response to stress (GO:0006950), and transport processes (GO:0006810).

The second cluster is downregulated and contains 340 genes and 17 transcription factors. Three motifs were significantly enriched in the promoter regions of genes in this cluster: `AAATAATT`, `TTAATTAA`, and `AAAATAAAAAWAAAAAWR`. This cluster is associated with *protein modification* (GO:0036211), *signal transduction and cell communication* (GO:0007165, GO:0007154), *response to hormone* (GO:0009725), and *response to light stimulus* (GO:0009416).

The third cluster, defined primarily by co-expression interactions, is also downregulated and includes 124 genes and 16 transcription factors, six of which act as central hubs (greater than 40 edges): *PGSC0003DMG401002532* (GATA), *PGSC0003DMG400025614* (HD-ZIP), *PGSC0003DMG401022600* (bHLH), *PGSC0003DMG400008834* (MYB), *PGSC0003DMG400030228* (ERF), and *PGSC0003DMG400001339* (MYB). This cluster is functionally enriched for protein phosphorylation (GO:0006468), developmental processes (GO:0032502), and cell wall organization or biogenesis (GO:0071554).

The enriched biological processes in heat GRN reflect well-known heat stress response: protein catabolism and proteolysis suggesting degradation of damaged proteins, oxidative stress response and RNA metabolism indicating cellular stress adaptation, ubiquitin-dependent pathways highlighting the need of proteins quality control, and transport processes which could facilitate metabolite and signal movement [62,63]. Identified TFs families such as AP2/ERF, NAC, WRKY, GATA, bHLH, MYB) are well-known in plant abiotic stress signaling. For instance, AP2/ERF and NAC families regulate stress-responsive gene expression widely under heat and drought, while WRKY factors often modulate oxidative stress and defense pathways [62–64]. The reconstructed GRN under drought stress captures the integration of drought signaling, hormonal regulation, structural and metabolic adaptation consistent with established molecular drought response in potato [65–67]. Under drought stress, involvement of TFs from such families as HD-ZIP, ERF, MYB, GATA indicate complex transcriptional networks controlling drought tolerance, hormone responses, protein metabolism, phosphorylation, and cell wall remodeling. These processes align with known drought adaptation mechanisms in potato and other plants [62]. Overall, the reconstructed GRNs reveal coordinated transcriptional programs integrating metabolic, hormonal, and developmental shifts that underpin potato's drought and heat acclimation mechanisms.

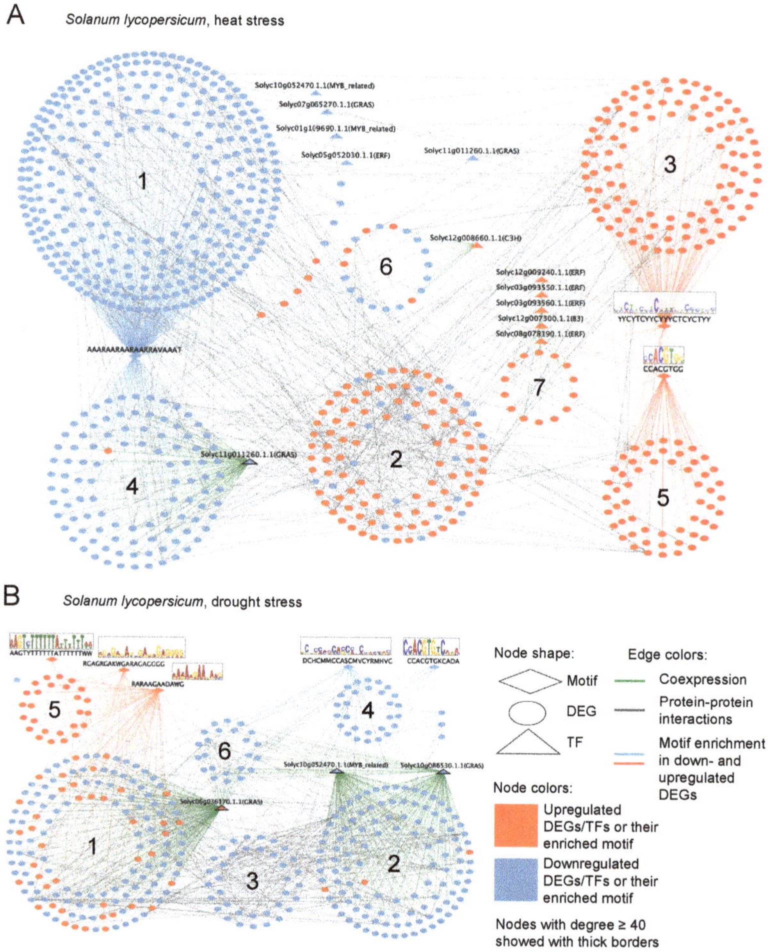

Figure 6. Integrated GRNs reconstructed for *Solanum lycopersicum* under (**A**) heat and (**B**) drought stress conditions. Nodes are colored by expression pattern: red for upregulated and blue for downregulated genes. Node shapes denote element type: circles for differentially expressed genes (DEGs), triangles for transcription factors (TFs), and rhombi for TF binding site motifs. Edge colors represent evidence type: green for co-expression, gray for protein–protein interactions (PPIs), and red or blue for motif enrichment in up- or downregulated DEGs, respectively.

2.5.2. *Solanum lycopersicum*

The reconstructed GRN of *S. lycopersicum* under **heat stress** (Figure 6A) comprised 764 genes and 1335 interactions, including 149 co-expression edges, 625 motif-associated connections, and 561 protein–protein interactions.

The first cluster consists of downregulated DEGs and includes 335 genes and four transcription factors. Genes in this cluster share significant enrichment of a promoter motif AAARAARAARAARRAVAAAT. Functionally, this cluster is enriched in metabolic processes such as *small molecule metabolism* (GO:0044281), *carbohydrate metabolism* (GO:0005975), *lipid metabolism* (GO:0006629), and *terpenoid metabolism* (GO:0016114).

The second cluster contains 111 genes linked predominantly by protein–protein interactions according to STRING data. No specific transcription factors or regulatory motifs were identified for this cluster. It exhibits mixed regulation, with a predominance of upregulated genes (90 out of 111). Functionally, several genes in this cluster are associated with *biosynthetic processes* (GO:0009058), *gene expression* (GO:0010467), and *protein folding* (GO:0006457).

The third cluster, comprising 111 genes, is primarily upregulated. Genes in this cluster share a significantly enriched promoter motif YYCYTCYYCYYYCTCYCTYY. Seven genes within this cluster are associated with *RNA modification* (GO:0009451).

The fourth cluster, similar to the first one, is downregulated and linked to the same regulatory motif AAARAARAARAARRAVAAAT. However, it shows pronounced co-expression with the transcription factor *Solyc11g011260.1.1* (GRAS family). Genes in this cluster are enriched for *small molecule metabolism* (GO:0044281), *cellular response to chemical stimulus* (GO:0070887), and *cellular homeostasis* (GO:0019725).

The reconstructed network of *S. lycopersicum* under **drought stress** (Figure 6B) comprised 387 genes and 891 interactions, including 420 co-expression, 165 motif-enriched, and 306 protein–protein associations.

The first cluster exhibits mixed regulation and contains 151 genes co-expressed with the transcription factor *Solyc06g036170.1.1* (GRAS). The majority of genes in this cluster are downregulated (110 DEGs). Functional enrichment indicates involvement in metabolic processes including *small molecule metabolism* (GO:0044281), *carbohydrate metabolism* (GO:0005975), and *amino acid metabolism* (GO:0006520).

The second cluster is downregulated and comprises 116 genes co-expressed with two transcription factors: *Solyc10g052470.1.1* (MYB-related) and *Solyc10g086530.1.1* (GRAS). This cluster is enriched for genes associated with *phosphorus metabolism* (GO:0006793) and *photosynthesis* (GO:0015979).

The reconstructed GRN under heat stress highlights the suppression of metabolic pathways (carbohydrate, lipid, terpenoid, and small molecule). Additionally, downregulation linked with cellular response to chemical stimuli and homeostasis, suggesting a shift in basal metabolism and stress adaptation which coexpressed with a GRAS family TF. Upregulated clusters include genes related to biosynthesis, gene expression, protein folding, and RNA modification, reflecting active stress mitigation and repair responses under heat stress. This regulatory landscape fits with the known patterns of gene expression in tomato during heat stress [68–70]. Under drought stress, tomato shows a predominantly downregulated GRN with clusters coexpressed with TFs from GRAS and MYB-related families which linked to metabolism (carbohydrate, amino acid, phosphorus, small molecule) and photosynthesis. Overall, the drought GRN reflects tomato's adaptive transcriptional suppression of growth and metabolic processes mediated by TFs to preserve water and ensure survival [71,72]. In summary, tomato's heat stress response GRN demonstrates a balance of downregulated metabolic functions with activated protective and repair pathways, while the drought response relies on downregulation of metabolism and photosynthesis through MYB and GRAS TFs to optimize water usage and to enhance the stress endurance.

2.5.3. Orthologs in Reconstructed Gene Networks

Taken together, the integrated network analysis reveals both shared molecular mechanisms underlying drought and heat responses and distinct evolutionary strategies of *S. tuberosum* and *S. lycopersicum*. The modular and hierarchical structure of these networks provides a systems-level perspective on stress adaptation and points to potential regulatory hubs for future functional studies.

A comparative analysis of the heat-responsive gene network components (961 genes in *S. tuberosum* and 764 genes in *S. lycopersicum*) identified 72 orthologous genes (see Supplementary Table S7), of which only three exhibited opposite regulation between the two species under heat stress. Among these orthologs, 41 genes were upregulated, representing several major functional groups.

- **Heat shock proteins:** three heat shock cognate 70 kDa proteins, two heat shock protein 83, and two additional heat shock proteins, which likely contribute to protein folding and protection under elevated temperature.
- **Transcriptional and RNA metabolism regulators:** two DNA-directed RNA polymerases, a serine/arginine-rich splicing factor, a DEAD-box ATP-dependent RNA helicase, a peptidyl–prolyl isomerase, and an AAR2 protein family member, indicating active post-transcriptional control of gene expression during heat stress.
- **Signaling components:** PERK1 kinase, WRKY transcription factor C, ethylene response factor ERF4, and a calmodulin-binding protein, reflecting the integration of hormonal and calcium-dependent signaling cascades in heat adaptation.

In contrast, 28 orthologous genes were downregulated, encompassing several categories of metabolic and organelle-associated functions.

- **Protein metabolism genes:** ubiquitin carrier protein, RING zinc finger protein, cysteine protease, protein phosphatase, heat shock protein DnaJ, protein translocase, and transitional endoplasmic reticulum ATPase, suggesting a reduction in protein turnover and degradation.
- **Other metabolic processes:** S-adenosyl-methionine-sterol-C-methyltransferase, delta-8 sphingolipid desaturase, fructose-1,6-bisphosphatase, glycerate dehydrogenase, amino acid transporter, 2-deoxyglucose-6-phosphate phosphatase, and starch synthase VI, which may indicate a shift in primary metabolism toward energy conservation.
- **Organelle-associated proteins:** chloroplast ferredoxin I, protein of the chloroplast import apparatus 2, DNA-directed RNA polymerase 2B, and mitochondrial small heat shock protein, pointing to altered plastid and mitochondrial functions under prolonged heat exposure.

Similarly, the comparison of drought-responsive GRNs (993 genes in *S. tuberosum* and 387 genes in *S. lycopersicum*) identified 33 orthologous genes, with only one showing opposite regulation. The majority of these orthologous DEGs (29) were downregulated, encompassing several key functional groups:

- **Photosynthesis- and chloroplast-related genes:** two chlorophyll a/b binding proteins, thylakoid soluble phosphoprotein, tetrapyrrole-binding protein, chloroplast fructose-1,6-bisphosphatase I, fructose-bisphosphate aldolase, PTAC16, and nitrite reductase, reflecting the suppression of photosynthetic activity under drought stress.
- **Cell wall and extracellular matrix metabolism:** arabinogalactan peptide 14, methionine-rich arabinogalactan, pectate lyase, and glycosyltransferase, suggesting the remodeling of the cell wall to prevent water loss.
- **Signaling and regulatory components:** serine/threonine protein kinase, P21-rho-binding domain-containing protein, extracellular calcium-sensing receptor, and

ubiquitin-protein ligase BRE1, which indicate modulation of intracellular signaling and stress perception.

- **Amino acid and energy metabolism:** aspartate kinase, proline-rich protein, and ATP-binding protein, reflecting adaptive shifts in nitrogen and energy metabolism.
- **Lipid and membrane metabolism:** delta-9 desaturase, desaturase, and plastidial delta-12 oleate desaturase, likely associated with maintaining membrane fluidity under dehydration.
- **Antioxidant-related genes:** two germins and dihydrolipoyl dehydrogenase, implying reduced reactive oxygen species (ROS) detoxification capacity during drought.

Only three orthologous genes were upregulated under drought stress: *Triacylglycerol lipase*, *ATP-binding protein*, and *Ent-kaurene oxidase 2*, which may participate in lipid remodeling and hormonal signaling (gibberellin biosynthesis) in response to water deficit.

Overall, the presence of a shared core of stress-responsive orthologs between the two species indicates a conserved molecular framework for abiotic stress adaptation, while species-specific differences in regulatory patterns and pathway engagement reflect divergent evolutionary strategies shaped by their physiological and ecological contexts.

For instance, in heat stress responses, the expression patterns of heat shock proteins and antioxidant enzymes differ between tomato cultivars and desert species, like *Rhazya stricta* [73]. A recent comparison of the transcriptome response of *Solanum lycopersicum* and its wild relative, *Solanum pennellii* revealed that approximately 43% of orthologous genes displayed species-specific expression patterns in response to drought [74]. Even among different *Solanum tuberosum* cultivars, a total of approximately 7000 species-specific DEGs were identified in response to cold stress, indicating that cultivar-specific regulatory networks differ substantially [75]. Therefore, it is not surprising that we found approximately 10% overlap of core gene networks to be orthologous, based on the high evidence of high species specificity of transcriptomic response and the heterogeneous data available for such analysis.

Thus, the comparative analysis allowed us to describe the genetic basis for the observed general trends in resource redistribution from growth and active metabolism to maintaining plant viability. Multiple pairs of orthologous genes were identified that unidirectionally change their expression in response to the same stress factor. Under heat stress, we observe massive activation of proteostasis components (such as HSPs and others), post-transcriptional control, and stress signaling for the synthesis of protective macromolecules. Simultaneously, components associated with protein metabolism and organelle activity are suppressed, confirming the principle of "economy and protection." Under drought, components associated with photosynthesis, cell wall metabolism, and some antioxidant defenses are massively suppressed, while signaling and lipid metabolism are modulated to adapt to dehydration. The virtually complete absence of opposing regulation of orthologous genes, which are among the top 1000 differentially expressed genes in each species, emphasizes the universality of this basic strategy. Beyond this conserved framework, species-specific features were discovered: potato modulates intracellular communication and transport to a greater extent, while tomato more actively restructures primary metabolism. Regarding differential responses depending on the type of stress, a stronger activation of the oxidative response in potato can be noted in heat.

3. Discussion

Understanding orthologous gene relationships across *Solanum* species is essential for translating genomic and transcriptomic knowledge from well-studied model crops to their lesser-known relatives. Orthogroup analysis provides a robust framework to identify conserved gene families and evolutionary patterns that shape abiotic stress responses [76].

Recent comparative genome-wide studies across the genus have revealed both conserved and lineage-specific gene families involved in stress signaling, metabolism, and development [60,77–79]. These findings collectively lay the groundwork for integrating functional genomics with evolutionary biology to dissect how gene networks evolve under selective pressures imposed by the environment. Currently, extensive omics resources are available not only for cultivated potato and tomato (*S. tuberosum* and *S. lycopersicum*), but also for several of their wild relatives (*S. pinnatisectum, S. chacoense, S. jamesii, S. penelli*, and others) [60,74], which serve as valuable reservoirs of alleles conferring abiotic stress tolerance and highlight both the evolutionary and applied potential of comparative studies across the genus.

Within this broader context, our reconstructed GRNs illuminate both shared and species-specific aspects of abiotic stress response. The cross-species comparison highlights a dual pattern: on the one hand, the specificity of stress-induced transcriptional regulation within each species, and on the other, a remarkable degree of conservation among a subset of orthologous genes. Out of 105 orthologous genes identified across the networks, 101 exhibited the same direction of regulation under abiotic stress conditions. Such high concordance points to the presence of a conserved regulatory core that may function as an ancestral scaffold for stress adaptation. Notably, this core includes several transcription factors with well-established roles in plant stress signaling, including *WRKY transcription factor C* (*PGSC0003DMG400001434*/*Solyc02g088345*), *Ethylene response factor ERF4* (*PGSC0003DMG400016004*/*Solyc05g052030*), and a *DNA-binding protein* (*PGSC0003DMG402009541*/*Solyc08g006240*).

At the same time, network topology and hub composition reveal important species-specific differences. In *S. tuberosum*, the coexpression network under heat stress contained multiple hub transcription factors (*PGSC0003DMG400001434* (WRKY), *PGSC0003DMG400027904* (AP2), *PGSC0003DMG400009378* (G2-like), *PGSC0003DMG400003296* (NF-YC), and *PGSC0003DMG400009353* (GATA)), suggesting complex regulatory cross-talk between transcriptional modules. Under drought stress, the central nodes shifted toward other TF families, including *PGSC0003DMG400030228* (ERF), *PGSC0003DMG401002532* (GATA), *PGSC0003DMG400025614* (HD-ZIP), *PGSC0003DMG401022600* (bHLH), and *PGSC0003DMG400001339* (MYB), indicating dynamic reorganization of regulatory control depending on stress type. In contrast, the *S. lycopersicum* networks exhibited a more compact structure, with a single hub under heat stress (*Solyc11g011260*, GRAS) and three under drought (*Solyc06g036170*, GRAS; *Solyc10g086530*, GRAS; *Solyc10g052470*, MYB-related). Together with the overrepresented promoter motifs shared between the species, these findings point to a multi-layered transcriptional regulation involving diverse TF families, consistent with our previous results [46]. However, it is worth noting that our study predicted possible associations between genes and transcription factors, based on three complementary sources of evidence: coexpression, protein–protein interactions, and enrichment of regulatory motifs in their promoter regions. Therefore, the putative roles of these transcription factors in mediating drought and heat responses require experimental validation. Thus, while both crops activate overlapping sets of protective pathways, the architecture and hierarchy of transcriptional control appear to have diverged during evolution and domestication.

A closer look at antioxidant systems regulation shows both specialized and shared strategies that *S. lycopersicum* and *S. tuberosum* use when facing abiotic stresses. When tomato plants experience heat stress, they increase expression of certain electron transport components, particularly 2Fe-2S ferredoxin-type and chloroplastic ferredoxin-1 proteins, along with gamma-glutamylcyclotransferase. This suggests an adaptive approach to managing photosynthetic electron flow while maintaining glutathione homeostasis during thermal stress [80]. This compartmentalized response fits with the fundamental principle that plant antioxidant systems are highly organized across different cellular spaces, with chloroplasts serving as the main ROS production sites requiring specialized, locally-acting defense mechanisms [81]. Interestingly, many classical ROS-scavenging enzymes, including glutathione reductase, multiple thioredoxins, and COQ6, are actually downregulated during heat stress. This might indicate a shift from eliminating ROS toward using ROS as signaling molecules that help fine-tune stress responses [82]. This nuanced regulation appears evolutionarily conserved, as shown by orthologous genes behaving similarly across species. For example, chloroplastic ferredoxin I (*Solyc01g103920* and *PGSC0003DMG400004532*) is consistently upregulated under heat in both tomato and potato, highlighting its fundamental role in redox buffering during thermal challenges. In contrast, drought stress triggers a different antioxidant profile: both species strongly increase expression of specific peroxidases, *Solyc02g092580* in tomato and *PGSC0003DMG401001731* in potato, emphasizing peroxidase-mediated hydrogen peroxide clearance as a central drought tolerance mechanism [83]. The stress-specific nature of these responses becomes even clearer when looking at repressed genes, which show minimal overlap between heat and drought conditions. This suggests plants have distinct redox remodeling programs for different environmental challenges [84]. Notably, our analysis reveals three evolutionarily conserved, stress-responsive orthologs with indirect antioxidant functions, germin-like proteins and dihydrolipoyl dehydrogenase, that are consistently downregulated under drought across both species. This pattern could represent strategic metabolic trade-offs that prioritize resource allocation toward immediate fight against ROS and maintaining full defense capacity [85]. Therefore, heat stress seems to favor redox signaling and electron redistribution pathways, while drought emphasizes enzymatic ROS detoxification. This sophisticated stress-specific tuning likely underlies the different growth resilience and ROS management observed between species facing distinct environmental challenges.

In the context of our comparative genomic and transcriptomic analyses, we also observed marked evolutionary divergence among paralogous genes in *S. lycopersicum*, *S. pennellii*, and *S. tuberosum* (Figure 4C), contrasted by strong evolutionary conservation among orthologs (Figure 4D), particularly between *S. lycopersicum* and *S. pennellii*. The number of stress-induced orthogroups varied by a factor of 1.5–2 between these species (Figure 4A), while the overlap of orthologs within the most relevant DEG-based regulatory networks of heat and drought response between *S. lycopersicum* and *S. tuberosum* did not exceed 7–10% (Supplementary Table S7). Furthermore, pronounced differences were detected in the composition of regulatory motifs and transcription factor families contributing to these networks. Taken together, these results indicate that genomic divergence within Solanaceae and the diversification of their molecular regulatory systems both contribute substantially, and possibly to a comparable extent, to the evolution of stress response mechanisms at the transcriptomic level. Therefore, a comprehensive systems-level approach that integrates regulatory, functional, and evolutionary perspectives represents the most promising strategy for uncovering the molecular foundations of stress adaptation in this plant family.

Placing our findings in a broader evolutionary context, recent pan-genomic analyses of the *Solanum* genus have demonstrated that extensive gene duplication and subsequent paralogue diversification are key drivers of transcriptional and functional divergence among related crops [52]. Even between the closely related species *S. tuberosum* and *S. lycopersicum*, such paralogous expansions can obscure genotype-to-phenotype relationships, providing a likely explanation for the species-specific transcriptional architectures observed in our study. Complementary evidence from the wild tomato *S. chilense* further suggests that drought-responsive gene networks differ in evolutionary age and transcriptional stability [86]. Core regulatory modules associated with cell cycle control are evolutionarily ancient and highly conserved, while metabolic and signaling networks are of more recent origin and display greater variability. Together, these insights indicate that the evolution of abiotic stress responses in *Solanum* involves both structural diversification through gene duplication and functional divergence in the timing and dynamics of stress-related network evolution. This combination of conserved regulatory backbones and flexible peripheral modules likely underlies the balance between robustness and adaptability that characterizes abiotic stress responses in *S. tuberosum* and *S. lycopersicum*.

When viewed in the context of earlier transcriptomic studies, our results reinforce and extend existing models of stress-induced regulation in Solanaceae. In agreement with prior work, heat stress in *S. tuberosum* is characterized by the upregulation of genes related to photosynthesis, cell wall modification, RNA processing, and protein turnover [87], whereas drought stress induces genes involved in carbohydrate metabolism, flavonoid and lipid biosynthesis, and secondary metabolism [88]. For *S. lycopersicum*, functional annotations of downregulated DEGs under heat stress emphasize cellular and metabolic processes, while heat shock proteins display pronounced activation [89]. Similarly, during drought response, tomato exhibits coordinated transcriptomic and proteomic regulation involving genes associated with general stress responses, abiotic stimuli, and oxidative balance [90]. Altogether, these findings suggest that *S. tuberosum* and *S. lycopersicum* share a conserved regulatory backbone but utilize distinct network architectures to achieve stress resilience, reflecting both their evolutionary divergence and contrasting physiological strategies.

From a broader perspective, the integration of orthology inference, network reconstruction, and meta-analysis presented here provides a new comparative framework for understanding abiotic stress adaptation in the Solanaceae. By resolving stress-responsive transcriptional networks at the cross-species level, our study bridges molecular and evolutionary scales of regulation, linking conserved orthologous modules to lineage-specific regulatory innovations. This integrative approach highlights the potential of systems-level analyses to uncover evolutionary constraints and plasticity within complex stress response networks.

In summary, our study establishes a comparative and evolutionary framework for linking gene network organization to abiotic stress adaptation within the Solanaceae. By integrating large-scale transcriptomic data, orthology analysis, and systems-level reconstruction, we identify conserved molecular signatures and reveal how evolutionary diversification of regulatory networks underpins species-specific strategies of drought and heat tolerance. These results advance our understanding of the molecular evolution of stress responses in plants and provide a solid foundation for future research in functional genomics and adaptive breeding of Solanaceae crops.

4. Materials and Methods

4.1. Pipeline for Identification and Systematic Analysis of Key Differentially Expressed Genes in Solanum tuberosum and Solanum lycopersicum Under Drought and Heat Stress

For the comprehensive large-scale analysis, the data were retrieved and processed through the following interconnected stages shown in the Figure 7.

4.2. Search and Selection of Relevant Transcriptomic Experiments for Analysis

Gene Expression Omnibus (GEO) was screened to find RNA-Seq datasets related to drought/water deficit and heat/high temperature stresses. To qualify for inclusion in the meta-analysis, an expression dataset had to meet the following criteria: (1) it's raw RNA-seq reads are deposited in a public data repository; (2) it is collected under controlled conditions with replicates; (3) it is provided with metadata and method description. As a result, the total of 450 samples from 21 experiments for *S. tuberosum* and *S. lycopersicum* were selected for further analysis. All included studies, along with associated information, are listed in Table 1.

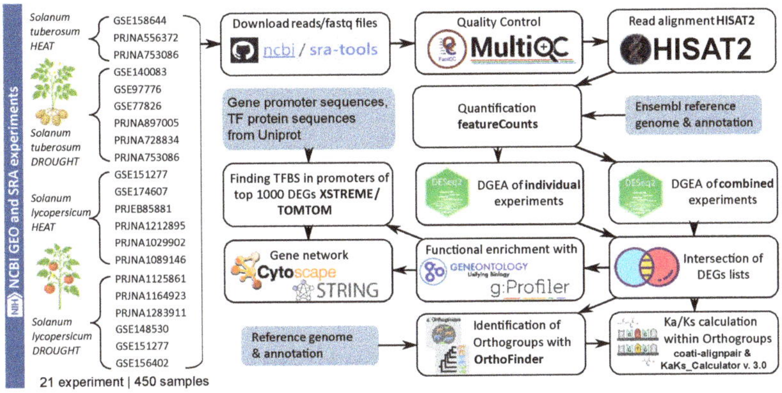

Figure 7. Overview of the collected datasets and the complete workflow used in the large-scale meta-analysis of drought and heat stress responses in *S. tuberosum* and *S. lycopersicum*. The schematic illustrates the main steps of the analysis, including data acquisition, quality control, read alignment, expression quantification, differential gene expression (DEG) analysis, functional enrichment of Gene Ontology (GO) terms, promoter motif analysis, and the construction of gene co-expression networks.

Table 1. Included studies, along with associated information.

GEO/Project ID	Number of Samples [1]	Reference
Solanum tuberosum		
heat stress:		
GSE158644	12	[39]
PRJNA556372	18	[38]
PRJNA753086	18	—
drought stress:		
GSE140083	24	[33]
GSE97776	54	[34]
GSE77826	48	[35]
PRJNA897005	20	[36]
PRJNA728834	21	[37]
PRJNA753086	17	—
Solanum lycopersicum		
heat stress:		
GSE151277	15	[41]
GSE174607	12	[91]
PRJEB85881	98	[92]
PRJNA1212895	6	—
PRJNA1029902	6	—
PRJNA1089146	6	[93]
drought stress:		
GSE148530	8	[42,43]
GSE151277	18	[41]
GSE156402	12	[94]
PRJNA1125861	27	[95]
PRJNA1164923	4	[96]
PRJNA1283911	6	—

[1] Indicates the number of samples included in the analysis (i.e., some samples were omitted due to low coverage or other issues).

4.3. RNA-Seq Data Processing

The raw reads were downloaded in FastQC format from the NCBI SRA repository (https://www.ncbi.nlm.nih.gov/geo/, accessed on 10 October 2025) using prefetch and fasterq-dump from SRA Toolkit (https://github.com/ncbi/sra-tools, accessed on 10 October 2025, version 3.1.1, NCBI, Bethesda, MD, USA). Reads quality was checked using FastQC (https://www.bioinformatics.babraham.ac.uk/projects/fastqc/, accessed on 10 October 2025, version 0.11.9, Babraham Institute, Cambridge, Great Britain) and results were aggregated with MultiQC v1.27 [97]. FastQC reports indicated that the quality of the raw sequencing data submitted to the SRA was adequate for downstream analysis; therefore, no pre-processing or trimming of the raw expression data was required.

For *Solanum tuberosum* Solanum_tuberosum.SolTub_3.0.dna_sm.toplevel.fa reference genome sequences file and Solanum_tuberosum.SolTub_3.0.57.chr.gtf gene annotation files were obtained from the Ensembl Plants database in FASTA and GFF formats. For *Solanum lycopersicum* Solanum_lycopersicum.SL3.0.dna.toplevel.fa and Solanum_lycopersicum.SL3.0.57.chr.gff3 were retrieved from the Ensembl Plants database for reference genome and gene annotation files, respectively (https://ftp.ensemblgenomes.ebi.ac.uk/pub/plants/release-57/, accessed on 10 October 2025).

Reads were mapped to the reference genomes of *S. tuberosum* and *S. lycopersicum* using Hisat2 [98] (https://daehwankimlab.github.io/hisat2/, accessed on 10 October 2025, version 2.2.1, Johns Hopkins University, Baltimore, MD, USA) with -no-unal and -no-mixed flags, and alignments were converted to BAM with samtools v.1.9 [99]. Aligned reads were quantified using featureCounts [100] with the following parameters: assign reads to all their overlapping meta-features (-O), set the minimum mapping quality score to 10 (-Q), set attribute type to 'Parent' (-g).

4.4. Differential Gene Expression Analysis

Two complementary approaches were applied to identify differentially expressed genes (DEGs) associated with drought and heat stress responses. (1) In the first approach, DEGs were identified within each individual experiment, and the resulting DEG lists were combined and ranked according to the frequency with which a gene appeared as a DEG across multiple transcriptomic experiments. (2) In the second approach, samples from all relevant experiments were merged into four combined datasets (two stress types $\times$ two species), and batch effects associated with experiment identifiers ("GSE/Project" variable) were adjusted by including this factor in the design formula. DEGs were then identified from the combined datasets.

For the first approach, pairwise comparisons were performed separately within each experiment to avoid confounding by batch effects (e.g., comparisons were made independently for different cultivars or stress durations).

For the second approach, the following design formula was applied to the DESeq2 model: design = GSE + stress, where the "GSE" term accounts for variation between projects, and the "stress" term captures the biological effect of interest. Since DESeq2 [101] uses a generalized linear model (GLM) to fit read counts and iteratively estimate β coefficients for each experimental factor, the effect of the unwanted variable ("GSE") is controlled by fitting its own β coefficient. The β coefficient corresponding to the stress factor thus represents the estimated $\log_2$ fold change (LFC).

In both approaches, DEGs were identified using the negative binomial model implemented in DESeq2 [101] on the raw gene count matrix generated by featureCounts. The contrasts were defined as "Heat vs Control" and "Drought vs Control." A $\log_2$ fold change $\geq$0.32 threshold was applied during hypothesis testing. The independent filtering of low-count genes was left at its default setting. To mitigate inflation of LFC values for lowly expressed genes, LFC shrinkage was performed using the adaptive shrinkage estimator (`type = "ashr"`). Genes with an *s*-value (estimated false sign rate) below 0.005 and an absolute $\log_2$ fold change ≥ 0.32 were considered significantly differentially expressed.

For downstream analyses, the intersection of DEGs identified by both approaches was used. For each species–stress combination, the top 1000 DEGs (top 500 upregulated and top 500 downregulated genes) from this intersection were selected. Genes were then ranked by the absolute value of their LFC within each occurrence group. Each group was defined by the number of pairwise comparisons in which a gene was identified as a DEG across individual experiments. For example, in *S. tuberosum* under drought stress, the top 500 upregulated DEGs included genes that were consistently detected as differentially expressed in multiple independent comparisons (from 19 down to 13). Within each occurrence group, genes were then sorted by decreasing absolute LFC, so that the most strongly upregulated genes appeared first.

4.5. Enrichment Analysis and Functional Annotation

Functional enrichment analysis of lists of differentially expressed genes was done with the `gprofiler2` R library (https://doi.org/10.32614/CRAN.package.gprofiler2, accessed on 10 October 2025). Genes without functional annotation (having the "gene of unknown function" mark in the annotation file) were searched in non-redundant NCBI protein database using `blastp`.

Functional enrichment analysis was performed using the `clusterProfiler` package in R (v. 4.4.1) to identify overrepresented Gene Ontology (GO) terms among the top differentially expressed genes (DEGs) in each species–stress combination.

4.6. Promoter Analysis and Motif Identification

Promoter analysis was performed to identify conserved cis-regulatory motifs potentially involved in the transcriptional regulation of drought- and heat-responsive genes in *S. tuberosum* and *S. lycopersicum*. Upstream promoter sequences were extracted from reference genome annotations using `bedtools flank` and `bedtools getfasta` [58], based on gene coordinates from Ensembl Plants release 57 (https://ftp.ensemblgenomes.ebi.ac.uk/pub/plants/release-57/, accessed on 10 October 2025).

Motif discovery was carried out using the XSTREME tool from the MEME Suite, and identified motifs were annotated with TOMTOM (https://meme-suite.org/). Only motifs with a *q*-value < 0.05 were retained for further analysis.

To associate the identified motifs with putative transcription factors (TFs), protein sequences of known TFs were retrieved from UniProt and aligned against the *S. tuberosum* and *S. lycopersicum* proteomes using `blastp` (parameters: `max_target_seqs` = 10, `evalue` = 0.01). BLAST v2.15.0+ results were filtered according to the following thresholds: minimum sequence identity of 50%, maximum difference in aligned sequence length of 30%, and minimum coverage of 50% for both query and target sequences. The resulting matches were used to infer potential regulatory associations between conserved promoter motifs and candidate stress-responsive transcription factors in both species.

4.7. Orthogroup Identification and Evolutionary Analysis

Comparative genomic analysis was performed to identify orthologous gene groups and to characterize evolutionary relationships among Solanaceae species, including wild potato accessions. Predicted proteomes from 11 plant genomes were analyzed, comprising eight reference Solanaceae genomes obtained from the PLAZA database (https://bioinformatics.psb.ugent.be/plaza.dev, accessed on 10 October 2025) and three wild potato genomes (*Solanum neorossii* PG6243, *S. pinnatisectum* PG1013, and *S. burkartii* PG4005 reported in [60]).

Protein-coding sequences were extracted using `gffread` v. 0.12.8 [102] from genome FASTA and GFF annotation files. Transcripts lacking valid start or stop codons, or containing internal stop codons, were removed, and overlapping isoforms were merged into single loci to obtain nonredundant proteome datasets.

Orthologous relationships were inferred using `OrthoFinder` v. 3.0.1b1 [59] with default parameters. Orthogroup clustering produced both standard and hierarchical orthogroup (HOG) assignments, and a species tree was inferred from concatenated multiple-sequence alignments of single-copy orthologs generated during the OrthoFinder run. Summary statistics, including total gene number, proportion of genes assigned to orthogroups, and number of species-specific orthogroups, were calculated automatically and verified using in-house R scripts (R v. 4.3.1) with the `tidyverse`, `ggtree`, and `cowplot` packages.

Selective constraints among orthologous genes were assessed using pairwise nonsyno nymous-to-synonymous substitution rate ratios (K_a/K_s) calculated with `KaKs_Calculator` v. 3.0 [103]. Coding sequences from each orthogroup were aligned with `coati-alignpair` (https://github.com/CartwrightLab/coati, accessed on 10 October 2025) using the Muse and Gaut codon model and the Gotoh alignment algorithm. Only hierarchical orthogroups containing genes from all six representative Solanaceae genomes (*Capsicum annuum, Nicotiana tabacum, Petunia axillaris, Solanum lycopersicum, S. pennellii,* and *S. tuberosum*) were retained for downstream analysis. Gene pairs with $K_s > 0.01$ or coding sequence length shorter than 300 bp were excluded to minimize artifacts due to saturation and low-quality alignments.

The resulting K_a/K_s distributions were used to evaluate patterns of purifying and neutral selection across orthogroups. All data processing, statistical analyses, and visualization were performed in the same R environment. Details on analyzed genomes, orthogroup composition, and sequence data sources are provided in Supplementary Files S3 and S4.

4.8. Gene Network Reconstruction, Analysis, and Visualization

GRNs were reconstructed for the top 500 upregulated and top 500 downregulated differentially expressed genes (DEGs) identified in *Solanum tuberosum* under drought stress and in *Solanum lycopersicum* under heat stress. For *S. lycopersicum* under drought conditions, a total of 440 DEGs were identified, including 369 downregulated genes, all of which were included in the network reconstruction. The smaller number of DEGs in this case reflects the limited sample size and the lower statistical power of available transcriptomic datasets for drought-treated *S. lycopersicum*, which constrained the detection of significant expression changes.

Three sources of evidence were integrated to build the networks:

1. **Motif-based regulatory associations.** Statistically enriched promoter motifs specifically associated with subsets of up- and downregulated genes were identified using `MEME/XSTREME` (see Section 4.6). These motifs were used to infer potential transcriptional regulatory links.
2. **Coexpression relationships.** Gene coexpression networks were reconstructed using the `WGCNA` package [104]. Adaptive correlation thresholds were applied depending on the number of experiments within each subset [47]:
    - *S. lycopersicum* (drought): threshold `0.4`, 6 experiments;
    - *S. lycopersicum* (heat): threshold `0.4`, 6 experiments;
    - *S. tuberosum* (drought): threshold `0.4`, 6 experiments;
    - *S. tuberosum* (heat): threshold `0.9`, 3 experiments.

    `WGCNA` was performed with the following settings: an unsigned network (`networkType = ''unsigned''`), soft-thresholding power chosen by the pickSoftThreshold function to attain scale-free topology, unsigned TOM (`TOMType = ''unsigned''`), average linkage hierarchical clustering for gene dendrogram construction, dynamic module detection (`cutreeDynamic`) with `deepSplit = 2` and a minimum module size of 30 genes, and module merging based on eigengene correlation > 0.75.

3. **Protein–protein interactions (PPIs).** Experimentally validated and database-curated interactions were retrieved from `STRING` (https://string-db.org/, accessed on 10 October 2025) [105]. Protein identifiers were imported using the `proteins by sequences` option, with the following parameters: evidence sources were set to `experiments` and `databases`, and the minimum interaction score was defined as `medium confidence (0.400)`.

All three interaction layers were merged into integrated GRNs and visualized in `Cytoscape` v. `3.10.3` (https://cytoscape.org/, accessed on 10 October 2025) [106]. Initial layouts were generated using the `Edge-weighted Spring Embedded` algorithm, followed by `Circular Layout` optimization to enhance cluster visualization. Clusters were defined based on connection types and prioritized as follows: (1) coexpression edges, (2) motif-based regulatory associations, and (3) PPIs for genes lacking direct regulatory links. Clusters were numbered in descending order according to the number of genes they contained. Functional enrichment analysis for each cluster was performed using Gene Ontology Biological Process terms in `ShinyGO` v. `0.85` (https://bioinformatics.sdstate.edu/go/, accessed on 10 October 2025) [107]. The complete list of enriched Gene Ontology terms identified across all four experimental conditions is provided in Supplementary File S6.

To identify orthologous genes within the reconstructed regulatory networks, we used the Ensembl BioMart Plants service (https://plants.ensembl.org/info/data/biomart/index.html, accessed on 10 October 2025) [108] and the *Ensembl Plant Genes 62* dataset.

5. Conclusions

In this study, we integrated transcriptomic and comparative genomic analyses to identify key genes and regulatory mechanisms underlying drought and heat stress responses in *Solanum tuberosum* and *Solanum lycopersicum*. By applying two complementary differential expression strategies, individual experiment analysis and combined dataset meta-analysis, we generated a robust and reproducible set of stress-responsive genes while minimizing experiment-specific variability. The integration of these approaches revealed both species-specific and conserved transcriptional responses: tomato exhibited a more focused heat-responsive expression profile, whereas potato displayed broader transcriptional activation under drought conditions. The functional annotation of differentially expressed genes demonstrated the activation of cellular protection mechanisms, including chaperone-mediated protein folding, oxidative stress mitigation, and crosstalk with immune signaling pathways. Downregulated genes were enriched in metabolic and photosynthetic processes, indicating energy reallocation toward stress tolerance and survival. Analysis of promoter motifs and transcription factor (TF) classes highlighted the central roles of bZIP, bHLH, DOF, and BBR/BPC families in regulating heat- and drought-responsive transcriptional programs, underscoring the multilayered organization of regulatory networks in Solanaceae. Comparative genomic analysis established the evolutionary framework for these transcriptional patterns. Potato genomes exhibited lineage-specific gene duplications and a higher proportion of species-specific orthogroups compared to tomato and other Solanaceae species, suggesting that recent duplication events contributed to the diversification of stress adaptation pathways. Together, these results indicate that both evolutionary diversification and transcriptional plasticity contribute to the stress resilience of Solanaceae species. Overall, our findings provide an integrated view of the molecular and evolutionary mechanisms underlying drought and heat stress responses in two major Solanaceae crops. This work establishes a reference framework for identifying candidate genes, regulatory elements, and conserved adaptive modules that can inform future functional studies and crop improvement strategies targeting environmental stress resilience. Our proposed framework for analyzing gene and gene network evolution enables large-scale prediction of key genes important for breeding and hybridization. We note the potential for further in-depth study of stress-induced orthogroups and the accumulation of substitutions within them, which will complement the overall picture of evolutionary trends. The obtained results reveal complex patterns in evolution and open new avenues for further research.

Supplementary Materials: The following supporting information can be downloaded at: https://www.mdpi.com/article/10.3390/plants14243851/s1, Supplementary File S1: Summary of the RNA-seq experiments used in this study. The metadata include accession identifiers, number of samples, sequencing layout and platform, plant cultivar, tissue type, growth conditions, duration of stress treatment, and links to the corresponding article/source of the data. Supplementary File S2: Differential gene expression analysis (DEA) results performed on Solanum tuberosum and Solanum lycopersicum datasets under drought and heat stress conditions. The file includes 12 worksheets organized as follows: 4 sheets present results of DEA for pairwise comparisons within individual experiments; 4 sheets contain results of DEA for combined datasets representing meta-analysis across experiments; and 4 sheets summarize the intersections of differentially expressed genes identified by both individual and combined analyses. Supplementary File S3: Genomes used for orthology inference. This table lists the genomes used as input for orthology inference with OrthoFinder, and provides links to the sources of reference proteoms. Supplementary File S4: Orthology inference results. This file corresponds to the OrthoFinder N0.tsv output, summarizing orthogroup and hierarchical orthogroup assignments across 11 analyzed species. Each row represents an orthogroup, and columns correspond to species included in the analysis. Entries list gene identifiers belonging to each orthogroup. Supplementary File S5: Nonsynonymous/synonymous substitution rate (K_a/K_s) substitution rate analysis. This file reports the K_a/K_s calculated using KaKs_Calculator v3.0 for a subset of six selected genomes. K_a and K_s values were computed for all pairwise alignments within orthogroups (gene composition of orthogroups can be found in Supplementary File S4). Supplementary File S6: Lists of Gene Ontology enriched terms in *S. tuberosum* and *S. lycopersicum* clusters of GRNs in response to heat and drought. Supplementary File S7: Lists of presented orthologous in reconstructed GRN in response to drought and heat stress in *S. tuberosum* and *S. lycopersicum*.

Author Contributions: Conceptualization E.I.B., A.V.B., U.S.Z. and A.V.D.; methodology, E.I.B., A.V.B., U.S.Z. and A.V.D.; formal analysis, E.I.B., A.V.B., U.S.Z. and A.V.D.; investigation, E.I.B., A.V.B., U.S.Z. and A.V.D.; writing—original draft preparation, E.I.B., A.V.B., U.S.Z. and A.V.D.; writing—review and editing, E.I.B., A.V.B., U.S.Z. and A.V.D.; visualization, E.I.B. and A.V.B.; data curation, E.I.B.; Project administration, E.I.B.; supervision, A.V.D. All authors have read and agreed to the published version of the manuscript.

Funding: This research was funded by Russian Science Foundation, grant number 24-24-00452 (https://rscf.ru/project/24-24-00452/, accessed on 10 October 2025). The APC was funded by Russian Science Foundation, grant number 24-24-00452.

Data Availability Statement: The original contributions presented in this study are included in the article/Supplementary Material. Further inquiries can be directed to the corresponding author(s).

Acknowledgments: We thank the Department of High Performance Computing at Siberian Federal University for assistance with the use of the high-performance computing cluster. We would like to thank editor and reviewers who revised this manuscript providing insightful comments.

Conflicts of Interest: The authors declare no conflicts of interest. The funders had no role in the design of the study; in the collection, analyses, or interpretation of data; in the writing of the manuscript; or in the decision to publish the results.

Abbreviations

The following abbreviations are used in this manuscript:

GEO	Gene Expression Omnibus
ROS	reactive oxygen species
DEGs	differentially expressed genes
GRN	gene regulatory network
ABA	abscisic acid
LFC	log_2 fold change
TF	transcription factor
GO	Gene Ontology
HOGs	hierarchical orthologous groups
HSF	heat shock transcription factors

References

1. The Intergovernmental Panel on Climate Change (IPCC). *Climate Change 2007: The Physical Science Basis*; Agenda; The Intergovernmental Panel on Climate Change (IPCC): Geneva, Switzerland, 2007; Volume 6, 333p.
2. Solomon, S.; Qin, D.; Manning, M.; Chen, Z.; Marquis, M.; Averyt, K.; Tignor, M.; Miller, H. Summary for Policymakers. In *Climate Change*; Cambridge University Press: Cambridge, UK, 2007; pp. 1–18.
3. Schafleitner, R.; Ramirez, J.; Jarvis, A.; Evers, D.; Gutierrez, R.; Scurrah, M. Adaptation of the potato crop to changing climates. In *Crop Adaptation to Climate Change*; Wiley: Hoboken, NJ, USA, 2011; pp. 287–297. [CrossRef]
4. Simelton, E.; Fraser, E.D.; Termansen, M.; Benton, T.G.; Gosling, S.N.; South, A.; Arnell, N.W.; Challinor, A.J.; Dougill, A.J.; Forster, P.M. The socioeconomics of food crop production and climate change vulnerability: A global scale quantitative analysis of how grain crops are sensitive to drought. *Food Secur.* **2012**, *4*, 163–179. [CrossRef]
5. Haverkort, A.; Verhagen, A. Climate change and its repercussions for the potato supply chain. *Potato Res.* **2008**, *51*, 223–237. [CrossRef]
6. Monneveux, P.; Ramírez, D.A.; Pino, M.T. Drought tolerance in potato (*S. tuberosum* L.): Can we learn from drought tolerance research in cereals? *Plant Sci.* **2013**, *205*, 76–86. [CrossRef]
7. Levy, D.; Veilleux, R.E. Adaptation of potato to high temperatures and salinity-a review. *Am. J. Potato Res.* **2007**, *84*, 487–506. [CrossRef]
8. Joshi, M.; Fogelman, E.; Belausov, E.; Ginzberg, I. Potato root system development and factors that determine its architecture. *J. Plant Physiol.* **2016**, *205*, 113–123. [CrossRef]
9. Iwama, K.; Yamaguchi, J. Abiotic stresses. In *Handbook of Potato Production, Improvement, and Postharvest Management*; CRC Press: Boca Raton, FL, USA, 2006; Chapter 7, pp. 231–278. [CrossRef]
10. Liu, C.; Li, Y.; Liu, Y.; Kear, P.; Feng, Y.; Wang, L.; Wang, D.; Luo, M.; Li, J. Effect of elevated temperature and CO_2 on growth of two early-maturing potato (*Solanum tuberosum* L.) varieties. *Clim. Smart Agric.* **2025**, *2*, 100034. [CrossRef]
11. Lee, Y.H.; Sang, W.G.; Baek, J.K.; Kim, J.H.; Shin, P.; Seo, M.C.; Cho, J.I. The effect of concurrent elevation in CO_2 and temperature on the growth, photosynthesis, and yield of potato crops. *PLoS ONE* **2020**, *15*, e0241081. [CrossRef] [PubMed]
12. Deblonde, P.; Ledent, J.F. Effects of moderate drought conditions on green leaf number, stem height, leaf length and tuber yield of potato cultivars. *Eur. J. Agron.* **2001**, *14*, 31–41. [CrossRef]
13. Kumar, S.N.; Govindakrishnan, P.; Swarooparani, D.; Nitin, C.; Surabhi, J.; Aggarwal, P. Assessment of impact of climate change on potato and potential adaptation gains in the Indo-Gangetic Plains of India. *Int. J. Plant Prod.* **2015**, *9*, 151–170. [CrossRef]
14. Eiasu, B.K.; Soundy, P.; Hammes, P.S. Response of potato (*Solarium tuberosum*) tuber yield components to gel-polymer soil amendments and irrigation regimes. *N. Z. J. Crop Hortic. Sci.* **2007**, *35*, 25–31. [CrossRef]
15. Schafleitner, R.; Gutierrez, R.; Espino, R.; Gaudin, A.; Pérez, J.; Martínez, M.; Domínguez, A.; Tincopa, L.; Alvarado, C.; Numberto, G.; et al. Field screening for variation of drought tolerance in *Solanum tuberosum* L. by agronomical, physiological and genetic analysis. *Potato Res.* **2007**, *50*, 71–85. [CrossRef]
16. Gregory, L. Physiology of tuberization in plants. (Tubers and tuberous roots.). In *Differenzierung und Entwicklung/Differentiation and Development*; Springer: Berlin/Heidelberg, Germany, 1965; pp. 1328–1354. [CrossRef]
17. Slater, J. The effect of night temperature on tuber initiation of the potato. *Eur. Potato J.* **1968**, *11*, 14–22. [CrossRef]
18. Ahn, Y.J.; Claussen, K.; Zimmerman, J.L. Genotypic differences in the heat-shock response and thermotolerance in four potato cultivars. *Plant Sci.* **2004**, *166*, 901–911. [CrossRef]

19. Guillemette, A.M.; Hernández Casanova, G.; Hamilton, J.P.; Pokorná, E.; Dobrev, P.I.; Motyka, V.; Rashotte, A.M.; Leisner, C.P. The physiological and molecular responses of potato tuberization to projected future elevated temperatures. *Plant Physiol.* **2025**, *197*, kiae664. [CrossRef] [PubMed]
20. Rykaczewska, K. The impact of high temperature during growing season on potato cultivars with different response to environmental stresses. *Am. J. Plant Sci.* **2013**, *4*, 2386–2393. [CrossRef]
21. Minhas, J.; Rawat, S.; Govindakrishnan, P.; Kumar, D. Possibilities of enhancing potato production in non-traditional areas. *Potato J.* **2011**, *38*, 14–17.
22. Asfaw, A.; Bonierbale, M.; Khan, M. Integrative breeding strategy for making climate-smart potato varieties for sub-Saharan Africa. In *Potato and Sweetpotato in Africa: Transforming the Value Chains for Food and Nutrition Security*; CABI: Wallingford, UK, 2015; pp. 134–142. [CrossRef]
23. Khan, Q.; Wang, Y.; Xia, G.; Yang, H.; Luo, Z.; Zhang, Y. Deleterious effects of heat stress on the tomato, its innate responses, and potential preventive strategies in the realm of emerging technologies. *Metabolites* **2024**, *14*, 283. [CrossRef] [PubMed]
24. Xu, J.; Wolters-Arts, M.; Mariani, C.; Huber, H.; Rieu, I. Heat stress affects vegetative and reproductive performance and trait correlations in tomato (*Solanum lycopersicum*). *Euphytica* **2017**, *213*, 156. [CrossRef]
25. Raja, V.; Qadir, S.U.; Alyemeni, M.N.; Ahmad, P. Impact of drought and heat stress individually and in combination on physio-biochemical parameters, antioxidant responses, and gene expression in Solanum lycopersicum. *3 Biotech* **2020**, *10*, 208. [CrossRef]
26. Mubarok, S.; Jadid, N.; Widiastuti, A.; Derajat Matra, D.; Budiarto, R.; Lestari, F.W.; Nuraini, A.; Suminar, E.; Pradana Nur Rahmat, B.; Ezura, H. Parthenocarpic tomato mutants, iaa9-3 and iaa9-5, show plant adaptability and fruiting ability under heat-stress conditions. *Front. Plant Sci.* **2023**, *14*, 1090774. [CrossRef]
27. Fragkostefanakis, S.; Mesihovic, A.; Simm, S.; Paupière, M.J.; Hu, Y.; Paul, P.; Mishra, S.K.; Tschiersch, B.; Theres, K.; Bovy, A.; et al. HsfA2 Controls the Activity of Developmentally and Stress-Regulated Heat Stress Protection Mechanisms in Tomato Male Reproductive Tissues. *Plant Physiol.* **2016**, *170*, 2461–2477. [CrossRef]
28. Fragkostefanakis, S.; Simm, S.; El-Shershaby, A.; Hu, Y.; Bublak, D.; Mesihovic, A.; Darm, K.; Mishra, S.K.; Tschiersch, B.; Theres, K.; et al. The repressor and co-activator HsfB1 regulates the major heat stress transcription factors in tomato. *Plant Cell Environ.* **2019**, *42*, 874–890. [CrossRef]
29. Hu, X.; Zhu, L.; Zhang, Y.; Xu, L.; Li, N.; Zhang, X.; Pan, Y. Genome-wide identification of C2H2 zinc-finger genes and their expression patterns under heat stress in tomato (*Solanum lycopersicum* L.). *PeerJ* **2019**, *7*, e7929. [CrossRef]
30. Zhou, R.; Yu, X.; Ottosen, C.; Zhang, T.; Wu, Z.; Zhao, T. Unique miRNAs and their targets in tomato leaf responding to combined drought and heat stress. *BMC Plant Biol.* **2019**, *20*, 107. [CrossRef]
31. Alayafi, A.A.M. Exogenous ascorbic acid induces systemic heat stress tolerance in tomato seedlings: Transcriptional regulation mechanism. *Environ. Sci. Pollut. Res.* **2019**, *27*, 19186–19199. [CrossRef] [PubMed]
32. Mukhtar, T.; Rehman, S.U.; Smith, D.L.; Sultan, T.; Seleiman, M.F.; Alsadon, A.A.; Amna; Ali, S.; Chaudhary, H.J.; Solieman, T.; et al. Mitigation of Heat Stress in Solanum lycopersicum L. by ACC-deaminase and Exopolysaccharide Producing *Bacillus cereus*: Effects on Biochemical Profiling. *Sustainability* **2020**, *12*, 2159. [CrossRef]
33. Barra, M.; Meneses, C.; Riquelme, S.; Pinto, M.; Lagüe, M.; Davidson, C.; Tai, H.H. Transcriptome profiles of contrasting potato (*Solanum tuberosum* L.) genotypes under water stress. *Agronomy* **2019**, *9*, 848. [CrossRef]
34. Pieczynski, M.; Wyrzykowska, A.; Milanowska, K.; Boguszewska-Mankowska, D.; Zagdanska, B.; Karlowski, W.; Jarmolowski, A.; Szweykowska-Kulinska, Z. Genomewide identification of genes involved in the potato response to drought indicates functional evolutionary conservation with Arabidopsis plants. *Plant Biotechnol. J.* **2018**, *16*, 603–614. [CrossRef]
35. Sprenger, H.; Kurowsky, C.; Horn, R.; Erban, A.; Seddig, S.; Rudack, K.; Fischer, A.; Walther, D.; Zuther, E.; Köhl, K.; et al. The drought response of potato reference cultivars with contrasting tolerance. *Plant Cell Environ.* **2016**, *39*, 2370–2389. [CrossRef]
36. Alvarez-Morezuelas, A.; Barandalla, L.; Ritter, E.; Ruiz de Galarreta, J.I. Transcriptome analysis of two tetraploid potato varieties under water-stress conditions. *Int. J. Mol. Sci.* **2022**, *23*, 13905. [CrossRef] [PubMed]
37. Jian, H.; Sun, H.; Liu, R.; Zhang, W.; Shang, L.; Wang, J.; Khassanov, V.; Lyu, D. Construction of drought stress regulation networks in potato based on SMRT and RNA sequencing data. *BMC Plant Biol.* **2022**, *22*, 381. [CrossRef]
38. Xi, Y.; Han, X.; Zhang, Z.; Joshi, J.; Borza, T.; Aqa, M.M.; Zhang, B.; Yuan, H.; Wang-Pruski, G. Exogenous phosphite application alleviates the adverse effects of heat stress and improves thermotolerance of potato (*Solanum tuberosum* L.) seedlings. *Ecotoxicol. Environ. Saf.* **2020**, *190*, 110048. [CrossRef]
39. Park, J.S.; Park, S.J.; Kwon, S.Y.; Shin, A.Y.; Moon, K.B.; Park, J.M.; Cho, H.S.; Park, S.U.; Jeon, J.H.; Kim, H.S.; et al. Temporally distinct regulatory pathways coordinate thermo-responsive storage organ formation in potato. *Cell Rep.* **2022**, *38*, 110579. [CrossRef] [PubMed]
40. Fragkostefanakis, S.; Simm, S.; Paul, P.; Bublak, D.; SCHARF, K.D.; Schleiff, E. Chaperone network composition in S olanum lycopersicum explored by transcriptome profiling and microarray meta-analysis. *Plant Cell Environ.* **2015**, *38*, 693–709. [CrossRef]

41. Wang, Q.; Zhan, X. Elucidating the role of SlBBX31 in plant growth and heat-stress resistance in tomato. *Int. J. Mol. Sci.* **2024**, *25*, 9289. [CrossRef]
42. Zhao, T.; Wu, T.; Pei, T.; Wang, Z.; Yang, H.; Jiang, J.; Zhang, H.; Chen, X.; Li, J.; Xu, X. Overexpression of SlGATA17 promotes drought tolerance in transgenic tomato plants by enhancing activation of the phenylpropanoid biosynthetic pathway. *Front. Plant Sci.* **2021**, *12*, 634888. [CrossRef] [PubMed]
43. Zhao, T.; Wu, T.; Zhang, J.; Wang, Z.; Pei, T.; Yang, H.; Li, J.; Xu, X. Genome-wide analyses of the genetic screening of C2H2-type zinc finger transcription factors and abiotic and biotic stress responses in tomato (*Solanum lycopersicum*) based on RNA-Seq data. *Front. Genet.* **2020**, *11*, 540. [CrossRef] [PubMed]
44. Todaka, D.; Quynh, D.T.N.; Tanaka, M.; Utsumi, Y.; Utsumi, C.; Ezoe, A.; Takahashi, S.; Ishida, J.; Kusano, M.; Kobayashi, M.; et al. Application of ethanol alleviates heat damage to leaf growth and yield in tomato. *Front. Plant Sci.* **2024**, *15*, 1325365. [CrossRef]
45. Benny, J.; Pisciotta, A.; Caruso, T.; Martinelli, F. Identification of key genes and its chromosome regions linked to drought responses in leaves across different crops through meta-analysis of RNA-Seq data. *BMC Plant Biol.* **2019**, *19*, 194. [CrossRef]
46. Bobrovskikh, A.V.; Zubairova, U.S.; Bondar, E.I.; Lavrekha, V.V.; Doroshkov, A.V. Transcriptomic data meta-analysis sheds light on high light response in *Arabidopsis thaliana* L. *Int. J. Mol. Sci.* **2022**, *23*, 4455. [CrossRef]
47. Bobrovskikh, A.V.; Zubairova, U.S.; Doroshkov, A.V. Identification of Key Differentially Expressed Genes in *Arabidopsis thaliana* Under Short-and Long-Term High Light Stress. *Int. J. Mol. Sci.* **2025**, *26*, 7790. [CrossRef]
48. Admas, T.; Jiao, S.; Pan, R.; Zhang, W. Pan-omics insights into abiotic stress responses: Bridging functional genomics and precision crop breeding. *Funct. Integr. Genom.* **2025**, *25*, 128. [CrossRef] [PubMed]
49. Khalil, A.; Fatima, K.; Sadaqat, M.; Tahir ul Qamar, M. Disease Resistance at the Molecular Level: Omics-Based Approaches for Tomato Defense. In *Omics Approaches for Tomato Yield and Quality Trait Improvement*; Springer: Singapore, 2025; pp. 323–345. [CrossRef]
50. Mascher, M.; Jayakodi, M.; Shim, H.; Stein, N. Promises and challenges of crop translational genomics. *Nature* **2024**, *636*, 585–593. [CrossRef] [PubMed]
51. Schreiber, M.; Jayakodi, M.; Stein, N.; Mascher, M. Plant pangenomes for crop improvement, biodiversity and evolution. *Nat. Rev. Genet.* **2024**, *25*, 563–577. [CrossRef] [PubMed]
52. Benoit, M.; Jenike, K.M.; Satterlee, J.W.; Ramakrishnan, S.; Gentile, I.; Hendelman, A.; Passalacqua, M.J.; Suresh, H.; Shohat, H.; Robitaille, G.M.; et al. *Solanum* pan-genetics reveals paralogues as contingencies in crop engineering. *Nature* **2025**, *640*, 135–145 [CrossRef]
53. Pan, C.; Yang, D.; Zhao, X.; Jiao, C.; Yan, Y.; Lamin-Samu, A.T.; Wang, Q.; Xu, X.; Fei, Z.; Lu, G. Tomato stigma exsertion induced by high temperature is associated with the jasmonate signalling pathway. *Plant Cell Environ.* **2019**, *42*, 1205–1221. [CrossRef]
54. Liu, S.; Yuan, X.; Wang, Y.; Wang, H.; Wang, J.; Shen, Z.; Gao, Y.; Cai, J.; Li, D.; Song, F. Tomato stress-associated protein 4 contributes positively to immunity against necrotrophic fungus Botrytis cinerea. *Mol. Plant-Microbe Interact.* **2019**, *32*, 566–582. [CrossRef]
55. Suzuki, N.; Katano, K. Coordination between ROS regulatory systems and other pathways under heat stress and pathogen attack. *Front. Plant Sci.* **2018**, *9*, 490. [CrossRef]
56. Chen, S.; Liu, A.; Zhang, S.; Li, C.; Chang, R.; Liu, D.; Ahammed, G.J.; Lin, X. Overexpression of mitochondrial uncoupling protein conferred resistance to heat stress and Botrytis cinerea infection in tomato. *Plant Physiol. Biochem.* **2013**, *73*, 245–253. [CrossRef]
57. Lee, K.; Kang, H. Recent insights into the physio-biochemical and molecular mechanisms of low temperature stress in tomato. *Plants* **2024**, *13*, 2715. [CrossRef]
58. Quinlan, A.R.; Hall, I.M. BEDTools: A flexible suite of utilities for comparing genomic features. *Bioinformatics* **2010**, *26*, 841–842. [CrossRef]
59. Emms, D.M.; Kelly, S. OrthoFinder: Phylogenetic orthology inference for comparative genomics. *Genome Biol.* **2019**, *20*, 238. [CrossRef]
60. Tang, D.; Jia, Y.; Zhang, J.; Li, H.; Cheng, L.; Wang, P.; Bao, Z.; Liu, Z.; Feng, S.; Zhu, X.; et al. Genome evolution and diversity of wild and cultivated potatoes. *Nature* **2022**, *606*, 535–541. [CrossRef]
61. Kumar, S.; Stecher, G.; Suleski, M.; Hedges, S.B. TimeTree: A resource for timelines, timetrees, and divergence times. *Mol. Biol. Evol.* **2017**, *34*, 1812–1819. [CrossRef] [PubMed]
62. Gul, S.; Gul, H.; Shahzad, M.; Ullah, I.; Shahzad, A.; Khan, S.U. Comprehensive analysis of potato (*Solanum tuberosum*) PYL genes highlights their role in stress responses. *Funct. Plant Biol.* **2024**, *51*, FP24094. [CrossRef] [PubMed]
63. Hamoud, Y.A.; Noman, M.; Yang, X.; Ahmed, T.; Alwutayd, K.M.; Qin, H.; Shaghaleh, H. Integrated multi-omics and transcriptomic analysis reveals key regulatory networks of heat stress responses in potato. *Plant Physiol. Biochem.* **2025**, *228*, 110204. [CrossRef] [PubMed]
64. Zhu, X.; Duan, X.; Majeed, Y.; Luo, J.; Guan, N.; Zheng, H.; Zou, H.; Jin, H.; Chen, Z.; Zhang, Y. Functional analysis of StWRKY75 gene in response to heat stress in potato (*Solanum tuberosum* L.). *Front. Plant Sci.* **2025**, *16*, 1617625. [CrossRef]

65. van Muijen, D.; Anithakumari, A.; Maliepaard, C.; Visser, R.G.; van der Linden, C.G. Systems genetics reveals key genetic elements of drought induced gene regulation in diploid potato. *Plant Cell Environ.* **2016**, *39*, 1895–1908. [CrossRef]
66. Qin, T.; Ali, K.; Wang, Y.; Dormatey, R.; Yao, P.; Bi, Z.; Liu, Y.; Sun, C.; Bai, J. Global transcriptome and coexpression network analyses reveal cultivar-specific molecular signatures associated with different rooting depth responses to drought stress in potato. *Front. Plant Sci.* **2022**, *13*, 1007866. [CrossRef]
67. Wen, S.; Jian, H.; Shang, L.; Kear, P.J.; Zhang, M.; Li, Y.; Yuan, P.; Lyu, D. Comprehensive transcriptional regulatory networks in potato through chromatin accessibility and transcriptome under drought and salt stresses. *Plant J.* **2025**, *121*, e70081. [CrossRef]
68. Balyan, S.; Rao, S.; Jha, S.; Bansal, C.; Das, J.R.; Mathur, S. Characterization of novel regulators for heat stress tolerance in tomato from Indian sub-continent. *Plant Biotechnol. J.* **2020**, *18*, 2118–2132. [CrossRef] [PubMed]
69. Mizoi, J.; Todaka, D.; Imatomi, T.; Kidokoro, S.; Sakurai, T.; Kodaira, K.S.; Takayama, H.; Shinozaki, K.; Yamaguchi-Shinozaki, K. The ability to induce heat shock transcription factor-regulated genes in response to lethal heat stress is associated with thermotolerance in tomato cultivars. *Front. Plant Sci.* **2023**, *14*, 1269964. [CrossRef]
70. Graci, S.; Barone, A. Tomato plant response to heat stress: A focus on candidate genes for yield-related traits. *Front. Plant Sci.* **2024**, *14*, 1245661. [CrossRef]
71. Shu, J.; Zhang, L.; Liu, G.; Wang, X.; Liu, F.; Zhang, Y.; Chen, Y. Transcriptome analysis and metabolic profiling reveal the key regulatory pathways in drought stress responses and recovery in tomatoes. *Int. J. Mol. Sci.* **2024**, *25*, 2187. [CrossRef]
72. Ma, Y.; Li, Y.; Wang, F.; Qing, Q.; Deng, C.; Wang, H.; Song, Y. Screening and Identification of Drought-Tolerant Genes in Tomato (*Solanum lycopersicum* L.) Based on RNA-Seq Analysis. *Plants* **2025**, *14*, 1471. [CrossRef]
73. Rurek, M.; Smolibowski, M. Variability of plant transcriptomic responses under stress acclimation: A review from high throughput studies. *Acta Biochim. Pol.* **2024**, *71*, 13585. [CrossRef] [PubMed]
74. Contreras-Riquelme, J.S.; Contreras, M.; Moyano, T.C.; Sjoberg, R.; Jimenez-Gomez, J.; Alvarez, J.M. Desert-adapted tomato Solanum pennellii exhibit unique regulatory elements and stress-ready transcriptome patterns to drought. *PLoS ONE* **2025**, *20*, e0324724. [CrossRef] [PubMed]
75. Li, X.; Zheng, Z.; Zhou, Y.; Yang, S.; Su, W.; Guo, H.; Ye, G.; Wang, J. Metabolome and transcriptome analyses reveal molecular responses of two potato (*Solanum tuberosum* L.) cultivars to cold stress. *Front. Plant Sci.* **2025**, *16*, 1543380. [CrossRef]
76. Zhang, C.; Nie, X.; Kong, W.; Deng, X.; Sun, T.; Liu, X.; Li, Y. Genome-wide identification and evolution analysis of the gibberellin oxidase gene family in six gramineae crops. *Genes* **2022**, *13*, 863. [CrossRef]
77. Huang, J.; Xu, W.; Zhai, J.; Hu, Y.; Guo, J.; Zhang, C.; Zhao, Y.; Zhang, L.; Martine, C.; Ma, H.; et al. Nuclear phylogeny and insights into whole-genome duplications and reproductive development of Solanaceae plants. *Plant Commun.* **2023**, *4*. [CrossRef]
78. Kyriakidou, M.; Anglin, N.L.; Ellis, D.; Tai, H.H.; Strömvik, M.V. Genome assembly of six polyploid potato genomes. *Sci. Data* **2020**, *7*, 88. [CrossRef]
79. Achakkagari, S.; Bozan, I.; Camargo-Tavares, J.; McCoy, H.; Portal, L.; Soto, J.; Bizimungu, B.; Anglin, N.; Manrique-Carpintero, N.; Lindqvist-Kreuze, H.; et al. The phased Solanum okadae genome and Petota pangenome analysis of 23 other potato wild relatives and hybrids. *Sci. Data* **2024**, *11*, 454. [CrossRef]
80. Liu, J.; Tang, X.; Zhang, H.; Wei, M.; Zhang, N.; Si, H. Transcriptome analysis of potato leaves under oxidative stress. *Int. J. Mol. Sci.* **2024**, *25*, 5994. [CrossRef]
81. Bobrovskikh, A.; Zubairova, U.; Kolodkin, A.; Doroshkov, A. Subcellular compartmentalization of the plant antioxidant system: An integrated overview. *PeerJ* **2020**, *8*, e9451. [CrossRef] [PubMed]
82. Mittler, R.; Zandalinas, S.I.; Fichman, Y.; Van Breusegem, F. Reactive oxygen species signalling in plant stress responses. *Nat. Rev. Mol. Cell Biol.* **2022**, *23*, 663–679. [CrossRef] [PubMed]
83. Su, P.; Sui, C.; Niu, Y.; Li, J.; Wang, S.; Sun, F.; Yan, J.; Guo, S. Comparative transcriptomic analysis and functional characterization reveals that the class III peroxidase gene *TaPRX-2A* regulates drought stress tolerance in transgenic wheat. *Front. Plant Sci.* **2023**, *14*, 1119162. [CrossRef]
84. Pardo-Hernández, M.; Arbona, V.; Simon, I.; Rivero, R.M. Specific ABA-independent tomato transcriptome reprogramming under abiotic stress combination. *Plant J.* **2024**, *117*, 1746–1763. [CrossRef]
85. Monson, R.K.; Trowbridge, A.M.; Lindroth, R.L.; Lerdau, M.T. Coordinated resource allocation to plant growth–defense tradeoffs. *New Phytol.* **2022**, *233*, 1051–1066. [CrossRef]
86. Wei, K.; Sharifova, S.; Zhao, X.; Sinha, N.; Nakayama, H.; Tellier, A.; Silva-Arias, G.A. Evolution of gene networks underlying adaptation to drought stress in the wild tomato *Solanum chilense*. *Mol. Ecol.* **2024**, *33*, e17536. [CrossRef] [PubMed]
87. Liu, B.; Kong, L.; Zhang, Y.; Liao, Y. Gene and metabolite integration analysis through transcriptome and metabolome brings new insight into heat stress tolerance in potato (*Solanum tuberosum* L.). *Plants* **2021**, *10*, 103. [CrossRef]
88. Aliche, E.B.; Gengler, T.; Hoendervangers, I.; Oortwijn, M.; Bachem, C.W.; Borm, T.; Visser, R.G.; van der Linden, C.G. Transcriptomic responses of potato to drought stress. *Potato Res.* **2022**, *65*, 289–305. [CrossRef]
89. Su, H.Z.; Ma, S.Y.; Ma, X.H.; Song, Y.; Wang, X.M.; Cheng, G.X. Transcriptome analyses show changes in heat-stress related gene expression in tomato cultivar 'Moneymaker'under high temperature. *J. Plant Biochem. Biotechnol.* **2023**, *32*, 328–337. [CrossRef]

90. Liu, M.; Zhao, G.; Huang, X.; Pan, T.; Chen, W.; Qu, M.; Ouyang, B.; Yu, M.; Shabala, S. Candidate regulators of drought stress in tomato revealed by comparative transcriptomic and proteomic analyses. *Front. Plant Sci.* **2023**, *14*, 1282718. [CrossRef] [PubMed]
91. Liu, Y.; Feng, Z.; Zhu, W.; Liu, J.; Zhang, Y. Genome-wide identification and characterization of cysteine-rich receptor-like protein kinase genes in tomato and their expression profile in response to heat stress. *Diversity* **2021**, *13*, 258. [CrossRef]
92. Sahu, A.; Psaroudakis, D.; Rolletschek, H.; Neumann, K.; Borisjuk, L.; Himmelbach, A.; Pinninti, K.; Töpfer, N.; Szymanski, J. panomiX: Investigating Mechanisms Of Trait Emergence Through Multi-Omics Data Integration. *Plant Phenomics* **2025**, 100131. [CrossRef]
93. Rosenkranz, R.R.; Vraggalas, S.; Keller, M.; Sankaranarayanan, S.; McNicoll, F.; Löchli, K.; Bublak, D.; Benhamed, M.; Crespi, M.; Berberich, T.; et al. A plant-specific clade of serine/arginine-rich proteins regulates RNA splicing homeostasis and thermotolerance in tomato. *Nucleic Acids Res.* **2024**, *52*, 11466–11480. [CrossRef]
94. Ogden, A.J.; Bhatt, J.J.; Brewer, H.M.; Kintigh, J.; Kariuki, S.M.; Rudrabhatla, S.; Adkins, J.N.; Curtis, W.R. Phloem exudate protein profiles during drought and recovery reveal abiotic stress responses in tomato vasculature. *Int. J. Mol. Sci.* **2020**, *21*, 4461. [CrossRef]
95. Bortolami, G.; de Werk, T.A.; Larter, M.; Thonglim, A.; Mueller-Roeber, B.; Balazadeh, S.; Lens, F. Integrating gene expression analysis and ecophysiological responses to water deficit in leaves of tomato plants. *Sci. Rep.* **2024**, *14*, 29024. [CrossRef]
96. Ng, W.Q.; Rejab, N.A.; Taheri, S.; Teo, C.H. Elucidating the roles of small open reading frames towards drought stress in Solanum lycopersicum. *Crop Breed. Appl. Biotechnol.* **2025**, *25*, e510625112. [CrossRef]
97. Ewels, P.; Magnusson, M.; Lundin, S.; Käller, M. MultiQC: Summarize analysis results for multiple tools and samples in a single report. *Bioinformatics* **2016**, *32*, 3047–3048. [CrossRef] [PubMed]
98. Kim, D.; Langmead, B.; Salzberg, S.L. HISAT: A fast spliced aligner with low memory requirements. *Nat. Methods* **2015**, *12*, 357–360. [CrossRef] [PubMed]
99. Danecek, P.; Bonfield, J.K.; Liddle, J.; Marshall, J.; Ohan, V.; Pollard, M.O.; Whitwham, A.; Keane, T.; McCarthy, S.A.; Davies, R.M.; et al. Twelve years of SAMtools and BCFtools. *Gigascience* **2021**, *10*, giab008. [CrossRef]
100. Liao, Y.; Smyth, G.K.; Shi, W. featureCounts: An efficient general purpose program for assigning sequence reads to genomic features. *Bioinformatics* **2014**, *30*, 923–930. [CrossRef]
101. Love, M.I.; Huber, W.; Anders, S. Moderated estimation of fold change and dispersion for RNA-seq data with DESeq2. *Genome Biol.* **2014**, *15*, 550. [CrossRef]
102. Pertea, G.; Pertea, M. GFF utilities: GffRead and GffCompare. *F1000Research* **2020**, *9*, ISCB Comm J-304. [CrossRef]
103. Zhang, Z. KaKs_Calculator 3.0: Calculating selective pressure on coding and non-coding sequences. *Genom. Proteom. Bioinform.* **2022**, *20*, 536–540. [CrossRef] [PubMed]
104. Langfelder, P.; Horvath, S. WGCNA: An R package for weighted correlation network analysis. *BMC Bioinform.* **2008**, *9*, 559. [CrossRef]
105. Szklarczyk, D.; Nastou, K.; Koutrouli, M.; Kirsch, R.; Mehryary, F.; Hachilif, R.; Hu, D.; Peluso, M.E.; Huang, Q.; Fang, T.; et al. The STRING database in 2025: Protein networks with directionality of regulation. *Nucleic Acids Res.* **2025**, *53*, D730–D737. [CrossRef]
106. Shannon, P.; Markiel, A.; Ozier, O.; Baliga, N.S.; Wang, J.T.; Ramage, D.; Amin, N.; Schwikowski, B.; Ideker, T. Cytoscape: A software environment for integrated models of biomolecular interaction networks. *Genome Res.* **2003**, *13*, 2498–2504. [CrossRef] [PubMed]
107. Ge, S.X.; Jung, D.; Yao, R. ShinyGO: A graphical gene-set enrichment tool for animals and plants. *Bioinformatics* **2020**, *36*, 2628–2629. [CrossRef]
108. Smedley, D.; Haider, S.; Ballester, B.; Holland, R.; London, D.; Thorisson, G.; Kasprzyk, A. BioMart–biological queries made easy. *BMC Genom.* **2009**, *10*, 22. [CrossRef] [PubMed]

plants

Article

Metabolic Responses of *Amaranthus caudatus* Roots and Leaves to Zinc Stress

Natalia Osmolovskaya [1,*,†], Tatiana Bilova [1,2,†], Anastasia Gurina [1], Anastasia Orlova [2], Viet D. Vu [1,3], Stanislav Sukhikh [4], Tatiana Zhilkina [5], Nadezhda Frolova [2,*], Elena Tarakhovskaya [1,6], Anastasia Kamionskaya [5] and Andrej Frolov [2]

1 Department of Plant Physiology and Biochemistry, St. Petersburg State University, 199034 St. Petersburg, Russia; bilova.tatiana@gmail.com (T.B.); anastasia.gurina@list.ru (A.G.); vuvietdung01031982@gmail.com (V.D.V.); elena.tarakhovskaya@gmail.com (E.T.)
2 Laboratory of Analytical Biochemistry and Biotechnology, K.A. Timiryazev Institute of Plant Physiology of the Russian Academy of Science, 127276 Moscow, Russia; orlova@ifr.moscow (A.O.); frolov@ifr.moscow (A.F.)
3 Coast Branch—Vietnam Russian Tropical Center, Nha Trang City 650000, Vietnam
4 Laboratory of Microbiology and Biotechnology, Immanuel Kant Baltic Federal University, 236041 Kaliningrad, Russia; ssukhikh@kantiana.ru
5 Federal Research Centre Fundamentals of Biotechnology of the Russian Academy of Science, 119071 Moscow, Russia; timofeeva.bio@gmail.com (T.Z.); akamio@fbras.ru (A.K.)
6 Vavilov Institute of General Genetics, St. Petersburg Branch, Russian Academy of Sciences, 199034 St. Petersburg, Russia
* Correspondence: n.osmolovskaya@spbu.ru (N.O.); frolovanadja@yandex.ru (N.F.)
† These authors contributed equally to this work.

Abstract: In recent decades, heavy metal pollution has become a significant environmental stress factor. Plants are characterized by high biochemical plasticity and can adjust their metabolism to ensure survival under a changing environment. Here we report, to our knowledge, the first gas chromatography-mass spectrometry (GC-MS)-based metabolomics study of Zn-induced stress responses in *Amaranthus caudatus* plants. The study was performed with root and leaf aqueous methanolic extracts after their lyophilization and sequential derivatization with methoxylamine hydrochloride and *N*-methyl-*N*-(trimethylsilyl)trifluoroacetamide. In total, 419 derivatives were detected in the samples, and 144 of them could be putatively annotated. The metabolic shifts in seven-week-old *A. caudatus* plants in response to a seven-day treatment with 300 μmol/L $ZnSO_4{\cdot}7H_2O$ in nutrient solution were organ-specific and more pronounced in roots. Most of the responsive metabolites were up-regulated and dominated by sugars and sugar acids. The revealed effects could be attributed to the involvement of these metabolites in osmotic regulation, antioxidant protection and Zn^{2+} complexation. A 59-fold up-regulation of gluconic acid in roots distinctly indicated enhanced glucose oxidation due to oxidative stress upon the Zn treatment. Gluconic acid might be further employed in Zn^{2+} complexation. Pronounced Zn-induced up-regulation of salicylic acid in roots and shoots suggested a key role of this hormone in stress signaling and activation of Zn stress tolerance mechanisms. Overall, our study provides the first insight into the general trends of Zn-induced biochemical rearrangements and main adaptive metabolic shifts in *A. caudatus*.

Keywords: *Amaranthus caudatus*; complexation of Zn^{2+} ions; GC-MS-based metabolomics; gluconate; heavy metals; metabolic adjustment; osmotic regulation; ROS scavenging; salicylate; Zn stress

1. Introduction

In nature, plants are constantly exposed to a variety of abiotic stresses, including soil contamination with heavy metals (HM) due to the progressing anthropogenic activities [1]. An excess of HM in soils ultimately leads to pronounced physiological and metabolic disturbances in crops, accompanied by a decrease in their productivity and quality [2,3]. The damaging effects of HMs are generally explained by the high reactivity and toxicity of their ions, which can be primarily manifested in their direct interactions with proteins and enhanced production of ROS [4,5]. However, plants exhibit remarkable tolerance and high potential for adaptation to environmental stressors due to their impressive biochemical plasticity [6]. Crop plants essentially differ in their tolerance to HMs in the soil, which relies on a broad range of adaptive mechanisms, including those implemented at the biochemical level [7]. Stress-induced rearrangement of cellular metabolism, also referred to as a metabolic adjustment, represents one of the most important adaptive strategies [8]. The stress-related dynamics of some selected metabolites can be addressed by standardized chromatographic methods [9]. However, in a more efficient way, stress-induced metabolic shifts can be characterized by the state-of-the-art untargeted metabolomics approaches [10,11], which provide access to the molecular mechanisms of the stress response [12].

Currently, metabolomics is recognized as the method of choice to study the complex mechanisms underlying plant responses to HM stress [13]. Because of the complexity of the plant metabolome, its full coverage cannot be achieved with only one instrumental platform [14]. Among the full array of comprehensive profiling methods, GC-MS and LC-MS are complementary powerful metabolomics techniques providing an analysis of primarily metabolites (most often as trimethylsilyl—TMS derivatives) and secondary metabolites, respectively [14,15]. To date, the metabolomics studies of the HM stress responses have focused mostly on the impact of Cd in a limited selection of plants, namely *Arabidopsis thaliana* [16,17], *Brassica juncea* [18], *Cynodon dactylon* [19], *Raphanus sativus* [15] and *Lycopersicon esculentum* [20], whereas the effects of other HMs, including such a widely spread element as Zn, have been only minimally addressed so far [21,22]. Zn naturally presents in the environment as a trace metal [23] and is an essential micronutrient mandatory for plant metabolism and growth [24,25]. For most agricultural soils, total Zn contents are usually in the range of 10–300 ppm, while the concentrations of bioavailable Zn^{2+} in bulk soil solutions are much lower and typically do not exceed 4.0 µmol/L [26]. However, in Zn-polluted soils, the content of this metal can increase dramatically, affecting principal physiological parameters and ultimately inducing strong toxicity responses in plants [24,25]. Multiple toxic manifestations of Zn excess in plants caused by disorders of photosynthesis, respiration, water status, mineral nutrition and redox balance were comprehensively reviewed recently by Kaur and Garg [25]. According to studies conducted in solution culture, the threshold concentrations of Zn^{2+}, inducing phytotoxic symptoms, ranged from 25–50 µmol/L to 250–500 µmol/L, strongly depending on plant species, developmental stage, duration of exposure and environmental conditions [23–25]. The narrow gap between the levels corresponding to Zn essentiality and toxicity represents an important feature of Zn physiology in various plant species [25].

Till now, characterization of the metabolic shifts induced by Zn stress has relied mostly on routine methods of targeted biochemical analysis. Such experiments typically covered only selected primary metabolites and addressed just a limited number of plant species [9,27,28]. Thus, the application of 0.5 mmol/L Zn^{2+} to *Triticum aestivum* and *Lactuca sativa* plants induced accumulation of a characteristic set of stress-related metabolites (proline, glycine betaine, soluble sugars, and free amino acids) [9,28]. This observation might indicate the involvement of these compounds in osmotic adjustment, which is critical for plant survival under Zn stress [9,28]. However, the same Zn dosage applied to *Brassica oleracea* resulted in pronounced down-regulation of proline and glycine betaine, while the content of GABA in plant tissues increased [9]. In sugar beet, overexposure to Zn induced essential shifts in carboxylate metabolism, which were manifested with up-regulation of organic acids, most strongly pronounced for citrate and malate [27]. This observation is in agreement with the fact that these highly efficient HM-chelators underlie one of the key mechanisms of HM tolerance in plants [29–31]. Obviously, implementation of the untargeted metabolomics approach would provide much better insight into the plant metabolic responses associated with Zn stress and the mechanisms conferring plant tolerance to this metal.

Due to their pronounced tolerance to environmental stresses and well-characterized potential for HM phytoremediation, *Amaranthus* species attract a special interest in the context of Zn stress [32–34], including Zn phytoextraction [35] and Zn phytostabilization [36]. These features of amaranth can be explained by its outstanding phenotypic plasticity, extreme adaptability to adverse environmental conditions and ability for efficient growth on agriculturally unfavorable territories [37]. Moreover, due to its high nutritional quality [38] and rich pattern of biologically active secondary metabolites [39,40], amaranth has been recognized as "a new millennium crop of nutraceutical values" [37]. Regarding the primary metabolism, amaranth species are featured with C4 type of photosynthesis [41] and enhanced production of oxalic acid [42,43]. However, to date, this genus is poorly characterized in terms of metabolic rearrangements under HM stress conditions, and comprehensive metabolomics studies of Zn-induced responses in amaranth are still lacking.

Therefore, here, we present a gas chromatography-mass spectrometry (GC-MS)-based metabolomics study addressing adaptive metabolic adjustment as a primary response of *A. caudatus* roots, young and mature leaves, to Zn stress. Such an approach seems to be adequate for a better understanding of the mechanisms underlying HM tolerance at the whole-plant level, as well as possible age-related changes in the dynamics of metabolic adjustment accompanying the tolerance onset. To address the stress-related dynamics of low-molecular-weight metabolites, we decided on GC-MS of their trimethylsylilated (TMS) derivatives. Due to the high chromatographic resolution, sensitivity and specificity of this technique, efficient and reliable detection and quantification of the metabolites critically involved in the stress-related metabolic adjustment (mono- and disaccharides, polyols, carboxylic acids, amino acids, etc.) could be achieved. We assume that these metabolites could act as osmoprotectors, chelating agents and regulators, maintaining the balance between biosynthetic and energy metabolic pathways under Zn stress.

2. Results

2.1. Physiological Responses of A. caudatus Plants to Zn-Exposure

A. caudatus plants grown in a hydroponic culture at the age of six weeks were exposed for a week to 300 μmol/L $ZnSO_4$ added to a nutrient solution. Over this period, the plants did not show any serious alterations in root and leaf morphology in comparison to the untreated controls (Supplementary Information S1, Figure S1(1,2)). The only visible Zn-related effect could be observed for the young leaves, which turned light green upon the treatment (Figure S1(1A)). This, however, was not the case with the mature (i.e., fully expanded) leaves, which remained dark green even after Zn application. The non-destructive measurements of chlorophyll content, accomplished with the 3rd mature leaf, also did not reveal any Zn-induced enhancement of chlorophyll degradation. However, application of the Zn stress resulted in a low (approximately 6%) but significant decrease in the efficiency of photosystem II (PSII). On the other hand, the Zn treatment had no effect on the water budget of the amaranth plants: both stomatal conductivity and leaf relative water content

(LRWC) did not show any alterations associated with Zn application (Figure S1(3A,4A)). These patterns of the plant morphological and physiological responses to Zn application were confirmed in the second independent experiment (Figure S1(2,3B,4B)).

As no inter-experimental differences in the phenotypes and physiological responses could be observed in two independent plant experiments (Figure S1(1–4), Supplementary Information S2, Table S2(1–4)), dynamics of metal content in tissues and levels of individual low-molecular weight metabolites upon the Zn stress application were measured for only one of them.

2.2. Patterns of Low-Molecular Weight Metabolites Detected by GC-MS in A. caudatus Leaves and Roots and Their Dynamics Upon Treatment with Zn^{2+}

Analysis of the low-molecular-weight metabolites (represented mostly by primary compounds and low-molecular-weight phenolics) in the leaves and roots of *A. caudatus* was accomplished with gas chromatography-electron ionization-quadrupole-mass spectrometry (GC-EI-Q-MS). Despite the high potential of LC-MS for analysis of both primary and secondary metabolites, we decided here on GC-MS as this method allows targeting the compounds involved in the stress-protective metabolic adjustment (amino acids and carboxylic acids, polyols and sugars) in the most efficient way. In this context, GC-MS appeared to be completely sufficient for achieving the objectives of this study. The resulting patterns of metabolites were first compared at the qualitative level, i.e., the principal differences in the leaf and root responses to Zn stress were addressed.

In total, 419 individual trimethylsilyl (TMS) and methyloxime (MEOX)/TMS derivatives (i.e., chromatographic signals which we define here as features) were detected in the GC-MS data acquired for the leaf and root extracts of control and Zn^{2+}-treated plants (Supplementary Information S1, Table S1(1)). Some annotated metabolites were represented by several features, which could result from different contributions of MEOX and TMS groups or sin/anti isomerism of sugar TMS-metoximes. Among them, 26 features could be unambiguously identified by electron ionization mass spectral (EI-MS) similarity search and co-elution with 21 authentic standards, whereas the other 118 features were putatively structurally annotated by EI-MS and retention index (RI) similarity search against available EI-MS public libraries (NIST, GMD) and in-house spectral libraries. Since in the GC-MS chromatograms some analytes were represented with several peaks of isomers, TMS or MEOX/TMS derivatives, the overall number of structurally annotated metabolites was 128. These metabolites could be assigned to the following classes: organic acids (di- and tricarboxylic acids of TCA cycle, other hydroxyacids and dicarboxylic acids, sugar acids, phenolic acids), short- and long-chain fatty acids, N-metabolites (amides, amines, proteinogenic and non-proteinogenic amino acids, purines/pyrimidines, nucleosides), sugars, polyols, phosphate-conjugated metabolites (first of all, sugar phosphates), and terpenes.

Annotation of the further 107 analytes relied on the presence of characteristic fragment ions ($m/z \pm 0.5$ Da) in the EI mass spectra which might serve as indicators for the following substance classes [44]: fragment ions of m/z 174 and 100 prospectively indicated amino group-containing compounds; m/z 103, 160, 217, 319 were characteristic for C5- and C6-monosaccharides; m/z 319 and 204 for sugar alcohols, m/z 318 and 319 for stereoisomers of inositol, m/z 333, 292 and 319 for sugar acids, m/z 361, 437 or/and 451 for di- and oligosaccharides; m/z 299, 315, 357 or/and 387 for organic phosphates. An essential part (41%) of the metabolite pattern included yet uncharacterized metabolite features, which did not yield any match in public and in-house EI-MS databases. Most of these metabolites were detected in all inter-group paired comparisons (i.e., leaves and roots from Zn^{2+}-treated vs. untreated plants). However, about 24 metabolites (boric acid, urea, 1-nitroso-3,5-dinitro-hexahydro-1,3,5-triazine, two unknown amines and 19 further unknowns) were missing (i.e., not detected under our experimental conditions) in one or more experimental groups. In particular, boric acid was not detected in young leaves and roots of the control plants, whereas urea was not detected in young leaves of Zn^{2+}-treated plants.

The relative abundances of individual metabolites detected in young leaves and roots of *A. caudatus* plants in the absence and presence of Zn stress were visualized by hierarchical clustering with heat map representation (Figure 1a). Although no significant stress-associated changes in leaf relative water contents, root and leaf wet and dry weights could be observed (Supplementary Information S1, Figure S1(4,5) and Supplementary Information S2, Table S2(4–6)), the metabolic responses in both leaves and roots demonstrated clear stress-induced patterns dominated with organic acids (predominantly sugar acids), sugars and, to some extent, amino acids (Tables 1 and 2). Since all these compounds are well-known stress metabolites, the stress state of Zn-treated plants can further be considered as unambiguously confirmed.

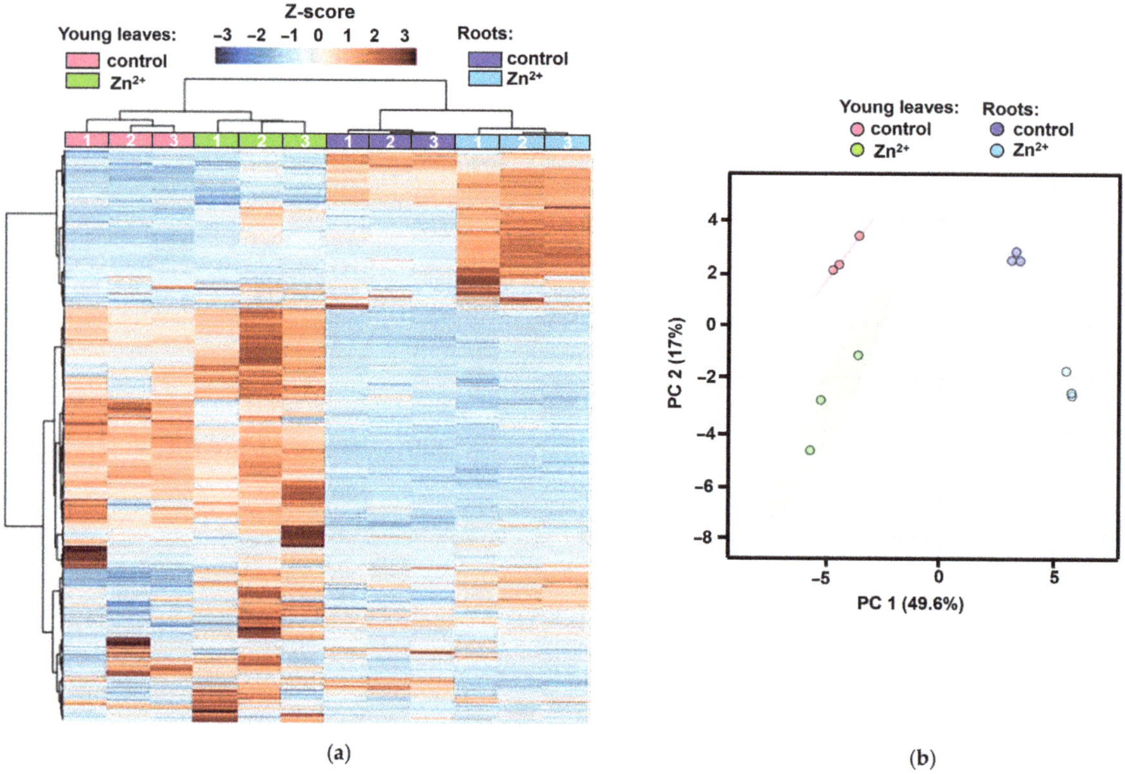

Figure 1. Results of hierarchical clustering with heat map representation (**a**) and principal component analysis (PCA), score plot built for the first two principal components, (**b**) accomplished for abundances of all 419 individual metabolites (features) detected as trimethylsilyl (TMS) and TMS-methyl oxime (MEOX) derivatives by untargeted GC-EI-Q-MS of aq. methanolic extracts of *A. caudatus* young leaves and roots. The PCA and Heatmap were built in MetaboAnalyst 4.0 (https://www.metaboanalyst.ca, accessed on 16 March 2020).

On the other hand, the absence of any significant changes in organ weights and LRWC clearly indicates that we succeeded in achieving moderate stress, as was anticipated. This conclusion was strongly supported by the data on Zn^{2+} tissue contents (Figure 2, Supplementary Information S2, Table S2(7)). Indeed, the obtained data clearly indicated that Zn was efficiently absorbed by treated plants and accumulated in roots approximately 28-fold compared to untreated controls (3531 vs. 125 μg/g DW). Zn content was also significantly increased in the mature and young leaves of the Zn-treated plants, although these changes were less pronounced in comparison to roots and were only 12- and 6-fold, respectively (168 vs. 14 and 479 vs. 81 μg/g, respectively). Thus, the patterns of the Zn-dependently up-regulated metabolites were in agreement with the observed stress response. The Zn-induced metabolic response of both leaves and roots was dominated by a pronounced increase in abundance of multiple features, which were well-represented in both organs. However, as can be seen from the corresponding heatmap (Figure 1a), the stress response in leaves was more diverse in terms of the number of affected metabolites. Nevertheless, although the root response to stress was less diverse, it was characterized by lower intra-group variability in comparison to leaves. This observation was confirmed by the principal component analysis (PCA) performed with the same dataset. Thus, as can be seen from the corresponding score plot built for the two first principal components (PC1 and PC2, Figure 1b), all experimental groups (both organs and treatments) were clearly separated with the percentage of explained variance for the constructed model 49.6 and 17.0% for PC1 and PC2, respectively. In agreement with the results of hierarchical clustering, all leaf groups showed higher dispersion in score plots and, therefore, lower confidence of stress-related alterations in the metabolome. In contrast, the stress-related metabolic shifts observed in roots were comparable with inter-organ differences.

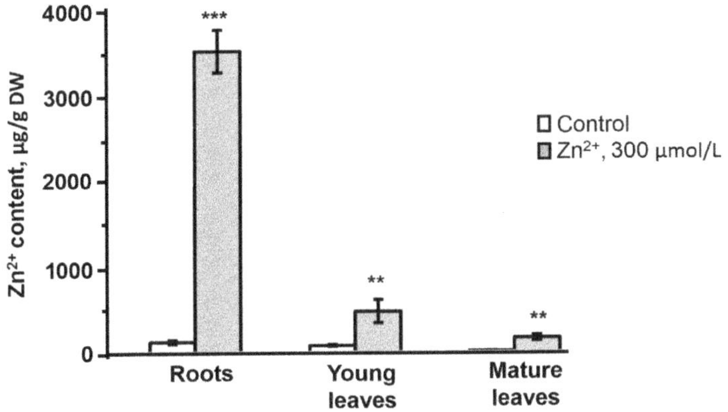

Figure 2. Impact of exogenous Zn^{2+} (300 μmol/L) on the contents of Zn^{2+} in roots, young and mature leaves harvested from seven-week-old *A. caudatus* plants ($n = 3$) grown in hydroponic nutrient solution in the presence (Zn^{2+}-group) and absence (control group) of 300 μmol/L $ZnSO_4$ for 1 week. ** and *** denote *p*-value (Student's test) <0.01 and <0.001, respectively, for significance of the differences observed in Zn^{2+} content between Zn-treated groups and untreated controls.

2.3. *Relative Quantification of the Zn-Responsive Primary Metabolites in the Leaves and Roots of A. caudatus*

For further characterization of the metabolic rearrangements in *A. caudatus* plants in response to Zn^{2+} treatment, organ-specific paired comparisons were considered. Volcano plots and heatmaps were constructed for visualization of significantly affected metabolites in young leaves (Figure 3) and in roots (Figure 4). The volcano plots represent the features differentially ($\geq$1.5-fold, $p \leq 0.05$) abundant in Zn-treated *A. caudatus* young leaves and roots in comparison to the same organs in the control plants. Thereby, the confidence (*p*-value) and magnitude of the alterations (fold change, FC) are visualized.

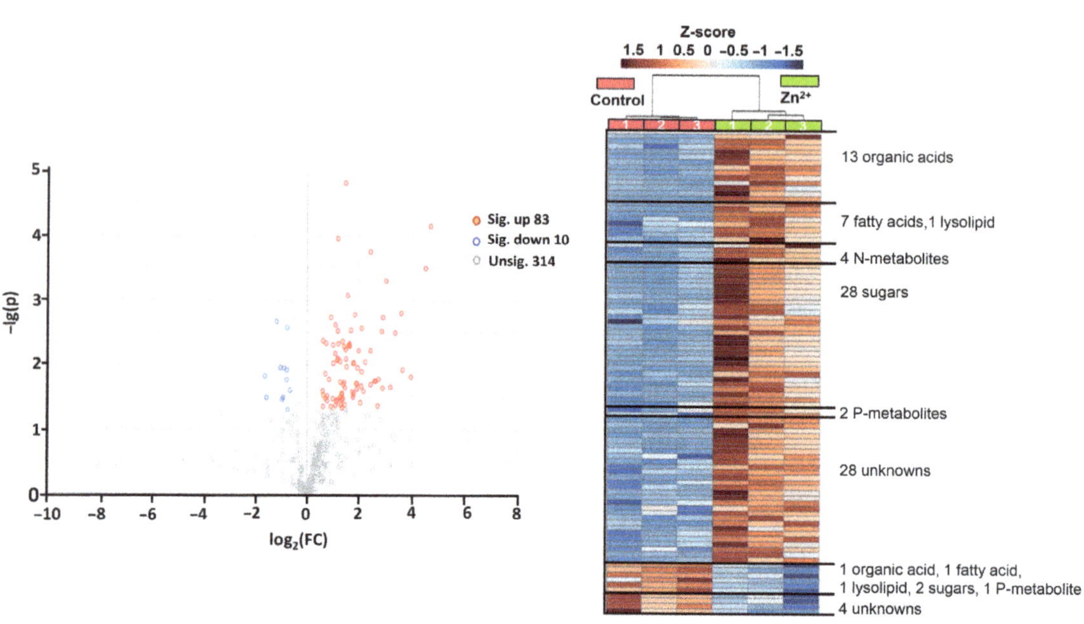

(**a**) (**b**)

Figure 3. Metabolic responses of *A. caudatus* young leaves to a seven-day-long treatment with 300 µmol/L $ZnSO_4$ (n = 3) in comparison to untreated controls (n = 3). (**a**) Volcano plot representing the numbers of the features differentially ($\geq$1.5-fold, $p \leq 0.05$) abundant in *A. caudatus* young leaves in comparison to young leaves of the control plants. Colored dot marks represent the features that demonstrated significantly higher (Sig. up, red) or lower (Sig. down, blue) abundance in Zn^{2+}-treated young leaves in comparison to the controls; grey dots correspond to the features that did not show any significant alteration in young leaves in response to Zn^{2+}-treatment. (**b**) Hierarchical clustering with heat map representation of 83 and 10 metabolites whose tissue contents were significantly ($p \leq 0.05$) higher and lower, respectively, in samples of the Zn^{2+}-treated plants. The metabolites were grouped according to their chemical classes. The list of Zn^{2+}-regulated metabolites is presented in Table 1 and Supplementary Information S1, Table S1(2). The volcano plot and the heatmap were constructed in MetaboAnalyst 4.0 (https://www.metaboanalyst.ca, accessed on 16 March 2020).

In total, 93 metabolites were differentially abundant in young leaves of the Zn-stressed plants. Among them, 83 analytes showed higher and 10 lower abundance, compared to the control. Based on the heatmap representation, the Zn^{2+} -responsive metabolites could be grouped by their chemical classes (Figure 3b). The metabolites, which were more abundant in the young leaves, were represented by 13 organic acids, seven fatty acids, 28 sugars, four N-metabolites, two organic phosphates, one lysolipid and 28 unknown metabolites. Metabolites decreasing their content upon Zn treatment were represented with one organic acid, one fatty acid, two sugars, one organic phosphate, one lysolipid and four unknown compounds (see Table 1 and Supplementary Information S1, Table S1(2) for identified features and the total list of differentially abundant leaf metabolites).

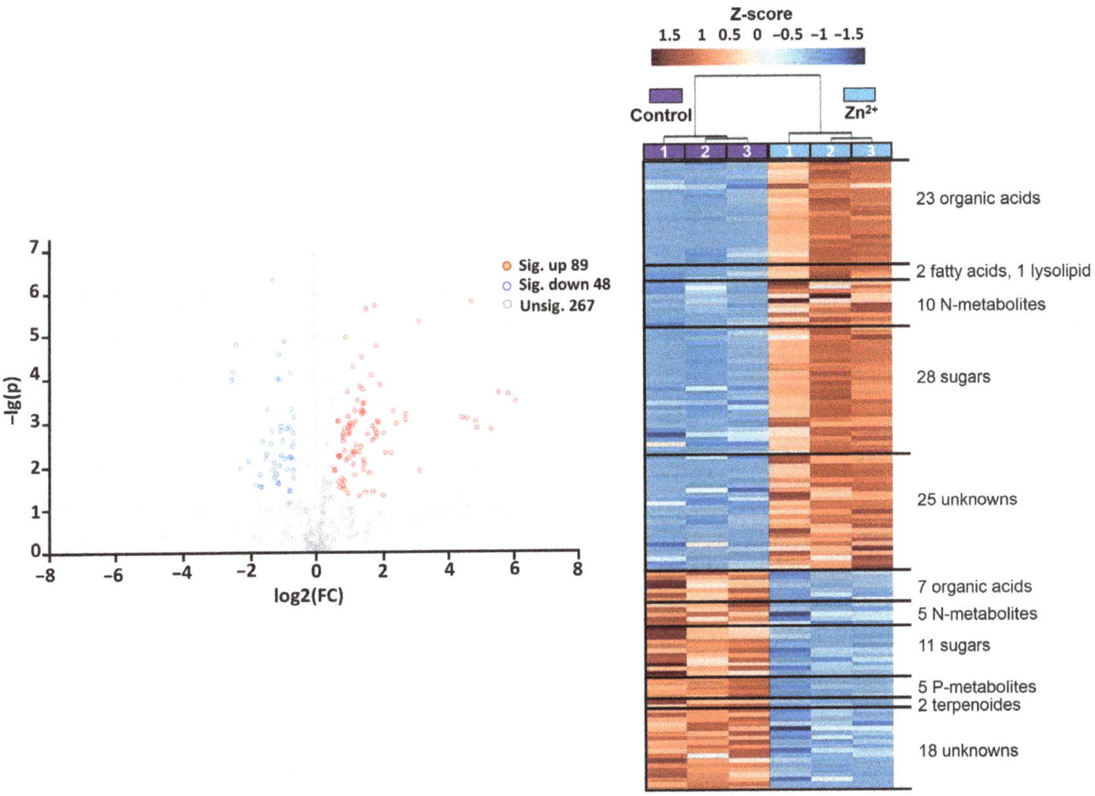

(**a**) (**b**)

Figure 4. Metabolic responses of *A. caudatus* roots to a seven-day-long treatment with 300 μmol/L $ZnSO_4$ (n = 3) in comparison to untreated controls (n = 3). (**a**) Volcano plot representing the numbers of features differentially ($\geq$1.5-fold, $p \leq 0.05$) abundant in *A. caudatus* roots in comparison to the roots of the control plants. Colored dot marks represent the features that demonstrated significantly higher (Sig. up, red) or lower (Sig. down, blue) abundances in Zn^{2+}-treated roots in comparison to the controls; grey dots correspond to the features that did not show any significant regulation in roots in response to Zn^{2+}-treatment. (**b**) Hierarchical clustering with heat map representation of the 89 and 48 metabolites whose tissue contents were significantly ($p \leq 0.05$) higher and lower, respectively, in the samples of Zn^{2+}-treated plants. The metabolites were grouped according to their chemical classes. The list of Zn^{2+}-regulated metabolites is presented in Table 2 and Supplementary Information S1, Table S1(3). The volcano plot and the heatmap were constructed in MetaboAnalyst 4.0 (https://www.metaboanalyst.ca, accessed on 16 March 2020).

The total number of root metabolites, which were differentially ($\geq$1.5-fold, $p \leq 0.05$) abundant in Zn^{2+}-treated plants in comparison to the control ones, was 137, i.e., 47% higher than in leaves. Among them, 89 metabolites were more abundant and 48 less abundant (Figure 4a). As can be seen in Figure 4a and Supplementary Information S1, Table S1(3), differential abundance of multiple metabolites could be confirmed with higher confidence in roots than in leaves ($p \leq 0.001$ vs. $p \leq 0.05$). Also, in contrast to the leaf data, the lowest p-values, observed in the up- and down-regulated metabolite groups were comparable for roots. The group of up-regulated root metabolites contained more organic acids (23), carbohydrates (28) and N-metabolites (10), but fewer fatty acids (2) compared to young leaves (Figures 3b and 4b). Also, no phosphates could be identified among up-regulated root metabolites (see Tables 2 and S1(3) for identified features and the total list of differentially abundant root metabolites). Overall, sugars and organic acids were the most representative metabolite classes in both young leaves and roots of the Zn-stressed amaranth plants.

Table 1. Zn^{2+}-regulated structurally annotated metabolites in the young leaves of *A. caudatus*.

#	Metabolite Feature [a]	Derivatives [b]	RI [c]	*m/z* [d]	FC [e]	*p* [f]
Metabolites demonstrating higher abundances in Zn-treated leaves in comparison to untreated ones						
1	Glyoxylic acid	1MEOX, 1TMS	1142.6	218	22	0.015
2	**3-Hydroxy-3-methylglutaric acid**	3TMS	1598.9	247	3.6	0.002
3	**Gluconic acid δ-lactone**	4TMS	1878.5	319	2.8	0.008
4	Ribonic acid-1,4-lactone	3TMS	1645.9	117	6.2	0.018
5	Lyxonic acid-1,4-lactone	3TMS	1729.7	217	1.8	0.017
6	Arabinonic acid-1,4-lactone	3TMS	1622.2	217	2	0.005
7	**Gluconic acid**	6TMS	1990.7	333	10	0.003
8	**Shikimic acid**	4TMS	1808.5	204	2.5	0.041
9	**Salicylic acid**	2TMS	1504.2	267	23	≤0.001/0.026
10	Octanoic acid	1TMS	1269.1	117	>100	0.0003
11	Stearic acid	1TMS	2223.3	341	3	0.001
12	Arachidic acid	1TMS	2392.2	369	2.8	≤0.001/0.005
13	Myristic acid	1TMS	1846.1	285	2.6	0.02
14	Oleic acid	1TMS	2205.5	339	2	0.034
15	Heptadecanoic acid	1TMS	2134.3	327	2.1	0.002
16	**Behenic acid (docosanoic acid)**	1TMS	2547.0	397	2	0.009
17	1-Monostearateglycerol	2TMS	2651.2	399	2.2	0.036
18	**Pyroglutamic acid (5-oxoproline)**	1TMS	1496.4	84	5.5	0.006
19	**Ethanolamine**	3TMS	1232.1	174	4.2	0.038
20	***N*-acetyl-serine**	2TMS	1503.1	116	12	0.002
21	**5-Methylcytosine**	2TMS	1534.0	254	7.4	0.023
22	**Fructofuranose, peak 1**	5TMS	1807.0	217	4.3	0.013
23	**Fructose, peak 2**	1MEOX, 5TMS	1942.9	217	4.3	0.021
24	Fructofuranose, peak 2	5TMS	1798.7	217	3.9	0.011
25	**Fructose, peak 1**	1MEOX, 5TMS	1934.8	217	3.8	0.024
26	**Glucose, peak 1**	1MEOX, 5TMS	1990.5	319	3.7	0.019
27	**Glucose, peak 2**	1MEOX, 5TMS	2007.2	319	2.6	0.033
28	**Mannose**	1MEOX, 5TMS	1984.3	319	3.5	0.028
29	**Galactose**	1MEOX, 5TMS	1979.6	319	2.1	0.042
30	**Myo-inositol**	6TMS	2076.9	305	1.9	0.044
31	Sucrose	8TMS	2540.8	361	6.6	0.043
32	Phosphoric acid monomethyl ester	2TMS	1185.5	241	1.7	0.036
Metabolites demonstrating lower abundances in Zn-treated leaves in comparison to untreated ones						
1	**Succinic acid**	2TMS	1316.6	247	2.0	0.012
2	Linoleic acid	1TMS	2192.7	337	1.7	0.003
3	2-Oleoylglycerol	2TMS	2629.7	129	1.7	0.012
4	**Glycerol-3-phosphate**	4TMS	1758.8	299	1.7	0.017

[a] Structurally annotated metabolites arranged by the following chemical classes: organic acids, fatty acids, lysolipids, N-metabolites, monosaccharides, polyols, disaccharides, organic phosphates. Numbers 1 or 2 next to the metabolite features with the same annotation indicate the peak (isomer) number of the metabolite. [b] The numbers of trimethylsilyl (TMS) and methyloxime (MEOX) groups. [c] RI, Kovach retention index. [d] The *m/z* value refers to the fragment selected as compound-characteristic for quantification by integration of peak areas at characteristic extracted ion chromatograms. [e] FC, fold change (at least 1.5-fold) in metabolite relative abundances in Zn-treated young leaves compared to controls. [f] *t*-test *p*-value ≤ 0.05. The symbol "/" indicates the *p*-values calculated without and with false discovery rate (FDR) correction by the Benjamini–Hochberg method and did not exceed the *p*-value threshold of 0.05 after application of the FDR correction. The total list of Zn-regulated differentially abundant metabolites, which includes structurally annotated metabolite features (36 species), presented in the table, the unknown features (32 species) and features (25 species) annotated only to a specific chemical class, is presented in Supplementary Information S1, Table S1(2). Bold fond marks metabolites which were also Zn-regulated in roots.

Table 2. Zn^{2+}-regulated structurally annotated metabolites in the roots of *A. caudatus*.

#	Metabolite Features [a]	Derivatives [b]	RI [c]	*m/z* [d]	FC [e]	*p* [f]
Metabolites demonstrating higher abundances in Zn-treated roots in comparison to untreated ones						
1	Malonic acid	2TMS	1211.3	233	2.1	0.002
2	Glyceric acid	3TMS	1330.4	292	2.5	≤0.001/0.004
3	Citric acid	4TMS	1814.1	273	2.7	0.0035/0.01
4	Citramalic acid	3TMS	1468.2	247	1.8	0.002/0.01
5	3,4-Dihydroxybutyric acid γ-lactone	-	1371.4	247	3.7	≤0.001/0.007
6	Adipic acid	2TMS	1504.9	111	1.9	0.023
7	Threonic acid	4TMS	1558.5	292	3.8	≤0.001/0.007
8	Erythronic acid	4TMS	1540.9	292	2.0	≤0.001/0.003
9	α-Hydroxyglutaric acid	3TMS	1573.8	247	2.4	0.003/0.014
10	**3-Hydroxy-3-methylglutaric acid**	3TMS	1598.9	247	2.4	≤0.001/0.007
11	**Gluconic acid**	6TMS	1990.7	333	59	≤0.001/0.003
12	**Gluconic acid δ-lactone**	4TMS	1878.5	319	25	≤0.001/0.007
13	Xylonic acid-1,4-lactone	3TMS	1629.5	117	2.8	≤0.001/0.005
14	**Shikimic acid**	4TMS	1808.5	204	2.7	≤0.001/0.005
15	**Salicylic acid**	2TMS	1504.2	267	27	≤0.001/≤ 0.001
16	2-Hydroxysebacic acid	2TMS	2525.2	317	5.2	≤0.001/0.005
17	**Behenic acid**	1TMS	2547.0	397	1.7	0.006/0.02
18	1-Monooleoylglycerol	2TMS	2632.9	129	1.7	0.006/0.02
19	Oleic acid amide	1TMS	2366.7	338	9.6	0.013
20	Proline [+CO_2]	2TMS	1577.6	142	6.8	≤0.001
21	Alanine	3TMS	1357.0	188	2.9	0.04
22	Leucine	2TMS	1271.7	158	2.0	0.042
23	*N,N*-Dimethylglycine	1TMS	1040.4	58	1.9	0.004/0.02
24	**N-acetyl-serine**	2TMS	1503.1	116	1.5	0.012/0.035
25	**5-Methylcytosine**	2TMS	1534.0	254	3.7	≤0.001/≤0.001
26	Uridine	3TMS	2389.4	217	2.3	0.005/0.02
27	Adenine—derivative	2TMS	2096.2	264	1.8	0.0018
28	Arabinose 1	1MEOX, 4TMS	1755.0	307	2.3	≤0.001
29	Arabinose 2	1MEOX, 4TMS	1758.9	307	2.2	0.005/0.02
30	Arabino-hexos-2-ulose (2-ketoglucose)	4TMS	1477.2	234	2.3	0.02/0.044
31	**Fructofuranose**	5TMS	1807.0	217	2.8	≤0.001/0.005
32	**Fructose, peak 1**	1MEOX, 5TMS	1934.8	217	2.8	≤0.001/0.004
33	**Fructose, peak 2**	1MEOX, 5TMS	1942.9	217	2.7	≤0.001/0.006
34	**Glucose, peak 1**	1MEOX, 5TMS	1990.5	319	4.0	≤0.001/0.002
35	**Glucose, peak 2**	1MEOX, 5TMS	2007.2	319	2.0	≤0.001/0.006
36	**Mannose**	1MEOX, 5TMS	1984.3	319	3.0	≤0.001/≤0.001
37	**Galactose**	1MEOX, 5TMS	1979.6	319	9.0	≤0.001/≤0.001
38	**Myo-inositol**	6TMS	1979.6	319	2.0	0.001/0.008
39	Ribitol	5TMS	1720.7	217	2.8	≤0.001/0.003
40	2-*O*-Glycerol-α-*D*-galactopyranoside	6TMS	2283.0	204	2.2	0.001/0.008
41	Trehalose	8TMS	2675.5	361	2.7	≤0.001/≤0.001
Metabolites demonstrating lower abundances in Zn-treated roots in comparison to untreated ones						
1	Oxalic acid	2TMS	1145.5	190	1.6	0.004/0.015
2	**Succinic acid**	2TMS	1316.6	247	1.5	≤0.001/0.006
3	Methylmaleic acid	2TMS	1348.9	259	2.3	0.02/0.05
4	*trans*-Caffeic acid	3TMS	2137.6	219	4.0	0.007/0.02
5	Itaconic acid	2TMS	1342.9	183	2.4	0.012/0.035
6	**Ethanolamine**	3TMS	1232.1	174	2.8	0.007/0.023
7	**Pyroglutamic acid**	1TMS	1496.4	84	2.2	0.024
8	Oleic acid amide	-	2323.2	59	1.5	0.02/0.048
9	**Glycerol-3-phosphate**	4TMS	1758.8	299	2.2	≤0.001/0.002
10	Glycerophosphoglycerol	5TMS	2181.2	357	5.1	≤0.001/≤0.001
11	*myo*-Inositol phosphate	7TMS	2361.7	318	1.8	0.001/0.008
12	Phytol	1TMS	2159.7	143	2.3	0.008/0.03
13	β-Sitosterol	1TMS	3087.5	486	3.0	0.003/0.012

[a] Structurally annotated metabolites arranged by the following chemical classes: organic acids, fatty acids, lysolipids, N-metabolites, monosaccharides, polyols, disaccharides, and organic phosphates. Numbers 1 or 2 next

to the metabolite features with the same annotation indicate the peak (isomer) number of the metabolite. [b] The numbers of trimethylsilyl (TMS) and methyloxime (MEOX) groups. [c] RI, Kovach retention index. [d] The m/z value refers to the fragment selected as compound-characteristic for quantification by integration of peak areas at characteristic extracted ion chromatograms. [e] FC, fold change (at least 1.5-fold) in metabolite relative abundances in Zn-treated roots compared to controls. [f] t-test p-value ≤ 0.05. The symbol "/" indicates the p-values calculated without and with false discovery rate (FDR) correction by the Benjamini–Hochberg method and did not exceed the p-value threshold of 0.05 after application of the FDR correction. Bold fond marks metabolites which were also Zn-regulated in young leaves.

As can be seen from Figure 3a, most of the differentially abundant metabolites in young leaves were up-regulated upon Zn treatment. Octanoic acid was the most affected compound among them (FC > 100, Table 1). The others were glyoxylic (FC 22), salicylic (FC 23) and gluconic (FC 10) acids. The other sugar acids (besides gluconic) and their lactons demonstrated a lower abundance increase (1.8–6.2-fold) in response to Zn^{2+} treatment, whereas disaccharides were 5.4–9.1-fold up-regulated. Zn^{2+}-related accumulation of monosaccharides, in particular galactose, glucose, mannose and fructose, was less pronounced and did not exceed 2.1–4.3-fold. Essential up-regulation was observed also for four N-metabolites—*N*-acetyl-serine (12.1-fold), 5-methylcytosine (7.4-fold), pyroglutamic acid (5.5-fold) and ethanolamine (4.2-fold). In contrast, only a few metabolites decreased their content in amaranth tissues upon Zn^{2+} treatment: succinic acid (2.0-fold), linoleic acid, 2-oleoylglycerol and glycerol-3-phosphate (each 1.7-fold, Table 1).

As can be seen from Figure 4a, the Zn^{2+} response in roots of *A. caudatus* was more balanced, although still dominated by up-regulation. The most pronounced abundance increase was observed for several sugar acids, especially for non-identified sugar acid RI1988 (69-fold), gluconic acid (59-fold) and gluconic acid-δ-lacton (25-fold) (Tables S1(3) and 2). A strong up-regulation (27-fold) was also observed for salicylic acid. Thus, the Zn-induced increase in content of the above-mentioned compounds in roots was higher than in leaves. This fact might indicate stronger involvement of roots in the metabolic response to Zn^{2+} treatment in our study. For other (not sugar-related) organic acids (malonic, glyceric, citric and threonic), much less pronounced up-regulation (2.0–3.8-fold) was observed. Also, a Zn-induced accumulation was observed for disaccharides (1.8–3.7-fold), including trehalose (2.7-fold), as well as for monosaccharides (FC 1.8–4.0), such as arabinose, fructose, glucose and mannose, with only galactose being much strongly up-regulated (9.0-fold). The content of several fatty acids, as well as amino acids (proline, alanine, leucine) and some other N-metabolites, also increased in roots of Zn-treated plants, with oleic acid amide (9.6-fold) and proline (6.8-fold) identified as the most responsive compounds in these classes.

The list of structurally annotated down-regulated root metabolites included 13 compounds (Table 2). The Zn-related decrease in abundance could be confirmed for five organic acids (oxalic, succinic, itaconic, methylmaleic and *trans*-caffeic), two N-metabolites, three P-metabolites and two terpenoids. The FC values for most of the compounds were in the range of 1.5–3.0, although two metabolites, namely *trans*-caffeic acid and glycerophosphoglycerol, were more strongly affected and demonstrated 4.0–5.1-fold down-regulation. Among the other 35 structurally unannotated features decreasing their relative abundances, most were represented with unknowns (18 species). The second most abundant compound class was that of carbohydrates, with eleven species demonstrating 1.6–4.7-fold down-regulation in roots of Zn-treated plants (Figure 4b and Table S1(3)).

To address the discrepancies and similarities in the Zn-stress responses of young leaves and roots, we compared the lists of significantly ($p \leq 0.05$) Zn-dependently regulated leaf and root metabolites, presented in Table 1 (Table S1(2)) and Table 2 (Table S1(3)). The comparison revealed 48 common features, most of which showed similar abundance dynamics in response to stress: 35 and 4 features, occurring both in roots and leaves, demonstrated concerted increase and decrease, respectively. Only nine metabolites showed different stress-related dynamics in the two compared plant organs, and among them, only two features (ethanolamine and pyroglutamic acid) were structurally annotated.

Among the features showing concerted stress-induced up-regulation, the following metabolites were structurally annotated: sugars and sugar-related substances (glucose, fructose, mannose, galactose, *myo*-inositol, gluconic acid), salicylic acid, shikimic acid, 3-hydroxy-3-methylglutaric acid, behenic acid, *N*-acetyl-serine and 5-methylcytosine. Some metabolites (salicylic acid, gluconic acid and its lactone, galactose) showed stronger Zn-dependent regulation in roots in comparison to leaves. Among the metabolites showing concerted Zn-induced abundance decrease in roots and leaves, succinic acid and glycerol-3-phosphate were structurally annotated.

To address the dynamics of Zn-dependently regulated metabolites in *A. caudatus* young leaves and roots, in more detail, all detected features can be classified in the following principal groups: (i) structurally annotated (36 and 54 features in young leaves and roots, respectively, Tables 1 and 2), (ii) annotated only to a specific compound class without further structure annotation (25 and 40 features, respectively, Supplementary Information S1, Table S1(2,3)) and (iii) unknowns (32 and 43 features, respectively, Table S1(2,3)). To reveal the metabolic pathways involved in the process of metabolic adjustment under Zn-stress (see below), the features assigned to the first group were further subjected to pathway analysis. The features representing the groups i and ii are discussed in more detail in the following sections.

2.4. Structure Elucidation of Di- and Oligosaccharides Based on RI and EI Mass Spectra

As di- and oligosaccharides represented one of the most strongly dominating chemical groups among the detected Zn-responsive metabolites of amaranth roots and leaves (16 in roots, 12 and 11 in young and mature leaves, respectively, Supplementary Information S1, Table S1(2–4)), the correctness of their annotation became a critically important aspect. Therefore, here we describe our annotation algorithm, which was applied to such compounds throughout the whole work. In the following paragraphs, we describe this universal logic in detail and exemplify it with a typical representative analyte—RI2864 di- or oligosaccharide (Table S1(3)), which demonstrated a 6.8-fold abundance increase in Zn-treated roots compared to control.

This metabolite was annotated as a di-/oligosaccharide by the similarity of its analytical behavior to other compounds of this class. The annotation was based on the following two criteria. The first is the similarity of EI mass spectra to those typical for disaccharides and trisaccharides (as can be illustrated by the EI-MS spectra from our in-house library, Supplementary Information S1, Figure S1(6)). The second is the elution within the characteristic RI window close to sucrose (Table S1(1), in-house library RI2541), which is considered to be a representative disaccharide and to maltotriose (GDM RI2887)—a typical trisaccharide. It is known that EI spectra of di- and oligosaccharides (MEOX/TMS derivatives of di- and oligosaccharides) demonstrate multiple common fragments with the spectra of MEOX/TMS derivatives of monosaccharides. However, they are featured with several characteristic signals which might serve as indicators of the glycosidic linkages between the monosaccharide residues. The cleavage of the glycosidic linkage of MEOX/TMS derivatives of the reducing disaccharide yields two characteristic structural moieties: a ring structure and an open chain moiety (a MEOX/TMS derivative). Under the EI conditions, both yield characteristic fragments with their signals clearly seen in the spectra [45].

As can be seen in Figure 5, the peak of RI2864 at t_R 45.31 showed much higher relative intensity in the total ion chromatogram (TIC) of the stressed root samples in comparison to the control ones (Figure 5a). The corresponding EI-MS spectrum (Figure 5b) delivered a rich pattern of signals. The base peak at m/z 204 appeared to be applicable for relative quantification by its extracted ion chromatogram (XIC) at m/z 204.0 ± 0.5 and t_R 45.31 (Figure 5c), i.e., integration of the area under the curve (Figure 5d). The assignment of RI2864 as a disaccharide relied on the presence of two characteristic fragments in the spectrum: m/z 363 corresponding to the pyranosyl moiety and m/z 273 corresponding to a TMS loss from it (marked red, Figure 5b). These fragments indicate the presence of rhamnopyranosyl moiety in the structure of the di- or oligosaccharide [45,46]. The open chain moiety (containing a methyl oxime group) is characterized by the fragment ion (of trace abundance) at m/z 538 (marked blue in Figure 5b). This fragment indicates that RI2864 might be a reducing sugar. The third group of signals (marked with black bold) at m/z 191, 204, 205, 217 and 319 represents the common characteristic fragments for carbohydrates [44]. Their structures are presented in Supplementary Information S1, Table S1(5). Thus, according to the EI spectrum and the literature data on the amaranth metabolism [47], we may conclude that RI2864 di- or oligosaccharide most likely corresponds to rutinose. This assignment was additionally supported by the absence of the fragment at m/z 361 and low intensity of the fragment signal at m/z 319: these fragments are characteristic of non-reducing disaccharides and are absent in reducing disaccharides like rutinose [46]. Moreover, the intensity ratio of the fragments at m/z 217 and m/z 204 was below one, which indicated the existence of a $1 \rightarrow 6$ glycosidic linkage in the disaccharide molecule, which is the case for rutinose. However, as the authentic standard of rutinose was not available, the compound assignment was treated as "putative identification". Annotation of two other di-/oligosaccharides is comprehensively described in Supplementary Information S1, Protocol S1-1 and illustrated in Figure S1(7,8).

2.5. Absolute Quantification of the Zn-Responsive Primary Metabolites in the Leaves and Roots of *A. caudatus* in Targeted GC-MS Experiments

To address the quantitative contribution of individual metabolites in the observed Zn-induced metabolic shifts, the absolute tissue contents of 29 principal metabolites representing several compound classes (carboxylic acids, amino acids, polyols, di- and monosaccharides) were addressed in the young leaves and roots of control and Zn-treated *A. caudatus* plants. For the targeted analysis, we decided on the external calibration method. Although this method is less expensive than the stable isotope dilution approach, it is material-, time- and work-consuming. Therefore, we decided to focus only on the metabolites that were earlier reported to be important in the mechanisms of HM tolerance in plants [9,29,31] and selected only 29 individual compounds for targeted analysis. Among these, 21 compounds (Table 3) were confirmed in young leaves and roots of control and Zn-treated *A. caudatus* plants. Additionally, it could be seen that most of the dissolved Glu standard formed pyroglutamate in solution. Therefore, pyroglutamate was used for the quantitation of Glu in plant samples.

The results of the absolute quantification survey conducted with external standardization are summarized in Table 3. As can be seen in the table, sucrose, as well as malic, benzoic, oxalic and pyroglutamic acids appeared to be the most abundant among the root and leaf primary metabolites selected for targeted analysis. Thus, their content in roots of untreated plants accounted for 7.6, 4.7, 9.1, 10.7 and 19.0 μmol/g DW, respectively. The levels of these metabolites in leaves of the untreated plants were 2–3 times lower. However, alanine showed approximately six-fold higher abundance in leaves compared to roots (7.9 vs. 1.4 μmol/g DW). Interestingly, the tissue content of the most abundant metabolites (malic, benzoic, oxalic and pyroglutamic acids and alanine) was only slightly and mostly insignificantly affected by Zn-induced stress. Indeed, more than a twofold accumulation could be observed only for pyroglutamate and alanine in young leaves and roots, respectively. The other analyzed metabolites were mostly up-regulated in the roots of Zn-treated plants, whereas this up-regulation was less pronounced in the leaves (Table 3). For several metabolites, the magnitude of the stress-induced responses was strikingly different between the plant organs. E.g., the relatively high content (7.6 µmol/g DW) of sucrose in roots was not changed upon the application of Zn^{2+} stress, whereas a relatively low amount of this sugar present in leaves (2.2 µmol/g DW) increased 7.6-fold under stress conditions.

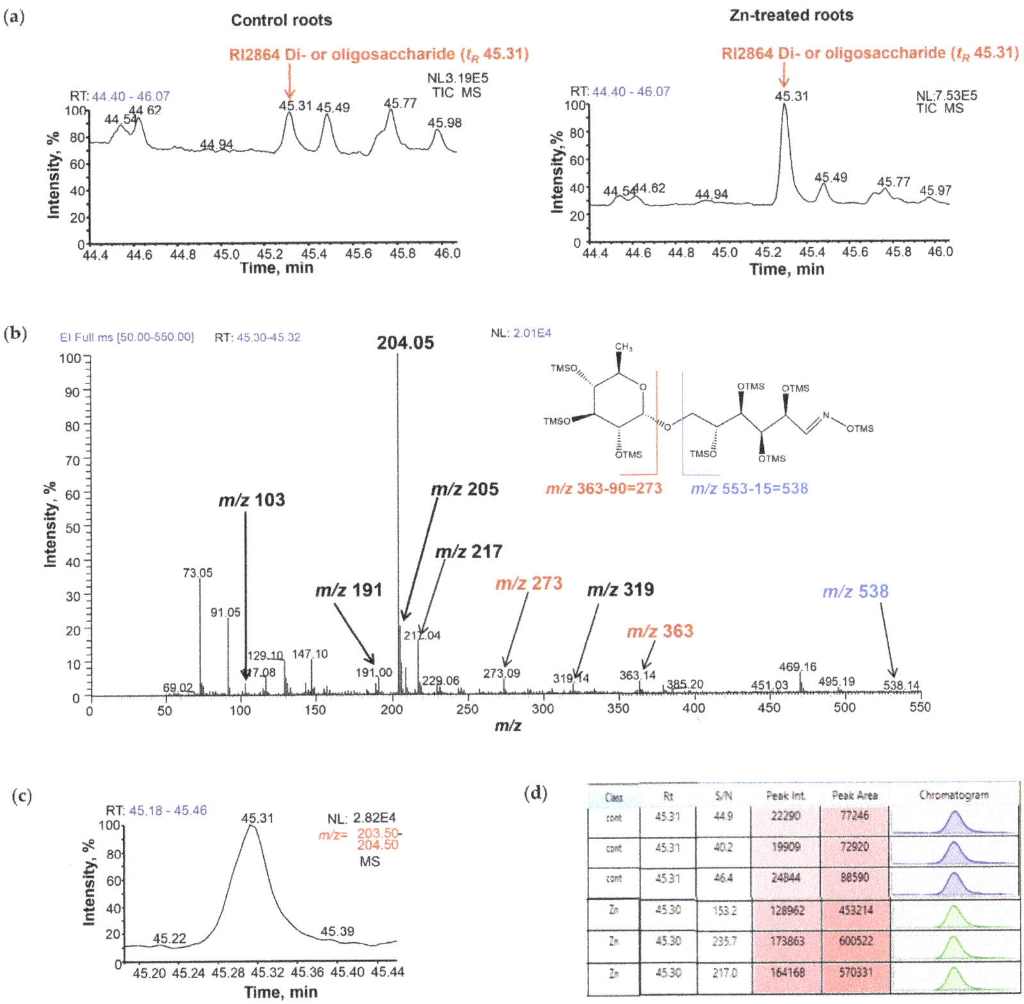

Figure 5. GC-MS information supporting annotation of a metabolite RI 2864 as an oligosaccharide, which showed a 6.8-fold abundance increase in Zn-treated *A. caudatus* roots as compared with the control. The analyte was eluted at the retention time (t_R) of 45.31 min (RI 2864) and was annotated to the class of di-/oligosaccharide with its preliminary identification as rutinose. (**a**) Total ion chromatograms (TICs) of control and Zn-treated *A. caudatus* root extracts presented for the t_R window of 44.4–46.1 min in which a peak of the RI2864 Di-/oligosaccharide located; (**b**) EI mass spectra of the metabolite (the fragments marked with red and blue arrows refer to the moieties of ring and open chain structure, respectively, those marked with black bold font presented fragments diagnostic for carbohydrates and showed in Supplementary Information S1, Table S1(5)); (**c**) Extracted ion chromatogram (XIC) for m/z 204 ± 0.5 (base peak in the EI spectrum) and t_R 45.31 of the RI2864 Di-/oligosaccharide; (**d**) A screenshot from MSDial program presented the RI2864 Di-/oligosaccharide XIC peak area integration values for three control and three Zn-treated samples.

Table 3. Tissue contents of the metabolites identified as differentially regulated in young leaves and roots of *A. caudatus* in response to Zn^{2+} treatment [a].

Metabolite	Average Content, µmol/g DW [b]								YL (Zn-Treated v. Cont.)		R (Zn-Treated v. Cont.)	
	YL Cont		YL Zn		R Cont		R Zn					
	µmol/g DW	±StD	µmol/g DW	±StD	µmol/g DW	±StD	µmol/g DW	±StD	FC [c]	*p* [d]	FC	*p*
Metabolites demonstrating an increase in tissue contents in response to Zn^{2+} treatment in comparison to control												
Young leaf and root metabolites												
Galactose (1MEOX, 5TMS)	0.05	0.01	0.10	0.03	0.05	0.004	0.38	0.02	**↑ 2.0**	**0.04**	**↑ 7.8**	**≤0.001**
Glucose (1MEOX, 5TMS)	0.26	0.02	0.76	0.2	0.93	0.06	3.3	0.28	**↑ 2.9**	**0.02**	**↑ 3.5**	**≤0.001**
myo-Inositol (6TMS)	0.31	0.06	0.59	0.2	0.36	0.03	0.74	0.08	**↑ 1.9**	**0.04**	**↑ 2.0**	**0.001**
Young leaf metabolites												
Pyroglutamic acid (1&2TMS)	10.0	6.1	25.0	7.2	19.0	1.9	18.0	3.8	**↑ 2.5**	**0.05**	↓ ≤1.3	0.46
Sucrose (8TMS)	2.2	0.7	17.0	9.1	7.6	0.3	8.2	0.83	**↑ 7.6**	**0.05**	↑ ≤1.3	0.31
Root metabolites												
Malonic acid (2TMS)	0.74	0.11	0.8	0.2	0.5	0.04	0.9	0.10	1.0	0.90	**↑ 1.9**	**0.002**
Citric acid (4TMS)	0.31	0.08	0.4	0.2	0.2	0.003	0.4	0.05	↑ 1.3	0.51	**↑ 2.1**	**0.003**
Erythronic acid (4TMS)	2.5	0.2	3.0	0.7	1.0	0.08	2.0	0.09	↑ ≤1.3	0.38	**↑ 1.9**	**≤0.001**
Alanine (2&3TMS)	7.9	3.3	3.8	0.9	1.4	0.07	2.9	0.47	↓ 2.1	0.11	**↑ 2.1**	**0.005**
Valine (2TMS)	0.59	0.20	0.4	0.09	0.5	0.03	1.1	0.34	↓ 1.6	0.15	**↑ 2.1**	**0.051**
Leucine (2TMS)	0.80	0.05	0.6	0.2	1.2	0.13	2.1	0.47	↓ ≤1.3	0.21	**↑ 1.7**	**0.036**
Proline (2TMS)	0.09	0.15	0.1	0.04	0.2	0.14	0.9	0.03	↑ ≤1.3	0.86	**↑ 5.1**	**0.001**
Glycerol (3TMS)	1.4	0.1	2.0	0.8	3.0	0.10	4.3	0.11	↑ 1.4	0.25	**↑ 1.4**	**≤0.001**
Arabinose (1MEOX, 4TMS)	0.21	0.02	0.2	0.05	0.07	0.01	0.14	0.02	1.0	0.87	**↑ 1.9**	**0.004**
Metabolites demonstrating a decrease in tissue contents in response to Zn^{2+} treatment in comparison to control												
Young leaf and root metabolites												
Succinic acid (2TMS)	4.2	0.14	2.2	0.7	2.6	0.1	1.7	0.13	**↓ 1.9**	**0.01**	**↓ 1.6**	**0.001**
Young leaf metabolites												
Benzoic acid (1TMS)	10.0	0.5	8.1	0.2	9.1	2.5	10.0	0.25	**↓ 1.3**	**0.002**	↑ ≤1.3	0.58
Root metabolites												
Oxalic acid (2TMS)	4.9	2.1	4.6	1.3	10.7	0.6	7.3	0.79	↓ 1.1	0.85	**↓ 1.5**	**0.004**
Metabolites without significant changes in content in response to Zn^{2+} treatment in comparison to control												
Fumaric acid (2TMS)	0.44	0.04	0.4	0.1	0.8	0.03	0.8	0.05	↓ ≤1.3	0.37	1.0	0.93
Malic acid (2TMS)	2.2	0.2	2.3	1.1	4.7	0.1	5.6	0.56	↑ ≤1.3	0.80	↑ ≤1.3	0.057
Aconitic acid (3TMS)	1.4	0.5	1.1	0.4	0.05	0.02	0.1	0.13	↓ ≤1.3	0.55	↑ 2.2	0.484
Isoleucine (1&2TMS)	0.16	0.03	0.1	0.06	0.4	0.04	0.5	0.07	↓ 1.7	0.17	↑ 1.3	0.062
Urea (2TMS)	3.7	5.8	0.3	0.00	1.5	1.3	3.9	2.12	↓ 11.8	0.37	↑ 2.5	0.18

[a] The analysis relied on the targeted GC-EI-Q-MS assay; [b] The average contents of metabolites in µmol/g DW (dry weight) and their standard deviations (StD) found in young leaves (YL cont) and roots (R cont) of control plants as well as in the organs of the plants exposed to Zn^{2+} excess, YL Zn and R Zn, respectively, are presented. [c] FC, fold change (not less than 1.3-fold) in the metabolite content in Zn-treated samples in comparison to controls. Arrows indicate increased (↑) or decreased (↓) contents of individual metabolites in the Zn-treated sample in comparison to the controls. [d] *T*-test, *p*-value. Statistically significant (*t*-test, $p \leq 0.05$) differences are marked in bold font. FC value 1.0 indicates no difference between the compared Zn-treated and untreated samples.

Galactose, glucose and *myo*-inositol demonstrated a concerted increase in their content in both organs, with the most striking stress-related difference observed for galactose (7.8-fold in roots). The most abundant representatives of organic acids in both young leaves and roots were benzoic, oxalic, malic, and succinic acids. Among them, only succinate demonstrated concerted dynamics of Zn^{2+} stress-related response, i.e., a 1.9- and 1.6-fold decrease in its abundance in young leaves and roots, respectively ($p \leq 0.01$, Table 3). Benzoic acid showed a slight (approximately 30%) abundance decrease in leaves and oxalic acid—a 50% decrease in roots of Zn-treated plants, whereas the content of malic acid demonstrated no significant changes in response to Zn^{2+} stress in both organs.

2.6. Annotation of Zn-Responsive Metabolic Pathways in A. caudatus Leaves and Roots

To address the role of individual metabolic pathways in the observed Zn-related responses, the data on stress-dependently regulated (≥1.5-fold, *t*-test $p \leq 0.05$) primary metabolites annotated in *A. caudatus* young leaves (36 features, Table 1) and roots (54 features, Table 2) were subjected to the pathway analysis. This analysis relied on the *Arabidopsis thaliana* pathway library (deposited by KEGG on-line platform from 03.2020). The results of the combined pathway enrichment analysis (global test) and pathway topology analysis (relative-betweenness centrality test) allowed identification of the most confident pathways strongly contributing in the Zn-related metabolic responses in the young leaves (Figure 6a, Supplementary Information S1, Table S1(6) and Supplementary Information S3, Part S1) and in roots (Figure 6b, Supplementary Information S1, Table S1(7), Supplementary Information S3, Part S2) of *A. caudatus*.

(a)

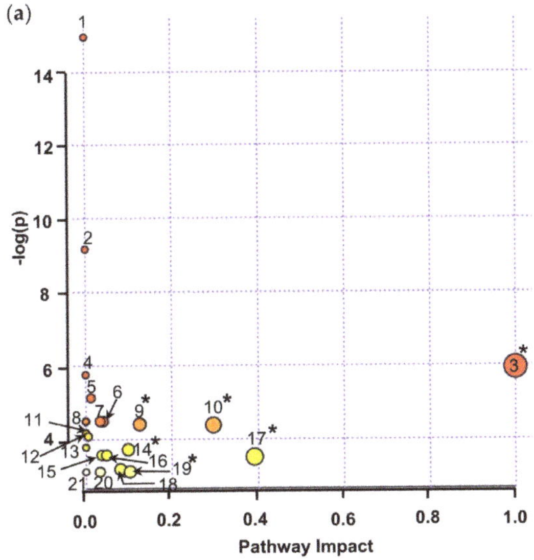

#	Pathway Name
1	Biosynthesis of unsaturated fatty acids
2	Fatty acid biosynthesis
3	Linoleic acid metabolism
4	Pentose phosphate pathway
5	Glutathione metabolism
6	Citrate cycle (TCA cycle)
7	Sulfur metabolism
8	Ala, Asp and Glu metabolism Propanoate and butanoate metabolism
9	Cutin, suberine and wax biosynthesis
10	Glyoxylate and dicarboxylate metabolism
11	Purine metabolism Gly, Ser and Thr metabolism
12	Glycerolipid metabolism
13	Amino sugar and nucleotide sugar metabolism
14	Glycerophospholipid metabolism
15	Fructose and mannose metabolism
16	Galactose metabolism
17	Starch and sucrose metabolism
18	Phe, Tyr and Trp biosynthesis
19	Inositol phosphate metabolism
20	Phosphatidylinositol signaling system
21	Ascorbate and aldarate metabolism

(b)

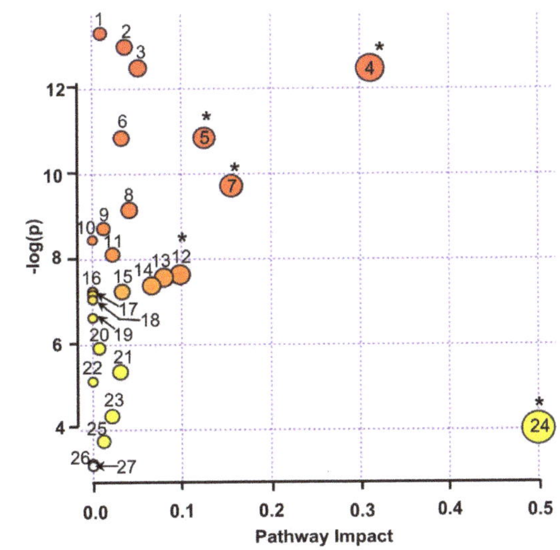

#	Pathway Name
1	Glycerolipid metabolism
2	Fructose and mannose metabolism
3	Glyoxylate and dicarboxylate metabolism
4	Starch and sucrose metabolism
5	Inositol phosphate metabolism
6	Phosphatidylinositol signalling system
7	Citrate cycle (TCA cycle)
8	Galactose metabolism
9	Amino sugar and nucleotide sugar metabolism
10	Pentose phosphate pathway
11	Gly, Ser and Thr metabolism
12	Glycerophospholipid metabolism
13	Phe, Tyr and Trp biosynthesis
14	Arg and Pro metabolism
15	Sulfur metabolism
16	Propanoate and butanoate metabolism
17	Aminoacyl-tRNA biosynthesis
18	Ala, Asp and Glu metabolism
19	Ascorbate and aldarate metabolism
20	Steroid biosynthesis
21	Pyrimidine metabolism
22	Cutin, suberine and wax biosynthesis
23	Val, Leu and Ile biosynthesis
24	C5-Branched dibasic acid metabolism
25	Glutathione metabolism
26	Selenocompound metabolism Carbon fixation in photosynthetic organisms
27	Val, Leu and Ile degradation

Figure 6. Pathway analysis of the Zn-regulated ($\geq$1.5-fold, *t*-test $p \leq 0.05$) metabolites of *A. caudatus* young leaves (**a**) and roots (**b**) annotated by untargeted GC-EI-Q-MS. The pathway analysis relied on the *Arabidopsis thaliana* pathway library (deposited on the KEGG on-line platform from 03.2020) and presents the results from combined pathway enrichment analysis (global test) and pathway topology

analysis (relative-betweenness centrality) to highlight the most confident pathways related to the Zn^{2+}-stress condition. Indicators such as higher circle position along the *Y*-axis and more intense red color of the circles indicate higher significance of the observed differences, whereas the other two indicators, such as circle size and position along the *X*-axis (pathway impact), indicate the impact of the annotated metabolites in the corresponding pathways. The pathways (marked numerically) listed in the corresponding tables on the right panels are ranked according to the results of pathway enrichment analysis. The symbol * marks pathways (their names are indicated by grey font) with their pathway impact value ≥ 0.1, which are considered as potential main contributors to metabolic response to Zn-exposure. More details for the pathway analysis are provided in Supplementary Information S1, Table S1(6,7) for young leaves and roots, respectively. To address the regulated metabolic pathways separately for the Zn-dependent high- and low-abundant ($\geq$1.5-fold, *t*-test $p \leq 0.05$) metabolites, please refer to the Supplementary Information S3, Parts S1 and S2. The Pathway analyses were performed in MetaboAnalyst 4.0 (https://www.metaboanalyst.ca, accessed on 19 March 2020) and details on the analyses are provided in Supplementary Information S4.

As can be seen from the results, the number of pathways involved in Zn-induced metabolic responses in roots was higher than in leaves (27 vs. 21, Supplementary Information S1, Table S1(6,7)). Moreover, based on the comparison of the false discovery rate (FDR)-adjusted (Holm–Bonferroni method) *p*-values, annotation of the root Zn-inducible pathways was more confident. It might be a direct consequence of the higher confidence of the stress-induced changes observed in roots (Figure 6b). However, most of the pathways for both organs were represented with only one hit (15 and 16 for leaves and roots, respectively, Table S1(6,7)). Galactose and fatty acid metabolism were featured with the highest numbers of confidently matched leaf metabolites, whereas galactose and glyoxylate/dicarboxylate metabolism were confirmed as the most represented in roots. Finally, the highest impact values ($\geq$0.1) were calculated for linoleic acid, starch and sucrose, glyoxylate and dicarboxylate metabolism, cutin, suberine and wax biosynthesis, glycerophospholipid metabolism and inositol phosphate metabolism in leaves, and for C5-branched dibasic acid metabolism, TCA cycle, starch and sucrose, glycerophospholipid and inositol phosphate metabolism in roots. Thus, the pathways which contributed most to the Zn-induced response in *A. caudatus* leaves and roots were represented mainly with sugars, sugar- and TCA-related acids and fatty acids, while the pathways of nitrogen metabolism were less involved and featured with low (<0.1) pathway impact values (Supplementary Information S1, Table S1(6,7)).

3. Discussion

3.1. Dynamics of the Metabolite Patterns in the Integrated Plant Response to Zn Stress

Metabolomics is an efficient tool in the studies of the plant responses to abiotic stress; in particular, for revealing the stress-induced metabolic rearrangements and adjustments in the plant metabolic network [48]. Although the untargeted metabolomics approach is widely used in the analysis of osmotic and drought stress response [10], such studies in the field of HM stress are still rare and focus mainly on the effects of Cd and Pb [15,19]. Meanwhile, contamination of agricultural soils with Zn^{2+} ions becomes an important environmental problem [25], and the metabolic shifts associated with Zn^{2+} toxicity require proper characterization. In the first step, it seems logical to characterize the stress-related metabolite patterns in a "static" mode using one time point and the conditions referred to in literature as Zn-stress. Therefore, here, for the first time, we employed the state-of-the-art GC-EI-Q-MS technique to characterize the metabolic responses in the roots and leaves of seven-week-old *A. caudatus* plants to a seven-day-long exposure to 300 µmol/L $ZnSO_4 \cdot 7H_2O$ in the nutrient solution. The choice of the current Zn concentration relied on the available literature data [24,25,49], including those for amaranth [35], which indicated that in various plants, Zn^{2+} concentrations in the range of 50–500 µmol/L triggered clear manifestations of a moderate metal stress accompanied by a well-defined adaptive response.

We showed that exposure to Zn^{2+} ions caused considerable changes in the metabolite profiles of different organs of *A. caudatus*. Indeed, 33 and 22% of the total number of MEOX/TMS derivatives detected in roots and young leaves, respectively, were differentially abundant in this experiment (Figure 1, Supplementary Information S1, Table S1(1)). Interestingly, the additional analyses performed with mature leaves revealed a lower number of differentially abundant metabolites, accounting for only 12% of the total list of detected compounds (Figure S1(9) and Table S1(4)).

The chemical patterns of the Zn^{2+} stress responses in young and mature leaves and roots were dominated by the up-regulated metabolites. This trend was more pronounced in leaves, regardless of their age, than in roots (Figure 3b, Supplementary Information S1, Figure S1(9) and Table S1(2–4)). Considering the metabolite dynamics in all three organs (roots and leaves of different ages) might provide access to the spatial dimension of metabolomics, i.e., to the distribution of metabolites at the level of the whole plant to ensure Zn^{2+} stress tolerance in *A. caudatus*.

The pathway analysis, accomplished for the differentially abundant leaf and root metabolites with unambiguous structural annotation, allowed revealing the metabolic processes involved in the Zn-induced responses. The pathways with the highest impact were mostly related to sugar, organic acid and lipid metabolism, whereas the metabolism of nitrogen-containing compounds was less affected (Table S1(6–8)). Although many of the revealed pathways were represented by only one hit, the pathway analysis procedure ensures the reliability of such results. Hits are based on the p-values calculated from the enrichment analysis, and the pathway impact (PI) values directly depend on the t-test p-values of the hit metabolites and their location/properties of the corresponding enzyme reactions in the pathways. Both these parameters rely on t-statistics. Thus, if a particular enzyme regulates the whole pathway, then the corresponding metabolite up- or downstream of that enzyme would define the impact of the whole pathway. Therefore, the lower pathway p-value and higher pathway impact value indicate that this pathway may have a higher contribution in the whole observed plant response.

Some of the hits obtained from the pathway analysis are common for several related pathways (Table S1(6–8)), implying cross-talk between different metabolic routes. The fact that the same compounds may be involved in different biochemical processes reflects the high complexity of the plant metabolome network. However, these metabolites may have different PI values in different pathways, and the pathway where the compound has the highest PI is expected to contribute most to the plant response to Zn-stress; e.g., among all confident pathways, the "starch and sucrose metabolism" demonstrated the highest impact for young leaves, roots, and mature leaves of *A. caudatus*.

Notably, the metabolites annotated as sugars and demonstrating higher abundance in Zn-treated plants compared to the control ones were much more diverse in young leaves and roots than in mature leaves. By contrast, the number of Zn-dependently accumulated organic acids was twice as high in roots as in leaves of any age (Figures 3b and 4b and Supplementary Information S1, Figure S1(9B)). This is in agreement with the results of the pathway analysis (especially those based on the root data), which indicated sugar- and organic acid-related pathways as the potential main contributors in the Zn-related stress response (Figure 6, Supplementary Information S1, Table S1(6–8) and Figure S1(10), Supplementary Information S3). Induction of organic acids and soluble sugars as "stress metabolites" is a general feature common for plant responses to different stresses [50], including Cd^{2+} stress [15–17]. However, in relation to Zn^{2+} stress, this aspect is still insufficiently addressed.

Therefore, based on the number of differentially abundant metabolites, their FC-values, and the data of the pathway enrichment analysis, we may conclude that up-regulation of sugars and organic acids is a key feature in the metabolic signature of the amaranth response to the Zn^{2+} stress.

3.2. Di- and Monosaccharides in the Response of Amaranth Plants to Zn Stress

The involvement of sugars in stress responses is attributed mainly to their role as osmolytes and osmoprotectants in maintaining cell turgor and stabilizing membranes of the plant cell [51]. We suggest that this was also the case in our study, as the trend towards a decrease in fresh (but not dry) biomass gain was shown for Zn-exposed plants (Supplementary Information S1, Figure S1(5)). Sugars are also known as ROS scavengers [50–52]. Moreover, due to their higher ROS reactivity and lower amenability to hydroxyl radical-related damage, disaccharides and sugar alcohols are recognized as more efficient antioxidants than monosaccharides [53,54]. In addition, sugars may contribute sufficiently to maintaining the redox environment in the cell [55]. Improvement of their availability under stress conditions may stimulate the pentose phosphate pathway (PPP) reactions and thus boost NADPH biosynthesis. Enhanced generation of NADPH was shown to dramatically increase the ROS scavenging capacity in plant cells [56].

The intracellular accumulation of Zn^{2+} ions might cause redox imbalance and overproduction of ROS due to interaction with redox proteins and disturbance of electron transport mechanisms [4,25]. This is recognized as a trigger for intracellular accumulation of soluble sugars (especially oligosaccharides) and their engagement in ROS scavenging [25]. Indeed, we observed up to nine- and seven-fold up-regulation of oligosaccharides in Zn-stressed *A. caudatus* young leaves and roots, respectively (Table S1(2,3)). This was accompanied by the sugar up-regulation up to five-fold in mature leaves (Table S1(4)). Soluble sugars may be formed as a result of stress-induced degradation of storage polysaccharides and/or enhancement of gluconeogenesis and the glyoxylate cycle [57]. However, among the annotated disaccharides increasing their abundance upon the Zn treatment, only sucrose in young and mature leaves and trehalose in roots were unambiguously identified by co-elution with authentic standards, whereas no exact structures could be annotated to the remaining unknown di- and oligosaccharides (Table S1(2–4)). Currently, a stress-induced increase in sucrose accumulation is considered to be associated with the activation of the enzymes involved in sugar metabolism, as well as up-regulation of sucrose transporters [58,59]. Our results revealed a more pronounced increase in sucrose content in the young leaves of treated plants compared to mature leaves (Table 3, Supplementary Information S1, Table S1(9)). Such an imbalance is unlikely to be caused solely by greater enhancement of in situ sucrose production in young leaves, which are typically regarded as mostly heterotrophic sinks dependent on phloem sugar supply [60,61]. Therefore, enhancement in sucrose translocation from mature (source) to young (sink) leaves should be considered as the second possible mechanism. A similar stress-induced redistribution of carbohydrates was previously shown for arabidopsis leaves [62]. At the systemic level, these results may indicate the stress-induced enhancement of metabolite allocation to the sinks as an adaptive response [62]. This phenomenon has recently been reported as an element of plant responses to specific abiotic stresses [63].

Zn-induced up-regulation of monosaccharides (Table S1(2–4)) is in agreement with the metabolic adjustment events reported in the leaves of *Camellia sinensis*, where these metabolites were assumed to be involved in osmoprotection [21]. The presence of pentoses in the Zn^{2+}-associated up-regulation patterns is also in agreement with the earlier published assumption that HMs (Cd^{2+} in particular) can trigger the switch of carbohydrate oxidative catabolism from glycolysis to the PPP [18]. A two-fold up-regulation of *myo*-inositol in the roots and young leaves of Zn-treated *A. caudatus* might be a part of the general stress response—this effect was earlier reported for *A. hypochondriacus* subjected to viral infection and was interpreted as a signaling event [64].

In this study, galactose appeared to be the most Zn^{2+}-responsive sugar in *A. caudatus* roots (Table 2). Galactose is known to play an essential role as a component of cell wall polysaccharides [65]. Roots can respond to trace metals by actively remodeling their metabolism and up-regulating specific biosynthetic pathways [66]. Thus, the walls of epidermal and cortical root cells were shown to thicken in the presence of Zn^{2+} excess, which was explained by enhancement of pectin biosynthesis [67,68]. Due to the presence of multiple carboxyl groups in the galacturonyl monomers of pectins, these polymers can efficiently scavenge Zn^{2+} ions, and wall thickening might essentially increase the capacity of root cells for Zn^{2+} binding [69]. This feature of pectins underlies one of the most efficient mechanisms of HM detoxification and HM-stress tolerance [70]. It is assumed that this mechanism might explain 30–40% of the total Zn^{2+} scavenging capacity of the root cells [69]. This corresponds well with our AAS data (Figure 2), which indicates that the major part of the absorbed Zn^{2+} was deposited in amaranth roots, thus preventing Zn^{2+} translocation to the shoot. Given the possible role of galactose as a precursor in ascorbic acid biosynthesis [65] via the *L*-galactose pathway [71], an increase in galactose content in the amaranth roots may also be associated with the involvement of ascorbate in the response to abiotic stress [72].

In addition to the above-mentioned aspects, galactose could contribute to the synthesis of galactolipids. Activation of the galactolipid metabolism was reported in maize and arabidopsis roots under phosphate deficiency conditions [73] and in the leaves of *N. caerulescens* treated with excess Zn [74]. This mechanism is considered to be an adaptive strategy, which involves remodeling of membrane lipids in response to phosphate deprivation: substitution of phospholipids with galactolipids [73,75]. Depletion of the available phosphate pools in the Zn-treated amaranth root cells may occur due to the direct Zn chelation by phosphate or/and increased demands for energy metabolism. This, in turn, might promote the synthesis of galactolipids instead of phospholipids.

Thus, the patterns of Zn-induced up-regulation in the metabolomes of *A. caudatus* roots and leaves may indicate the following roles of sugars in the HM-stress response: (i) osmotic protection, (ii) allocation of metabolites (in particular, sucrose) from mature leaves to young leaves, (iii) ROS scavenging (relying predominantly on oligosaccharides), (iv) switching the carbohydrate oxidative catabolism from glycolysis to the PPP (that is manifested with characteristic dynamics of pentoses), signaling (accomplished by *myo*-inositol), (v) remodeling of cell walls and, most likely, membranes to enhance the Zn-binding capacity of these structures (galactosyl residues in membrane lipids, pectins and other polymers of the cell wall).

3.3. Organic Acids in the Response of Amaranth Plants to Zn Stress

Organic acids represent the second compound class in the metabolome of *A. caudatus*, which was strongly affected by Zn^{2+} treatment. Among these compounds, salicylic acid (SA) was especially strongly up-regulated in young leaves and roots (Tables 1 and 2). This metabolite is one of the key plant hormones involved in activation of diverse stress responses [76–78], including HM-induced metabolic shifts such as the enhancement of redox metabolism and activation of antioxidant defense, in particular, up-regulation of secondary metabolism and biosynthesis of ROS-scavenging osmoprotectors [78,79].

Although plants' responses to HM stress are generally well characterized, Zn-induced biochemical alterations are still understudied. It was suggested that SA can induce alternative oxidase (AOX), which contributes to redox and metabolic homeostasis and signaling in plant cells [80]. Thus, SA might be involved in both ROS production and scavenging [81]. In particular, AOX1 was found to modulate oxidative stress triggered by Cd^{2+} exposure [17]. However, to date, this mechanism of stress tolerance has not been experimentally confirmed for other HMs. The strong induction of SA in *A. caudatus* roots and leaves exposed to Zn^{2+} (Tables 1 and 2) suggests that this metabolite may be involved in AOX up-regulation. This assumption is further supported by a significant increase in the content of shikimic acid (which is known as the key intermediate in the biosynthesis of aromatic acids) in young leaves and roots (Tables 1 and 2) [82]. The pathways of the SA biosynthesis in plants are still not fully understood. Currently, two pathways are discussed: the first relies on cinnamic acid and phenylalanine ammonia-lyase (PAL), while the second originates from isochorismate (IC) and relies on IC synthase [83]. The second pathway is typical for plants and is characteristic, for example, of arabidopsis [84].

The observed Zn-dependent up-regulation of shikimic acid in young leaves and roots is in good agreement with the stress-related dynamics of phenylpropanoids in mature leaves. In particular, the increased levels of *p*- and *o*-coumaric acids (Supplementary Information S1, Table S1(1–4) might indicate the enhancement of lignin biosynthesis, which underlies the improvement of cell wall rigidity [85]. On the other hand, in roots, the metabolites of this group (e.g., caffeic acid, Table 2) showed mostly a down-regulation pattern in response to Zn stress. Caffeic acid is an important intermediate in lignin biosynthesis, a pathway which is known to be involved in the HM-response [86]. A similar decrease in caffeic acid content in response to application of Cu, Cd and Pb was reported for tomatoes [87]. Interestingly, the opposite effect was shown for *Matricaria chamomilla* under Ni-exposure [88]. It cannot be ruled out that the decrease in the level of caffeic acid is associated with a strong enhancement of SA biosynthesis in *A. caudatus* exposed to Zn stress and the switching of the phenylpropanoid pathway to its production.

Gluconic acid and five unknown sugar acids demonstrated a dramatic increase in their content in roots of Zn-exposed *A. caudatus*, whereas a similar response in leaves was less pronounced (Supplementary Information S1, Table S1(2,3)). Accumulation of gluconic acid was earlier reported in radish roots in response to application of Cd^{2+} and Pb^{2+} [15] and in leaves of corns upon the application of Cu^{2+} [89]. An especially strong (up to 34-fold) increase in gluconate content was shown in leaves of Zn hyperaccumulator *N. caerulescens* upon the application of 0.5 mmol/L Zn^{2+} [74]. The ability of this metabolite to form stable complexes with Cd^{2+} has been intensively discussed. However, reliable assignment of the Zn^{2+} coordination chemistry in planta is challenging [74].

While the shifts in plant metabolism underlying the above-described responses remain largely unknown, the pathways of gluconate biosynthesis have been well studied in yeast. In these organisms, the biosynthesis of gluconate relies on the oxidation of glucose in the so-called $NADP^+$-dependent glucose dehydrogenase—gluconate shunt of PPP [90]. Since this reaction does not involve glucose-6-phosphate dehydrogenase (G6PDH), the regulatory enzyme controlling the rate of the entire cycle, it can be considered as an alternative route for the glucose entry into the PPP. A similar shunt is believed to function in plants [91], although the specific mechanisms for its Zn-dependent activation and its relevance for other aldonic acids still need investigation. Meanwhile, G6PDH was shown to be a Cd^{2+}- [92] and Zn^{2+}-inducible metabolite [93]. Thus, assuming an ubiquitous character of this phenomenon, both pathways of gluconate biosynthesis might be involved in the plant response to Zn stress.

It should be emphasized that non-enzymatic autoxidation of glucose might contribute to the development of metal-induced oxidative stress. Thus, enhanced ROS generation and associated overproduction of reactive carbonyl compounds (RCCs) can be expected under stress conditions [94]. These highly reactive molecules readily interact with biopolymers [95]. Although the resulting conjugates were confirmed *in vivo* [48], their physiological effects in plants still remain unknown [96]. Our data indicate stronger accumulation of gluconate in roots of *A. caudatus* compared to leaves (Tables 1 and 2), which suggests the involvement of gluconic acid in Zn^{2+} chelation directly in roots.

Organic acids of the TCA cycle (malate, citrate) and oxalate are involved in the well-known mechanism of HM tolerance in plants [29–31]. It was shown that citrate is more efficient than malate in the chelating of Zn^{2+} ions [29]. However, in response to Zn-stress, the tissue level of both acids in young leaves did not change, while only a slight increase in their content was detected in roots and mature leaves. Oxalate is known to considerably contribute to Zn^{2+} complexation in plants with high tissue content of this acid [97], such as *A. caudatus* [30]. In our study, we showed that oxalate was one of the abundant metabolites in all tested organs of amaranth (Table 3, Supplementary Information S1, Table S1(9)). However, Zn application did not result in any changes in the content of soluble oxalate in young leaves and caused a slight decrease in its level in mature leaves and roots. The latter observation could be attributed to the formation of insoluble Zn^{2+} oxalate complexes in the walls of root cells [97]. This phenomenon can lead to a sharp decrease in oxalate concentration in tissue liquids. At the analytical level, it might even result in recognition of this metabolite as "below detection limit" by the conventional GC-MS employed here.

To date, the HM-related shifts in the tissue pools of organic acids have been reported only for the specific case of Cd-induced plant stress [17,19]. This aspect was mostly discussed in terms of the stress-induced activation of mitochondrial AOX [17]. The dynamics of some TCA cycle intermediates observed here, together with similar data on Cd^{2+} stress [17,19], suggest Zn-dependent stimulation of the mitochondrial activity via enhancement of ATP, NAD(P)H and antioxidant biosynthesis. Our data are also in line with the recently published assumption of Kaur and Garg [25]. The authors proposed that plant respiration can be enhanced under high Zn^{2+} levels that might result in the overproduction of multiple organic acids via the TCA cycle. This might underlie the onset of Zn^{2+} tolerance, since most of the accumulated acids can be involved in the sequestration of excess Zn^{2+} in vacuoles. Moreover, Zn-associated changes in organic acid content (i.e., up-regulation of malate and citrate and down-regulation of succinate) are consistent with the concept of stress-induced metabolic adjustment through the switch of the TCA cycle from a "closed" to an "open" configuration with the formation of malate and citrate "valves" provoked by increase in cell oxidative status [98]. Since amaranth is a C4 plant, its pools of C3 and C4 organic acids represent reserves of energy (ATP) and reducing equivalents (NADPH) in adaptation to environmental stresses [99].

Strong up-regulation of glyoxylic acid in young leaves (Table 1) can be explained in the context of the following considerations. This toxic intermediate of the glyoxylate cycle is detoxified in planta with the formation of oxalate and may thus be considered as a putative precursor for oxalate synthesis in oxalate-rich plants [100]. As this mechanism can be relevant for amaranth, it needs to be studied in more detail in the context of Zn^{2+} stress. In addition, glyoxylate might impact the prospective glycation-based regulatory pathways in plants [52].

Overall, the accumulation of salicylic, shikimic, and gluconic acids, some TCA cycle intermediates, oxalate and glyoxylate observed in this study might be a result of Zn-dependent activation of the phenylpropanoid pathway, gluconate shunt, PPP, cell respiration and glyoxylate cycle. These organic acids are involved in the processes of redox signaling and homeostasis (salicylic acid), ATP and NADPH synthesis, chelating and sequestrating of Zn excess into vacuoles (gluconate, citrate, malate), and cell wall rigidification (phenylpropanoids).

3.4. Fatty Acids in the Response of Amaranth Plants to Zn Stress

Fatty acids (FAs) represent another group of amaranth metabolites affected by Zn^{2+} stress. Most of the differentially accumulated FAs were saturated (octanoic, behenic, stearic, arachidic, myristic and heptadecanoic acids), whereas unsaturated FAs were represented only by oleic acid and linoleic acids. Many saturated acids accumulated in young leaves of zinc-stressed plants, while linoleic acid showed the opposite trend (Table 1). These results are consistent with the data for spinach leaves [101], which demonstrated an increase in saturated FAs accompanied by a decrease in unsaturated FAs in response to increasing doses of Cd^{2+}.

Among FAs, the octanoic acid was most strongly affected, displaying >100-fold accumulation in young leaves. The more pronounced effect of Zn^{2+} on FA contents in young leaves compared to other organs might be explained by the role of octanoic acid in (i) *de novo* fatty acid biosynthesis in mitochondria and (ii) biosynthesis of lipoic acid, an important cofactor of mitochondrial oxo-acid dehydrogenase complexes and glycine decarboxylase [102]. Furthermore, the up-regulation of long-chain saturated FAs (e.g., stearic and myristic acids) might indicate the enhancement of wax and cutin biosynthesis, which is known to be more active in growing tissues [103].

The pathway analysis confirmed a more efficient stress-induced involvement of FAs in the pathways of lipid, cutin, wax, and suberin biosynthesis in young leaves compared to roots (Figure 6). Although the pathway analysis accomplished for up- and down-regulated intermediates of lipid biosynthesis gave the most confident results for roots, the impact of FAs in these pathways (especially the linoleic acid metabolism) was still higher for the young leaves (Figure 6; Supplementary Information S1, Table S1(2,3); Supplementary Information S3, Parts S1 and S2). In this regard, a slight decrease in the level of linoleic acid (C18:2) in young leaves of *A. caudatus* could be caused by its enhanced involvement in the synthesis of cutin under Zn^{2+} stress [104].

To summarize, the increase in the content of saturated FAs in response to the accumulation of Zn^{2+} in tissues may be considered as an adaptive feature of amaranth plants. These metabolic rearrangements might contribute to the deposition of cutin and wax in the walls of leaf cells, but also might improve the rigidity of leaf and root cell membranes to make them less Zn^{2+}-permeable.

3.5. Nitrogen Metabolites in the Response of Amaranth Plants to Zn Stress

The dynamics of nitrogen-containing metabolites were intensively discussed in the context of plant responses to HM stress [9,105] with special attention paid to the role of proline as a key molecule contributing to stress tolerance. This metabolite is universally recognized as one of the most important plant osmolytes, a stress-associated signaling molecule, a potent HM chelator and ROS scavenger [105,106]. Thus, a two-fold increase in proline content was shown for Zn-stressed *Vigna unguiculata* seedlings [107] and up to four-fold for *Sinapis alba* seedlings [108]. However, in some plants, Zn^{2+} appeared to be less efficient in triggering proline accumulation compared to other HMs [9,28].

Our study revealed a Zn-induced increase in the levels of proteinogenic amino acids such as leucine, alanine and proline only in roots, but not in young and mature leaves of *A. caudatus* (Tables 1 and 2, Supplementary Information S1, Table S1(4)). These results correspond well with the pattern of metabolic responses reported for *Arabidopsis thaliana* under Cd^{2+} stress [17]. Keunen and co-workers attributed the up-regulation of N-containing metabolites to the primary role of roots as the first contact site for HMs in plants, and discussed it in the context of related signaling pathways and protective mechanisms.

Regarding the zinc effect, it has been reported that 0.5 mmol/L Zn^{2+} induced the accumulation of proline in the leaves of stress-sensitive *Lactuca sativa* plants but not in stress-tolerant species such as *Brassica oleracea* [9]. Thus, the up-regulation of proline in plant tissues might be considered as a symptom of Zn stress rather than a marker of tolerance to Zn toxicity. So far, the genus *Amaranthus* has not been addressed in terms of the proline response to Zn-induced stress. However, this aspect was considered in the context of salinity stress. Thus, Wouyou and co-workers reported that the salt-induced increase in proline content in a tolerant cultivar of *A. cruentus* could be expressed only in roots [109]. Therefore, an increase in the proline content in Zn-treated *A. caudatus* plants (which was manifested mostly in roots) could be considered as an adaptive response driven by stress-induced metabolic adjustments.

An unexpected result was the Zn-induced increase in the content of pyroglutamic acid (PG) in young amaranth leaves, accompanied by a decrease in its level in the roots (Tables 1 and 2). Currently, the biosynthesis of this cyclic lactam in plants is discussed in the context of its production via glutathione (GSH) reduction-oxidation and degradation [110]. Since GSH metabolism is affected under HM stress, changes in PG dynamics may be observed [110]. Moreover, PG was suggested to be a putative precursor of glutamate, and it was hypothesized that this compound is a major contributor to the steady-state glutamate level in plant leaves and is a main reservoir for glutamate and glutathione [111]. Thus, the conversion of PG to proline under stress conditions seems likely [112]. However, the formation of PG can also be a method-related artifact. Indeed, at least partly, the measured PG levels can be underlined by the spontaneous conversion of glutamic acid to PG under high temperature conditions of the GC injector.

Ethanolamine is one more analyte to be discussed. This compound attracted our attention due to a Zn-dependent up-regulation, which was observed; however, this was only in young leaves (Table 1). Despite its well-known importance in stress response [113], the volatility of this metabolite raises doubts about the possibility of its reliable detection by our GC-MS protocol. Indeed, most likely, this compound might be evacuated during freeze drying and/or evaporation under reduced pressure (which are the ultimate steps of the sample preparation procedure). Therefore, we suggest that ethanolamine appears as a method-related artifact, a degradation product of thermolabile N-metabolites, which readily degrade under high temperature conditions of the splitless-split injector.

Three non-proteinogenic amino acid derivatives, namely *N*-acetyl-serine, 5-methyl-cytosine and *N,N*-dimetylglycine, were up-regulated in the cells of Zn-treated amaranth plants. This was also the case for adenine and several essential nucleosides—guanosine and uridine (Tables 1 and 2, Supplementary Information S1, Table S1(4)). Most of these metabolites are known as effectors and secondary messengers directly involved in the transduction of intracellular signals and hormone regulation of stress responses in plants [114,115]. The patterns and the degree of this up-regulation in Zn-treated amaranth depended on the leaf age (Table 1, Supplementary Information S1, Table S1(4)). Generally, results of the pathway analysis indicate that nitrogen metabolism is more strongly involved in the Zn-induced responses in roots than in leaves of *A. caudatus*. Nevertheless, young leaves were also considerably affected (Figure 6, Supplementary Information S1, Figure S1(10), Supplementary Information S3). Notably, pronounced up-regulation of *N*-acetyl-serine and 5-methyl-cytosine in young leaves might imply the enhanced methylation of the corresponding substrates in response to Zn exposure. *N*-acetyl-serine may originate from its positioning isomer, *O*-acetyl-serine, and both these isomers are involved in the biosynthesis of the sulfur-containing amino acids [115]. Therefore, the increase in the abundance of *N*-acetyl-serine shown in our research might relate to the active turnover of *S*-adenosylmethionine, a universal donor of methyl groups. An increase in the content of 5-methyl-cytosine might indicate the activation of DNA methylation under Zn exposure. It is known that activation of DNA methylation is involved in epigenetic regulation underlying the tolerance to different stress conditions [116], including the HM stress [117,118].

Thus, alterations in N-metabolism in Zn-exposed plants might indicate involvement of (i) proline-related signaling, (ii) ROS scavenging in roots, (iii) induction of PG (and/or glutamate) biosynthesis in young leaves and roots and (iv) DNA methylation in Zn-induced stress tolerance of *A. caudatus*.

Finally, regarding the possible cross-talk between the biochemical processes and individual metabolites discussed above, it is important to note that the accumulation of salicylic acid might be a crucial event in the metabolic response of *A. caudatus* roots and young leaves to Zn stress. This signaling molecule may serve as a key node controlling various metabolic pathways. The primary function of the plant hormone is regulating the induction of antioxidant systems to balance increased ROS production under stressful conditions [81]. Not less importantly, SA can regulate sugar metabolism, triggering soluble sugar accumulation in leaves and roots [119]. These sugars function as osmotic regulators, ROS scavengers, and carbon and energy substrates. Moreover, SA is able to sustain a high energy level in plant cells by redirecting sugar substrates from glycolysis to the PPP, thus providing NADPH, which is necessary to cope with oxidative stress and is required for various biosynthetic processes [120]. The key intermediate of the PPP, 6-phosphogluconic acid, may be a precursor of gluconic acid, a chelator of Zn^{2+} ions. SA may also trigger the synthesis of proline in roots [121], activate the phenylpropanoid pathway leading to lignin biosynthesis in mature leaves [122] and probably might induce alterations in lipid metabolism leading to cutin and wax synthesis in young leaves similar to those shown in fruits [123]. Most of the putative SA effects enhancing stress tolerance in plants were studied under stresses other than HM exposure. Thus, a strong SA up-regulation in leaves and roots of Zn-treated plants found in our work may open new perspectives to further examine the role of this hormone as a stress metabolite. An overview of the Zn-related metabolic responses of *A. caudatus* leaves and roots is presented in Figure 7.

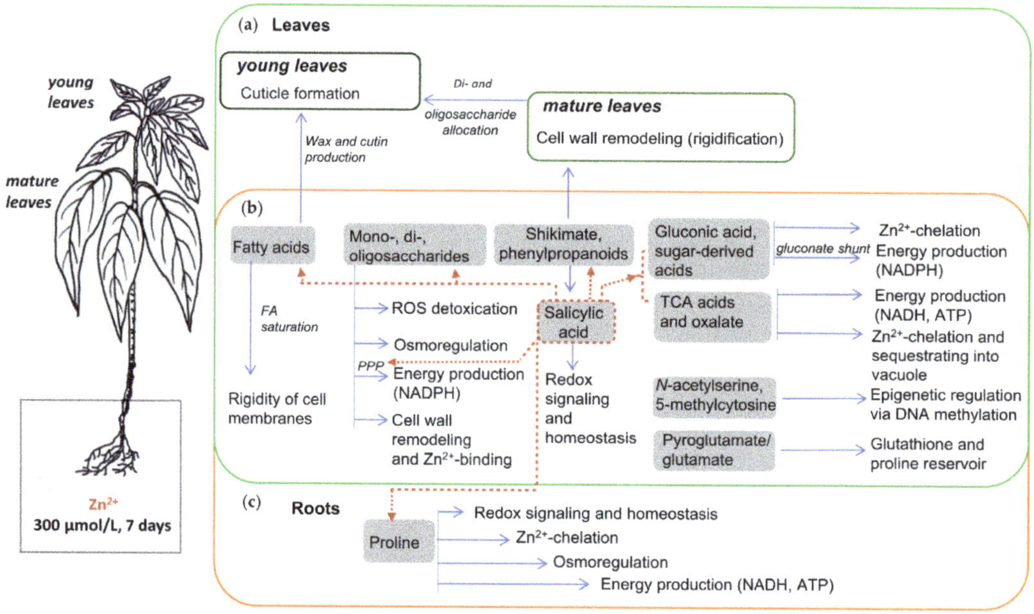

Figure 7. *Amaranthus caudatus* metabolic shifts and possible biological responses induced by Zn-treatment (Zn^{2+} 300 µmol/L, 7 days). The leaf-specific (**a**), leaf and root common (**b**) and root-specific

(c) metabolic responses to the Zn excess exposure. Red dashed line and arrows highlight the proposed on the basis of literature [119–123] central role of salicylic acid and its regulations in the metabolic response of *A. caudatus*.

4. Materials and Methods

4.1. Reagents

Unless stated otherwise, materials were obtained from the following manufacturers: Conlac GmbH (Leipzig, Germany): hexane (puriss p. a.); Macherey-Nagel GmbH and Co KG (Düren, Germany): *N*-methyl-*N*-(trimethylsilyl)trifluoroacetamide (MSTFA, MS grade); Reanal (Budapest, Hungary): *L*-aspartic acid, 2-oxoglutaric acid; Vekton (Saint-Petersburg, Russia): methanol (LC grade). All other chemicals were purchased from Merck KGaA (Darmstadt, Germany). Water was purified in-house with a water conditioning and purification system Millipore Milli-Q Gradient A10 system (resistance 18 mΩ/cm, Merck Millipore, Darmstadt, Germany).

4.2. Plant Culturing and Zn^{2+} Stress Application

Amaranthus caudatus L.-variety Karwa dauta plants were used in the study. The seeds were obtained from the collection of vegetable crops of Vavilov All-Russia Institute of Plant Genetic Resources (Saint Petersburg, Russia). The seeds were surface sterilized with 3% (*v*/*v*) H_2O_2 solution for 20 min, washed with deionized water and germinated in containers filled with calcined quartz sand. The plants were grown in two independent experiments with an identical setup (n = 9 per treatment group, in total 27 plants in nine vessels) under a 16:8 day/night regimen, 70–75% relative humidity, and day/night temperatures of 24/18 °C. Light was provided by fluorescent lamps with a wavelength range of 320–780 nm and a photosynthetic photon flux density of 120 μmol m^{-2} s^{-1}.

During the first seven days after germination, the seedlings were watered with a ten-fold diluted nutrient solution (0.1 n.s., Supplementary Information S1, Protocol S1-2) three times a day. For the next three weeks, the concentration of the solution was increased on a weekly basis to 0.2 n.s., 0.5 n.s. and 1.0 n.s, respectively. The four-week-old plants were transferred to new 3 L hydroponic vessels filled with the full-strength (1.0 n.s.) nutrient solution (the setup relied on three plants per vessel).

After two weeks of culturing in the hydroponic system, the vessels with six-week-old plants with well-developed (i.e., mature) leaves that could be easily distinguished from young leaves were randomly split into three equal groups (n = 3). Plants from the first group were harvested before stress application for a separate biomass assessment of their roots, young and mature leaves. The second group of plants was designated as "Zn-treated". These plants were subjected to Zn^{2+} stress for one week, which was accomplished by supplementation of 300 μmol/L $ZnSO_4{\cdot}7H_2O$ in the nutrient solution. This choice for Zn^{2+} concentration and the Zn^{2+} stress duration was based on the available literature data [24,25,49], including those for amaranth [35]. The plants of the third group were referred to as "controls" and remained untreated. The experiments were set in three biological replicates, i.e., three vessels with a total of nine plants in each group. The Zn-treated and control plants were characterized by an array of physiological parameters (stomatal conductivity, chlorophyll content, photosystem II activity, leaf relative water contents) prior (Day 0) and after the Zn-stress exposure (Day 7), and harvested afterwards for further biomass assessment, atomic absorption spectroscopy and metabolomics experiments.

4.3. Physiological Assays

Chlorophyll content, photosystem II activity and stomatal conductivity were assessed in a nondestructive way as described in [124–126], respectively, using the 3rd (counting from the plant top) fully expanded mature leaf of each experimental and control plant. The measurements relied on the portable devices: chlorophyll meter atLEAF (FT Green LLC, Wilmington, DE, USA), fluorometer Junior-PAM (HeinzWalz GmbH, Effeltrich, Germany), porometer SC-1 (Delta-T Devices Ltd., Cottbus, Germany), respectively, according to the manufacturer's instructions. The obtained atLEAF units were used to calculate chlorophyll content ($mg \cdot cm^2$) [124]. The leaf relative water contents (LRWC) were as follows: LRWC (%) = (fresh weight − dry weight) × 100%/fresh weight. For the determination of this parameter for young and mature leaves, the 4th young (i.e., not fully expanded) leaf and 4th mature (fully expanded) leaf, respectively, as indicated in Supplementary Information S1, Table S1(4), were collected from each plant.

4.4. Plant Fresh and Dry Biomass Determination

Roots, young and mature leaves of six- and seven-week-old plants were collected separately. The roots were sequentially rinsed for 5 min with 0.1 mmol/L $CaCl_2$ and deionized water. Afterwards, the roots, young and mature leaves were weighed, oven-dried at 105 °C for 1 h and dried at 70 °C for the following 24 h. Thus, the fresh and dry weights were determined prior to and after the Zn^{2+} application.

4.5. Atomic Absorption Spectroscopy (AAS)-Based Analysis of Zn in Plant Organs

The contents of Zn in roots, young and mature leaves of Zn-treated and control plants were determined in the relevant dried and ground plant samples by atomic absorption spectroscopy (AAS, Shimadzu AA-7000, Kyoto, Japan). Plant samples (100 mg) were placed in heat-resistant glass flasks and digested with 5 mL of concentrated $HNO_3/HClO_4$ 4:1 (*v*/*v*) at 160 °C (electric stove), followed by subsequent dilution to 100 mL with deionized water and further Zn analysis [127].

The fresh, dry weights and metal contents were assessed for pooled samples (three leaf or root samples to build one experimental replicate, in total n = three per group).

4.6. GC-MS-Based Metabolite Profiling

Profiling of polar primary metabolites relied on the in-house established protocol, including (i) extraction of plant material with aq. methanol, (ii) subsequent derivatization of the extracted primary metabolites and (iii) their analysis by gas chromatography-electron ionization-quadrupole-mass spectrometry (GC-EI-Q-MS) as described elsewhere [128] with a few modifications indicated in Supplementary Information S1, Protocol S1-3. The prepared samples (1 µL) containing derivatized metabolites were analyzed by a GC2010 gas chromatograph coupled online to a quadrupole mass selective detector, Shimadzu GCMS QP201, operating under the instrumental settings summarized in Table S1(10). An assessment of the method performance relied on quality controls (QCs, i.e., aliquots of the pool prepared by mixing all individual extracts), which were included in the sequence for the GC-MS analysis [128]. The quality of the acquired chromatograms was assessed by verification of the baseline regularity, background MS noise, as well as the symmetry, width and height of peaks. To obtain qualitative information about individual metabolites, the chromatograms were processed by AMDIS (version 2.66 from 08.08.2008, www.amdis.net, accessed on 5 March 2020). This software was used to accomplish deconvolution of mass spectra, peak picking, Kovach retention indices (RI) calculation by retention times (t_R) of C_8–C_{20} alkanes and annotation of analytes. The further analysis relied on the untargeted approach, i.e., unbiased TIC-based identification of all signals with the signal-to-noise ratio (S/N) $\geq$ 3. For this, the trimethylsilyl (TMS) and methyl oxime (MEOX)-TMS derivatives underlying all detected individual chromatographic peaks (corresponding to individual MEOX/TMS derivatives, further defined as features) were annotated by t_R, RI and EI-MS data—the results of spectral similarity search with the NIST Search engine against available electron ionization (EI)-MS spectral libraries such as National Institute of Standards and Technology (NIST, https://webbook.nist.gov/chemistry/, accessed on 10 March 2020), Golm Metabolome Database (GMD, http://gmd.mpimp-golm.mpg.de/, accessed on 10 March 2020), and in-house spectral library. The similarity between the experimental spectra and the reference EI-MS entries in external libraries was assessed by the match factor. Its values, such as >800, 799–700, 699–600, 599–500 and <500, were interpreted as excellent, very good, good, satisfactory and unsatisfactory results, respectively (Table S1(1) and Figure S1(11)). Relative quantification was accomplished via direct comparison of individual analyte abundances, derived as integral areas of corresponding peaks in extracted ion chromatograms (XIC, $m/z \pm 0.5$ Da) for representative intense signals at specific t_R XIC. The analyte quantification procedure was accomplished with Xcalibur™ and LCquan™ (version 2.5.6, ThermoFisher Scientific Inc., Bremen, Germany) and MSDial software (https://systemsomicslab.github.io/compms/msdial/main.html, accessed on 10 March 2020), which performs alignment of chromatograms by t_R of analytes and the integration of analyte peak areas. The quantitation results (i.e., integrated peak areas detected in each sample) were organized into a digital matrix and were statistically processed by using the Metaboanalyst 4.0 (https://www.metaboanalyst.ca/, accessed on 16 March 2020). Prior to the statistical analysis, the data were normalized by dry weight of samples and filtered to exclude the features not detected in ≥20% of the samples. For metabolites which were not detected in less than 20% of samples (and less than 30% per experimental group), imputation of missing values was performed by the random forest algorithm with the help of MetImp 1.2 on-line tool [129]. The quantitative metabolomics data were analyzed by principal component analysis (PCA). To visualize changes in the levels of individual assigned metabolites, hierarchical clustering analysis with heatmap representation was accomplished.

Targeted absolute quantitative analysis relied on external standardization with 29 authentic standards (oxalic acid, malonic acid, succinic acid, tartaric acid, malic acid, aconitic acid, citric acid, fumaric acid, benzoic acid, ascorbic acid, erythronic acid, glycerol, arabinose, glucose, galactose, *myo*-inositol, sucrose, urea, Ala, Trp, Ile, Leu, Asn, Asp, Glu, Pro, Val, Ser, Thr) prepared as a total mix serially diluted in the range from 0.2 µmol/L to 0.2 mmol/L. The statistical interpretation and bioinformatics post-processing of the acquired quantitative data relied on MetaboAnalyst 5.0 (https://www.metaboanalyst.ca accessed on 10 September 2021) online platform [130].

4.7. Metabolic Pathway Analysis

Annotation of individual metabolites to Zn-responsive pathways relied on the *Arabidopsis thaliana* pathway library (deposited by KEGG on-line platform from 03.2020). The pathway analysis was accomplished with MetaboAnalyst 4.0 software (https://www.metaboanalyst.ca, accessed on 19 March 2020) and employed the combination of pathway enrichment analysis by the globaltest method and pathway topology analysis [131]. As the pathway enrichment analysis relies on the compound-specific abundance values, only significantly (t-test $p \leq 0.05$) differentially abundant structurally annotated metabolites were included in the input list. The pathway topology analysis gives access to the pathways, in which the changes in more important positions of the enzyme network more significant impact on the observed differences. To assess the node centrality (i.e., to estimate node importance), the betweenness centrality method was applied.

4.8. Statistics

The quantitative metabolomics data were analyzed by the methods of multivariate (principal component analysis, PCA) and univariate (volcano plot) statistics. Results of the PCA are presented as score plots built for the first two principal components. The plots were obtained for abundances of individual metabolites detected in the form of TMS and TMS-MEOX derivatives by untargeted GC-EI-Q-MS of *A. caudatus* aq. methanolic extracts, which were prepared from young leaves and roots. The PCA score plots were built in MetaboAnalyst 4.0 (https://www.metaboanalyst.ca, accessed on 16 March 2020). The results of the univariate statistical analysis (presented as figures and tables) were expressed as mean $\pm$ standard deviation (StD) of three biological replicates. Significance of the differences in the contents of individual metabolites between Zn-treated groups and untreated controls was estimated by *t*-test (Student's test). Thereby, for the untargeted metabolite profiling, the threshold values for the *p*-value and fold change (FC) were 0.05 and 1.5, respectively, as they are often in metabolomics studies [132]. For the targeted analysis, whose main focus is highly sensitive, precise absolute quantification of selected metabolites, the FC threshold was decreased to 1.3. To address the reliability of the observed Zn^{2+}-related changes in the abundances of individual metabolites, the false discovery rates (FDRs) were assessed at $p \leq 0.05$ with the Benjamini–Hochberg method for all comparisons [133].

The results obtained by the targeted approach were expressed in µmol/g DW as mean $\pm$ standard deviation (StD) of three biological replicates (each pooled from three individual leaves or root samples). All calculations were performed within the linear dynamic ranges (LDRs, $R^2 \geq 0.95$) derived in additional dilution experiments ($n = 3$) with individual authentic standards. The resulting individual calibration curves were built on a logarithmic scale for the analyte peak areas as a function of the applied concentration (0.2, 0.5, 1, 2, 5, 10, 20, 50, 100 and 200 µmol/L, Supplementary Information S1, Figure S1(12)).

5. Conclusions

In this study, we provided a first insight into the metabolic responses of *A. caudatus* plants to moderate Zn-stress. Our metabolomics survey revealed strong changes in the contents of multiple primary metabolites in Zn-stressed plants. The more pronounced changes in the metabolic profiles of amaranth roots compared to leaves indicate a clear organ specificity of the stress-induced biochemical rearrangements. At least in part, this effect may be explained by the crucial role of roots in Zn uptake and detoxication in plants. Sugars and organic acids were the principal up-regulated metabolites in all organs of *A. caudatus*, which may indicate their key role during the early stages of the plant's biochemical adaptation to Zn-related toxicity. The specific dynamics of sugar metabolites in different organs may reflect the activation of source-sink relocation of reserve di- and oligosaccharides. The involvement of mono- and disaccharides in the stress response can be attributed to their role in ROS scavenging and osmoregulation. The observed increase in galactose content in the roots suggests that root cell walls contribute to the scavenging of Zn^{2+} ions. However, an increased demand for galactose to maintain the synthesis of membrane galactolipids under stress cannot be ruled out. The higher responsivity of organic acids to Zn^{2+} in roots might be attributed to their active participation in Zn^{2+} chelation. The extremely high (59-fold) up-regulation of gluconic acid in roots, which can be caused either by direct glucose oxidation or by the putative gluconate shunt of the pentose phosphate pathway, may prevent free Zn^{2+} transfer from roots to leaves. The impact of the TCA cycle metabolites in stress-induced metabolic adjustments was much less pronounced. The accumulation of salicylic acid both in young leaves and roots suggests a possible role of this signaling molecule in the activation of Zn-stress tolerance mechanisms in *A. caudatus* plants. Increased saturated fatty acids in Zn-stressed plants may cause rigidification of cell membranes in both leaves and roots, as well as cuticle formation in young leaves. Finally, the up-regulation of the phenylpropanoid pathway intermediates may underlie the enhancement of lignin biosynthesis and stiffening of the cell walls.

Among N-metabolites, proline was the most affected compound in the roots of *A. caudatus*. The accumulation of *N*-acetylserine and 5-metylcitosine in roots and leaves may be an indicator of DNA-methylation and epigenetic regulation induced by exposure to Zn^{2+}. In general, our results showed that adaptation of *A. caudatus* to Zn-stress was accompanied by considerable metabolic shifts involving a large set of metabolic pathways. Since biochemical rearrangements can be carried out at both posttranslational and transcriptional levels, in the next step of our study, we examined the proteomic response of *A. caudatus* to Zn exposure.

Supplementary Materials: The following supporting information can be downloaded at: https://www.mdpi.com/article/10.3390/plants14142119/s1, Supplementary Information S1: Protocols, Figures and Tables; Supplementary Information S2: Measurements and calculations of chlorophyll content (Table S2(1)), photosystem II activity (Table S2(2)), stomatal conductivity (Table S2(3)), LRWC (Table S2(4)), fresh weight (Table S2(5)), dry weight (Table S2(6)), Zn^{2+} content (Table S2(7)); Supplementary Information S3: Metabolic Pathways revealed by the Pathway Analysis; Supplementary Information S4: Pathway Analysis detailed information.

Author Contributions: N.O. proposed the idea of the study and wrote the manuscript draft, T.B., A.K., S.S. and A.F. designed the laboratory experiment; T.B., E.T. and N.F. accomplished processing and post-processing of metabolomics data, V.D.V., A.G. and T.Z. grew plants; V.D.V., A.G. and T.Z. accomplished plant Zn-stress experiments and collected plant material. A.G. and T.Z. measured physiological parameters. V.D.V. and T.B. performed the extraction of primarily metabolites; A.O. performed GC-MS analyses for thermostable primarily metabolome characterization; N.O., T.B., E.T., A.K., N.F., S.S. and A.F. contributed to writing the final manuscript draft and critical discussion; N.O., T.B., E.T. and A.F. supervised the final version of the manuscript. All authors have read and agreed to the published version of the manuscript.

Funding: Funding for this study was from the Russian Scientific Foundation (grants # 21-74-30003 and 23-44-00101).

Data Availability Statement: All relevant data are available within the article and Supplementary Materials. The raw GC-MS data files related to the research are available on the NIH Common Fund's National Metabolomics Data Repository (NMDR) website, the Metabolomics Workbench, https://www.metabolomicsworkbench.org, accessed on 29 January 2024, where it has been assigned Project ID PR001910. The data can be accessed directly via its Project DOI: 10.21228/M8TT6W. Additionally, all the initial analytical data obtained when estimating the peak areas on chromatograms are also available and could be provided by request.

Acknowledgments: The authors acknowledge Saint-Petersburg State University for a research project 118201839 and the Chemical Analysis and Materials Research Centre and Center of Molecular and Cell Technologies for technical support under agreement C-111/30(01/1-69-46-SPbU). We are thankful to Vavilov All-Russia Institute of Plant Genetic Resources (Saint Petersburg, Russia) for providing *Amaranthus caudatus* L. seeds variety Karwa dauta, which were used in the study. The infrastructural support from the Ministry of Science and Higher Education of the Russian Federation (theme #122042700044-6) is acknowledged. The work relied on the experimental climate control facility in the Institute of Bioengineering, Research Center of Biotechnology, Russian Academy of Science, U-73547. The technical support provided by Maria Soloveva, Valeria Mashkina and Ludmila Kuchaeva is acknowledged.

Conflicts of Interest: The authors declare no conflicts of interest.

Abbreviations

AOX, alternative oxidase; DW, dry weight; EI, electron ionization; EI-MS, electron ionization mass spectrum; FA, fatty acid; FC, fold change; FDR, false discovery rate; G6PDH, glucose-6-phosphate dehydrogenase; GC-EI-Q-MS, gas chromatography-electron ionization-quadrupole mass spectrometry; GDM, Golm Metabolome Database; GSH, glutathione; HM, heavy metals; HMDB, Human

Metabolome Database; KEGG, Kyoto Encyclopedia of Genes and Genomes; LDR, linear dynamic ranges; LRWC, leaf related water content; MEOX-TMS, methyl oxime-trimethylsilyl; MSTFA, *N*-methyl-*N*-(trimethylsilyl) trifluoroacetamide; n.s., nutrient solution; N-metabolites, nitrogen containing metabolites; NIST, National Institute of Standards and Technology; PC, principal component; PCA, principal component analysis; PG, pyroglutamic acid; P-metabolites, phosphate containing metabolites; PPP, pentose phosphate pathway; RI, retention index; ROS, reactive oxygen species; SA, Salicylic acid; StD, standard deviation; TCA cycle, tricarboxylic acid cycle; TIC, total ion current; TMS, trimethylsilyl; t_R, retention time; XIC, extracted ion chromatograms.

References

1. Ghori, N.-H.; Ghori, T.; Hayat, M.Q.; Imadi, S.R.; Gul, A.; Altay, V.; Ozturk, M. Heavy metal stress and responses in plants. *Int. J. Environ. Sci. Technol.* **2019**, *16*, 1807–1828. [CrossRef]
2. Goyal, D.; Yadav, A.; Prasad, M.; Singh, T.B.; Shrivastav, P.; Ali, A.; Dantu, P.K.; Mishra, S. Effect of Heavy Metals on Plant Growth: An Overview. In *Contaminants in Agriculture: Sources, Impacts and Management*; Naeem, M., Ansari, A.A., Gill, S.S., Eds.; Springer International Publishing: Cham, Switzerland, 2020; pp. 79–101. ISBN 978-3-030-41552-5.
3. El-Sappah, A.H.; Zhu, Y.; Huang, Q.; Chen, B.; Soaud, S.A.; Abd Elhamid, M.A.; Yan, K.; Li, J.; El-Tarabily, K.A. Plants' molecular behavior to heavy metals: From criticality to toxicity. *Front. Plant Sci.* **2024**, *15*, 1423625. [CrossRef]
4. Sharma, S.S.; Dietz, K.-J. The relationship between metal toxicity and cellular redox imbalance. *Trends Plant Sci.* **2009**, *14*, 43–50. [CrossRef] [PubMed]
5. Hossain, M.A.; Liu, F.; Burritt, D.J.; Fujita, M.; Huang, B. *Priming-Mediated Stress and Cross-Stress Tolerance in Crop Plants;* Academic Press: Cambridge, MA, USA, 2020; ISBN 978-0-12-817893-5.
6. Gratani, L. Plant Phenotypic Plasticity in Response to Environmental Factors. *Adv. Bot.* **2014**, *2014*, e208747. [CrossRef]
7. Singh, M.; Kumar, J.; Singh, S.; Singh, V.P.; Prasad, S.M.; Singh, M.P.V.V.B. Adaptation Strategies of Plants against Heavy Metal Toxicity: A Short Review. *Biochem. Pharmacol.* **2015**, *4*, 161. [CrossRef]
8. Zhang, H.; Zhu, J.; Gong, Z.; Zhu, J.-K. Abiotic stress responses in plants. *Nat. Rev. Genet.* **2022**, *23*, 104–119. [CrossRef]
9. Paradisone, V.; Barrameda-Medina, Y.; Montesinos-Pereira, D.; Romero, L.; Esposito, S.; Ruiz, J.M. Roles of some nitrogenous compounds protectors in the resistance to zinc toxicity in *Lactuca sativa* cv. Phillipus and *Brassica oleracea* cv. Bronco. *Acta Physiol. Plant.* **2015**, *37*, 137. [CrossRef]
10. Patel, M.K.; Pandey, S.; Kumar, M.; Haque, M.I.; Pal, S.; Yadav, N.S. Plants Metabolome Study: Emerging Tools and Techniques. *Plants* **2021**, *10*, 2409. [CrossRef]
11. Allwood, J.W.; Williams, A.; Uthe, H.; van Dam, N.M.; Mur, L.A.J.; Grant, M.R.; Pétriacq, P. Unravelling Plant Responses to Stress—The Importance of Targeted and Untargeted Metabolomics. *Metabolites* **2021**, *11*, 558. [CrossRef]
12. Carrera, F.P.; Noceda, C.; Maridueña-Zavala, M.G.; Cevallos-Cevallos, J.M. Metabolomics, a Powerful Tool for Understanding Plant Abiotic Stress. *Agronomy* **2021**, *11*, 824. [CrossRef]
13. Jamla, M.; Khare, T.; Joshi, S.; Patil, S.; Penna, S.; Kumar, V. Omics approaches for understanding heavy metal responses and tolerance in plants. *Curr. Plant Biol.* **2021**, *27*, 100213. [CrossRef]
14. Jorge, T.F.; Rodrigues, J.A.; Caldana, C.; Schmidt, R.; van Dongen, J.T.; Thomas-Oates, J.; António, C. Mass spectrometry-based plant metabolomics: Metabolite responses to abiotic stress. *Mass Spectrom. Rev.* **2016**, *35*, 620–649. [CrossRef] [PubMed]
15. Wang, Y.; Xu, L.; Shen, H.; Wang, J.; Liu, W.; Zhu, X.; Wang, R.; Sun, X.; Liu, L. Metabolomic analysis with GC-MS to reveal potential metabolites and biological pathways involved in Pb & Cd stress response of radish roots. *Sci. Rep.* **2015**, *5*, 18296. [CrossRef]
16. Sun, X.; Zhang, J.; Zhang, H.; Ni, Y.; Zhang, Q.; Chen, J.; Guan, Y. The responses of *Arabidopsis thaliana* to cadmium exposure explored via metabolite profiling. *Chemosphere* **2010**, *78*, 840–845. [CrossRef]
17. Keunen, E.; Florez-Sarasa, I.; Obata, T.; Jozefczak, M.; Remans, T.; Vangronsveld, J.; Fernie, A.R.; Cuypers, A. Metabolic responses of *Arabidopsis thaliana* roots and leaves to sublethal cadmium exposure are differentially influenced by ALTERNATIVE OXIDASE1a. *Environ. Exp. Bot.* **2016**, *124*, 64–78. [CrossRef]
18. D'Alessandro, A.; Taamalli, M.; Gevi, F.; Timperio, A.M.; Zolla, L.; Ghnaya, T. Cadmium Stress Responses in *Brassica juncea*: Hints from Proteomics and Metabolomics. *J. Proteome Res.* **2013**, *12*, 4979–4997. [CrossRef]
19. Xie, Y.; Hu, L.; Du, Z.; Sun, X.; Amombo, E.; Fan, J.; Fu, J. Effects of Cadmium Exposure on Growth and Metabolic Profile of Bermudagrass [*Cynodon dactylon* (L.) Pers.]. *PLoS ONE* **2014**, *9*, e115279. [CrossRef]
20. Hédiji, H.; Djebali, W.; Cabasson, C.; Maucourt, M.; Baldet, P.; Bertrand, A.; Boulila Zoghlami, L.; Deborde, C.; Moing, A.; Brouquisse, R.; et al. Effects of long-term cadmium exposure on growth and metabolomic profile of tomato plants. *Ecotoxicol. Environ. Saf.* **2010**, *73*, 1965–1974. [CrossRef]

21. Zhang, Y.; Wang, Y.; Ding, Z.; Wang, H.; Song, L.; Jia, S.; Ma, D. Zinc stress affects ionome and metabolome in tea plants. *Plant Physiol. Biochem. PPB* **2017**, *111*, 318–328. [CrossRef]
22. Zemanová, V.; Pavlíková, D.; Novák, M.; Hnilička, F. The Dual Role of Zinc in Spinach Metabolism: Beneficial × Toxic. *Plants* **2024**, *13*, 3363. [CrossRef]
23. Kopittke, P.M.; Blamey, F.P.C.; Asher, C.J.; Menzies, N.W. Trace metal phytotoxicity in solution culture: A review. *J. Exp. Bot.* **2010**, *61*, 945–954. [CrossRef] [PubMed]
24. Tsonev, T.; Lidon, F. Zinc in plants—An overview. *Emir. J. Food Agric.* **2012**, *24*, 322–333.
25. Kaur, H.; Garg, N. Zinc toxicity in plants: A review. *Planta* **2021**, *253*, 129. [CrossRef] [PubMed]
26. Broadley, M.R.; White, P.J.; Hammond, J.P.; Zelko, I.; Lux, A. Zinc in plants. *New Phytol.* **2007**, *173*, 677–702. [CrossRef]
27. Sagardoy, R.; Morales, F.; Rellán-Álvarez, R.; Abadía, A.; Abadía, J.; López-Millán, A.F. Carboxylate metabolism in sugar beet plants grown with excess Zn. *J. Plant Physiol.* **2011**, *168*, 730–733. [CrossRef]
28. Li, X.; Yang, Y.; Jia, L.; Chen, H.; Wei, X. Zinc-induced oxidative damage, antioxidant enzyme response and proline metabolism in roots and leaves of wheat plants. *Ecotoxicol. Environ. Saf.* **2013**, *89*, 150–157. [CrossRef]
29. Anjum, N.A.; Hasanuzzaman, M.; Hossain, M.A.; Thangavel, P.; Roychoudhury, A.; Gill, S.S.; Rodrigo, M.A.M.; Adam, V.; Fujita, M.; Kizek, R.; et al. Jacks of metal/metalloid chelation trade in plants—An overview. *Front. Plant Sci.* **2015**, *6*, 192.
30. Osmolovskaya, N.; Vu, D.V.; Kuchaeva, L. The role of organic acids in heavy metal tolerance in plants. *Biol. Commun.* **2018**, *63*, 9–16. [CrossRef]
31. Panchal, P.; Miller, A.J.; Giri, J. Organic acids: Versatile stress-response roles in plants. *J. Exp. Bot.* **2021**, *72*, 4038–4052. [CrossRef]
32. Bosiacki, M.; Kleiber, T.; Kaczmarek, J. Evaluation of Suitability of *Amaranthus Caudatus* L. and *Ricinus Communis* L. in Phytoextraction of Cadmium and Lead from Contaminated Substrates. *Arch. Environ. Prot.* **2013**, *39*, 47–59. [CrossRef]
33. Watanabe, T.; Murata, Y.; Osaki, M. Amaranthus Tricolor Has the Potential for Phytoremediation of Cadmium-Contaminated Soils. *Commun. Soil Sci. Plant Anal.* **2009**, *40*, 3158–3169. [CrossRef]
34. Ko, C.-H.; Chang, F.-C.; Wang, Y.-N.; Chung, C.-Y. Extraction of Heavy Metals from Contaminated Soil by Two *Amaranthus* spp. *CLEAN-Soil Air Water* **2014**, *42*, 635–640. [CrossRef]
35. Lukatkin, A.S.; Bashmakov, D.I.; Al Harbawee, W.E.Q.; Teixeira da Silva, J.A. Assessment of physiological and biochemical responses of *Amaranthus retroflexus* seedlings to the accumulation of heavy metals with regards to phytoremediation potential. *Int. J. Phytoremediat.* **2021**, *23*, 219–230. [CrossRef]
36. Hunková, J.; Lisinovičová, M.; Lancikova, V.; Szabóová, M. A comparative analysis of heavy metal stress responses in different grain amaranth cultivars. *ResearchGate* **2025**, *14*, 100619. [CrossRef]
37. Rastogi, A.; Shukla, S. Amaranth: A new millennium crop of nutraceutical values. *Crit. Rev. Food Sci. Nutr.* **2013**, *53*, 109–125. [CrossRef]
38. Achigan-Dako, E.G.; Sogbohossou, O.E.D.; Maundu, P. Current knowledge on *Amaranthus* spp.: Research avenues for improved nutritional value and yield in leafy amaranths in sub-Saharan Africa. *Euphytica* **2014**, *197*, 303–317. [CrossRef]
39. Ragasa, C.Y.; Austria, J.P.M.; Subosa, A.F.; Torres, O.B.; Shen, C.-C. Chemical Constituents of *Amaranthus viridis*. *Chem. Nat. Compd.* **2015**, *51*, 146–147. [CrossRef]
40. Joshi, D.C.; Sood, S.; Hosahatti, R.; Kant, L.; Pattanayak, A.; Kumar, A.; Yadav, D.; Stetter, M.G. From zero to hero: The past, present and future of grain amaranth breeding. *Theor. Appl. Genet.* **2018**, *131*, 1807–1823. [CrossRef]
41. Ludwig, M. The Roles of Organic Acids in C4 Photosynthesis. *Front. Plant Sci.* **2016**, *7*, 647.
42. Gélinas, B.; Seguin, P. Oxalate in grain amaranth. *J. Agric. Food Chem.* **2007**, *55*, 4789–4794. [CrossRef]
43. Osmolovskaya, N.G.; Dung, V.V.; Kudryashova, Z.K.; Kuchaeva, L.N.; Popova, N.F. Effect of Cadmium on Distribution of Potassium, Calcium, Magnesium, and Oxalate Accumulation in *Amaranthus cruentus* L. Plants. *Russ. J. Plant Physiol.* **2018**, *65*, 553–562. [CrossRef]
44. Harvey, D.J.; Vouros, P. Mass spectrometric fragmentation of trimethylsilyl and related alkylsilyl derivatives. *Mass Spectrom. Rev.* **2020**, *39*, 105–211. [CrossRef] [PubMed]
45. Boldizsár, I.; Füzfai, Z.; Molnár-Perl, I. Characteristic fragmentation patterns of trimethylsilyl and trimethylsilyl-oxime derivatives of plant disaccharides as obtained by gas chromatography coupled to ion-trap mass spectrometry. *J. Chromatogr. A* **2011**, *1218*, 7864–7868. [CrossRef] [PubMed]
46. Füzfai, Z.; Boldizsár, I.; Molnár-Perl, I. Characteristic fragmentation patterns of the trimethylsilyl and trimethylsilyl-oxime derivatives of various saccharides as obtained by gas chromatography coupled to ion-trap mass spectrometry. *J. Chromatogr. A* **2008**, *1177*, 183–189. [CrossRef] [PubMed]
47. Kalinova, J.; Dadakova, E. Rutin and total quercetin content in amaranth (*Amaranthus* spp.). *Plant Foods Hum. Nutr. (Dordr. Neth.)* **2009**, *64*, 68–74. [CrossRef]
48. Paudel, G.; Bilova, T.; Schmidt, R.; Greifenhagen, U.; Berger, R.; Tarakhovskaya, E.; Stöckhardt, S.; Balcke, G.U.; Humbeck, K.; Brandt, W.; et al. Osmotic stress is accompanied by protein glycation in *Arabidopsis thaliana*. *J. Exp. Bot.* **2016**, *67*, 6283–6295. [CrossRef]

49. Chen, H.; Song, L.; Zhang, H.; Wang, J.; Wang, Y.; Zhang, H. Cu and Zn Stress affect the photosynthetic and antioxidative systems of alfalfa (*Medicago sativa*). *J. Plant Interact.* **2022**, *17*, 695–704. [CrossRef]
50. Keunen, E.; Peshev, D.; Vangronsveld, J.; Van Den Ende, W.; Cuypers, A. Plant sugars are crucial players in the oxidative challenge during abiotic stress: Extending the traditional concept. *Plant Cell Environ.* **2013**, *36*, 1242–1255. [CrossRef]
51. Pommerrenig, B.; Ludewig, F.; Cvetkovic, J.; Trentmann, O.; Klemens, P.A.W.; Neuhaus, H.E. In Concert: Orchestrated Changes in Carbohydrate Homeostasis Are Critical for Plant Abiotic Stress Tolerance. *Plant Cell Physiol.* **2018**, *59*, 1290–1299. [CrossRef]
52. Shumilina, J.; Kusnetsova, A.; Tsarev, A.; Janse van Rensburg, H.C.; Medvedev, S.; Demidchik, V.; Van den Ende, W.; Frolov, A. Glycation of Plant Proteins: Regulatory Roles and Interplay with Sugar Signalling? *Int. J. Mol. Sci.* **2019**, *20*, 2366. [CrossRef]
53. Peshev, D.; Vergauwen, R.; Moglia, A.; Hideg, E.; Van den Ende, W. Towards understanding vacuolar antioxidant mechanisms: A role for fructans? *J. Exp. Bot.* **2013**, *64*, 1025–1038. [CrossRef]
54. Gangola, M.P.; Ramadoss, B.R. Chapter 2—Sugars Play a Critical Role in Abiotic Stress Tolerance in Plants. In *Biochemical, Physiological and Molecular Avenues for Combating Abiotic Stress Tolerance in Plants*; Wani, S.H., Ed.; Academic Press: Cambridge, MA, USA, 2018; pp. 17–38. ISBN 978-0-12-813066-7.
55. Couée, I.; Sulmon, C.; Gouesbet, G.; El Amrani, A. Involvement of soluble sugars in reactive oxygen species balance and responses to oxidative stress in plants. *J. Exp. Bot.* **2006**, *57*, 449–459. [CrossRef] [PubMed]
56. Bolouri-Moghaddam, M.R.; Le Roy, K.; Xiang, L.; Rolland, F.; Van den Ende, W. Sugar signalling and antioxidant network connections in plant cells. *FEBS J.* **2010**, *277*, 2022–2037. [CrossRef] [PubMed]
57. Wu, D.; Cai, S.; Chen, M.; Ye, L.; Chen, Z.; Zhang, H.; Dai, F.; Wu, F.; Zhang, G. Tissue Metabolic Responses to Salt Stress in Wild and Cultivated Barley. *PLoS ONE* **2013**, *8*, e55431. [CrossRef]
58. Du, Y.; Zhao, Q.; Chen, L.; Yao, X.; Zhang, W.; Zhang, B.; Xie, F. Effect of drought stress on sugar metabolism in leaves and roots of soybean seedlings. *Plant Physiol. Biochem. PPB* **2020**, *146*, 1–12. [CrossRef]
59. Thomas, A.; Beena, R.; Laksmi, G.; Soni, K.B.; Alex, S.; Viji, M.M. Changes in sucrose metabolic enzymes to water stress in contrasting rice genotypes. *Plant Stress* **2022**, *5*, 100088. [CrossRef]
60. Turgeon, R. The Sink-Source Transition in Leaves. *Annu. Rev. Plant Physiol. Plant Mol. Biol.* **1989**, *40*, 119–138. [CrossRef]
61. Geiger, D. Plant Sucrose Transporters from a Biophysical Point of View. *Mol. Plant* **2011**, *4*, 395–406. [CrossRef]
62. Dong, S.; Zhang, J.; Beckles, D.M. A pivotal role for starch in the reconfiguration of 14C-partitioning and allocation in *Arabidopsis thaliana* under short-term abiotic stress. *Sci. Rep.* **2018**, *8*, 9314. [CrossRef]
63. Paul, M.J.; Watson, A.; Griffiths, C.A. Linking fundamental science to crop improvement through understanding source and sink traits and their integration for yield enhancement. *J. Exp. Bot.* **2020**, *71*, 2270–2280. [CrossRef]
64. Srivastava, S.; Bisht, H.; Sidhu, O.P.; Srivastava, A.; Singh, P.C.; Pandey, R.M.; Raj, S.K.; Roy, R.; Nautiyal, C.S. Changes in the metabolome and histopathology of *Amaranthus hypochondriacus* L. in response to *Ageratum enation* virus infection. *Phytochemistry* **2012**, *80*, 8–16. [CrossRef]
65. Zhang, G.-Y.; Liu, R.-R.; Zhang, C.-Q.; Tang, K.-X.; Sun, M.-F.; Yan, G.-H.; Liu, Q.-Q. Manipulation of the Rice L-Galactose Pathway: Evaluation of the Effects of Transgene Overexpression on Ascorbate Accumulation and Abiotic Stress Tolerance. *PLoS ONE* **2015**, *10*, e0125870. [CrossRef]
66. Krzesłowska, M. The cell wall in plant cell response to trace metals: Polysaccharide remodeling and its role in defense strategy. *Acta Physiol. Plant.* **2011**, *33*, 35–51. [CrossRef]
67. Glińska, S.; Gapińska, M.; Michlewska, S.; Skiba, E.; Kubicki, J. Analysis of *Triticum aestivum* seedling response to the excess of zinc. *Protoplasma* **2016**, *253*, 367–377. [CrossRef]
68. Doyama, K.; Yamaji, K.; Haruma, T.; Ishida, A.; Mori, S.; Kurosawa, Y. Zn tolerance in the evergreen shrub, *Aucuba japonica*, naturally growing at a mine site: Cell wall immobilization, aucubin production, and Zn adsorption on fungal mycelia. *PLoS ONE* **2021**, *16*, e0257690. [CrossRef]
69. Caldelas, C.; Weiss, D.J. Zinc Homeostasis and isotopic fractionation in plants: A review. *Plant Soil* **2017**, *411*, 17–46. [CrossRef]
70. Krzesłowska, M.; Rabęda, I.; Basińska, A.; Lewandowski, M.; Mellerowicz, E.J.; Napieralska, A.; Samardakiewicz, S.; Woźny, A. Pectinous cell wall thickenings formation—A common defense strategy of plants to cope with Pb. *Environ. Pollut.* **2016**, *214*, 354–361. [CrossRef]
71. Castro, J.C.; Castro, C.G.; Cobos, M. Genetic and biochemical strategies for regulation of L-ascorbic acid biosynthesis in plants through the L-galactose pathway. *Front. Plant Sci.* **2023**, *14*, 1099829. [CrossRef]
72. Xiao, M.; Li, Z.; Zhu, L.; Wang, J.; Zhang, B.; Zheng, F.; Zhao, B.; Zhang, H.; Wang, Y.; Zhang, Z. The Multiple Roles of Ascorbate in the Abiotic Stress Response of Plants: Antioxidant, Cofactor, and Regulator. *Front. Plant Sci.* **2021**, *12*, 598173. [CrossRef]
73. Wang, F.; Ding, D.; Li, J.; He, L.; Xu, X.; Zhao, Y.; Yan, B.; Li, Z.; Xu, J. Characterisation of genes involved in galactolipids and sulfolipids metabolism in maize and Arabidopsis and their differential responses to phosphate deficiency. *Funct. Plant Biol.* **2020**, *47*, 279–292. [CrossRef] [PubMed]
74. Foroughi, S.; Baker, A.J.M.; Roessner, U.; Johnson, A.A.T.; Bacic, A.; Callahan, D.L. Hyperaccumulation of zinc by *Noccaea caerulescens* results in a cascade of stress responses and changes in the elemental profile. *Metallomics* **2014**, *6*, 1671–1682. [CrossRef]

75. Härtel, H.; Dormann, P.; Benning, C. DGD1-independent biosynthesis of extraplastidic galactolipids after phosphate deprivation in Arabidopsis. *Proc. Natl. Acad. Sci. USA* **2000**, *97*, 10649–10654. [CrossRef]
76. Agnihotri, A.; Gupta, P.; Dwivedi, A.; Seth, C.S. Counteractive mechanism (s) of salicylic acid in response to lead toxicity in *Brassica juncea* (L.) Czern. cv. Varuna. *Planta* **2018**, *248*, 49–68. [CrossRef]
77. Yu, L.-L.; Liu, Y.; Zhu, F.; Geng, X.-X.; Yang, Y.; He, Z.-Q.; Xu, F. The enhancement of salt stress tolerance by salicylic acid pretreatment in *Arabidopsis thaliana*. *Biol. Plant* **2020**, *64*, 150–158. [CrossRef]
78. Yang, H.; Fang, R.; Luo, L.; Yang, W.; Huang, Q.; Yang, C.; Hui, W.; Gong, W.; Wang, J. Uncovering the mechanisms of salicylic acid-mediated abiotic stress tolerance in horticultural crops. *Front. Plant Sci.* **2023**, *14*, 1226041. [CrossRef]
79. Sharma, A.; Sidhu, G.P.S.; Araniti, F.; Bali, A.S.; Shahzad, B.; Tripathi, D.K.; Brestic, M.; Skalicky, M.; Landi, M. The Role of Salicylic Acid in Plants Exposed to Heavy Metals. *Molecules* **2020**, *25*, 540. [CrossRef]
80. Saha, B.; Borovskii, G.; Panda, S.K. Alternative oxidase and plant stress tolerance. *Plant Signal. Behav.* **2016**, *11*, e1256530. [CrossRef]
81. Poór, P. Effects of Salicylic Acid on the Metabolism of Mitochondrial Reactive Oxygen Species in Plants. *Biomolecules* **2020**, *10*, 341. [CrossRef]
82. Chen, Z.; Zheng, Z.; Huang, J.; Lai, Z.; Fan, B. Biosynthesis of salicylic acid in plants. *Plant Signal. Behav.* **2009**, *4*, 493–496. [CrossRef] [PubMed]
83. Zhong, Q.; Hu, H.; Fan, B.; Zhu, C.; Chen, Z. Biosynthesis and Roles of Salicylic Acid in Balancing Stress Response and Growth in Plants. *Int. J. Mol. Sci.* **2021**, *22*, 11672. [CrossRef]
84. Torrens-Spence, M.P.; Bobokalonova, A.; Carballo, V.; Glinkerman, C.M.; Pluskal, T.; Shen, A.; Weng, J.-K. PBS3 and EPS1 Complete Salicylic Acid Biosynthesis from Isochorismate in Arabidopsis. *Mol. Plant* **2019**, *12*, 1577–1586. [CrossRef]
85. Liu, Q.; Luo, L.; Zheng, L. Lignins: Biosynthesis and Biological Functions in Plants. *Int. J. Mol. Sci.* **2018**, *19*, 335. [CrossRef]
86. Riaz, U.; Kharal, M.A.; Murtaza, G.; Zaman, Q.U.; Javaid, S.; Malik, H.A.; Aziz, H.; Abbas, Z. Prospective Roles and Mechanisms of Caffeic Acid in Counter Plant Stress: A Mini Review. *Pak. J. Agric. Res.* **2019**, *32*, 8–19. [CrossRef]
87. Kisa, D.; Kayir, O.; Saglam, N.; Şahin, S.; Öztürk, L.; Elmastaş, M. Changes of phenolic compounds in tomato associated with the heavy metal stress. *J. Nat. Appl. Sci.* **2019**, *2*, 35–43.
88. Kovácik, J.; Klejdus, B.; Backor, M. Phenolic metabolism of *Matricaria chamomilla* plants exposed to nickel. *J. Plant Physiol.* **2009**, *166*, 1460–1464. [CrossRef]
89. Zhao, L.; Huang, Y.; Keller, A.A. Comparative Metabolic Response between Cucumber (*Cucumis sativus*) and Corn (*Zea mays*) to a Cu(OH)2 Nanopesticide. *J. Agric. Food Chem.* **2018**, *66*, 6628–6636. [CrossRef]
90. Corkins, M.E.; Wilson, S.; Cocuron, J.-C.; Alonso, A.P.; Bird, A.J. The gluconate shunt is an alternative route for directing glucose into the pentose phosphate pathway in fission yeast. *J. Biol. Chem.* **2017**, *292*, 13823–13832. [CrossRef]
91. Kornecki, J.F.; Carballares, D.; Tardioli, P.W.; Rodrigues, R.C.; Berenguer-Murcia, Á.; Alcántara, A.R.; Fernandez-Lafuente, R. Enzyme production of D-gluconic acid and glucose oxidase: Successful tales of cascade reactions. *Catal. Sci. Technol.* **2020**, *10*, 5740–5771. [CrossRef]
92. Yang, Y.; Fu, Z.; Su, Y.; Zhang, X.; Li, G.; Guo, J.; Que, Y.; Xu, L. A cytosolic glucose-6-phosphate dehydrogenase gene, ScG6PDH, plays a positive role in response to various abiotic stresses in sugarcane. *Sci. Rep.* **2014**, *4*, 7090. [CrossRef]
93. Van Assche, F.; Cardinaels, C.; Clijsters, H. Induction of enzyme capacity in plants as a result of heavy metal toxicity: Dose-response relations in *Phaseolus vulgaris* L., treated with zinc and cadmium. *Environ. Pollut. (Barking, Essex: 1987)* **1988**, *52*, 103–115. [CrossRef]
94. Ávila, F.; Schmeda-Hirschmann, G.; Silva, E. The Major Chromophore Arising from Glucose Degradation and Oxidative Stress Occurrence during Lens Proteins Glycation Induced by Glucose. *Molecules* **2017**, *23*, 6. [CrossRef]
95. Antonova, K.; Vikhnina, M.; Soboleva, A.; Mehmood, T.; Heymich, M.-L.; Leonova, T.; Bankin, M.; Lukasheva, E.; Gensberger-Reigl, S.; Medvedev, S.; et al. Analysis of Chemically Labile Glycation Adducts in Seed Proteins: Case Study of Methylglyoxal-Derived Hydroimidazolone 1 (MG-H1). *Int. J. Mol. Sci.* **2019**, *20*, 3659. [CrossRef]
96. Soboleva, A.; Frolova, N.; Bureiko, K.; Shumilina, J.; Balcke, G.; Zhukov, V.; Tikhonovich, I.; Frolov, A. Dynamics of Reactive Carbonyl Species in Pea Root Nodules in Response to Polyethylene Glycol (PEG)-Induced Osmotic Stress. *Int. J. Mol. Sci.* **2022**, *23*, 2726. [CrossRef]
97. Rauser, W.E. Structure and function of metal chelators produced by plants: The case for organic acids, amino acids, phytin, and metallothioneins. *Cell Biochem. Biophys.* **1999**, *31*, 19–48. [CrossRef]
98. Igamberdiev, A.U.; Eprintsev, A.T. Organic Acids: The Pools of Fixed Carbon Involved in Redox Regulation and Energy Balance in Higher Plants. *Front. Plant Sci.* **2016**, *7*, 1042.
99. Igamberdiev, A.U.; Bykova, N.V. Role of organic acids in the integration of cellular redox metabolism and mediation of redox signalling in photosynthetic tissues of higher plants. *Free Radic. Biol. Med.* **2018**, *122*, 74–85. [CrossRef]

100. Zarei, A.; Brikis, C.J.; Bajwa, V.S.; Chiu, G.Z.; Simpson, J.P.; DeEll, J.R.; Bozzo, G.G.; Shelp, B.J. Plant Glyoxylate/Succinic Semialdehyde Reductases: Comparative Biochemical Properties, Function during Chilling Stress, and Subcellular Localization. *Front. Plant Sci.* **2017**, *8*, 1399. [CrossRef]
101. Zemanová, V.; Pavlík, M.; Pavlíková, D.; Kyjaková, P. Changes in the contents of amino acids and the profile of fatty acids in response to cadmium contamination in spinach. *Plant Soil Environ.* **2015**, *61*, 285–290. [CrossRef]
102. Wada, H.; Shintani, D.; Ohlrogge, J. Why do mitochondria synthesize fatty acids? Evidence for involvement in lipoic acid production. *Proc. Natl. Acad. Sci. USA* **1997**, *94*, 1591–1596. [CrossRef]
103. Koch, K.; Barthlott, W. Plant Epicuticular Waxes: Chemistry, Form, Self-Assembly and Function. *Nat. Prod. Commun.* **2006**, *1*, 1934578X0600101123. [CrossRef]
104. He, M.; Ding, N.-Z. Plant Unsaturated Fatty Acids: Multiple Roles in Stress Response. *Front. Plant Sci.* **2020**, *11*, 562785.
105. Sharma, S.S.; Dietz, K.-J. The significance of amino acids and amino acid-derived molecules in plant responses and adaptation to heavy metal stress. *J. Exp. Bot.* **2006**, *57*, 711–726. [CrossRef]
106. Singh, S.; Singh, P.; Tomar, R.S.; Sharma, R.A.; Singh, S.K. Proline: A Key Player to Regulate Biotic and Abiotic Stress in Plants. In *Towards Sustainable Natural Resources: Monitoring and Managing Ecosystem Biodiversity*; Rani, M., Chaudhary, B.S., Jamal, S., Kumar, P., Eds.; Springer International Publishing: Cham, Switzerland, 2022; pp. 333–346. ISBN 978-3-031-06443-2.
107. Singh, J.; Hembram, P.; Basak, J. Potential of *Vigna unguiculata* as a Phytoremediation Plant in the Remediation of Zn from Contaminated Soil. *Am. J. Plant Sci.* **2014**, *5*, 1156–1162. [CrossRef]
108. Repkina, N.; Nilova, I.; Kaznina, N. Effect of Zinc Excess in Substrate on Physiological Responses of *Sinapis alba* L. *Plants* **2023**, *12*, 211. [CrossRef]
109. Wouyou, A.; Prodjinoto, H.; Zanklan, A.S.; Vanpee, B.; Lutts, S.; Gandonou, C.B. Implication of Ions and Organic Solutes Accumulation in Amaranth (*Amaranthus cruentus* L.) Salinity Resistance. *Am. J. Plant Sci.* **2019**, *10*, 2335–2353. [CrossRef]
110. Dorion, S.; Ouellet, J.C.; Rivoal, J. Glutathione Metabolism in Plants under Stress: Beyond Reactive Oxygen Species Detoxification. *Metabolites* **2021**, *11*, 641. [CrossRef]
111. Kumar, A.; Bachhawat, A.K. Pyroglutamic acid: Throwing light on a lightly studied metabolite. *Curr. Sci.* **2012**, *102*, 288–297.
112. Jiménez-Arias, D.; García-Machado, F.J.; Morales-Sierra, S.; Luis, J.C.; Suarez, E.; Hernández, M.; Valdés, F.; Borges, A.A. Lettuce plants treated with L-pyroglutamic acid increase yield under water deficit stress. *Environ. Exp. Bot.* **2019**, *158*, 215–222. [CrossRef]
113. Moussa, H.R.; Selem, E.E.-S.M.; Ghramh, H.A. Ethanolamine affects physiological responses of salt-treated jute plants. *Int. J. Veg. Sci.* **2019**, *25*, 581–589.
114. Van Damme, T.; Blancquaert, D.; Couturon, P.; Van Der Straeten, D.; Sandra, P.; Lynen, F. Wounding stress causes rapid increase in concentration of the naturally occurring 2′,3′-isomers of cyclic guanosine- and cyclic adenosine monophosphate (cGMP and cAMP) in plant tissues. *Phytochemistry* **2014**, *103*, 59–66. [CrossRef]
115. Watanabe, M.; Chiba, Y.; Hirai, M.Y. Metabolism and Regulatory Functions of O-Acetylserine, S-Adenosylmethionine, Homocysteine, and Serine in Plant Development and Environmental Responses. *Front. Plant Sci.* **2021**, *12*, 643403. [CrossRef]
116. Abdulraheem, M.I.; Xiong, Y.; Moshood, A.Y.; Cadenas-Pliego, G.; Zhang, H.; Hu, J. Mechanisms of Plant Epigenetic Regulation in Response to Plant Stress: Recent Discoveries and Implications. *Plants* **2024**, *13*, 163. [CrossRef]
117. Gallo-Franco, J.J.; Sosa, C.C.; Ghneim-Herrera, T.; Quimbaya, M. Epigenetic Control of Plant Response to Heavy Metal Stress: A New View on Aluminum Tolerance. *Front. Plant Sci.* **2020**, *11*, 602625.
118. Fasani, E.; Giannelli, G.; Varotto, S.; Visioli, G.; Bellin, D.; Furini, A.; DalCorso, G. Epigenetic Control of Plant Response to Heavy Metals. *Plants* **2023**, *12*, 3195. [CrossRef]
119. Dong, C.-J.; Wang, X.-L.; Shang, Q.-M. Salicylic acid regulates sugar metabolism that confers tolerance to salinity stress in cucumber seedlings. *Sci. Hortic.* **2011**, *129*, 629–636. [CrossRef]
120. Chen, Y.; Sun, J.; Lin, H.; Lin, M.; Lin, Y.; Wang, H.; Hung, Y.-C. Salicylic acid reduces the incidence of *Phomopsis longanae* Chi infection in harvested longan fruit by affecting the energy status and respiratory metabolism. *Postharvest Biol. Technol.* **2020**, *160*, 111035. [CrossRef]
121. Regulatory Role of Proline in Heat Stress Tolerance: Modulation by Salicylic Acid. In *Plant Signaling Molecules*; Woodhead Publishing: Cambridge, UK, 2019; pp. 437–448.
122. Wang, Z.; Guo, J.; Luo, W.; Niu, S.; Qu, L.; Li, J.; Chen, Y.; Li, G.; Yang, H.; Lu, D. Salicylic Acid Cooperates with Lignin and Sucrose Signals to Alleviate Waxy Maize Leaf Senescence Under Heat Stress. *Plant Cell Environ.* **2025**, *48*, 4341–4355. [CrossRef]
123. Jiang, B.; Liu, R.; Fang, X.; Tong, C.; Chen, H.; Gao, H. Effects of salicylic acid treatment on fruit quality and wax composition of blueberry (*Vaccinium virgatum* Ait). *Food Chem.* **2022**, *368*, 130757. [CrossRef]
124. Richardson, A.D.; Duigan, S.P.; Berlyn, G.P. An Evaluation of Noninvasive Methods to Estimate Foliar Chlorophyll Content. *New Phytol.* **2002**, *153*, 185–194.
125. Murchie, E.H.; Lawson, T. Chlorophyll fluorescence analysis: A guide to good practice and understanding some new applications. *J. Exp. Bot.* **2013**, *64*, 3983–3998. [CrossRef]

126. Monteith, J.L.; Campbell, G.S.; Potter, E.A. Theory and performance of a dynamic diffusion porometer. *Agric. For. Meteorol.* **1988**, *44*, 27–38. [CrossRef]
127. Hina, B.; Rizwani, G.-H.-; Naseeb, U.; Huma, A.; Hyder, Z. Application of Atomic Absorption Spectroscopy to determine Mineral and Heavy Metal distribution level of Medicinal Plants. *J. Anal. Tech. Res.* **2023**, *5*, 26–32.
128. Leonova, T.; Popova, V.; Tsarev, A.; Henning, C.; Antonova, K.; Rogovskaya, N.; Vikhnina, M.; Baldensperger, T.; Soboleva, A.; Dinastia, E.; et al. Does Protein Glycation Impact on the Drought-Related Changes in Metabolism and Nutritional Properties of Mature Pea (*Pisum sativum* L.) Seeds? *Int. J. Mol. Sci.* **2020**, *21*, 567. [CrossRef]
129. Wei, R.; Wang, J.; Su, M.; Jia, E.; Chen, S.; Chen, T.; Ni, Y. Missing Value Imputation Approach for Mass Spectrometry-based Metabolomics Data. *Sci. Rep.* **2018**, *8*, 663. [CrossRef]
130. Pang, Z.; Chong, J.; Zhou, G.; de Lima Morais, D.A.; Chang, L.; Barrette, M.; Gauthier, C.; Jacques, P.-É.; Li, S.; Xia, J. MetaboAnalyst 5.0: Narrowing the gap between raw spectra and functional insights. *Nucleic Acids Res.* **2021**, *49*, W388–W396. [CrossRef]
131. Goeman, J.J.; Bühlmann, P. Analyzing gene expression data in terms of gene sets: Methodological issues. *Bioinformatics* **2007**, *23*, 980–987. [CrossRef] [PubMed]
132. Cochran, C.; Martin, K.; Rafferty, D.; Choi, J.; Leontyev, A.; Shetty, A.; Kurup, S.; Puthanveetil, P. Untargeted Metabolomics Based Prediction of Therapeutic Potential for Apigenin and Chrysin. *Int. J. Mol. Sci.* **2023**, *24*, 4066. [CrossRef]
133. Benjamini, Y.; Hochberg, Y. Controlling the False Discovery Rate: A Practical and Powerful Approach to Multiple Testing. *J. R. Stat. Soc. Ser. B (Methodol.)* **1995**, *57*, 289–300. [CrossRef]

plants

MDPI

Article

Metabolic Responses to the Zinc Stress in the Roots and Leaves of *Amaranthus caudatus*: The Proteomics View

Anastasia Gurina [1,2,†], Tatiana Bilova [1,2,*,†], Daria Gorbach [2], Alena Soboleva [2], Nataliia Stepanova [3], Olga Babich [4], Christian Ihling [5], Anastasia Kamionskaya [3], Natalia Osmolovskaya [1] and Andrej Frolov [2,*]

1 Department of Plant Physiology and Biochemistry, St. Petersburg State University, 199034 St. Petersburg, Russia; anastasia.gurina@list.ru (A.G.); n.osmolovskaya@spbu.ru (N.O.)

2 Laboratory of Analytical Biochemistry and Biotechnology, K.A. Timiryazev Institute of Plant Physiology, Russian Academy of Science, 127276 Moscow, Russia; daria.gorba4@yandex.ru (D.G.); oriselle@yandex.ru (A.S.)

3 Federal Research Centre Fundamentals of Biotechnology, Russian Academy of Science, 119071 Moscow, Russia; akamio@fbras.ru (A.K.)

4 Research and Educational Center "Applied Biotechnologies", Immanuel Kant Baltic Federal University, 236041 Kaliningrad, Russia; olich.43@mail.ru

5 Department of Pharmaceutical Chemistry and Bioanalytics, Institute of Pharmacy, Martin-Luther Universität Halle-Wittenberg, 06120 Halle, Germany; christian.ihling@pharmazie.uni-halle.de

* Correspondence: bilova.tatiana@gmail.com (T.B.); afrolov0375@yandex.ru (A.F.)

† These authors contributed equally to this work.

Abstract: Zinc excess (Zn stress) could lead to deleterious effects in plants such as enhanced ROS production, inhibition of photosynthetic machinery, and impairment of nutrient uptake. Hence, we aimed to investigate the complexity of metabolic responses to Zn stress in *Amaranthus caudatus* young and mature leaves, as well as in roots by means of proteomics. Our previous metabolomics research has indicated potential involvement of gluconate and salicylate in Zn tolerance mechanisms. However, proteomics study of metabolic adjustments underlying Zn stress tolerance can give additional insight to the issue, as a lot of enzymes are known to be affected by the excess of transitional metals. The results obtained through bottom-up proteomics were complementary to our earlier metabolomics data and, furthermore, enlightened other important details in the metabolic response of *A. caudatus* plants to the applied Zn stress. In particular, the significant involvement of redox-related enzymes was shown, especially for the roots, and their possible interactions with salicylate and jasmonate signaling could be proposed. Furthermore, Zn^{2+}-induced changes in roots and young leaves strongly affected sugar metabolism, enhanced protein quality control system, while mature leaves were characterized by remarkable decrease in subunits of photosynthetic electron transport complexes. Thus, this work emphasizes massive metabolic reprogramming aimed to reinforce root defense responses while supporting young leaves with sugar metabolites. Mass spectrometry proteomics data are available via ProteomeXchange with identifier PXD069557.

Keywords: *Amaranthus caudatus*; zinc stress; bottom-up proteomics; metabolic adjustment

1. Introduction

Zinc (Zn) is one of the eight essential micronutrients [1] acting as a cofactor of multiple enzymes involved in signal transduction, protein biosynthesis, maturation and structure integrity, DNA replication and repair, detoxication of reactive oxygen species (ROS) [2,3]. In agreement with this, more than 2000 proteins in the *Arabidopsis thaliana* proteome were predicted to have Zn-binding cites [4] that might indicate high responsivity of plant proteome to Zn.

In plants, Zn uptake relies on the root absorption of Zn(II) cations, which are sufficiently abundant in soils, although in poorly soluble state [5]. Therefore, the equilibrium concentrations of free Zn^{2+} in the soil solution are quite low (typically not exceeding a few micromoles per L) [6] and essentially depend on the physicochemical properties of the soil, secretion activity of the plant roots, and metabolic features of the rhizosphere microbial community [2]. In agreement with this, plants typically accumulate Zn at relatively low levels (30–100 μg/g dry weight) [7]. These metabolically adequate contents can be efficiently managed by constitutive detoxification systems [8]. Accumulation of Zn(II) ions above these values results in overwhelming the plant metal detoxification capacity and development of heavy metal stress, which is typically accompanied by clear signs of toxicity [4].

Despite limited availability of Zn for its uptake by plant roots, several scenarios might lead to enhanced solubilization of the soil Zn pool [2]. Thus, acidified soils favor enhanced liberation of Zn from insoluble complexes and impact on development of Zn toxicity due to accumulation of this micronutrient in tissues [9]. Moreover, increased Zn contents might be underlied by prolonged application of phosphate-based fertilizers and by the use of sewage sludge [9]. Not less importantly, stressors such as drought or flooding can positively affect Zn availability and thereby favor higher Zn^{2+} concentrations in the soil solution [10]. Finally, accumulation of Zn in the plant tissues might be underlied by industrial activities and enhanced soil pollution [11].

From the point of plant physiology, high tissue Zn contents represent an abiotic stressor triggering a well-defined pattern of physiological responses: [9] decrease in seed germination rates, retardation of seedling growth [12], chlorosis in young leaves [13], necrosis and reddening lesions in mature leaves [14], decrease in photosynthetic efficiency [15], inhibition of nutrient [15,16] and water uptake [17], reduced stomatal activity [15]. To an essential extent, these effects can be attributed to overproduction of reactive oxygen species (ROS) and development of oxidative stress [18,19], which is, at least partly, underlied by impaired redox reactions in the electron transport chains (ETCs) of the chloroplasts and mitochondria [20]. Another toxicity mechanism might be associated with the loss of protein functionality due to the displacement of metal ions (particularly Fe^{2+}/Fe^{3+}, Mg^{2+}, and Mn^{2+}) by Zn^{2+} from their functional sites of enzymes [3].

Fortunately, plants possess well-established mechanisms of biochemical adaptation to Zn stress, such as the activation of antioxidant defense, enhanced synthesis of transport proteins involved in the uptake, partitioning and sequestration of Zn ions, as well as redirection of the central metabolism to cover the associated energy demand [18]. However, plants essentially differ in efficiency of the principal mechanism behind the tolerance to the Zn stress—in ability to rapidly compartmentalize Zn into vacuoles via transporters [21] and scavenging of Zn^{2+} in the cytosol by chelation with organic acids [22], amino acids (cysteine) [23], and glutathione [19].

The genus *Amaranthus* is not only a valuable source of nutrients and biologically active metabolites [24], but is also a highly adaptive crop [25] and a versatile tool for phytoremediation of Zn-contaminated agricultural soils [26]. In our previous work [27] we addressed the metabolic rearrangements triggered by the Zn stress in the roots and leaves of *A. caudatus* plants, and described the accompanying adaptive metabolic shifts therein. However, to address these metabolic changes in the most comprehensive way, the expressional changes at the proteome level (i.e., stress-induced dynamics of effector enzymes, transporters and ROS scavenging systems) need to be understood as well.

In this context, proteomics represents a versatile tool for disclosing the molecular mechanisms behind the Zn toxicity and plant tolerance to Zn stress, as was shown by Fukao et al. (2011) in *Arabidopsis thaliana* seedlings, Šimon et al. (2021) in *Oryza sativa* roots, and by Lucini et al. (2015) in *Lactuca sativa* leaves using gel-based and gel-free bottom-up approaches [28–30]. A study on *A. thaliana* seedlings grown under Zn excess can serve as a model to illustrate the significance of proteomics data in understanding the development of adaptive physiological reactions or Zn toxicity symptoms [30]. The seedling toxicity symptom such as chlorosis was associated with an increase in abundance of a few proteins known to be responsive to iron deficiency, indicating that Zn displaces iron from active sites of enzymes involved in chlorophyll synthesis. Another *A. thaliana* Zn toxicity symptom is a reduction in seedling root growth, associated with decrease in abundance of a few subunits of V-ATPase which are responsible for cell expansion and Zn^{2+} ion sequestration into vacuole [30]. However, despite the impressive body of the data accumulated so far, such a promising food crop as *A. caudatus* was still not addressed by the proteomics approach in respect of Zn stress.

Therefore, here we employ LC-MS-based bottom-up proteomics to address the metabolic responses of the *A. caudatus* root and leaves to moderate Zn stress. Thereby, we aimed to identify the Zn-responsive proteins in the roots, young and mature leaves of the *A. caudatus* plants exposed to Zn stress for seven days, as well as their comprehensive discussion in the context of the Zn^{2+}-induced adaptive metabolic shifts discovered in these organs earlier [27].

2. Results

2.1. Suppressing Exposure to Zn Caused Suppression of the Photosystem II Efficiency in the A. caudatus Leaves

After two weeks of growing under the hydroponic conditions, the six-week-old *A. caudatus* plants were exposed to 300 μmol/L $ZnSO_4 \times 7H_2O$ supplemented to the nutrient solution. A one-week stress application did not cause severe alterations in leaf and root morphology in comparison to the age-matched untreated controls. The visibly detectable changes were observed only for the young leaves, which became light green upon application of the Zn stress (Supplementary Information 1, Figure S1-1).

Physiological assays accomplished by means of non-invasive methods for the 3rd fully expanded leaf revealed no significant changes in stomatal conductance, leaf relative water contents (LRWC), and chlorophyll contents upon Zn stress application in comparison to control plants (Figure 1a,b). However, as could be seen from the decrease in the F_v/F_m values, photosystem II (PS II) efficiency appeared to be affected by stress conditions (Figure 1a,b). These results were also confirmed in the second independent experiment, which was accomplished to validate the initial findings (Figures S1-4, S1-2a and S1-3b).

2.2. Protein Isolation and Tryptic Digestion

The protein concentrations in the resulting young and mature leaves and roots extracts (assessed with the BCA assay) were in the range of 0.33–1.42 mg/mL, and the protein recoveries were 0.13–0.72 mg/g fresh weight (Table S1-1). Densitometric analysis of the electropherograms obtained with 5 μg of the total protein extracts loaded on each lane of the gels demonstrated comparable signal intensities per line across all gels normalized by the lines loaded with 5 μL protein ladder ($50.85 \times 10^4 \pm 68.19 \times 10^3$ arbitrary units, RSD = 13%), indicating sufficient precision of protein assay. The following proteolysis, accomplished with 25 μg of the individual protein extracts could be considered as complete or close to complete as no distinct protein bands were detectable in the electropherograms acquired with filter remnants, assuming sensitivity of the Coomassie staining better than

30 ng (Figures S1-5 and S1-6) [31]. Unfocused signals at the very beginning of the lines most likely could be attributed to non-protein negatively charged biopolymers or glycoproteins.

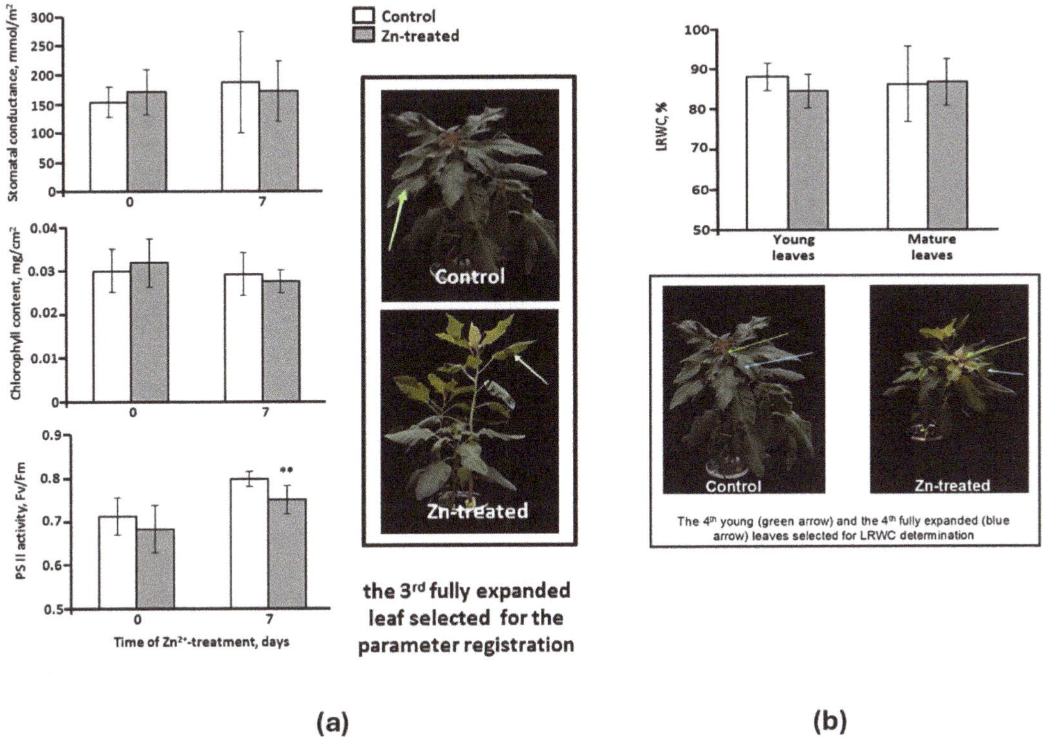

Figure 1. Impact of exogenous Zn^{2+} (300 µmol/L) on stomatal conductivity, chlorophyll content, photosystem II (PS II) activity (**a**), and leaf relative water content (**b**) of the third mature leaf (the leaf is indicated by green arrows at photos in the insert; numeration of the mature leaves were performed just beneath the plant top with young (i.e., not fully expanded) leaves) from six-week-old (before Zn^{2+}-exposure, 0 day) and seven-week-old *A.caudatus* plants (n = 9) (after Zn^{2+}-exposure, 7 day) grown in hydroponic nutrient solution. Gray columns indicate the Zn-treated group, namely plants whose physiological parameters were measured prior (0 day) and after the Zn-stress exposure (7 day). The Zn-stress exposure of the plants was performed by supplementation of 300 µmol/L $ZnSO_4$ in hydroponic nutrient solution for one week. White columns indicate the control group of plants which were grown during the whole experimental period (from 0 to 7 days) in hydroponic nutrient solution without the addition of 300 µmol/L $ZnSO_4$. F_v/F_m—variable fluorescence (F_v)/maximal fluorescence (F_m). Asterisks ** indicate significant difference between Zn-treated and control groups per time point at p (t-test) < 0.01. The experimental data obtained for a leaf of each plant are presented in Osmolovskaya et al. [27] Supplementary Information 2, Tables S2-1–S2-3).

2.3. Identification of Peptides and Annotation of Proteins

As the amaranth genome is still not established, for the database search we opted for the sequence of *C. quinoa*, which is the closest relative of *A. caudatus* with sequenced genome. The parallel search against the *A. thaliana* database (which contains more entries, although with a lower expected homology) yielded less numbers of confident matches. Therefore, the results of this search were not analyzed further.

The search against the *C. quinoa* protein sequence database resulted in the identification of 4608 peptides, comprising 1686 protein groups (non-redundant proteins) which represented 3163 possible individual proteins (Supplementary Information 2). To address the inter-organ differences in the absence and presence of the stress response, we accomplished inter-organ comparisons within each treatment group. This analysis, accomplished for the control group, yielded 2420 peptides comprising 873 non-redundant proteins, which could be assigned to 2205 possible individual proteins in all organs (Supplementary Information 1, Figure S1-7). Among these numbers, 444 peptides, 215 protein groups, and 663 individual proteins were common (18–30% of the total counts). Analogously, 2305 peptides, 856 protein groups, and 2120 proteins could be found in roots, young, and mature leaves of the Zn^{2+}-treated plants. Thereby, 26–32% of the total protein numbers appeared to be common for all organs (491, 231, and 681 peptides, protein groups and proteins, respectively, Figure S1-8).

At the next step, the paired stress-control comparisons were addressed for each of the organs. The analysis revealed 1738 and 2870 peptides discovered in roots and leaves (young and mature collectively), respectively. These peptides represented 1145 and 2018 individual proteins, falling in 629 and 1057 protein groups (Figures S1-9–S1-11). In total, 568 protein groups with 797 possible individual proteins were detected in roots both in control and stress-exposed groups, while 44 and 17 non-redundant proteins (with 189 and 159 accessions) were found in the control or Zn-treated plants (Figure S1-9a,b; for the identified peptides behind these annotations, see Figure S1-9c). Considering the identifications for young and mature leaves separately, 574 and 483 non-redundant proteins (with 1077 and 941 accessions) could be annotated to these groups, respectively (Figures S1-10a,b and S1-11a,b; for the identified peptides behind these annotations, see Figures S1-10c and S1-11c). The numbers of the protein groups and individual proteins common for the young leaves from the control and Zn-treated groups were 535 and 783, respectively. On the other hand, 8 non-redundant proteins (136 accessions) were found only in the control samples, while 31 of those (158 accessions) could be annotated solely to the stress-exposed group (Figure S-10a,b; for the corresponding identified peptides behind, see Figure S1-10c).

Finally, 466 protein groups were common for the mature leaves harvested in all treatment groups, while 7 and 10 of them were found only in the controls and Zn-exposed ones. These non-redundant proteins could be attributed to 690 potential accessions (with 132 and 119 of them found only in the control and Zn-exposed groups, Figure S-11a,b; for identified peptides behind, see Figure S1-11c).

2.4. Label-Free Quantification

At the first step, we addressed the quantitative differences within the whole dataset in glance. For this, one-way ANOVA was accomplished at the confidence level of $p \leq 0.05$ and the fold change (FC) threshold of 1.5 (both for increase and decrease in abundance in comparison to control). This multi-group comparison (in total 6 sample groups representing two treatments and three organs) revealed 355 non-redundant proteins, which were differentially expressed in at least 2 sample groups (Supplementary Information 3, Table S3-1).

Principal component analysis (PCA), accomplished for this differentially expressed part of the *A. caudatus* proteome, revealed clear separation in the organ-dependent manner with 81.2% of explained difference by the principal component 1 (PC1) visible in the corresponding scores plot (Figure 2a). Thus, the four leaf-related sample groups (representing the young and mature leaves collected from the control and the stress-exposed plants) clustered together and apart from the two root sample groups (control and stress).

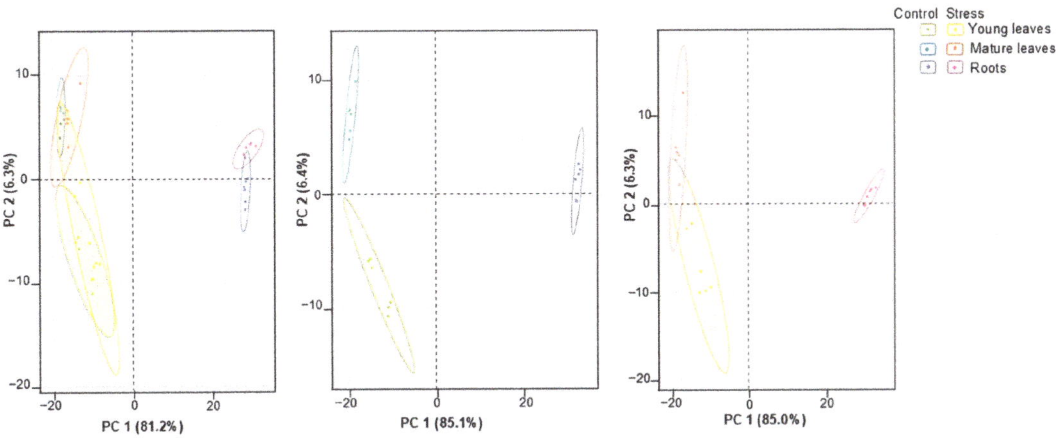

(a) (b) (c)

Figure 2. Principal component analysis with a scores plot performed for all treatment groups (roots, young, and mature leaves) with or without stress application (**a**), all groups of plants grown without Zn treatment (**b**), and all groups of plants grown under Zn stress conditions (**c**). PC—principal component; Control—plants grown without Zn treatment; Stress—plants grown under Zn stress (supplementation of 300 µmol/L Zn^{2+} to the hydroponic solution for 7 days).

The same pattern could be observed when the PCA tests were accomplished separately for all differentially expressed proteins annotated in the controls and in the Zn-exposed plants (360 and 375 differentially abundant non-redundant proteins, respectively; Tables S3-2 and S3-3). In both cases 85% of the total variance could be explained by PC1 (Figure 2b,c).

Based on corresponding loadings plots, the top 20 major contributors in these patterns could be assigned for each comparison (Figure S1-12). The analysis of these lists (Tables S1-2–S1-4) showed that the observed patterns of group separation were most likely defined by organ specificity and were not affected by the heavy metal treatment. It was especially relevant for the accessions expressed only in chloroplasts of the green parts.

Table S1-2 presents the five most responsive proteins (with their organ specificity) among all plant organs irrespective of Zn treatment, which were the major contributors in the observed differences. Thus, in roots of Zn-exposed plants, the germin-like protein appeared to be the positive contributor, whereas in the mature leaves of the control plants it was downregulated. UPF0603 followed an opposite pattern, being the most abundant in mature leaves and the least abundant in roots under control conditions. In contrast, the expression of the remaining three proteins was the highest in the control mature leaves but was the lowest in the roots of the Zn-exposed plants. The magnitude of these differences varied notably among the organs, ranging from 117.1 to 1461.8-fold.

Similarly, uncharacterized protein LOC110707738 isoform X1, UPF060, chloroplastic-like probable glutathione S-transferase parC isoform X1, chlorophyll *a-b* binding protein, and germin-like protein (72.9–infinite-fold) headlined the lists of the major contributors among all control groups (Table S1-3). Germin-like protein, peroxidase 12-like, NdhJ, putative pentatricopeptide repeat-containing protein, and UPF0603 (ranging from 214.4-fold to infinite-fold changes) were identified as the most significant contributors among all stress-treated groups (Table S1-4).

At the next step, hierarchical clustering analysis (HCA) with heatmap representation was accomplished for the whole sample set, which demonstrated comparable generalized results. The analysis revealed higher homogeneity of the root samples (obtained from both control and Zn-exposed plants), as could be seen by more distinct clustering of the corresponding groups at the column tree of the plot (Figure 3). On the other hand, the leaf samples (both young and mature) appeared to be distributed in a more or less uneven manner without any visible impact of the exposure to Zn (Figure 3).

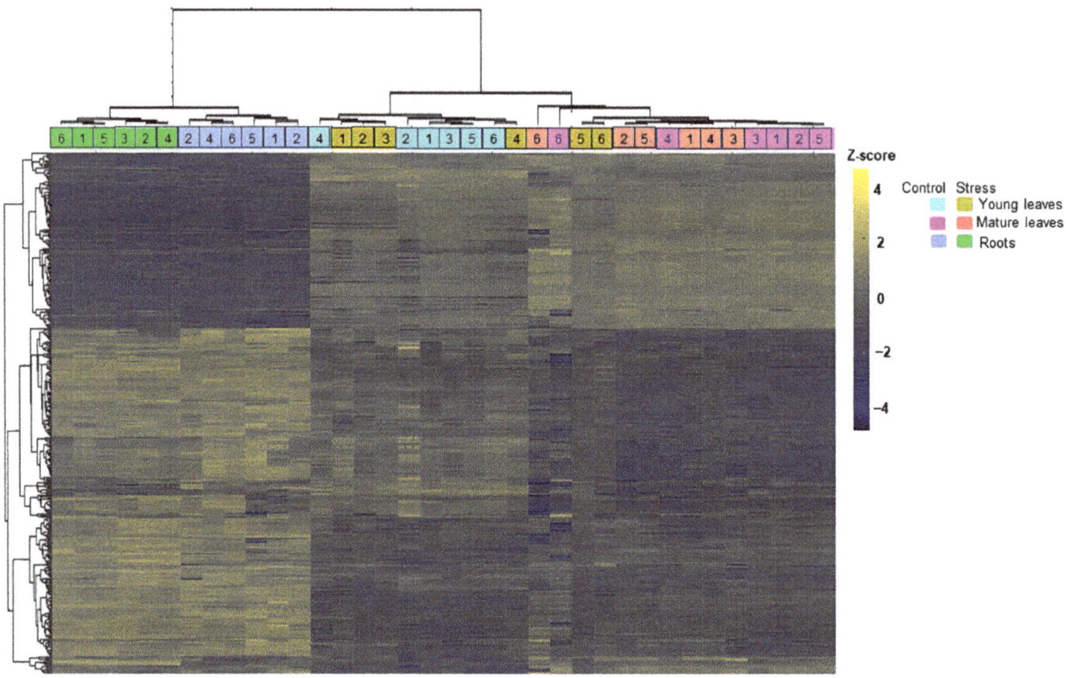

Figure 3. Hierarchical clustering with a heatmap representation showing expression patterns of individual samples performed for all the treatment groups (roots, young, and mature leaves) with or without stress application. Control—plants grown without Zn treatment; stress—plants grown under Zn stress (supplementation of 300 μmol/L Zn^{2+} to the hydroponic solution for 7 days). Figures indicate individual samples (n = 6) in each experimental group.

Processing of samples from control and Zn-exposed plants separately revealed the same pattern for the roots (Figure S1-13a,b). In contrast to the above-described results of the global HCA test, the control mature and young leaf samples also formed apparent organ-specific clusters (Figure S1-13a), which were more or less dispersed upon stress exposure (Figure S1-13b).

Although the multiple comparisons by multivariate statistics and one-way ANOVA provided quite an informative general overview of the organ- and stress-specificity of the observed protein dynamics, they are still too complex for making direct and unambiguous conclusions about specific aspects of the plant response to Zn stress. Therefore, to address the response of individual plant organs to Zn stress in the most comprehensive way, we considered individual stress vs. control-paired comparisons at the level of each organ (Supplementary Information 4). The corresponding PCA tests revealed distinct stress-dependent inter-group separation in the corresponding scores plots with 52.8%, 40.8%, and 60.2% of the total explained variance for the roots, young, and mature leaves, respectively (Figures S1-14a–S1-16a).

In more detail, 78 root proteins appeared to be differentially expressed upon seven days of exposure to the Zn stress among 629 unambiguously annotated, with 34 and 44 of them up- and down-regulated in the Zn-exposed plants, respectively (t-test: $p \leq 0.05$;

FC $\geq$ 1.5, FDR correction at $q \leq 0.05$; Tables S4-1 and S4-2). Analysis of the corresponding PCA data (and particularly the loadings plot) allowed the identification of the major contributors to the observed stress-related differences in the root differentially abundant proteins (Figure S1-14a,b; Table S1-5). Among them, the abundance of putative 12-oxophytodienoate reductase 11 isoform X1, probable glutathione S-transferase parC, chloroplast stem-loop binding protein of 41 kDa b, glutathione S-transferase-like protein, and probable inactive purple acid phosphatase 29 (PC1 0.09–0.15 absolute values; FC 1.6–27.7-fold) were significantly raised, while differential abundance level of proteins like plasma membrane ATPase 4-like, plasma membrane ATPase 4-like isoform X1, tubulin alpha-3 chain, ABC transporter F family member 3-like, and tubulin beta-5 chain-like were decreased (PC1 0.11–0.14; FC 1.7–3.5-fold). Further, heatmap cluster analysis was performed to highlight the differential accumulation patterns among proteins from the amaranth roots, where the clearest stress-related response could be observed (Figure S1-17).

Following the identification of the major patterns through PCA analysis, statistical evaluation further quantified the significance of differential protein accumulation. The statistical analysis thereby validated the observed clustering trends in the proteomics dataset. Increased level of protein abundances was observed for the probable glutathione S-transferase parC, putative 12-oxophytodienoate reductase 11 isoform X1, non-symbiotic hemoglobin 1-like, uncoupling protein 4-like, and S-transferase-like with the FC range of 10.6–39.1 (Table 1). Increased level of protein abundances was observed for such root proteins as peroxidase 3-like, ABC transporter F family member 3-like, aspartyl protease AED3-like, glutamine synthetase, and dihydrolipoyl dehydrogenase 2 (FC 2.3–5.0, Table 1).

For the young leaves, 40 proteins occurred as differentially expressed among 574 totally annotated, with 17 proteins demonstrating increased abundance and 23 proteins with decreased abundance (t-test: $p \leq 0.05$; FC $\geq$ 1.5, FDR correction at $q \leq 0.05$; Tables S4-3 and S4-4). Analysis of the PCA data by the first two principal components (PC1 and PC2) visualized via the corresponding scores and loadings plots allowed identification of the major contributors to the observed differences (Figure S1-15a,b; Table S1-10). Among them, CBS domain-containing protein CBSX3, dihydropyrimidine dehydrogenase ($NADP^+$), plasma membrane ATPase 4-like, peroxiredoxin-2E-1, dihydrolipoyl dehydrogenase (PC1 0.17–0.22 absolute values; FC 1.5–2.0-fold) abundance were increased, while uncharacterized protein LOC110707738 isoform X1, 60S ribosomal protein L10-like, 60S ribosomal protein L4-like, magnesium-protoporphyrin IX monomethyl ester cyclase, PsbC showed significant decrease in its abundance (PC1 0.09–0.21; FC 1.6–3.0-fold). Additionally, heatmap cluster analysis was conducted to reveal differential accumulation patterns among proteins from the young leaves, where the most distinct stress-related effects were observed (Figure S1-18).

T-test results were in good agreement with the multivariate statistics data results. Proteins which demonstrated the most pronounced abundance increase under Zn-stress conditions were represented by the receptor-like protein 12, aspartate aminotransferase, CBS domain-containing protein CBSX3, ferredoxin-NADP reductase and probable aldo-keto reductase 4 (1.9–2.7-fold), and magnesium-protoporphyrin IX monomethyl ester cyclase, chloroplastic-like, PsbC protein, 60S ribosomal protein L4-like, stearoyl-[acyl-carrier-protein] 9-desaturase, and porphobilinogen deaminase comprised top of the list of the entries with the decreased level of abundance after Zn exposure (1.9–3.1-fold) (Table 2).

Table 1. Top 20 most strongly regulated accessions among the 78 proteins constituting the part of amaranth root proteome, which was differentially expressed in response to Zn treatment.

Protein Name [a]	KEGG Accession	FC [b]	*p*	Bin [c]	Subcellular Localization
Proteins displaying a higher abundance under Zn stress in comparison to control					
probable glutathione S-transferase parC	XP_021729368.1	39.1	8.6×10^{-6}	9, 10, 26	cytoplasm, plastid, nucleus
putative 12-oxophytodienoate reductase 11 isoform X1	XP_021714783.1	27.6	7.3×10^{-7}	9	cytoplasm, peroxisome
non-symbiotic hemoglobin 1-like	XP_021731656.1	15	1.0×10^{-4}	10	cytoplasm, nucleus
mitochondrial uncoupling protein 4-like	XP_021725393.1	13.1	1.9×10^{-4}	2, 26	mitochondrion
glutathione S-transferase-like	XP_021763022.1	10.6	2.3×10^{-5}	10, 26	nucleus, cytoplasm, plasma membrane
probable inactive purple acid phosphatase 29	XP_021726654.1	10.5	5.0×10^{-5}	25	cytoplasm
alcohol dehydrogenase 1	XP_021760951.1	10.1	1.9×10^{-4}	3	cytoplasm
probable aldo-keto reductase 4	XP_021724433.1	8.2	5.0×10^{-5}	26	mitochondrion
bifunctional riboflavin biosynthesis protein RIBA 1, chloroplastic-like	XP_021713589.1	5.9	2.3×10^{-4}	7	plastid
LOW QUALITY PROTEIN: UPF0603 protein At1g54780, chloroplastic-like	XP_021772984.1	4.9	7.3×10^{-3}	1, 19	plastid
Proteins displaying a decreased abundance under Zn stress in comparison to control					
peroxidase 3-like	XP_021726359.1	5	3.7×10^{-4}	21	cell wall
ABC transporter F family member 3-like	XP_021757719.1	3.5	1.9×10^{-4}	17	cytoplasm, nucleus
aspartyl protease AED3-like	XP_021752910.1	2.9	4.0×10^{-4}	19, 26	cell wall
glutamine synthetase, chloroplastic	XP_021727743.1	2.4	1.4×10^{-3}	4, 25	plastid
dihydrolipoyl dehydrogenase 2, chloroplastic-like	XP_021730249.1	2.3	2.7×10^{-2}	5	plastid
peroxidase 72-like	XP_021714590.1	2.3	1.5×10^{-2}	21	cell wall
vacuolar-processing enzyme-like	XP_021716668.1	2.2	3.7×10^{-4}	19	vacuole
apoptosis-inducing factor 2-like	XP_021725188.1	2.1	4.4×10^{-3}	35	cytoplasm
40S ribosomal protein S27-2	XP_021762421.1	2.1	2.9×10^{-2}	17	ribosome, cytoplasm
plasma membrane ATPase 4-like	XP_021738895.1	2	6.2×10^{-5}	24, 26	plasma membrane

[a] Identification of peptides and annotation of proteins relied on the database search against amino acid sequences of *Chenopodium quinoa* (uploaded from KEGG protein database) accomplished with the SEQUEST algorithm; Mercator v4.5 was used for proteins annotation, subcellular localization of proteins was defined using BUSCA; [b] FC, fold change, for proteins that demonstrated an increase in abundance, FC(Zn stress/Control) indicates an ($\geq$1.5-fold) increase in their abundance under stress Zn compared to the control, and for proteins that showed a decrease in abundance, FC(Control/Zn stress) shows a ($\geq$1.5-fold) decrease in their abundance under Zn stress compared to the control. Thus, the FC values were adjusted for cases showing a decrease by presenting the inverse of the FC value; [c] Numbers correspond to the bins annotated with Mercator MapMan software (v4.5) (Table S1-6).

Mature leaf proteome accounted for 21 differentially accumulated accessions from the 483 annotated in total. Among these, 8 and 13 proteins appeared to be stress dependently up- and down-regulated, respectively (*t*-test: $p \leq 0.05$; FC $\geq$ 1.5, FDR correction at $q \leq 0.05$; Tables S4-5 and S4-6). PCA test results and loading plots for the mature leaf samples especially revealed eigenvalues of PC1 for the 20 major contributors (Figure S1-16a,b; Table S1-11). Thus, heat shock protein 83, probable glutathione S-transferase parC, betaine aldehyde dehydrogenase, annexin D2-like protein, citrate synthase comprised the group of the most differentially abundant proteins (PC1 0.14–0.22; FC 1.5–4.3-fold), while differential accumulation of PetA protein, catalase, photosystem II 10 kDa polypeptide, LRR receptor-like serine/threonine-protein kinase GSO1, 50S ribosomal protein L9 were significantly declined (PC1 0.21–0.27; FC 1.6–2.1-fold). Cluster analysis coupled with heatmap representation was carried out to emphasize the protein abundance level changes in the mature leaves, where the most pronounced stress-related responses were observed (Figure S1-19).

Table 2. Top 20 most strongly regulated accessions among the 40 proteins constituting the part of amaranth young leaf proteome, which was differentially expressed in response to Zn treatment.

Protein Name [a]	KEGG Accession	FC [b]	p	Bin [c]	Subcellular Localization
Proteins displaying a higher abundance under Zn stress in comparison to control					
receptor-like protein 12	XP_021726832.1	2.7	4.1×10^{-3}	26	plasma membrane
aspartate aminotransferase, cytoplasmic	XP_021723631.1	2.2	2.3×10^{-2}	4	cytoplasm
CBS domain-containing protein CBSX3, mitochondrial-like	XP_021763137.1	2.2	4.4×10^{-4}	10	mitochondrion
ferredoxin-NADP reductase, chloroplastic-like	XP_021727571.1	2.2	5.3×10^{-3}	1	chloroplast
probable aldo-keto reductase 4	XP_021724433.1	2.0	3.3×10^{-2}	26	mitochondrion
polygalacturonase inhibitor-like	XP_021724286.1	1.9	6.7×10^{-3}	26	cell wall
plasma membrane ATPase 4-like	XP_021723433.1	1.9	7.7×10^{-4}	24, 26	plasma membrane
enolase	XP_021743148.1	1.9	1.1×10^{-2}	3	cytoplasm
dihydropyrimidine dehydrogenase ($NADP^+$), chloroplastic-like	XP_021732698.1	1.8	3.8×10^{-4}	6	chloroplast
peroxiredoxin-2E-1, chloroplastic-like	XP_021743149.1	1.8	1.1×10^{-3}	10	chloroplast
Proteins displaying a decreased abundance under Zn stress in comparison to control					
magnesium-protoporphyrin IX monomethyl ester [oxidative] cyclase, chloroplastic-like	XP_021745614.1	3.1	1.7×10^{-3}	7	chloroplast
PsbC (chloroplast)	YP_009380126.1	2.6	7.0×10^{-3}	1	chloroplast
60S ribosomal protein L4-like	XP_021721538.1	2.1	6.5×10^{-4}	17	ribosome, cytoplasm
stearoyl-[acyl-carrier-protein] 9-desaturase, chloroplastic	XP_021751679.1	2.0	2.5×10^{-2}	4	chloroplast
porphobilinogen deaminase, chloroplastic-like	XP_021752880.1	1.9	8.1×10^{-3}	7	chloroplast
translocase of chloroplast 159, chloroplastic-like	XP_021763183.1	1.9	4.2×10^{-3}	23	chloroplast
granule-bound starch synthase 2, chloroplastic/amyloplastic-like	XP_021721398.1	1.8	1.1×10^{-2}	3	chloroplast
NdhH (chloroplast)	YP_009380186.1	1.7	2.3×10^{-2}	1	chloroplast
60S ribosomal protein L19-3-like	XP_021724307.1	1.7	8.4×10^{-3}	17, 26	nucleolus
40S ribosomal protein S30	XP_021774723.1	1.7	1.5×10^{-2}	17	ribosome, cytoplasm

[a] Identification of peptides and annotation of proteins relied on the database search against amino acid sequences of *Chenopodium quinoa* (uploaded from KEGG protein database) accomplished with the SEQUEST algorithm; Mercator v4.5 was used for proteins annotation, subcellular localization of proteins was defined using BUSCA; [b] FC, fold change, for proteins that demonstrated an increase in abundance, FC(Zn stress/Control) indicates an ($\geq$1.5-fold) increase in their abundance under stress Zn compared to the control, and for proteins that showed a decrease in abundance under Zn stress compared to the control, the FC values were adjusted by presenting the inverse of the FC value as FC(Control/Zn stress, $\geq$1.5); [c] Numbers correspond to the bins annotated with Mercator MapMan software (v4.5) (Table S1-6).

T-test results were consistent with the multivariate statistical data. The stress-induced proteins were represented by the heat shock protein 83, annexin D2-like protein, probable glutathione S-transferase, uncharacterized protein LOC110692268, and urease-like protein (1.7–4.2-fold). At the same time, callose synthase 10-like, PetA, NdhH, photosystem

II 10 kDa polypeptide, magnesium-chelatase subunit ChlI were the top 5 proteins with decreased abundance level under Zn treatment (1.7–2.6-fold) (Table 3).

Table 3. Top 18 most strongly regulated accessions among the 21 proteins constituting the part of amaranth root proteome, which was differentially expressed in response to Zn treatment.

Protein Name [a]	KEGG Accession	FC [b]	p	Bin [c]	Subcellular Localization
Proteins displaying a higher abundance under Zn stress in comparison to control					
heat shock protein 83	XP_021726736.1	4.2	5.1×10^{-6}	26	cytoplasm
annexin D2-like	XP_021714165.1	2.1	5.9×10^{-4}	22, 26	cytoplasm, plasma membrane
probable glutathione S-transferase parC	XP_021729368.1	2.1	7.2×10^{-6}	9, 10, 26	cytoplasm, chloroplast, nucleus
uncharacterized protein LOC110692268	XP_021724961.1	2.0	1.5×10^{-3}	-	extracellular space
low quality protein: urease-like	XP_021744615.1	1.8	6.1×10^{-3}	6	cytoplasm
betaine aldehyde dehydrogenase, chloroplastic	XP_021733695.1	1.6	4.6×10^{-5}	9, 26	chloroplast
citrate synthase, mitochondrial-like	XP_021731867.1	1.5	5.6×10^{-4}	2	mitochondrion
luminal-binding protein-like	XP_021763741.1	1.5	5.050^{-3}	19	endoplasmic reticulum
Proteins displaying a decreased abundance under Zn stress in comparison to control					
callose synthase 10-like	XP_021720217.1	2.6	1.0×10^{-2}	21	plasma membrane
PetA (chloroplast)	YP_009380143.1	2.1	1.6×10^{-4}	1	chloroplast
NdhH (chloroplast)	YP_009380186.1	1.8	1.1×10^{-2}	1	chloroplast
photosystem II 10 kDa polypeptide, chloroplastic	XP_021748287.1	1.8	1.9×10^{-3}	1	chloroplast
magnesium-chelatase subunit ChlI, chloroplastic-like	XP_021749286.1	1.7	8.9×10^{-3}	7	chloroplast
ribosomal protein S14 (chloroplast)	YP_009380128.1	1.7	1.7×10^{-2}	17	ribosome, chloroplast
LRR receptor-like serine/threonine-protein kinase GSO1	XP_021775210.1	1.7	3.1×10^{-3}	18, 26	plasma membrane
uncharacterized protein LOC110707738 isoform X1	XP_021741469.1	1.7	2.1×10^{-2}	-	chloroplast
50S ribosomal protein L9, chloroplastic	XP_021724417.1	1.6	3.7×10^{-3}	17	ribosome, chloroplast
cytochrome b6/f complex iron-sulfur subunit, chloroplastic	XP_021726772.1	1.6	5.7×10^{-3}	1	chloroplast

[a] Identification of peptides and annotation of proteins relied on the database search against amino acid sequences of *Chenopodium quinoa* (uploaded from KEGG protein database) accomplished with the SEQUEST algorithm; Mercator v4.5 was used for proteins annotation; subcellular localization of proteins was defined using BUSCA; [b] FC, fold change, for proteins that demonstrated an increase in abundance; FC(Zn stress/Control) indicates an (≥1.5-fold) increase in their abundance under stress Zn compared to the control, and for proteins that showed a decrease in abundance under Zn stress compared to the control, the FC values were adjusted by presenting the inverse of the FC value as FC(Control/Zn stress, ≥1.5); [c] Numbers correspond to the bins annotated with Mercator MapMan software (v4.5) (Table S1-6).

2.5. Functional Annotation of the Differentially Accumulated Proteins

Functional annotation of the differentially expressed proteins relied on the Mercator MapMan software (v 4.5) with the outputs being manually verified and curated based on the literature data. This was complemented by the advanced annotation workflow using open biological databases with an accession-based search (BRENDA Enzyme Database,

Gene Ontology, Kyoto Encyclopedia of Genes and Genomes, The Arabidopsis Information Resource, UniprotKB).

Representation of individual functional classes was rich in the *A. caudatus* root proteome both in terms of the functional diversity (i.e., the number of annotated bins) and the numbers of entries per bin. All the differentially abundant root proteins were distributed among 20 bins out of 28 available (Figure 4a; Supplementary Information 5, Tables S5-1 and S5-2), containing 78 entries in general. Thereby, multiple proteins could be attributed to more than one cellular function. For three entries, no specific molecular function could be assigned. Therefore, these species were annotated as unknown. Such diversity of the proteome response could be attributed mostly to the proteins which demonstrated a down-regulation trend in their response to Zn stress.

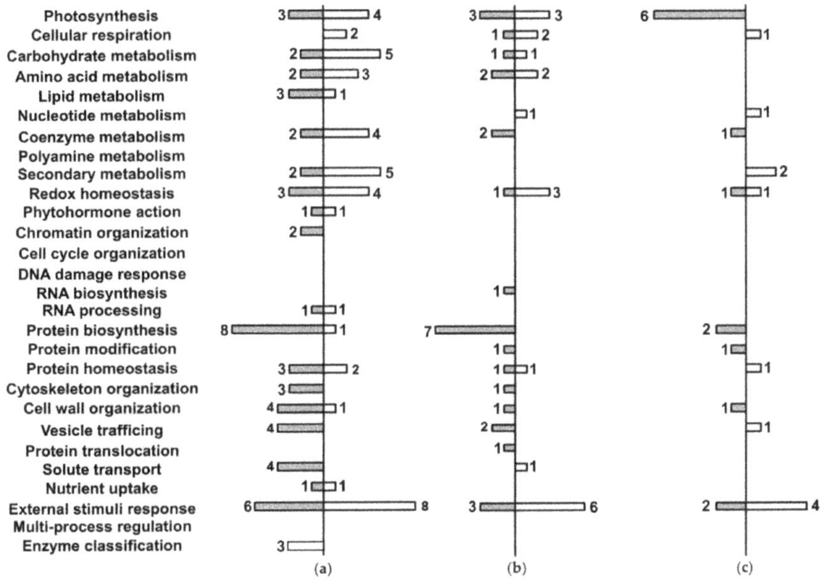

Figure 4. Functional annotation of the proteins identified as differentially expressed in the roots (**a**), young (**b**), and mature leaves (**c**) upon the exposure of the *A. caudatus* plants to 300 μmol/L Zn^{2+} supplemented to the hydroponic solution for 7 days (Zn stress) in comparison to the untreated controls. The bar labels indicate the numbers of proteins identified as up- (white) and down- (gray) regulated upon exposure to Zn stress. Protein annotation relied on Mercator MapMan (v4.5) software. Individual proteins comprising each functional class are listed in Supplementary Information 5.

Root proteins that demonstrated decrease in abundance when stress-induced comprised 18 functional classes (bins), while proteins with increased level of abundance were included in 16 bins. The latter part of the root proteome consisted of proteins involved in external stimuli response with eight entries (two glutathione S-transferases, uncoupling protein 4-like, probable aldo-keto reductase 4, heat shock protein 83) demonstrating a 2.4–39.1-fold abundance increase. Enzymes of the carbohydrate and secondary metabolism pathways also contributed to the observed differences with five proteins annotated within each functional class. The carbohydrate metabolism group included alcohol dehydrogenase 1 (10.1-fold), sucrose synthase-like (3.5-fold), phosphoglycerate kinase 3 (2.2-fold). Secondary metabolism protein functional class included probable glutathione S-transferase parC (39.1-fold), putative 12-oxophytodienoate reductase 11 (27.6-fold), BADH acyltransferase DCR-like (2.3-fold).

The external stimuli response functional class also dominated among root proteome with decreased level of protein abundance, where it was represented by six members—aspartyl protease AED3-like, several isoforms of the plasma membrane ATPase 4-like, and peroxidase 57-like, although in this case the stress-related proteins were less affected in comparison to the up-regulated ones (1.9–2.9 fold). Remarkably, in contrast to the proteins related to the stress response, the enzymes involved in protein biosynthesis were strongly suppressed under Zn stress conditions (eight entries in total). This group was represented by the ABC transporter F family member 3-like protein and different components of ribosomal machinery (1.5–3.5 fold-change).

The stress-related dynamics of the leaf proteins demonstrated different, although to some extent similar, patterns. Thereby, as can be seen from Figure 4, in leaves the Zn-induced changes were less pronounced in comparison to the roots. This effect could be seen both in terms of the numbers of the differentially expressed proteins and the degree of their abundance changes. Thus, only 40 differentially abundant species with the regulation degree not exceeding 3.1-fold were detected for the young leaves (in comparison to 78 differentially abundant proteins in roots, which demonstrated up to 39.1-fold changes).

The analysis of the young leaf proteome showed that 17 of 28 available functional classes (bins) appeared to be involved in the stress-related response in young leaves (Figure 4b, Tables S5-3 and S5-4). The down-regulated part of the leaf proteome was represented with 15 bins, whereas only 9 of them constituted its up-regulated part.

The proteins, discovered as Zn-dependently down-regulated in young leaves in response to the Zn treatment, dominated by the polypeptides involved in protein biosynthesis (seven entries). This functional class was represented by several ribosomal proteins and elongation factors TuB and 2-like (1.5–2.1-fold change) as the most responsive to the toxic metal. For the up-regulated part of the young leaf proteome, the group of the proteins responsive to external stimuli appeared to be the most abundant: in total, six proteins demonstrating a 1.7–2.7 abundance increase (receptor-like protein 12, probable aldo-keto reductase 4, polygalacturonase inhibitor-like, plasma membrane ATPase 4-like, and universal stress protein PHOS34-like) could be identified.

Finally, the proteome of the mature leaves appeared to be the least affected by Zn toxicity (Figure 4). Indeed, only 21 accessions representing only 12 bins (7 bins covering both increase and decrease in protein abundance level, with two representatives without specific function annotated) were differentially expressed in comparison to the untreated controls, which was approximately half of those discovered in young leaves (Figure 4c, Tables S5-5 and S5-6). Thereby, the proteins involved in responses to external stimuli (bin 26) appeared to be the most representative group of proteins with the increased abundance level, with four entries demonstrating up to 4.2-fold Zn-induced abundance change (heat shock protein 83, betaine aldehyde dehydrogenase, probable glutathione S-transferase parC, and annexin D2-like polypeptide). On the other hand, polypeptides involved in photosynthesis (six entries) were the most pronounced group with the decreased abundance level, although the degree of the abundance drop did not exceed 2.1-fold. The suppressed proteins of this class included PetA, NdhH, photosystem II 10 kDa polypeptide, cytochrome b6-f complex iron-sulfur subunit, NdhA.

Since this experiment revealed a relatively small number of Zn-responsive proteins (78, 40, and 21 for roots, young, and mature leaves, respectively), pathway enrichment analysis is unlikely to yield results for such small sets of differentially regulated proteins with fewer than 100 in each plant organ. Thus, further elucidation of important Zn-regulated metabolic pathways was based on numbers of Zn-responsive proteins assigned to a specific function. This highlighted at least six major protein functions: stress-response and redox metabolism, protein biosynthesis and homeostasis, photosynthesis, sugar metabolism, energy metabolism, and ion transport.

2.6. Prediction of Sub-Cellular Localization of Zn Stress-Exposed Differentially Accumulated Proteins

Prediction of the protein sub-cellular localization relied on BUSCA online tool with subsequent manual verification based on the literature data (Supplementary Information 6). This was further supported by annotation through an accession-based search in several open biological databases (BRENDA Enzyme Database, Gene Ontology, KEGG, The Arabidopsis Information Resource (TAIR), UniprotKB).

All accessions annotated as differentially accumulated in roots and leaves (both young and mature) were distributed among 17 cellular compartments, where cytosol (15–64% of proteins) and plastids (20–38% of proteins) tended to dominate in terms of the numbers of the individual proteins (Figure 5). The response to Zn stress in roots involved much higher diversity of compartments in comparison to the leaves, especially to the mature ones (Figure 5A,B). Remarkably, more than two-fold differences in the numbers of the up-regulated cytosolic proteins could be observed in the roots of the stress-exposed plants as compared to the young leaves collected from the same plants (Figure 5A,C, 39% and 15%, respectively).

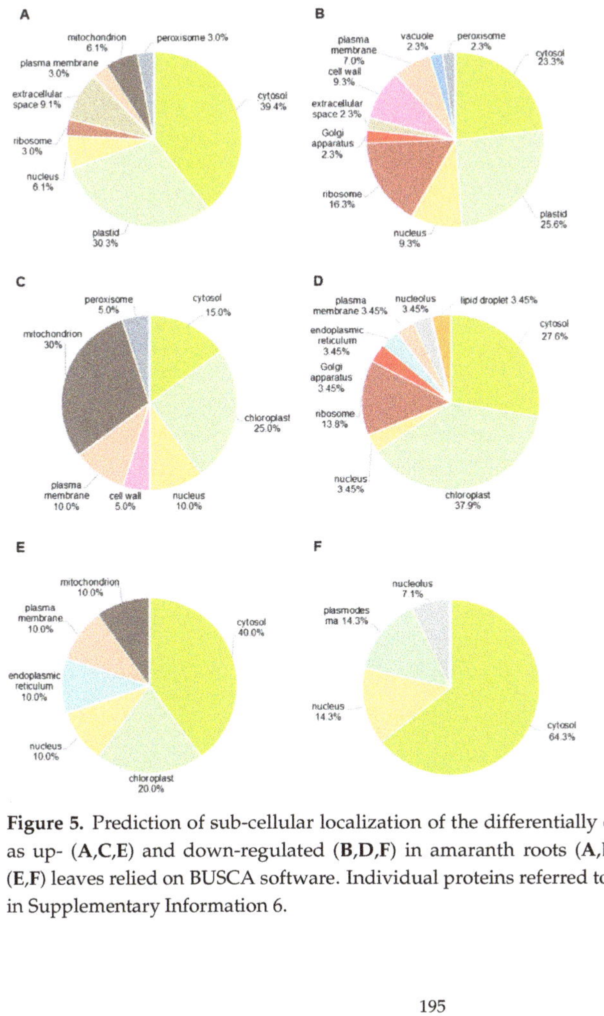

Figure 5. Prediction of sub-cellular localization of the differentially expressed proteins identified as up- (**A**,**C**,**E**) and down-regulated (**B**,**D**,**F**) in amaranth roots (**A**,**B**), young (**C**,**D**) and mature (**E**,**F**) leaves relied on BUSCA software. Individual proteins referred to each compartment are listed in Supplementary Information 6.

In contrast to roots, stress response in leaves was accompanied with the changes in only seven compartments, predominantly cytosol, chloroplast, and nucleus (Figure 5E,F). On the other hand, the lowest contribution was shown for peroxisomes (0–5% of proteins), vacuoles (0–2% of proteins), and nucleoli (0–7% of proteins). Interestingly, abundances of ribosomal proteins were mostly decreased in response to Zn exposure with the highest numbers of affected polypeptides observed in roots (Figure 5B, 16%). In contrast, mitochondrial proteins were mostly increased in abundance with the most pronounced effect observed in the young leaves (Figure 5C, 30%). Surprisingly, proteins annotated to the Golgi apparatus were found only in the down-regulated part of the proteome, but this compartment was represented by only a few polypeptides (Figure 5B,D, 2–3%).

3. Discussion

With the growing involvement of heavy metals in the industrial production and household, high capacity of multiple plant species to accumulate pollutants of this type [32] has attracted particular attention since it can be efficiently used for soil remediation purposes [33]. While for some heavy metals (e.g., cadmium) no constitutive physiological role in plants has been found so far, others are known as essential micronutrients, critically impacting catalytic and structural functions of cellular proteins [9]. Although Zn belongs to the latter group, enhancement of its root absorption leads to clear manifestation of toxicity signs [9] and development of a well-defined stress response [2,34,35]. Understanding particular molecular mechanisms behind the Zn-associated toxicity and accompanying stress responses is critically important for the development of new approaches to attenuate deleterious metal-related physiological effects and to establish efficient phytoremediation tools.

Here, we employ the comprehensive bottom-up proteomics approach to characterize the Zn-related changes in the proteomes of roots, young, and mature leaves of *A. caudatus* plants exposed to high Zn^{2+} concentrations during their growing in a hydroponics system. The roots and leaves differentially responded to the stressor, which could be seen by pronounced difference in the qualitative and quantitative patterns of differentially abundant proteins (DAPs). The organ-specificity of DAPs patterns can be explained in the context of the roles of these organs in Zn stress response. Indeed, as roots represent the first barrier on the way of soil solutes to the plant organism, the most pronounced Zn-induced shifts in the plant proteome (78 DAPs) were observed in this organ. Young leaves were the second affected plant organ. Their relatively intense response (forty DAPs) might be explained by their high metabolic activity. In agreement with this, mature leaves with relatively low metabolic activity appeared to be the least affected (Figure 3; Tables 1–3). Results of our previous metabolomics study [27] were in good agreement with the results of the proteomics experiment presented here.

3.1. *Plant Response to Zn^{2+} Was Accompanied by Characteristic Changes in Redox Metabolism and Stress-Inducible Proteome*

Protein functional class related to the external stimuli responses (bin 26) was one of the most represented in all organs addressed. Moreover, a pattern of the proteins involved in the constitutive and stress-inducible antioxidative defense mechanisms (bin 10 "Redox homeostasis") appeared to be affected in the metabolically active organs (roots and young leaves, Supplementary Information 5).

Enhancement in production of reactive oxygen species (ROS) is a universal reaction to any alterations in environment, which might be underlied by signaling events in terms of the stress response regulation [36] or stress-associated molecular damage due to overproduction of highly toxic radicals [37]. These reactions can be to some extent suppressed by the enzymatic antioxidant systems [38]. Here, the efficiency of these ROS-protective systems could be judged by the numbers of Zn-dependently up-regulated antioxidant enzymes, and by the degree of their up-regulation.

Thus, glutathione-*S*-transferase (GST), which was present in roots in two isoforms, showed a prominent up-regulation upon the exposure to Zn (Table S5-1). This effect can be explained by the role of this enzyme in the detoxification of highly reactive natural toxicants, such as lipid peroxides, by their conjugation with glutathione (GSH) [39]. On the other hand, this strong Zn-related increase in the relative abundance of GST can be explained by the direct involvement of GSH in detoxification of the Zn^{2+} ions which can be chelated directly by GSH to be transported in vacuole by GST [40,41].

This expressional response was in agreement with the shifts in the dynamics of the enzymes involved in the ascorbate-glutathione cycle: ascorbate peroxidase (APX, 1.7-fold down-regulated) and monodehydroascorbate reductase-like protein (MDHAR, 1.6-fold up-regulated, Tables S5-1 and S5-2) [42]. Such an increase in the abundance of the ascorbate-producing enzyme (MDHAR) might indicate enhancement in the ascorbic acid production and increase in its equilibrium concentration in cellular liquids. Ascorbate is not only a powerful scavenger of H_2O_2, but it plays a crucial role in the zeaxanthin synthesis (which is involved in dissipation of excess light energy under stress conditions) and reduction in catalytic transition metals in active centers of multiple ROS-detoxifying enzymes that is critical for maintaining their functionality [43].

On the other hand, decrease in the relative abundance of APX (present in two extracellular isoforms) and a concomitant rise in the levels of enzymes implicated in H_2O_2 production (several germin-like proteins) might indicate development of oxidative stress in amaranth roots (Tables S4-1 and S4-2). This transient increase in the tissue H_2O_2 contents might be a regulatory signal involved in the metabolic shifts associated with induction of protective phenylpropanoids and deposition of lignin in the cell wall [44].

It is known that H_2O_2 is a signaling agent employed in various defense responses and acting in a tight crosstalk with salicylate (SA) and jasmonate (JA) [45], activating the related signaling pathways, which are known to manage downstream activation of systemic tolerance mechanisms [46]. In line with this, the enzymes of JA metabolism were strongly affected by Zn stress. Thus, on one hand, probable linoleate 9S-lipoxygenase 5 (9-LOX5) and putative 12-oxophytodienoate reductase 11 (12-OPDAR), which are the key enzymes of the JA synthesis [47,48], appeared to be up-regulated in roots upon the exposure to Zn^{2+} (1.9- and 27.6- fold, respectively). On the other, peroxisomal 2-hydroxy-acid oxidase isoform X1, the principal enzyme of the JA metabolism, which is known to switch off the JA-signaling by hydroxylation of this hormone in its 12th position [49], was two-fold up-regulated in this organ. This concerted enhancement of JA biosynthesis and degradation might indicate strong involvement of H_2O_2-triggered JA-signaling in adaptation to Zn stress.

This assumption is supported by up-regulation of the root proteins, which are regulated by the JA- and SA-related pathways. For example, germin-like proteins, which positively impact on the tissue H_2O_2 levels [50] and play an important role in pathogen resistance (tightly regulated by JA-dependent mechanisms) [51], were found to be more than 1.5-fold up-regulated in roots. Further, non-symbiotic hemoglobin 1-like protein, which is known to be activated by JA and SA [52], was strikingly up-regulated in the roots (Table S5-1). This observation fitted well to the results of our metabolomics experiments [27], which indicated strong up-regulation of SA in roots and young leaves, highlighting a special role of SA in the mechanisms behind the plant tolerance to the Zn stress.

Similarly to the roots, the enzymes of the redox metabolism and an array of closely related proteins appeared to be the most responsive to the Zn exposure in the young leaves. This coordinated response suggests systemic regulation, where root-emerging redox shifts are transmitted with transporters, membrane receptors, and redox, signaling pathways to newly formed leaves [53]. The Zn-dependently regulated leaf proteome was predominantly represented by the proteins associated with response to external stimuli (bin 26, Tables S5-3 and S5-4, Figure 4b). A substantial part of these proteins was involved in signaling pathways and demonstrated differential patterns of the quantitative response. For example, leucine reached repeat (LRR)-containing receptor-like protein 12 and polygalacturonase inhibitor-like appeared to be Zn-dependently up-regulated, while the LRR-containing regulatory subunit of serine/threonine protein phosphatase 2A (PP2A) was down-regulated.

Currently, the LRR-containing proteins are the subject of extensive discussion with respect to various stressors, predominantly of the biotic nature [54,55]. The receptor-like protein 12 particularly, besides its function in stress protection, plays a pivotal role in the maintenance of meristems, serving as a CLV2-like peptide receptor in *A. thaliana* [56]. Polygalacturonase inhibitor protein is known to be induced by SA in response to biotic stress. It suppresses the activity of pathogen-secreted pectin-depolymerizing enzymes, which, in turn, limits the degree of cell-wall loosening [57]. Already at the beginning of this century, it was shown that pectin impacts binding heavy metal ions and prevents, thereby impacting their translocation into the cell [58]. PP2A is featured with broad specificity of substrates, and is known to be involved in auxin transport, ethylene and abscisic acid signaling, growth and development [59].

The described pattern of the Zn-induced protein expression shifts in young leaves might indicate the key role of the molecular signals transmitted therein from roots and mature leaves. The observed signature of signaling proteins might enable young leaves (as the most susceptible plant parts), to pre-adjust their metabolism for oncoming increase in tissue metal contents.

In contrast to the young leaves, the stress response of the mature leaf proteome dominated with a distinct pattern of up-regulated proteins as follows: heat shock protein 83 (HSP83), glutathione *S*-transferase parC (GST parC), and betaine aldehyde dehydrogenase (BADH). This response might highlight a general mechanism of the response to Zn stress, including maintaining the protein structure during/upon the stress exposure (HSP83) [60], conjugation of toxic compounds via GST [39], and accumulation of the osmolyte glycine betaine via BADH [61].

Annexin D2-like protein, a cytosolic polypeptide which was also found to be up-regulated in the mature leaves, is capable of relocating from cytosol to membranes (plasma membrane, ER, Golgi apparatus, vacuolar, nuclear, vesicles, etc. [62]) in response to enhanced cellular ROS production and increased cytosolic Ca^{2+} levels (Table S5-5) [63]. It is plausible to assume that both these stimuli are the parts of the regulatory signaling pathways including long-distance signal loops from roots. These effects, at least partly, are mediated by annexin D2, which likely activates defensive mechanisms [64].

Down-regulation of catalase more than 1.5-fold upon Zn treatment in mature leaves might be underlied by the signaling function of hydrogen peroxide, i.e., this might reduce H_2O_2 scavenging and thereby increase its equilibrium concentrations in tissue liquids. Catalase is known to be regulated by direct SA binding, which decreases the enzyme activity [65]. Thus, our observation is in agreement with the evidence for SA-repressed catalase mRNA expression described in the literature [66].

LRR receptor-like serine/threonine-protein kinase GSO1, the abundance of which was stress-dependently decreased in mature leaves, was intensively discussed in the context of root growth regulation and embryonic cuticle development [67,68]. Recently, GSO1 in combination with GSO2 was reported to be a positive regulator of cell proliferation [69]. Thus, it can be hypothesized that down-regulation of GSO1 in the mature amaranth leaves (Table S5-6) might be a signal linked to stress-related retardation of root and young leaf growth. This down-regulation of GSO1 protein likely designates stress-induced resource reallocation to prioritize defense processes under growth of new tissues [70].

3.2. Exposure to Zn^{2+} Affects Protein Biosynthesis and Homeostasis

Zn stress is known to cause protein damage, which is typically accompanied by their misfolding and aggregation [71]. In agreement with this, here we observed multiple alterations in the abundances of individual polypeptides associated with protein metabolism and structure maintenance (Figure 3). These shifts might be at least partly underlied by seven and five down-regulated ribosomal proteins (RPs) in amaranth roots and young leaves, respectively, and two transcript elongation factors (TEFs) in young leaves (Tables S5-2 and S5-4).

Now it is believed that besides their canonical structural role, RPs are implicated in additional specialized extraribosomal regulatory activities, which play important roles in the diversity of physiological processes [72]. Thus, it is proposed that some RPs could be capable of remodeling ribosomes for the synthesis of specific proteins that are required for a certain developmental stage or necessary for the response to defined biotic or abiotic stressor [73,74].

In addition, the regulatory function of RPs might be accomplished through interaction with specific proteins or/and metabolites—components of signaling pathways such as transcriptional factors and phytohormones [72]. Further, involvement of the TEF proteins in response to the applied Zn stress can be posited from suggestion that TEFs are implicated in transcriptional reprogramming, i.e., the process by which they regulate RNA polymerase II to generate transcripts adjusted for the specific demands of a given environment [75].

Another set of Zn-dependently regulated actors of protein metabolism was represented by the molecules involved in protein maturation and folding. Among them, stress-inducible HSP83 and proteases FTSH2 were up-regulated in the roots, whereas vacuolar-processing enzyme ClpA and aspartyl protease AED3-like protein appeared to be down-regulated (Tables S5-1 and S5-2). These results further suggest a multi-faceted protein protection mechanism, which includes protein refolding and protection against damage by heat shock proteins and proteases, and reducing the impact of impaired protein synthesis during ATP deficiency.

Among the accessions involved in proteolytic degradation in young leaves, the U-box domain containing 16-like protein was the only up-regulated protein from the group of ubiquitin ligases, which provided precisely tuned protein ubiquitination (Table S5-3) [76].

3.3. Zn Stress Results in Expressional Suppression of the Electron Transport Chain Proteins in the Chloroplast

Increased demand for energy and NAD(P)H supplementation is one of the most critical biochemical challenges associated with any stress response [77]. ATP and NAD(P)H are critically important for maintaining various metabolic processes including the activity of antioxidant system and correction of physiologically disadvantageous metabolic shifts [78].

The primary source of these energy-rich metabolites as well as organic carbon-rich substrates of energy metabolism is photosynthesis. However, the efficiency of this process in mature leaves is shown to be negatively affected by Zn stress [27]. Indeed, the excess of Zn might impair photosynthetic processes through disruption of chlorophyll synthesis by displacement of iron from active site of an enzyme that catalyze the formation of chlorophyll precursor, photochlorophilllide [30]; by substitution magnesium in the chlorophyll molecules [79]; and by causing direct damage of the proteins constituting electron transport chain (ETC) of the chloroplast [9].

In this study, we succeeded to obtain the first insight into these aspects at the proteome level. Thus, in agreement with the described scenario, we discovered the following six photosynthetic chloroplast ETC proteins which exhibited significant decrease in their abundance in mature leaves of Zn-exposed plants (Figure 4c): photosystems (PSs) I and II components (PsaA, PSII 10 kDa polypeptide), cytochrome proteins PetA and iron-sulfur subunit of cytochrome b_6/f complex, NADH-dehydrogenases NdhH and NdhA (Table S5-6). These results might indicate the enhanced degradation of the ETC proteins under Zn stress, which, in turn, can lead to disturbances in electron transport and ATP production processes. The overload of the ETC with electrons and related overproduction of ROS can further damage ETC proteins, especially those containing iron [80].

The young leaves of the Zn-exposed plants which showed visible chlorosis symptoms displayed another pattern of DAPs implicated in photosynthetic ETC. Specifically, PsbC, NdhH, and UPF0603 (*A. thaliana* homolog At1g54780, annotated as thylakoid lumen protein 18.3 in TAIR) were downregulated, while ferredoxin-$NADP^+$ reductase (FNR) and oxygen-evolving enhancer protein 2 (OEE2) were up-regulated (Tables S5-3 and S5-4). NdhH is a constituent of the protein complex implicated in cycled electron flow (CEF) around PSI [81]. This presents an alternative form of electron transport which is involved in ATP production under the conditions associated with high energy demand [82]. The down-regulation of NdhH in young leaves observed here might indicate the general impairment of photosynthetic complexes by Zn excess. The decline in the abundance of thylakoid lumen protein 18.3, regulating PSII reassembly, and PsbC, the core light-harvesting subunit of PSII, might stand for temporal cessation of PSII turnover in the stressful conditions by abovementioned direct Zn inhibition or as the result of oxidative damage [83].

The up-regulation of two other chloroplast proteins (FNR and OEE2) in young leaves under Zn stress probably reflects adaptive mechanisms against oxidative damage to maintain energy balance. Thus, an increased FNR abundance may suggest compensatory mechanisms accompanying disrupted efficiency of the photosynthetic machinery under Zn stress and compromised capacity of the Zn-exposed plant to supply NADPH equivalents for ROS counteraction and, most important, for sugars production in Calvin cycle [84]. OEE2 may serve to stabilize damaged PSII as its structural component [85].

In general, these findings indicate that exposure to high Zn^{2+} concentrations damages photosynthetic ETC or even prevents its establishment and chlorophyll biosynthesis in the young leaves causing their chlorosis. This, in turn, limits subsequent sugar synthesis. As a result of the decreased effectiveness of photosynthesis, young leaves have to compensate for the lack of monosaccharides by inducing their allocation from mature leaves.

3.4. Zn Stress Results in Rearrangement of Sugar and Energy Metabolism in A. caudatus Roots

In agreement with published data on the metabolic responses in heavy metal-exposed plants [86], our results indicate the inhibition of starch synthesis in the roots of *A. caudatus* plants grown under Zn stress. Indeed, fructose-1,6-bisphosphatase and glucose-1-phosphate adenylyltransferase—the two key enzymes involved in the early steps of starch biosynthesis—were found to be down-regulated in the Zn-affected root tissues.

Not less importantly, two sucrose synthase-like isoforms appeared to be Zn-induced in the amaranth roots (Tables S5-1 and S5-2). This enzyme is already known to be responsive to heavy metal ions with pronounced suppression of sucrose biosynthesis and activation of its degradation [87,88]. Thus, the results of our proteomics survey are in good agreement with observation of Li et al. Moreover, our proteomics data fit well with the results of the comprehensive metabolic profiling, accomplished for the same amaranth root material and clearly demonstrating a pronounced Zn-induced metabolic shift in favor of monosaccharides [27]. Some of the sugars might be involved in sugar signaling. Indeed, glucose produced by invertase may further maintain activity of hexokinase, one of major components of the sugar signaling cascades [89]. In fact, two hexokinase isoforms were annotated in amaranth roots. However, these enzymes did not exert Zn stress-associated expressional response (Tables S2-3 and S2-4, S4-1 and S4-2). Interestingly, fructokinase, the primary fructose-phosphorylating enzyme, responded to Zn stress by increasing the number of isomers from 10 found in roots of control plants to 15 found in roots of Zn-treated plants. It is known that some of fructokinase isoforms could be inhibited by fructose in concentration exceeding 0.5–1 mmol/L [90] and it is proposed that fructokinase might act as a fructose sensor [91]. Differential expression patterns of different fructokinase isoforms were found in other plants (rice, sunflower) in response to various abiotic stresses [92,93]. Thus, our data, together with the literature data, suggest that under unfavorable conditions, different isoforms of fructokinase may play an important role in regulating the amount of metabolized carbohydrates.

Furthermore, three proteins representing bins 2 and 3 (Cellular respiration and Carbohydrate metabolism, respectively), namely 2,3-bisphosphoglycerate-independent phosphoglycerate mutase, phosphoglycerate kinase 3, and alcohol dehydrogenase 1 (Tables S5-1 and S5-2), were found to be upregulated under Zn stress conditions. In the context of the above discussed data on the stress-induced dynamics of the enzymes involved in starch and sucrose degradation, it can be inferred that energy metabolism serves as the principal pathway for the utilization of the overproduced monosaccharides in the roots of the Zn-exposed plants.

The up-regulation of glycolytic enzymes observed here can also be triggered by a decline in cellular energy status (i.e., in the ATP/ADP ratio [94]), and is currently regarded as a protective survival strategy under stress [95]. These changes are also consistent with the sugar signaling model, as evidenced by the role of Sucrose Non-fermenting-Related Kinase 1 (SnRK1) in this process [96]. This multimeric protein serves as a sugar sensor that is activated under energy-deprived conditions, such as stress, to promote catabolic pathways and downstream enhancement of defense proteins expression/activity [97]. On the other hand, Target of Rapamycin (TOR) kinase functions as an antagonist to SnRK1, promoting anabolic growth and ribosome biogenesis coordinating nutrient availability and energy status [98]. The expression levels of SnRK1 or TOR remained unchanged in response to zinc exposure in any of the plant organs in the present experiment. Nonetheless, it has been reported that these proteins are subject to regulation at the post-translational level, primarily through the process of phosphorylation [99], as well as through interactions with sugar phosphates, such as trehalose-6-phosphate [100]. In addition, a multitude of downstream targets have been identified [101]. The upregulation of several glycolytic enzymes discussed earlier, sucrose-degrading enzymes (sucrose synthase-like isoforms), and redox-related enzymes with heat shock protein 83 (also discussed above) in roots is consistent with the hypothesis that a sugar-regulated signal is generated, involving SnRK1 activation, which in turn activates catabolic processes and defense responses via transcriptional factors [102]. This signal emanated primarily more likely in mature leaves as a source tissue. Phloem-mobile sucrose is synthesized mainly in mature leaves and transported to sink tissues such as roots and young leaves, where it serves as substrate for energy generation and can influence energy-sensing pathways such as SnRK1, particularly in roots under stress conditions [96]. Furthermore, up-regulation of key enzymes of pyruvate dehydrogenase complex and TCA cycle (dihydrolipoyl dehydrogenase and citrate synthase, respectively) in the young leaves may occur in response to higher energy demands upon heavy metal stress, as these enzymes play crucial roles in energy metabolism and are often activated under conditions of nutrient stress to enhance metabolic efficiency and energy production [103]. The observed downregulation of multiple ribosomal proteins in root tissue in response to Zn exposure suggests that TOR signaling was likely repressed. The aforementioned repression led to a coordinated decrease in ribosomal biogenesis and overall protein biosynthesis as energy-intensive process.

Remarkably, such concerted shifts in the levels of abovementioned proteins in roots related to sugar metabolism, glycolysis, and fermentation were negligible in young and mature leaves. Nevertheless, our metabolomics study, accomplished with the same plant material [27], revealed a Zn-induced increase in the relative contents of soluble sugars in these organs, especially in young leaves. This finding suggests that the translocation of sugars from mature leaves rather than their *de novo* synthesis might underlie their up-regulation in the young leaves. Interestingly, this fact could not be supported by differential expression of sugar transporters in none of the studied organs of the Zn-exposed plants, although several of them were annotated in our dataset, e.g., sugar carrier protein C-like and sugar transport protein 1-like, both identified in roots and leaves (Supplementary Information 2). Thus, the plausible reason for the increase in the sugar relative contents could be changes in the process of translation and post-translational regulation of their transporters.

3.5. The Excess of the Zn(II) Ions Is Sequestered into the Vacuole and Cell Wall by Means of Several Transport Protein Systems

Exposure of the *A. caudatus* plants to Zn^{2+} caused clear alterations in the expressional patterns of membrane transporters. Thus, three isoforms of the root plasma membrane (P-type) H^+-ATPase 4 and vacuolar (V-type) H^+-ATPase subunit B2 appeared to be up to 2-fold down-regulated, which might cause compromised ability of the plant root to manage ionic balance and translocation under stress conditions [30,104,105]. On the other hand, multiple V-type H^+-ATPases, annotated in the roots and leaves, did not show any stress-associated expressional response. This, however, does not ultimately indicate their unaffected functional state, as these V-type H^+-ATPases are known to be regulated on the post-translational level (by phosphorylation) under the conditions of heavy metal stress (Cd, Cu, and Ni) [106].

The changes in the activity of P- and V-type H^+-ATPases in roots (due to expressional or regulatory mechanisms) are critically important for suppression of Zn translocation to the shoot. This mechanism is implicated in the attenuation of Zn excess toxicity (i) by sequestering free Zn^{2+} ions or their chelates with low-molecular weight ligands from cytosol with the help of various types of Zn transporters (e.g., metal tolerance protein (MTP), vacuolar iron transporter (VIT), etc.) [9,107] or (ii) by promoting an accumulation of anionic metabolites (GSH, organic acids, amino acids, phytates, phosphate) in vacuoles, which further form complexes with Zn^{2+} [9,22]. This vacuolar sequestration of Zn^{2+} in root cells limits the ions translocation from roots to leaves and is regarded as one of HM tolerance mechanisms [108]. The proteomic datasets of amaranth roots and leaves studied here revealed no identification of any Zn transporters. Nevertheless, multiple representatives of ATP-binding cassette (ABC) transporters, a transporter group which was assumed to also be implicated in the translocation of free Zn^{2+} and their complexes across membranes into vacuoles [109], were annotated in all organs (Supplementary Information 2). However, no Zn-related alterations in their relative abundances could be observed in all organs (Supplementary Information 4). This fact might point to other mechanisms rather than transcriptional regulation that may be involved in control of ABC transporters. This might be the case here, as ABC transporters are often found to be post-translationally or epigenetically regulated [110,111]. The absence of the expressional response might be a species- or metal-specific feature. For example, transcriptional regulation of the ABC transporters was shown in tomato plants exposed to cadmium [112].

3.6. Consideration of the Zn-Induced Alterations in A. caudatus Proteome in the Context of the Accompanying Metabolic Adjustments Gives Access to the Mechanisms of Zn Stress Tolerance

Generally, metabolic adjustments observed in our previous study on the same plant model [27] in response to Zn stress were more pronounced (both in terms of the numbers of affected metabolites and magnitude of corresponding shifts in their relative abundance) in the roots and young leaves than in the mature leaves. Thereby, multiple sugars and organic acids (in particular sugar acids) were the major classes of compounds to be strongly up-regulated in all organs. Generalized information relating to the further discussed principal changes in primary metabolism that occurred in the roots and leaves of *A. caudatus* plants in response to Zn stress and supported by findings of this proteomic research is presented in Figure 6.

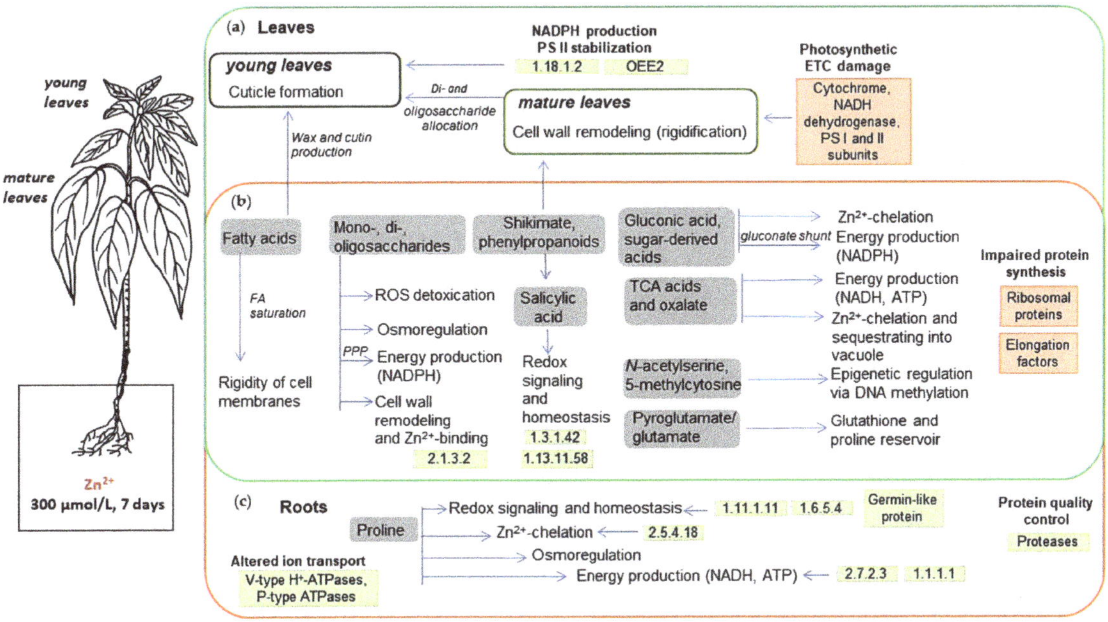

Figure 6. Consideration of the acquired proteomics data in the context of the metabolic shifts induced in the leaves and roots of the *A. caudatus* plants after their exposure to zinc in aq. culture (Zn^{2+}, 300 µmol/L, 7 days, described earlier by Osmolovskaya et al. [27]). The scheme presents the alterations specific for leaves (**a**), common for leaves and roots (**b**), and specific for roots (**c**). The proteomic findings observed here are indicated with green (up-regulation) and red (down-regulation) rectangles as follows: 1.1.1.1—alcohol dehydrogenase; 1.11.1.11—ascorbate peroxidase; 2.1.3.2—probable inactive purple acid phosphatase 29; 1.18.1.2—ferredoxin-NADP+ reductase; 2.5.1.18—glutathione S-transferase; 1.6.5.4—monodehydroascorbate reductase; 2.7.2.3—phosphoglycerate kinase; 1.3.1.42—putative 12-oxophytodienoate reductase 11 isoform X1; 1.13.11.58—probable linoleate 9S-lipoxygenase 5; OEE2—oxygen evolving enhancer protein 2.

Our proteomics data did not provide direct evidence of increased pectin synthesis, a common response to heavy metal stress in plants [113]. Indeed, increased pectin deposition in cell walls of plants upon Cd stress was associated with elevated expression of galacturonic acid transferase in tomato and potato, as well as with increased activity of pectin methylesterase [114]. Thus, both expression of key enzymes of pectin biosynthesis and pectin modifications (e.g., decreased methylesterification) influence heavy metal sequestration and, consequently, detoxification by cell wall. Thus, we can assume that regulation of this biopolymer does not rely mainly on the protein expression level in amaranth roots upon Zn stress. On the other hand, probable carbohydrate esterase At4g34215 likely belongs to SGNH-hydrolase superfamily of enzymes [115] and was down-regulated under Zn stress in the amaranth roots. Members of this superfamily are known for their ability to modify wide diversity of carbohydrates including pectins [116]. BAHD acyltransferase DCR-like up-regulated in roots could also be involved in pectin and cuticle remodeling [117]. The opposing regulation of these two cell-wall-modifying enzymes promotes the dynamic reconstruction of heteropolysaccharides in the root cell wall, improving Zn^{2+} chelation and counteracting their translocation into the shoot. Furthermore, the pattern of Zn-dependently up-regulated sugars dominated with galactose, which demonstrated a 9-fold increase in roots. This observation might be explained by the fact that galactose is putatively involved in the enhanced deposition of cell wall pectin, which has a high capacity for heavy metal binding [118].

Furthermore, galactose might be involved in the biosynthesis of galactolipids, which is known to be stimulated in response to phosphate starvation [119]. The deficit of phosphate might be underlied by enhanced involvement of this anion in chelation of Zn(II) ions under the conditions of their increased abundance in the plant tissues [120]. This assumption is confirmed by the fact that exposure of plants to high concentrations of Zn^{2+} results in low levels of phosphate in plants [121]. Our results also indicate the development of phosphate deficiency, which can be clearly seen from a 10-fold increase in the abundance of probable inactive purple acid phosphatase 29 (Table S5-1), which is a universally recognized indicator of this physiological state [122]. In the context of this finding, it is possible to assume that involvement of galactose in the biosynthesis of galactolipids can be activated by the substitution membrane phospholipids with galactolipids [123]. This is an adaptive strategy for membrane lipid remodeling during phosphate deprivation observed in several plants [119,123]. Additionally, LC-MS analysis of *N. caerulescens* leaves from Zn-treated plants revealed Zn-dependently altered galactolipid profiles, with a sharp increase in 12-oxo-phytodienoic acid (OPDA) contents, oxidized galactolipid species [124]. These changes fit well with the report of López-Orenes et al. on the heavy metal-induced SA-promoted oxylipin synthesis and the role of this oxylipin up-regulation in stress signaling and plant tolerance to abiotic stresses [125].

The role of SA in the plant response to the Zn exposure is another aspect to be discussed. In our earlier metabolomics survey [27] we observed a strong up-regulation of this phytohormone in young leaves and roots (23- and 27-fold, respectively). Although direct players involved in the SA biosynthetic pathway were not found in our proteomics dataset, multiple DAPs appeared to be involved in the processes, which, at least partly, rely on SA signaling. For example, SA is known to induce pathogenesis-related proteins, such as non-symbiotic hemoglobin 1-like protein in roots and polygalacturonase inhibitor-like in young leaves (Tables S4-1 and S4-3). Moreover, SA promotes biosynthesis of oxylipins [126], including jasmonates, which could be confirmed by the induction of putative 12-OPDAR and 9-LOX5 upon Zn stress in the roots (Table S4-1).

Increased (2–3.5-fold) expression of sucrose synthase in roots of Zn-treated plants (Table S4-1) might indicate stress-induced sucrose influx and its metabolism in roots. This enzyme together with hexose metabolizing hexokinases and fructokinases (see Section 4.4) supplies substrates for energy metabolism via their oxidation in glycolysis or PPP. This is in line with our previous metabolomics study which showed an increase in the levels of hexoses and, in particular, strong accumulation gluconic acid in roots and young leaves [27]. In plant cells gluconate acts as a powerful chelator of metal ions [127] and can be produced in the gluconate shunt of the pentose phosphate pathway (PPP) [128], described so far for bacteria, yeast, and fungi [129]. It is worth emphasizing that at the background of the ongoing oxidative stress, the gluconate formation, at least partly, may be caused by non-enzymatic glucose oxidation and thus contributes to metal-induced oxidative stress [27]. The presence of this phenomenon (i.e., enhanced ROS production) could be clearly seen by a 13-fold increase in the expression of mitochondrial uncoupling protein 4-like (Table S4-1), which discharges the mitochondrial membrane to reduce ROS production [130], and by a 2–5-fold reduction in the expression of five different peroxidases (Supplementary Information 3, Table S3-2). As a result, in conditions where sugar accumulation and oxidative stress occur simultaneously, generation of reactive carbonyl compounds (RCCs) might be activated, accompanied by the development of protein glycation and carbonyl stress [131]. This, in turn, may explain [132] the observed activation in expression of aldo-keto reductase (8-fold for probable aldo-keto reductase 4, Table S4-1) as a protein anti-glycation repair mechanism.

4. Materials and Methods

4.1. Plant Growth Conditions and Zn Stress Application

The detailed information on plant growth conditions was described in our previous work by Osmolovskaya et al. [27]. The study employed *Amaranthus caudatus* L. var. Karwa dauta. The seeds were provided by the Vavilov All-Russia Institute of Plant Genetic Resources, Saint Petersburg, Russia. The seeds were subjected to surface sterilization using a 3% (*v*/*v*) H_2O_2 solution for 20 min and rinsed with deionized water. The seeds were then germinated in containers filled with calcined quartz sand. The plants were cultivated in two separate experiments, each with an identical setup. Each experiment consisted of nine treatment groups, with a total of 27 plants distributed across nine vessels. The plants were grown under a controlled environment, with a 16 h day and 8 h night cycle, a relative humidity of 70–75%, and day/night temperatures of 24/18 °C. The plants were illuminated by fluorescent lamps (wavelength of 320–780 nm and a photosynthetic photon flux density of 120 µmol m^{-2} s^{-1}).

In the initial seven days after the seeds germinated, the seedlings were irrigated with a ten-fold diluted nutrient solution (0.1 n.s., see Supplementary Information 1, Protocol S1-1) three times a day. Over three weeks, the concentration of the solution was gradually increased to 0.2 n.s., 0.5 n.s., and 1.0 n.s., respectively. After four weeks, the plants were transplanted into new 3 L hydroponic containers filled with the full-strength (1.0 n.s.) nutrient solution, with three plants per container.

After two weeks of growth in the hydroponic setup, the containers with six-week-old plants, featuring fully developed leaves that could be clearly distinguished from younger ones, were randomly divided into three equal groups ($n = 3$). The plants in the first group were harvested prior to the application of stress, for a separate evaluation of their root, young, and mature leaf biomass. The second group of plants was designated as the "Zn-treated" group with addition of 300 micromoles per liter of $ZnSO_4 \times 7H_2O$ to the nutrient solution. These stress conditions were chosen based on the previous research, including studies on amaranth [26]. The plants in the third group were considered as the control group and were not subjected to any treatment. The experiments were conducted in three independent biological replicates, with a total of nine plants in each group. The plants treated with zinc and the control plants were evaluated for a range of physiological characteristics (stomatal conductance, chlorophyll content, photosystem II activity, leaf water content) before (Day 0) and after the zinc stress (Day 7), and were then harvested for proteomics.

4.2. Protein Isolation and Digestion

The total leaf and root protein fractions were isolated from approximately 200 and 250 mg of fresh-frozen grounded leaf and root material per sample, respectively, by the phenol extraction method as it was described by Frolov et al. [133]. The resulting dry acetone protein pellets were reconstituted in 4% (*w/v*) aqueous (aq.) sodium dodecyl sulfate (SDS) under continuous sonication in 5 min cycles with ice cooling. Protein concentrations were determined by BCA assay kit according to the manufacturer's protocol and cross-validated by sodium dodecyl sulfate polyacrylamide gel electrophoresis (SDS-PAGE) as described previously [134].

Aliquots of protein extracts (25 μg) were digested with trypsin according to the filter-aided sample preparation (FASP)-based protocol of Leonova et al. [135] with minimal changes (for details see Supplementary Information 1, Protocol S1-2). The completeness of protein digestion was verified with SDS-PAGE as described by Greifenhagen and co-workers [134]. The resulting tryptic hydrolysates were desalted and pre-cleaned with reverse-phase solid phase extraction (rpSPE) with six layers of C18 material-filled 200 μL tips (Merck KGaA, Darmstadt, Germany) according to the in-house established procedure [136]. The samples were dried under reduced pressure and stored at −20 °C until analysis.

4.3. Nano LC-MS/MS Experiments

Each of the samples containing dried tryptic hydrolysates were sequentially reconstituted in 60, 20, and 3% (*v*/*v*) aq. acetonitrile containing 0.1% (*v*/*v*) formic acid, and resulted samples were loaded on an Acclaim PepMap trap column (300 μm × 5 mm, 5 μm particle size, Thermo Fisher Scientific, Waltham, MA, USA) during 15 min at the flow rate of 30 μL/min. Peptides were separated on a RP C18 emitter column (PicoFrit, 75 μm × 250 mm, 15 μm tip diameter (New Objective) packed with ReproSil-Pur C18-AQ material, 1.9 μm, 120 Å, Dr. Maisch, Ammerbuch-Entringen, Germany) using an Ultimate 3000 RSLC nano chromatography system (Thermo Fisher Scientific, Waltham, MA, USA) coupled on-line to an Orbitrap XL mass spectrometer (Thermo Fisher Scientific, Waltham, MA, USA) operated in the positive ion mode via a nano-ESI source under the LC and MS settings specified in Tables S1-9 and S1-10, respectively. The nanoLC-ESI-Orbitrap-MS analysis relied on data-dependent acquisition (DDA) experiments, comprising in each cycle one survey Orbitrap-MS scan and dependent MS/MS scans for five the most abundant signals with the charge states from two to seven selected in the previous cycle.

4.4. Processing, Post-Processing, and Statistical Analysis of the Proteomics Data

The acquired LC-MS data were processed with Proteome Discoverer (v2.2, Thermo Fisher Scientific, Waltham, MA, USA) for identification of tryptic peptides and annotation of proteins (FDR correction at $q \leq 0.05$ and Percolator node to filter high-confident identifications; for other search settings, see Table S1-11), whereas label-free quantification relied on QIProgenesis software (Waters GmbH, Eschborn, Germany). To select proteins that are significantly differentially abundant, thresholds for fold change (FC) and FDR adjusted *p*-value were set at 1.5 as the lowest value and 0.05 as the highest value, respectively. With this, proteins that showed an increase greater than 1.5-fold in abundance under Zn stress compared to the control (FC(Zn-stress/Control) ≥ 1.5) were considered as up-accumulated; proteins that showed a decrease greater than 1.5-fold in abundance under Zn stress compared to the control (FC(Control/Zn-stress) ≥ 1.5) were considered to be down-accumulated. Thus, the FC values were adjusted for cases of proteins showing a decrease in abundance by presenting the inverse of the FC value. For identification and annotation of proteins *Arabidopsis thaliana* (Uniprot) and *Chenopodium quinoa* polypeptide sequence databases uploaded from Uniprot and Kyoto Encyclopedia of Genes and Genomes (KEGG), respectively, on 12 September 2023 were used. The quantitative data outputs from QIProgenesis software were post-processed in R programming environment (v4.0.2, packages as follows: factoextra, pheatmap, VennDiagram) following log_2-transformation. The differentially expressed proteins were annotated with Mercator MapMan platform (v4.5, https://plabipd.de/portal/mercator-sequence-annotation, accessed on 30 September 2023) to characterize functional roles and BUSCA online tool (http://busca.biocomp.unibo.it/, accessed on 1 October 2023) for sub-cellular localization prediction.

5. Conclusions

Comprehensive knowledge about the underlying mechanisms of plants adjustments and tolerance to heavy metals stress is critically important to achieve high efficiency of crops cultivation in polluted soils. In this regard, the integration of proteomics and metabolomics data is the best way to achieve this goal. The employed bottom-up proteomics approach revealed clear organ-specific responses to Zn exposure in *A. caudatus*, which were more pronounced in roots, indicating their crucial role as the first barrier on the way of pollutant uptake and distribution in plants. Our data indicated Zn-induced fine-tuned adaptive shifts in redox homeostasis, and enhancement of protein biosynthesis, repair/folding, and degradation systems. These adaptations, however, could not exclude the appearance of pronounced signs of toxicity and damage, like expressional suppression of ETC complexes in young and mature leaves.

Notably, the results of the proteomics study were consistent with the findings of the associated metabolomics research. Thus, the induction of SA-mediated protective reactions, as well as impact of SA on JA biosynthesis could be confirmed at the proteomics level. Furthermore, the strong up-regulation of gluconic acid (metal ion chelator in roots and young leaves) observed earlier was supported here by the induction of proteins forming soluble sugars for PPP which can also contribute to gluconate synthesis. Remarkably, while Zn-induced oxidative stress was evidenced by the upregulation of multiple antioxidative enzymes, a similar up-regulation was not observed for sugar transporters in roots and young leaves. Nevertheless, the annotation of several sugar transporters in these organs suggests the possibility of their regulation on post-translational level.

These findings may suggest that post-translational rather than transcriptional regulation is involved in fine-tuning control of the ion and sugar transport systems under Zn stress, which needs to be addressed in future studies.

To conclude, our results expand the existing knowledge on Zn-induced protein dynamics and give a better insight into the associated process of the metabolic adjustment as a part of the general response of plants to the Zn stress.

Supplementary Materials: The following supporting information can be downloaded at https://www.mdpi.com/article/10.3390/plants14213315/s1, Supplementary Information 1: Protocols, Figures, and Tables; Supplementary Information 2: Peptides, proteins, and protein groups (non-redundant proteins), identified in amaranth (*Amaranthus caudatus*) plants under Zn stress conditions; Supplementary Information 3: Proteins, differentially expressed (ANOVA) in amaranth (*A. caudatus*) plants under Zn stress conditions; Supplementary Information 4: Proteins, differentially expressed (*t*-test) in amaranth (*A. caudatus*) plants under Zn stress conditions; Supplementary Information 5: Functional classes of proteins, differentially expressed in amaranth (*A. caudatus*) plants under Zn stress conditions; Supplementary Information 6: Prediction of localization of the proteins, differentially expressed in amaranth (*A. caudatus*) plants under Zn stress conditions.

Author Contributions: A.F. and N.O. proposed the idea of the study; A.G. and T.B. wrote the manuscript draft; A.G., N.S. and A.K. grew plants, performed plant Zn-stress experiments, and collected plant material; A.G., A.S., A.K. and A.F. designed the laboratory experiment; A.G. and A.S. performed the protein extraction and digestion; C.I. performed nano-LC-MS/MS experiments; D.G., A.S. and A.G. accomplished processing, post-processing, and statistical analysis of proteomics data; A.G., T.B., N.O., O.B. and A.F. contributed to writing the final manuscript draft and critical discussion; N.O., T.B., O.B. and A.F. supervised the final version of the manuscript. All authors have read and agreed to the published version of the manuscript.

Funding: Funding for this study was from the Russian Scientific Foundation (grants # 23-14-00383 and # 22-16-00047P).

Data Availability Statement: The original data presented in the study are openly available via the ProteomeXchange Consortium in the PRIDE [137] database with a unique identifier PXD069557 and 10.6019/PXD069557.

Acknowledgments: We are thankful to Vavilov All-Russia Institute of Plant Genetic Resources (Saint Petersburg, Russia) for providing *Amaranthus caudatus* L. seeds variety Karwa dauta, which were used in the study. We also thank Tatiana Leonova for the help with deposition of proteomics data in the PRIDE data repository. The authors acknowledge Saint-Petersburg State University for a research project 118201839 and the Chemical Analysis and Materials Research center and Center of Molecular and Cell Technologies for technical support under agreement C-111/30(01/1-69-46-SPbU). The work relied on the experimental climate control facility in the Institute of Bioengineering, Research Center of Biotechnology, Russian Academy of Science, U-73547. The technical support provided by Mariia Soloveva, Valeria Mashkina, and Ludmila Kuchaeva is acknowledged. The infrastructural support from the Ministry of Science and Higher Education of the Russian Federation (theme #122042700044-6) is acknowledged.

Conflicts of Interest: The authors declare no conflicts of interest.

Abbreviations

9-LOX5, linoleate 9S-lipoxygenase 5; 12-OPDAR, 12-oxophytodienoate reductase; ABC, ammonium bicarbonate; APX, ascorbate peroxidase; BADH, betaine aldehyde dehydrogenase; BCA, bicinchoninic acid; CEF, cyclic electron flow; DAP, differentially abundant protein; DDA, data-dependent acquisition; ETC, electron transport chain; FC, fold change; FNR, ferredoxin-NADP$^+$ reductase; GSH, reduced glutathione; GST, Glutathione S-transferase; HCA, hierarchical clustering analysis; IAA, iodoacetamide; JA, jasmonic acid; KEGG, Kyoto Encyclopedia of Genes and Genomes; LRR, leucine reach repeats; MDHA, monodehydroascorbate reductase; MS/MS, tandem mass spectrometry; OEE2, oxygen-evolving enhancer protein 2; PCA, Principal component analysis; PP2A, protein phosphatase; PPP, pentose phosphate pathway; PS, photosystem; RC, reactive carbonyl; ROS, reactive oxygen species; RP, ribosomal protein; RSD, relative standard deviation; SA, salicylic acid; SDS, sodium dodecyl sulfate; SDS-PAGE, sodium dodecyl sulfate polyacrylamide gel electrophoresis; SnRK1, Sucrose Non-fermenting-Related Kinase 1; TAIR, The Arabidopsis Information Resource; TEF, transcript elongation factor; TOR, Target of Rapamycin.

References

1. Ali, A.; Bhat, B.A.; Rather, G.A.; Malla, B.A.; Ganie, S.A. Proteomic Studies of Micronutrient Deficiency and Toxicity. In *Plant Micronutrients: Deficiency and Toxicity Management*; Aftab, T., Hakeem, K.R., Eds.; Springer International Publishing: Cham, Switzerland, 2020; pp. 257–284, ISBN 978-3-030-49856-6.
2. Hamzah Saleem, M.; Usman, K.; Rizwan, M.; Al Jabri, H.; Alsafran, M. Functions and Strategies for Enhancing Zinc Availability in Plants for Sustainable Agriculture. *Front. Plant Sci.* **2022**, *13*, 1033092. [CrossRef]
3. Stanton, C.; Sanders, D.; Krämer, U.; Podar, D. Zinc in Plants: Integrating Homeostasis and Biofortification. *Mol. Plant* **2022**, *15*, 65–85. [CrossRef] [PubMed]
4. Broadley, M.R.; White, P.J.; Hammond, J.P.; Zelko, I.; Lux, A. Zinc in Plants. *New Phytol.* **2007**, *173*, 677–702. [CrossRef] [PubMed]

5. Claus, J.; Bohmann, A.; Chavarría-Krauser, A. Zinc Uptake and Radial Transport in Roots of Arabidopsis Thaliana: A Modelling Approach to Understand Accumulation. *Ann. Bot.* **2013**, *112*, 369–380. [CrossRef]
6. Krämer, U. Changing Paradigms for the Micronutrient Zinc, a Known Protein Cofactor, as a Signal Relaying Also Cellular Redox State. *Quant. Plant Biol.* **2025**, *6*, e7. [CrossRef]
7. White, P.J.; Broadley, M.R. Physiological Limits to Zinc Biofortification of Edible Crops. *Front. Plant Sci.* **2011**, *2*, 80. [CrossRef]
8. Hall, J.L. Cellular Mechanisms for Heavy Metal Detoxification and Tolerance. *J. Exp. Bot.* **2002**, *53*, 1–11. [CrossRef]
9. Kaur, H.; Garg, N. Zinc Toxicity in Plants: A Review. *Planta* **2021**, *253*, 129. [CrossRef]
10. Oyewo, O.A.; Adeniyi, A.; Bopape, M.F.; Onyango, M.S. Heavy Metal Mobility in Surface Water and Soil, Climate Change, and Soil Interactions. In *Climate Change and Soil Interactions*; Elsevier: Amsterdam, The Netherlands, 2020; pp. 51–88. ISBN 978-0-12-818032-7.
11. Vardhan, K.H.; Kumar, P.S.; Panda, R.C. A Review on Heavy Metal Pollution, Toxicity and Remedial Measures: Current Trends and Future Perspectives. *J. Mol. Liq.* **2019**, *290*, 111197. [CrossRef]
12. Amin, H.; Arain, B.A.; Jahangir, T.M.; Abbasi, A.R.; Abbasi, M.S.; Amin, F. Comparative Zinc Tolerance and Phytoremediation Potential of Four Biofuel Plant Species. *Int. J. Phytoremediat.* **2023**, *25*, 1014–1028. [CrossRef]
13. Wang, C.; Zhang, S.H.; Wang, P.F.; Hou, J.; Zhang, W.J.; Li, W.; Lin, Z.P. The Effect of Excess Zn on Mineral Nutrition and Antioxidative Response in Rapeseed Seedlings. *Chemosphere* **2009**, *75*, 1468–1476. [CrossRef] [PubMed]
14. Fontes, R.L.F.; Cox, F.R. Effects of Sulfur Supply on Soybean Plants Exposed to Zinc Toxicity. *J. Plant Nutr.* **1995**, *18*, 1893–1906. [CrossRef]
15. Zemanová, V.; Pavlíková, D.; Novák, M.; Hnilička, F. The Dual Role of Zinc in Spinach Metabolism: Beneficial × Toxic. *Plants* **2024**, *13*, 3363. [CrossRef]
16. Meng, Y.; Xiang, C.; Huo, J.; Shen, S.; Tang, Y.; Wu, L. Toxicity Effects of Zinc Supply on Growth Revealed by Physiological and Transcriptomic Evidences in Sweet Potato (*Ipomoea batatas* (L.) Lam). *Sci. Rep.* **2023**, *13*, 19203. [CrossRef]
17. Rucińska-Sobkowiak, R. Water Relations in Plants Subjected to Heavy Metal Stresses. *Acta Physiol. Plant* **2016**, *38*, 257. [CrossRef]
18. Balafrej, H.; Bogusz, D.; Triqui, Z.-E.A.; Guedira, A.; Bendaou, N.; Smouni, A.; Fahr, M. Zinc Hyperaccumulation in Plants: A Review. *Plants* **2020**, *9*, 562. [CrossRef]
19. Barrameda-Medina, Y.; Montesinos-Pereira, D.; Romero, L.; Blasco, B.; Ruiz, J.M. Role of GSH Homeostasis under Zn Toxicity in Plants with Different Zn Tolerance. *Plant Sci.* **2014**, *227*, 110–121. [CrossRef]
20. Szopiński, M.; Sitko, K.; Gieroń, Ż.; Rusinowski, S.; Corso, M.; Hermans, C.; Verbruggen, N.; Małkowski, E. Toxic Effects of Cd and Zn on the Photosynthetic Apparatus of the *Arabidopsis halleri* and Arabidopsis Arenosa Pseudo-Metallophytes. *Front. Plant Sci.* **2019**, *10*, 748. [CrossRef]
21. DalCorso, G.; Martini, F.; Fasani, E.; Manara, A.; Visioli, G.; Furini, A. Enhancement of Zn Tolerance and Accumulation in Plants Mediated by the Expression of Saccharomyces Cerevisiae Vacuolar Transporter ZRC1. *Planta* **2021**, *253*, 117. [CrossRef]
22. Sarret, G.; Saumitou-Laprade, P.; Bert, V.; Proux, O.; Hazemann, J.-L.; Traverse, A.; Marcus, M.A.; Manceau, A. Forms of Zinc Accumulated in the Hyperaccumulator *Arabidopsis halleri*. *Plant Physiol.* **2002**, *130*, 1815–1826. [CrossRef]
23. Zeng, X.-W.; Ma, L.Q.; Qiu, R.-L.; Tang, Y.-T. Effects of Zn on Plant Tolerance and Non-Protein Thiol Accumulation in Zn Hyperaccumulator *Arabis paniculata* Franch. *Environ. Exp. Bot.* **2011**, *70*, 227–232. [CrossRef]
24. Rastogi, A.; Shukla, S. Amaranth: A New Millennium Crop of Nutraceutical Values. *Crit. Rev. Food Sci. Nutr.* **2013**, *53*, 109–125. [CrossRef] [PubMed]
25. Riggins, C.W.; Barba de la Rosa, A.P.; Blair, M.W.; Espitia-Rangel, E. Editorial: Amaranthus: Naturally Stress-Resistant Resources for Improved Agriculture and Human Health. *Front. Plant Sci.* **2021**, *12*, 726875. [CrossRef] [PubMed]
26. Lukatkin, A.S.; Bashmakov, D.I.; Al Harbawee, W.E.Q.; Teixeira da Silva, J.A. Assessment of Physiological and Biochemical Responses of *Amaranthus retroflexus* Seedlings to the Accumulation of Heavy Metals with Regards to Phytoremediation Potential. *Int. J. Phytoremediat.* **2021**, *23*, 219–230. [CrossRef]
27. Osmolovskaya, N.; Bilova, T.; Gurina, A.; Orlova, A.; Vu, V.D.; Sukhikh, S.; Zhilkina, T.; Frolova, N.; Tarakhovskaya, E.; Kamionskaya, A.; et al. Metabolic Responses of *Amaranthus caudatus* Roots and Leaves to Zinc Stress. *Plants* **2025**, *14*, 2119. [CrossRef]
28. Lucini, L.; Bernardo, L. Comparison of Proteome Response to Saline and Zinc Stress in Lettuce. *Front. Plant Sci.* **2015**, *6*, 240. [CrossRef]
29. Šimon, M.; Shen, Z.-J.; Ghoto, K.; Chen, J.; Liu, X.; Gao, G.-F.; Jemec Kokalj, A.; Novak, S.; Drašler, B.; Zhang, J.-Y.; et al. Proteomic Investigation of Zn-Challenged Rice Roots Reveals Adverse Effects and Root Physiological Adaptation. *Plant Soil.* **2021**, *460*, 69–88. [CrossRef]
30. Fukao, Y.; Ferjani, A.; Tomioka, R.; Nagasaki, N.; Kurata, R.; Nishimori, Y.; Fujiwara, M.; Maeshima, M. iTRAQ Analysis Reveals Mechanisms of Growth Defects Due to Excess Zinc in Arabidopsis. *Plant Physiol.* **2011**, *155*, 1893–1907. [CrossRef]
31. Frolov, A.; Blüher, M.; Hoffmann, R. Glycation Sites of Human Plasma Proteins Are Affected to Different Extents by Hyperglycemic Conditions in Type 2 Diabetes Mellitus. *Anal. Bioanal. Chem.* **2014**, *406*, 5755–5763. [CrossRef]

32. Nouri, J.; Khorasani, N.; Lorestani, B.; Karami, M.; Hassani, A.H.; Yousefi, N. Accumulation of Heavy Metals in Soil and Uptake by Plant Species with Phytoremediation Potential. *Environ. Earth Sci.* **2009**, *59*, 315–323. [CrossRef]
33. Yan, A.; Wang, Y.; Tan, S.N.; Mohd Yusof, M.L.; Ghosh, S.; Chen, Z. Phytoremediation: A Promising Approach for Revegetation of Heavy Metal-Polluted Land. *Front. Plant Sci.* **2020**, *11*, 359. [CrossRef]
34. Nagajyoti, P.C.; Lee, K.D.; Sreekanth, T.V.M. Heavy Metals, Occurrence and Toxicity for Plants: A Review. *Environ. Chem. Lett.* **2010**, *8*, 199–216. [CrossRef]
35. Subba, P.; Mukhopadhyay, M.; Mahato, S.K.; Bhutia, K.D.; Mondal, T.K.; Ghosh, S.K. Zinc Stress Induces Physiological, Ultra-Structural and Biochemical Changes in Mandarin Orange (*Citrus reticulata* Blanco) Seedlings. *Physiol. Mol. Biol. Plants* **2014**, *20*, 461–473. [CrossRef]
36. Hasanuzzaman, M.; Bhuyan, M.H.M.B.; Parvin, K.; Bhuiyan, T.F.; Anee, T.I.; Nahar, K.; Hossen, M.S.; Zulfiqar, F.; Alam, M.M.; Fujita, M. Regulation of ROS Metabolism in Plants under Environmental Stress: A Review of Recent Experimental Evidence. *Int. J. Mol. Sci.* **2020**, *21*, 8695. [CrossRef]
37. Tripathy, B.C.; Oelmüller, R. Reactive Oxygen Species Generation and Signaling in Plants. *Plant Signal. Behav.* **2012**, *7*, 1621–1633. [CrossRef] [PubMed]
38. Das, K.; Roychoudhury, A. Reactive Oxygen Species (ROS) and Response of Antioxidants as ROS-Scavengers during Environmental Stress in Plants. *Front. Environ. Sci.* **2014**, *2*, 53. [CrossRef]
39. Skórzyńska-Polit, E.; Drążkiewicz, M.; Krupa, Z. Lipid Peroxidation and Antioxidative Response in Arabidopsis Thaliana Exposed to Cadmium and Copper. *Acta Physiol. Plant* **2010**, *32*, 169–175. [CrossRef]
40. Khan, M.; Samrana, S.; Zhang, Y.; Malik, Z.; Khan, M.D.; Zhu, S. Reduced Glutathione Protects Subcellular Compartments From Pb-Induced ROS Injury in Leaves and Roots of Upland Cotton (*Gossypium hirsutum* L.). *Front. Plant Sci.* **2020**, *11*, 412. [CrossRef]
41. Waśkiewicz, A.; Gładysz, O.; Szentner, K.; Goliński, P. Role of Glutathione in Abiotic Stress Tolerance. In *Oxidative Damage to Plants*; Elsevier: Amsterdam, The Netherlands, 2014; pp. 149–181, ISBN 978-0-12-799963-0.
42. Gill, S.S.; Anjum, N.A.; Hasanuzzaman, M.; Gill, R.; Trivedi, D.K.; Ahmad, I.; Pereira, E.; Tuteja, N. Glutathione and Glutathione Reductase: A Boon in Disguise for Plant Abiotic Stress Defense Operations. *Plant Physiol. Biochem.* **2013**, *70*, 204–212. [CrossRef]
43. Hasanuzzaman, M.; Bhuyan, M.H.M.B.; Anee, T.I.; Parvin, K.; Nahar, K.; Mahmud, J.A.; Fujita, M. Regulation of Ascorbate-Glutathione Pathway in Mitigating Oxidative Damage in Plants under Abiotic Stress. *Antioxidants* **2019**, *8*, 384. [CrossRef]
44. Xiao, T.; ShangGuan, X.; Wang, Y.; Tian, Z.; Peng, K.; Shen, Z.; Hu, Z.; Xia, Y. The Germin-like Protein OsGLP8-7 Is Involved in Lignin Synthesis for Acclimation to Copper Toxicity in Rice. *J. Plant Physiol.* **2024**, *303*, 154335. [CrossRef]
45. Tamás, L.; Mistrík, I.; Huttová, J.; Halušková, L.; Valentovičová, K.; Zelinová, V. Role of Reactive Oxygen Species-Generating Enzymes and Hydrogen Peroxide during Cadmium, Mercury and Osmotic Stresses in Barley Root Tip. *Planta* **2010**, *231*, 221–231. [CrossRef]
46. Leon, J.; Lawton, M.A.; Raskin, I. Hydrogen Peroxide Stimulates Salicylic Acid Biosynthesis in Tobacco. *Plant Physiol.* **1995**, *108*, 1673–1678. [CrossRef]
47. Qi, X.; Chen, M.; Liang, D.; Xu, Q.; Zhou, F.; Chen, X. Jasmonic Acid, Ethylene and ROS Are Involved in the Response of Cucumber (*Cucumis sativus* L.) to Aphid Infestation. *Sci. Hortic.* **2020**, *269*, 109421. [CrossRef]
48. León Morcillo, R.J.; Ocampo, J.A.; García Garrido, J.M. Plant 9-Lox Oxylipin Metabolism in Response to Arbuscular Mycorrhiza. *Plant Signal. Behav.* **2012**, *7*, 1584–1588. [CrossRef] [PubMed]
49. Miersch, O.; Neumerkel, J.; Dippe, M.; Stenzel, I.; Wasternack, C. Hydroxylated Jasmonates Are Commonly Occurring Metabolites of Jasmonic Acid and Contribute to a Partial Switch-off in Jasmonate Signaling. *New Phytol.* **2008**, *177*, 114–127. [CrossRef] [PubMed]
50. Ilyas, M.; Rasheed, A.; Mahmood, T. Functional Characterization of Germin and Germin-like Protein Genes in Various Plant Species Using Transgenic Approaches. *Biotechnol. Lett.* **2016**, *38*, 1405–1421. [CrossRef]
51. Mao, L.; Ge, L.; Ye, X.; Xu, L.; Si, W.; Ding, T. ZmGLP1, a Germin-like Protein from Maize, Plays an Important Role in the Regulation of Pathogen Resistance. *Int. J. Mol. Sci.* **2022**, *23*, 14316. [CrossRef]
52. Qu, Z.-L.; Zhong, N.-Q.; Wang, H.-Y.; Chen, A.-P.; Jian, G.-L.; Xia, G.-X. Ectopic Expression of the Cotton Non-Symbiotic Hemoglobin Gene GhHbd1 Triggers Defense Responses and Increases Disease Tolerance in Arabidopsis. *Plant Cell Physiol.* **2006**, *47*, 1058–1068. [CrossRef]
53. Oteiza, P.I. Zinc and the Modulation of Redox Homeostasis. *Free Radic. Biol. Med.* **2012**, *53*, 1748–1759. [CrossRef]
54. Dang, P.M.; Lamb, M.C.; Bowen, K.L.; Chen, C.Y. Identification of Expressed R-Genes Associated with Leaf Spot Diseases in Cultivated Peanut. *Mol. Biol. Rep.* **2019**, *46*, 225–239. [CrossRef] [PubMed]
55. Zhou, H.; Xie, Y.; Jiang, Y.; Nadeem, H.; Wang, Y.; Yang, N.; Zhu, H.; Tang, C. GhTLP1, a Thaumatin-like Protein 1, Improves Verticillium Wilt Resistance in Cotton via JA, ABA and MAPK Signaling Pathway-Plant Pathways. *Int. J. Biol. Macromol.* **2023**, *253*, 127388. [CrossRef] [PubMed]

56. Zrenner, R.; Verwaaijen, B.; Genzel, F.; Flemer, B.; Grosch, R. Transcriptional Changes in Potato Sprouts upon Interaction with Rhizoctonia Solani Indicate Pathogen-Induced Interference in the Defence Pathways of Potato. *Int. J. Mol. Sci.* **2021**, *22*, 3094. [CrossRef] [PubMed]
57. Rathinam, M.; Rao, U.; Sreevathsa, R. Novel Biotechnological Strategies to Combat Biotic Stresses: Polygalacturonase Inhibitor (PGIP) Proteins as a Promising Comprehensive Option. *Appl. Microbiol. Biotechnol.* **2020**, *104*, 2333–2342. [CrossRef]
58. Połeć-Pawlak, K.; Ruzik, R.; Lipiec, E.; Ciurzyńska, M.; Gawrońska, H. Investigation of Pb(II) Binding to Pectin in Arabidopsis Thaliana. *J. Anal. At. Spectrom.* **2007**, *22*, 968–972. [CrossRef]
59. País, S.M.; Téllez-Iñón, M.T.; Capiati, D.A. Serine/Threonine Protein Phosphatases Type 2A and Their Roles in Stress Signaling. *Plant Signal. Behav.* **2009**, *4*, 1013–1015. [CrossRef]
60. Neumann, D.; Lichtenberger, O.; Günther, D.; Tschiersch, K.; Nover, L. Heat-Shock Proteins Induce Heavy-Metal Tolerance in Higher Plants. *Planta* **1994**, *194*, 360–367. [CrossRef]
61. Fitzgerald, T.L.; Waters, D.L.E.; Henry, R.J. Betaine Aldehyde Dehydrogenase in Plants. *Plant Biol.* **2009**, *11*, 119–130. [CrossRef]
62. Konopka-Postupolska, D.; Clark, G. Annexins as Overlooked Regulators of Membrane Trafficking in Plant Cells. *Int. J. Mol. Sci.* **2017**, *18*, 863. [CrossRef]
63. Dufoo-Hurtado, M.D.; Huerta-Ocampo, J.Ã.; Barrera-Pacheco, A.; Barba De La Rosa, A.P.; Mercado-Silva, E.M. Low Temperature Conditioning of Garlic (*Allium sativum* L.) Â€œseedâ€ Cloves Induces Alterations in Sprouts Proteome. *Front. Plant Sci.* **2015**, *6*, 332. [CrossRef]
64. Mortimer, J.C.; Laohavisit, A.; Macpherson, N.; Webb, A.; Brownlee, C.; Battey, N.H.; Davies, J.M. Annexins: Multifunctional Components of Growth and Adaptation. *J. Exp. Bot.* **2008**, *59*, 533–544. [CrossRef] [PubMed]
65. Rüffer, M.; Steipe, B.; Zenk, M.H. Evidence against Specific Binding of Salicylic Acid to Plant Catalase. *FEBS Lett.* **1995**, *377*, 175–180. [CrossRef] [PubMed]
66. Sun, T.; Liu, F.; Wang, W.; Wang, L.; Wang, Z.; Li, J.; Que, Y.; Xu, L.; Su, Y. The Role of Sugarcane Catalase Gene ScCAT2 in the Defense Response to Pathogen Challenge and Adversity Stress. *Int. J. Mol. Sci.* **2018**, *19*, 2686. [CrossRef] [PubMed]
67. Chen, C.; He, G.; Li, J.; Perez-Hormaeche, J.; Becker, T.; Luo, M.; Wallrad, L.; Gao, J.; Li, J.; Pardo, J.M.; et al. A Salt Stress-activated GSO1-SOS2-SOS1 Module Protects the *Arabidopsis* Root Stem Cell Niche by Enhancing Sodium Ion Extrusion. *EMBO J.* **2023**, *42*, e113004. [CrossRef]
68. Creff, A.; Brocard, L.; Joubès, J.; Taconnat, L.; Doll, N.M.; Marsollier, A.-C.; Pascal, S.; Galletti, R.; Boeuf, S.; Moussu, S.; et al. A Stress-Response-Related Inter-Compartmental Signalling Pathway Regulates Embryonic Cuticle Integrity in Arabidopsis. *PLoS Genet.* **2019**, *15*, e1007847. [CrossRef]
69. Racolta, A.; Bryan, A.C.; Tax, F.E. The Receptor-like Kinases GSO1 and GSO2 Together Regulate Root Growth in *Arabidopsis* through Control of Cell Division and Cell Fate Specification. *Dev. Dyn.* **2014**, *243*, 257–278. [CrossRef]
70. Zhu, Q.; Feng, Y.; Xue, J.; Chen, P.; Zhang, A.; Yu, Y. Advances in Receptor-like Protein Kinases in Balancing Plant Growth and Stress Responses. *Plants* **2023**, *12*, 427. [CrossRef]
71. Tamás, M.; Sharma, S.; Ibstedt, S.; Jacobson, T.; Christen, P. Heavy Metals and Metalloids As a Cause for Protein Misfolding and Aggregation. *Biomolecules* **2014**, *4*, 252–267. [CrossRef]
72. Stępiński, D. Decoding Plant Ribosomal Proteins: Multitasking Players in Cellular Games. *Cells* **2025**, *14*, 473. [CrossRef]
73. Mazahar, M.; Achala, B.; Anusree, S.; Kirti, P.B. Ribosomal Proteins and Their Extra Ribosomal Functions in Abiotic Stress Tolerance of Plants. *Mod. Concepts Dev. Agron.* **2019**, *4*. [CrossRef]
74. Fakih, Z.; Germain, H. Implication of Ribosomal Protein in Abiotic and Biotic Stress. *Planta* **2025**, *261*, 85. [CrossRef]
75. Grasser, K.D. The Role of RNA Polymerase II Transcript Elongation Factors in Plant Stress Responses. *J. Exp. Bot.* **2025**, *76*, 2447–2454. [CrossRef] [PubMed]
76. Mao, X.; Yu, C.; Li, L.; Wang, M.; Yang, L.; Zhang, Y.; Zhang, Y.; Wang, J.; Li, C.; Reynolds, M.P.; et al. How Many Faces Does the Plant U-Box E3 Ligase Have? *Int. J. Mol. Sci.* **2022**, *23*, 2285. [CrossRef] [PubMed]
77. Chai, M.; Wei, P.; Chen, Q.; An, R.; Chen, J.; Yang, S.; Wang, X. NADK3, a Novel Cytoplasmic Source of NADPH, Is Required under Conditions of Oxidative Stress and Modulates Abscisic Acid Responses in Arabidopsis. *Plant J.* **2006**, *47*, 665–674. [CrossRef] [PubMed]
78. Valderrama, R.; Corpas, F.J.; Carreras, A.; Gómez-Rodríguez, M.V.; Chaki, M.; Pedrajas, J.R.; Fernández-Ocaña, A.; Del Río, L.A.; Barroso, J.B. The Dehydrogenase-mediated Recycling of NADPH Is a Key Antioxidant System against Salt-induced Oxidative Stress in Olive Plants. *Plant Cell Environ.* **2006**, *29*, 1449–1459. [CrossRef]
79. Moustakas, M.; Dobrikova, A.; Sperdouli, I.; Hanć, A.; Moustaka, J.; Adamakis, I.-D.S.; Apostolova, E. Photosystem II Tolerance to Excess Zinc Exposure and High Light Stress in *Salvia sclarea* L. *Agronomy* **2024**, *14*, 589. *Agronomy* **2024**, *14*, 589. [CrossRef]
80. Petrov, V.; Hille, J.; Mueller-Roeber, B.; Gechev, T.S. ROS-Mediated Abiotic Stress-Induced Programmed Cell Death in Plants. *Front. Plant Sci.* **2015**, *6*, 69. [CrossRef]
81. Zhang, C.; Shuai, J.; Ran, Z.; Zhao, J.; Wu, Z.; Liao, R.; Wu, J.; Ma, W.; Lei, M. Structural Insights into NDH-1 Mediated Cyclic Electron Transfer. *Nat. Commun.* **2020**, *11*, 888. [CrossRef]

82. Shuyskaya, E.; Rakhmankulova, Z.; Prokofieva, M.; Kazantseva, V.; Lunkova, N. Impact of Salinity, Elevated Temperature, and Their Interaction with the Photosynthetic Efficiency of Halophyte Crop *Chenopodium quinoa* Willd. *Agriculture* **2023**, *13*, 1198. [CrossRef]
83. Huihui, Z.; Xin, L.; Zisong, X.; Yue, W.; Zhiyuan, T.; Meijun, A.; Yuehui, Z.; Wenxu, Z.; Nan, X.; Guangyu, S. Toxic Effects of Heavy Metals Pb and Cd on Mulberry (*Morus alba* L.) Seedling Leaves: Photosynthetic Function and Reactive Oxygen Species (ROS) Metabolism Responses. *Ecotoxicol. Environ. Saf.* **2020**, *195*, 110469. [CrossRef]
84. Potters, G.; Horemans, N.; Jansen, M.A.K. The Cellular Redox State in Plant Stress Biology—A Charging Concept. *Plant Physiol. Biochem.* **2010**, *48*, 292–300. [CrossRef]
85. Bai, J.; Qin, Y.; Liu, J.; Wang, Y.; Sa, R.; Zhang, N.; Jia, R. Proteomic Response of Oat Leaves to Long-Term Salinity Stress. *Environ. Sci. Pollut. Res.* **2017**, *24*, 3387–3399. [CrossRef] [PubMed]
86. Yang, Y.; Wang, S.; Zhao, C.; Jiang, X.; Gao, D. Responses of Non-Structural Carbohydrates and Biomass in Plant to Heavy Metal Treatment. *Sci. Total Environ.* **2024**, *909*, 168559. [CrossRef] [PubMed]
87. Cai, S.-W.; Huang, W.-X.; Xiong, Z.-T.; Ye, F.-Y.; Ren, C.; Xu, Z.-R.; Liu, C.; Deng, S.-Q.; Zhao, J. Comparative Study of Root Growth and Sucrose-Cleaving Enzymes in Metallicolous and Non-Metallicolous Populations of Rumex Dentatus under Copper Stress. *Ecotoxicol. Environ. Saf.* **2013**, *98*, 95–102. [CrossRef] [PubMed]
88. Li, C.; Liu, Y.; Tian, J.; Zhu, Y.; Fan, J. Changes in Sucrose Metabolism in Maize Varieties with Different Cadmium Sensitivities under Cadmium Stress. *PLoS ONE* **2020**, *15*, e0243835. [CrossRef]
89. Varshney, V.; Singh, J.; Salvi, P. Sugar Signaling and Their Interplay in Mitigating Abiotic Stresses in Plant: A Molecular Perspective. In *Smart Plant Breeding for Field Crops in Post-Genomics Era*; Sharma, D., Singh, S., Sharma, S.K., Singh, R., Eds.; Springer Nature: Singapore, 2023; pp. 369–393, ISBN 978-981-19-8217-0.
90. German, M.A.; Asher, I.; Petreikov, M.; Dai, N.; Schaffer, A.A.; Granot, D. Cloning, Expression and Characterization of LeFRK3, the Fourth Tomato (*Lycopersicon esculentum* Mill.) Gene Encoding Fructokinase. *Plant Sci.* **2004**, *166*, 285–291. [CrossRef]
91. Granot, D.; Kelly, G.; Stein, O.; David-Schwartz, R. Substantial Roles of Hexokinase and Fructokinase in the Effects of Sugars on Plant Physiology and Development. *J. Exp. Bot.* **2014**, *65*, 809–819. [CrossRef]
92. Fulda, S.; Mikkat, S.; Stegmann, H.; Horn, R. Physiology and Proteomics of Drought Stress Acclimation in Sunflower (*Helianthus annuus* L.). *Plant Biol.* **2011**, *13*, 632–642. [CrossRef]
93. Guglielminetti, L.; Morita, A.; Yamaguchi, J.; Loreti, E.; Perata, P.; Alpi, A. Differential Expression of Two Fructokinases in *Oryza sativa* Seedlings Grown under Aerobic and Anaerobic Conditions. *J. Plant Res.* **2006**, *119*, 351–356. [CrossRef]
94. Bui, L.T.; Novi, G.; Lombardi, L.; Iannuzzi, C.; Rossi, J.; Santaniello, A.; Mensuali, A.; Corbineau, F.; Giuntoli, B.; Perata, P.; et al. Conservation of Ethanol Fermentation and Its Regulation in Land Plants. *J. Exp. Bot.* **2019**, *70*, 1815–1827. [CrossRef]
95. Jardine, K.J.; McDowell, N. Fermentation-mediated Growth, Signaling, and Defense in Plants. *New Phytol.* **2023**, *239*, 839–851. [CrossRef] [PubMed]
96. Wurzinger, B.; Nukarinen, E.; Nägele, T.; Weckwerth, W.; Teige, M. The SnRK1 Kinase as Central Mediator of Energy Signaling between Different Organelles. *Plant Physiol.* **2018**, *176*, 1085–1094. [CrossRef]
97. Li, G.; Zhao, Y. The Critical Roles of Three Sugar-Related Proteins (HXK, SnRK1, TOR) in Regulating Plant Growth and Stress Responses. *Hortic. Res.* **2024**, *11*, uhae099. [CrossRef]
98. Busche, M.; Scarpin, M.R.; Hnasko, R.; Brunkard, J.O. TOR Coordinates Nucleotide Availability with Ribosome Biogenesis in Plants. *Plant Cell* **2021**, *33*, 1615–1632. [CrossRef]
99. Crepin, N.; Rolland, F. SnRK1 Activation, Signaling, and Networking for Energy Homeostasis. *Curr. Opin. Plant Biol.* **2019**, *51*, 29–36. [CrossRef]
100. Tsai, A.Y.-L.; Gazzarrini, S. Trehalose-6-Phosphate and SnRK1 Kinases in Plant Development and Signaling: The Emerging Picture. *Front. Plant Sci.* **2014**, *5*, 119. [CrossRef]
101. Han, C.; Wang, H.; Shi, W.; Bai, M.-Y. The Molecular Associations between the SnRK1 Complex and Carbon/Nitrogen Metabolism in Plants. *New Crops* **2024**, *1*, 100008. [CrossRef]
102. Peixoto, B.; Baena-González, E. Management of Plant Central Metabolism by SnRK1 Protein Kinases. *J. Exp. Bot.* **2022**, *73*, 7068–7082. [CrossRef]
103. Angulo-Bejarano, P.I.; Puente-Rivera, J.; Cruz-Ortega, R. Metal and Metalloid Toxicity in Plants: An Overview on Molecular Aspects. *Plants* **2021**, *10*, 635. [CrossRef] [PubMed]
104. Fukao, Y.; Ferjani, A. V-ATPase Dysfunction under Excess Zinc Inhibits Arabidopsis Cell Expansion. *Plant Signal. Behav.* **2011**, *6*, 1253–1255. [CrossRef] [PubMed]
105. Li, J.; Guo, Y.; Yang, Y. The Molecular Mechanism of Plasma Membrane H+-ATPases in Plant Responses to Abiotic Stress. *J. Genet. Genom.* **2022**, *49*, 715–725. [CrossRef]
106. Janicka-Russak, M.; Kabala, K.; Burzynski, M.; Klobus, G. Response of Plasma Membrane H+-ATPase to Heavy Metal Stress in *Cucumis sativus* Roots. *J. Exp. Bot.* **2008**, *59*, 3721–3728. [CrossRef] [PubMed]

107. Ulhassan, Z.; Sheteiwy, M.S.; Khan, A.R.; Hamid, Y.; Azhar, W.; Hussain, S.; Salam, A.; Zhou, W. Zinc Toxicity in Plants: A Brief Overview on Recent Developments. In *Zinc in Plants*; Elsevier: Amsterdam, The Netherlands, 2025; pp. 77–93. ISBN 978-0-323-91314-0.
108. Verbruggen, N.; Hermans, C.; Schat, H. Molecular Mechanisms of Metal Hyperaccumulation in Plants. *New Phytol.* **2009**, *181*, 759–776. [CrossRef] [PubMed]
109. Do, T.H.T.; Martinoia, E.; Lee, Y.; Hwang, J.-U. 2021 Update on ATP-Binding Cassette (ABC) Transporters: How They Meet the Needs of Plants. *Plant Physiol.* **2021**, *187*, 1876–1892. [CrossRef] [PubMed]
110. I. Stolarczyk, E.; J. Reiling, C.; M. Paumi, C. Regulation of ABC Transporter Function Via Phosphorylation by Protein Kinases. *CPB* **2011**, *12*, 621–635. [CrossRef]
111. Yadav, S.; Kalwan, G.; Gill, S.S.; Jain, P.K. The ABC Transporters and Their Epigenetic Regulation under Drought Stress in Chickpea. *Plant Physiol. Biochem.* **2025**, *223*, 109903. [CrossRef]
112. Hashmi, S.S.; Lubna; Bilal, S.; Jan, R.; Asif, S.; Abdelbacki, A.M.M.; Kim, K.-M.; Al-Harrasi, A.; Asaf, S. Exploring the Role of ATP-Binding Cassette Transporters in Tomato (*Solanum lycopersicum*) under Cadmium Stress through Genome-Wide and Transcriptomic Analysis. *Front. Plant Sci.* **2025**, *16*, 1536178. [CrossRef]
113. Kosakivska, I.V.; Babenko, L.M.; Romanenko, K.O.; Korotka, I.Y.; Potters, G. Molecular Mechanisms of Plant Adaptive Responses to Heavy Metals Stress. *Cell Biol. Int.* **2021**, *45*, 258–272. [CrossRef]
114. Jia, H.; Wang, X.; Wei, T.; Zhou, R.; Muhammad, H.; Hua, L.; Ren, X.; Guo, J.; Ding, Y. Accumulation and Fixation of Cd by Tomato Cell Wall Pectin under Cd Stress. *Environ. Exp. Bot.* **2019**, *167*, 103829. [CrossRef]
115. Bitto, E.; Bingman, C.A.; McCoy, J.G.; Allard, S.T.M.; Wesenberg, G.E.; Phillips, G.N. The Structure at 1.6 Å Resolution of the Protein Product of the At4g34215 Gene from *Arabidopsis thaliana*. *Acta Crystallogr. D Biol. Crystallogr.* **2005**, *61*, 1655–1661. [CrossRef]
116. Anderson, A.C.; Stangherlin, S.; Pimentel, K.N.; Weadge, J.T.; Clarke, A.J. The SGNH Hydrolase Family: A Template for Carbohydrate Diversity. *Glycobiology* **2022**, *32*, 826–848. [CrossRef]
117. Panikashvili, D.; Shi, J.X.; Schreiber, L.; Aharoni, A. The Arabidopsis *DCR* Encoding a Soluble BAHD Acyltransferase Is Required for Cutin Polyester Formation and Seed Hydration Properties. *Plant Physiol.* **2009**, *151*, 1773–1789. [CrossRef]
118. Doyama, K.; Yamaji, K.; Haruma, T.; Ishida, A.; Mori, S.; Kurosawa, Y. Zn Tolerance in the Evergreen Shrub, Aucuba Japonica, Naturally Growing at a Mine Site: Cell Wall Immobilization, Aucubin Production, and Zn Adsorption on Fungal Mycelia. *PLoS ONE* **2021**, *16*, e0257690. [CrossRef]
119. Pfaff, J.; Denton, A.K.; Usadel, B.; Pfaff, C. Phosphate Starvation Causes Different Stress Responses in the Lipid Metabolism of Tomato Leaves and Roots. *Biochim. Biophys. Acta (BBA) Mol. Cell Biol. Lipids* **2020**, *1865*, 158763. [CrossRef]
120. Bouain, N.; Shahzad, Z.; Rouached, A.; Khan, G.A.; Berthomieu, P.; Abdelly, C.; Poirier, Y.; Rouached, H. Phosphate and Zinc Transport and Signalling in Plants: Toward a Better Understanding of Their Homeostasis Interaction. *J. Exp. Bot.* **2014**, *65*, 5725–5741. [CrossRef]
121. Ding, J.; Liu, L.; Wang, C.; Shi, L.; Xu, F.; Cai, H. High Level of Zinc Triggers Phosphorus Starvation by Inhibiting Root-to-Shoot Translocation and Preferential Distribution of Phosphorus in Rice Plants. *Environ. Pollut.* **2021**, *277*, 116778. [CrossRef] [PubMed]
122. Sabet, M.S.; Zamani, K.; Lohrasebi, T.; Malboobi, M.A.; Valizadeh, M. Functional Assessment of an Overexpressed Arabidopsis Purple Acid Phosphatase Gene (Atpap26) in Tobacco Plants. *Iran. J. Biotechnol.* **2018**, *16*, 31–41. [CrossRef] [PubMed]
123. Wang, F.; Ding, D.; Li, J.; He, L.; Xu, X.; Zhao, Y.; Yan, B.; Li, Z.; Xu, J. Characterisation of Genes Involved in Galactolipids and Sulfolipids Metabolism in Maize and Arabidopsis and Their Differential Responses to Phosphate Deficiency. *Funct. Plant Biol.* **2020**, *47*, 279. [CrossRef] [PubMed]
124. Foroughi, S.; Baker, A.J.M.; Roessner, U.; Johnson, A.A.T.; Bacic, A.; Callahan, D.L. Hyperaccumulation of Zinc by *Noccaea caerulescens* Results in a Cascade of Stress Responses and Changes in the Elemental Profile. *Metallomics* **2014**, *6*, 1671–1682. [CrossRef]
125. López-Orenes, A.; Martínez-Pérez, A.; Calderón, A.A.; Ferrer, M.A. Pb-Induced Responses in *Zygophyllum fabago* Plants Are Organ-Dependent and Modulated by Salicylic Acid. *Plant Physiol. Biochem.* **2014**, *84*, 57–66. [CrossRef]
126. Liu, W.; Park, S.-W. 12-Oxo-Phytodienoic Acid: A Fuse and/or Switch of Plant Growth and Defense Responses? *Front. Plant Sci.* **2021**, *12*, 724079. [CrossRef]
127. Kavita, B.; Shukla, S.; Naresh Kumar, G.; Archana, G. Amelioration of Phytotoxic Effects of Cd on Mung Bean Seedlings by Gluconic Acid Secreting Rhizobacterium Enterobacter Asburiae PSI3 and Implication of Role of Organic Acid. *World J. Microbiol. Biotechnol.* **2008**, *24*, 2965–2972. [CrossRef]
128. Kornecki, J.F.; Carballares, D.; Tardioli, P.W.; Rodrigues, R.C.; Berenguer-Murcia, Á.; Alcántara, A.R.; Fernandez-Lafuente, R. Enzyme Production of D -Gluconic Acid and Glucose Oxidase: Successful Tales of Cascade Reactions. *Catal. Sci. Technol.* **2020**, *10*, 5740–5771. [CrossRef]
129. Corkins, M.E.; Wilson, S.; Cocuron, J.-C.; Alonso, A.P.; Bird, A.J. The Gluconate Shunt Is an Alternative Route for Directing Glucose into the Pentose Phosphate Pathway in Fission Yeast. *J. Biol. Chem.* **2017**, *292*, 13823–13832. [CrossRef]

130. Ledesma, A.; De Lacoba, M.; Rial, E. The mitochondrial uncoupling proteins. *Genome Biol.* **2002**, *3*, reviews3015.1. [CrossRef]
131. Bilova, T.; Lukasheva, E.; Brauch, D.; Greifenhagen, U.; Paudel, G.; Tarakhovskaya, E.; Frolova, N.; Mittasch, J.; Balcke, G.U.; Tissier, A.; et al. A Snapshot of the Plant Glycated Proteome. *J. Biol. Chem.* **2016**, *291*, 7621–7636. [CrossRef]
132. Sengupta, D.; Naik, D.; Reddy, A.R. Plant Aldo-Keto Reductases (AKRs) as Multi-Tasking Soldiers Involved in Diverse Plant Metabolic Processes and Stress Defense: A Structure-Function Update. *J. Plant Physiol.* **2015**, *179*, 40–55. [CrossRef] [PubMed]
133. Frolov, A.; Bilova, T.; Paudel, G.; Berger, R.; Balcke, G.U.; Birkemeyer, C.; Wessjohann, L. Early Responses of Mature Arabidopsis Thaliana Plants to Reduced Water Potential in the Agar-Based Polyethylene Glycol Infusion Drought Model. *J. Plant Physiol.* **2017**, *208*, 70–83. [CrossRef] [PubMed]
134. Greifenhagen, U.; Frolov, A.; Blüher, M.; Hoffmann, R. Site-Specific Analysis of Advanced Glycation End Products in Plasma Proteins of Type 2 Diabetes Mellitus Patients. *Anal. Bioanal. Chem.* **2016**, *408*, 5557–5566. [CrossRef]
135. Leonova, T.; Ihling, C.; Saoud, M.; Frolova, N.; Rennert, R.; Wessjohann, L.A.; Frolov, A. Does Filter-Aided Sample Preparation Provide Sufficient Method Linearity for Quantitative Plant Shotgun Proteomics? *Front. Plant Sci.* **2022**, *13*, 874761. [CrossRef]
136. Frolov, A.; Shumilina, J.; Etemadi Afshar, S.; Mashkina, V.; Rhomanovskaya, E.; Lukasheva, E.; Tsarev, A.; Sulima, A.S.; Shtark, O.Y.; Ihling, C.; et al. Responsivity of Two Pea Genotypes to the Symbiosis with Rhizobia and Arbuscular Mycorrhiza Fungi—A Proteomics Aspect of the "Efficiency of Interactions with Beneficial Soil Microorganisms" Trait. *Int. J. Mol. Sci.* **2025**, *26*, 463. [CrossRef]
137. Perez-Riverol, Y.; Bandla, C.; Kundu, D.J.; Kamatchinathan, S.; Bai, J.; Hewapathirana, S.; John, N.S.; Prakash, A.; Walzer, M.; Wang, S.; et al. The PRIDE database at 20 years: 2025 update. *Nucleic Acids Res.* **2025**, *53*, D543–D553. [CrossRef]

plants

MDPI

Article

Defying Death: A Multi-Omics Approach to Understanding Desiccation Tolerance and Senescence in *Eragrostis nindensis*

Christine F. Madden [1], Brett Williams [2], Sagadevan Mundree [3], Sébastien Acket [4], Eric Ruelland [4], Henk W. M. Hilhorst [1] and Jill M. Farrant [1,*]

1 Department of Molecular and Cell Biology, University of Cape Town, Cape Town 7700, South Africa; chrissiemadden@gmail.com (C.F.M.); henk.hilhorst@uct.ac.za (H.W.M.H.)
2 School of Biology and the Environment, Queensland University of Technology, Brisbane, QLD 4001, Australia; b.williams@qut.edu.au
3 School of Agriculture and Food Sustainability, University of Queensland, Brisbane, QLD 4072, Australia; s.mundree@uq.edu.au
4 Unité Génie Enzymatique & Cellulaire, Université de Technologie de Compiègne, UMR CNRS 7025, F-60203 Compiègne, France; sebastien.acket@utc.fr (S.A.); eric.ruelland@utc.fr (E.R.)
* Correspondence: jill.farrant@uct.ac.za; Tel.: +27-823782499

Abstract: *Eragrostis nindensis* is a resurrection grass capable of surviving near-complete desiccation. We compared non-senescent leaf tissue (NST) and senescent leaf tissue (ST) to investigate the cellular and molecular basis of desiccation tolerance and senescence. NST recovered fully after drying, while ST failed to regain viability. Integrated transcriptomic (using RNA-Seq), lipidomic (using LC-MS), and ultrastructural (Transmission Electron Microscopical) analyses revealed that NST maintain RNA processing, protein folding, and translational activity during desiccation. Lipidomic data and ultrastructure showed preferential accumulation of polyunsaturated triacylglycerols (TAGs) and lipid droplets in NST, supporting membrane protection and energy buffering. In contrast, ST showed cellular collapse, reduced oleosin protein accumulation, and signatures of senescence. These findings highlight the importance of post-transcriptional and post-translational regulation, as well as lipid metabolism, in preserving cellular integrity during desiccation in this species.

Keywords: *Eragrostis nindensis*; desiccation tolerance; resurrection plants; senescence; RNA processing; lipid droplets; triacylglycerol; RNA sequencing

1. Introduction

Eragrostis nindensis is a polyploid grass capable of surviving near-complete protoplasmic water loss (>90% relative water content (RWC), equivalent to 0.1 g H_2O/g dry weight) and subsequently restoring full metabolic activity on rehydration. This phenomenon of vegetative desiccation tolerance is underpinned by complex molecular, metabolic, and structural adaptations [1–3]. Such plants are colloquially termed resurrection plants, and indeed while all survive drying to 10% RWC, there is a critical threshold of ca. 30% RWC for their survival and at which even drought-adapted species typically experience irreversible damage due to mechanical and oxidative stress [4–7].

Core desiccation responses across resurrection plants include the upregulation of early light inducible proteins (ELIPS), heat shock proteins (HSPs), late embryogenesis abundant (LEA) proteins, osmoprotectants (particularly sucrose, raffinose family oligosaccharides and trehalose), and various antioxidants [3,8–10]. However, species differ in the ratios and specific identities of these protectants, and in how the photosynthetic apparatus is managed during dehydration and recovery. Resurrection plants are broadly classified as either homoiochlorophyllous, retaining chlorophyll and thylakoid structure during drying, or poikilochlorophyllous, such as *E. nindensis*, which degrade chlorophyll and dismantle their thylakoids on drying, reconstituting the photosynthetic machinery on rehydration [11,12].

Earlier work on *E. nindensis* by Vander Willigen et al. [5,13,14] showed that leaf tissues differ in their desiccation response. Older leaves, termed senescent tissue (ST), senesce on desiccation, whereas younger ones do not and are termed non-senescent tissue (NST), retaining viability after drying. Both tissues undergo photosynthetic shutdown, but ST shows earlier membrane instability (electrolyte leakage at 50% RWC) and fails to maintain cellular integrity below 30% RWC [12–14]. In contrast, NST maintains cellular integrity through vacuole fragmentation and some degree of reversible wall folding to minimise plasmolysis and cythorrhesis.

Lipid metabolism, although underexplored, has emerged as a key determinant of desiccation tolerance. Lipids are central to membrane biogenesis, cell growth, and signalling [15,16], yet their thermal sensitivity makes them highly responsive to environmental stress [17]. Resurrection plants such as *Xerophyta humilis* [18], *Craterostigma plantagineum* [19], *Boea hygroscopica* [20], and *Paraisometrum mileense* [21] preserve membrane integrity during drying by increasing fatty acid (FA) unsaturation, particularly in thylakoids [2,22–24]. However, stress-induced lipid imbalance and reactive oxygen species (ROS) production can damage membranes, as ROS preferentially target unsaturated FAs [25,26]. Consequently, membrane remodelling and lipid-based protection are recurring features of desiccation tolerance [21,27]. A central component of this process is triacylglycerol (TAG), which sequesters toxic intermediates such as free fatty acids (FFAs), thereby mitigating oxidative damage and lipotoxicity [15]. TAG is synthesised in the endoplasmic reticulum and stored in lipid droplets (LDs), organelles stabilised by oleosins that regulate LD structure and size [28–32]. Although well studied in oilseeds, TAG and LDs also accumulate in vegetative tissues under abiotic stress [33,34], where they have been increasingly recognised as dynamic regulators of lipid homeostasis and stress signalling rather than passive storage sites [28,33]

This study integrates transcriptomic, ultrastructural, physiological, and lipidomic analyses to investigate desiccation responses in *E. nindensis*, focusing on the contrasting outcomes in NST and ST. It adds considerably to the knowledge of desiccation tolerance in *E. nindensis* and is the first study comparing multi-omics of non-senescent and senescent tissues of the same resurrection plant. By comparing gene expression, lipid composition, and LD dynamics between these tissues, we aim to uncover mechanisms that enable NST to resist senescence and recover after drying. Special attention is given to RNA and protein regulation pathways, membrane stabilisation, and the role of specific lipid species and LD-associated proteins. These insights may reveal molecular markers of desiccation tolerance and improve understanding of how lipid metabolism and gene regulation underpin desiccation tolerance (and suppression of drought-induced senescence) in this and potentially other resurrection plants.

2. Results

2.1. Physiological Changes Across a Desiccation–Rehydration Cycle

Initially, RWC remained relatively unchanged, maintaining high leaf water contents (ca. 75% RWC) for eight days after water was withheld (Figure 1A). Rapid water loss subsequently occurred, resulting in an air-dry state by day 13. Both NST and ST dried at similar rates. Plants were kept in an air-dry state for two weeks before being rehydrated by thorough watering (day 28). Only the NST made a full recovery after rehydration whereas the ST did not recover, resulting in leaf death (Figure 1B,D; Video S1). These findings correlate with previous studies on *E. nindensis* [14,35].

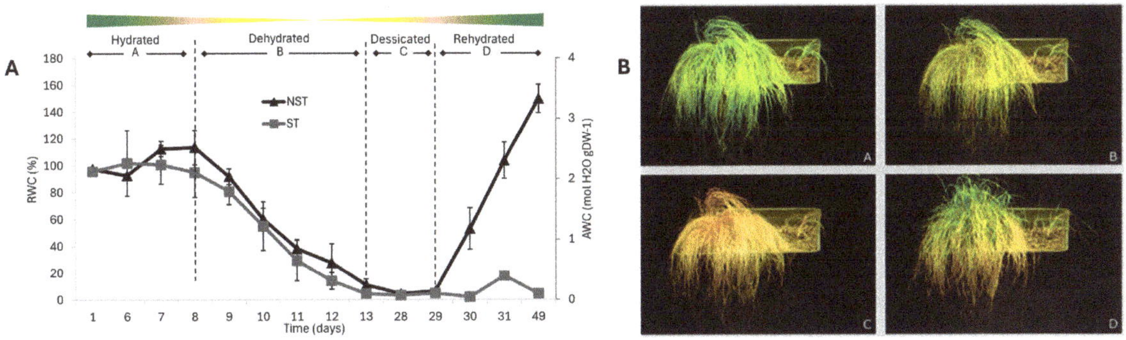

Figure 1. (**A**) Changes in relative water content (RWC) and absolute water content (AWC) over time in the resurrection plant *Eragrostis nindensis*, comparing desiccation-tolerant non-senescent tissue (NST) and desiccation-sensitive senescent tissue (ST) during drying and rehydration. Four physiological stages are indicated: A—Hydrated (100% RWC), B—Dehydrated (60–25% RWC), C—Desiccated (air-dry), and D—Rehydrated (12 h post-watering). The colour gradient illustrates chlorophyll loss during dehydration and its recovery upon rehydration. Breaks in the x-axis (days 1–6 and 13–28) are included to improve readability. Vertical bars represent standard errors (n = 6 plants, 3 technical replicates per plant). (**B**) Representative photographs of the four treatment stages corresponding to panel (**A**). Two-way ANOVA showed significant effects of time ($p < 0.0001$), tissue ($p < 0.0001$), and their interaction ($p < 0.0001$).

2.2. *Ultrastructural Changes During Drying and Rehydration*

2.2.1. Non-Senescent Tissue

E. nindensis is a C4 grass with typical Kranz anatomy, consisting of bundle sheath (BS) cells associated with vascular tissues, surrounded by palisade mesophyll cells [12]. As the BS cells are most informative regarding photosynthesis, they are discussed here. For mesophyll cell observations, see Supplementary Data (Figures S1 and S2). In general, the same trends are seen between the two tissue types, with the obvious exception of starch presence in mesophyll, which experienced considerably more cell wall folding during dehydration. Osmophilic body (OB) accumulation was less pronounced in mesophyll relative to BS cells.

In NST, fully hydrated BS cells had starch-filled chloroplasts arranged centrifugally, with well-formed stacked grana indicating active photosynthesis and starch production (Figure 2A). Upon drying to 60% RWC, the chloroplasts retained their structure and proximity to vascular bundles, though starch depletion was evident (Figure 2B). Vacuolation increased, and small, regularly sized lipid droplets (LDs) began accumulating along the plasma membrane.

At 40% RWC, BS cells were slightly compressed, with chloroplasts loosely congregated near the vascular bundles alongside fragmenting vacuoles. Several large OBs were present between chloroplasts (Figure 2C). Severe drying (25% RWC) caused dramatic changes. Cells became dominated by tightly packed vesicles, and chloroplasts showed complete thylakoid disassembly, characterised by thylakoid vesicles and clustered plastoglobuli. Several LDs accumulated densely along the cell wall and near plasmodesmata, suggesting a role in maintaining cell-to-cell communication, consistent with the role of LDs in intercellular trafficking [28,36]. The larger OBs seen at 40% RWC were not as evident (Figure 2D).

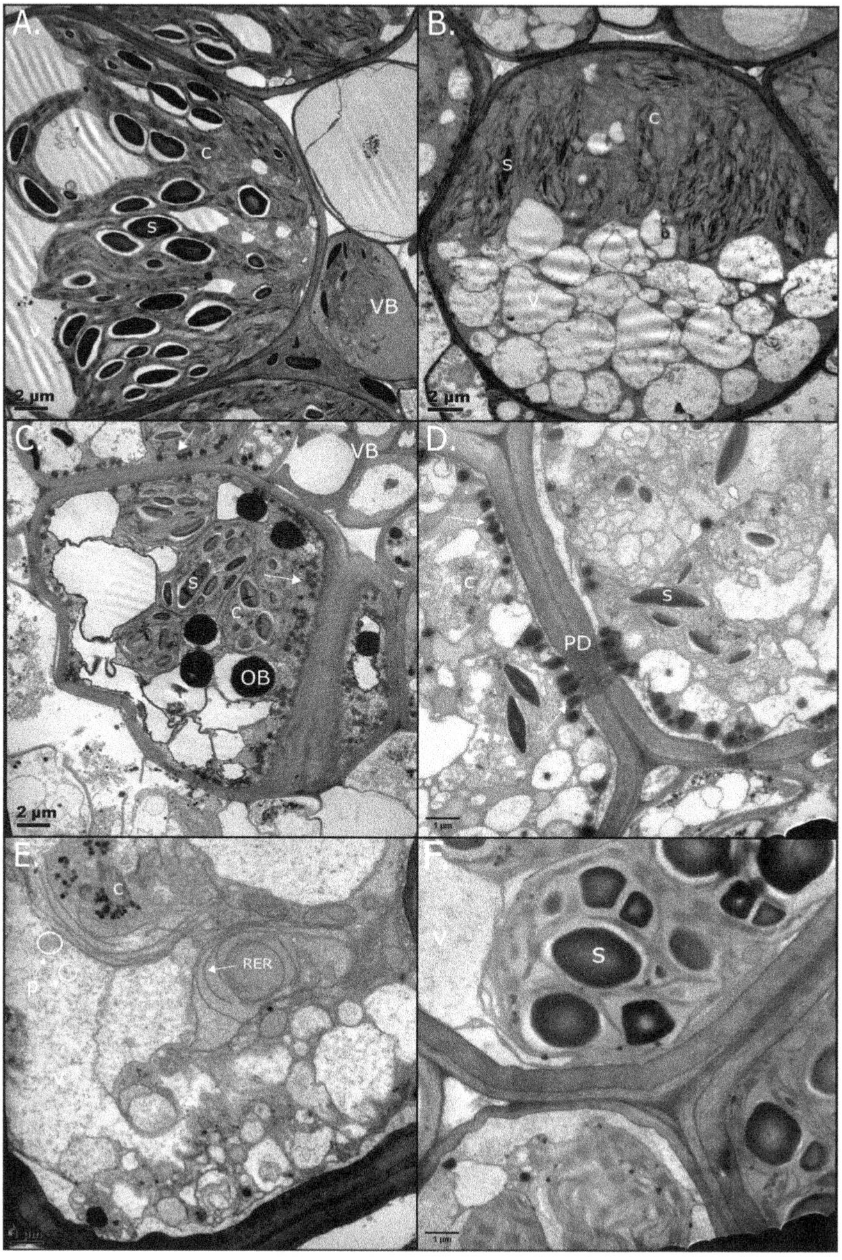

Figure 2. Transmission electron micrographs (TEMs) of bundle sheath (BS) cells from NST of *E. nindensis* during progressive drying (RWC) and rehydration (12 h). (**A**) At full hydration (100% RWC), chloroplasts (c) were structurally intact, actively photosynthesising, and accumulating starch (s). A vascular bundle (VB) is evident, affirming that this is a BS cell. (**B**) At 60% RWC, vacuoles (v) became fragmented, and low-density lipid droplets (LDs) appeared near the plasmalemma. (**C**) At 40% RWC, BS cell size was reduced but shape maintained; large osmophilic bodies (OBs) emerged and small lipid

droplets accumulated along the cell wall (arrow). (**D**) At 25% RWC, thylakoids were dismantled, but plasmodesmata (PD) remained intact, indicating preserved cell-to-cell communication. LD accumulation was prominent (arrow). OBs were not as evident. (**E**) In the desiccated state (<10% RWC), extensive ribosomal endoplasmic reticulum (RER; arrowed) was observed, implying active protein synthesis. Darkly stained clusters resembling polyribosomes (p, circles) were observed. (**F**) At 12 h post-rehydration, chloroplasts were largely reassembled, and starch (s) granules were again present.

In the air-dry, desiccated state, chloroplasts remained disassembled (Figure 2E), indicating that ultrastructural reorganisation for desiccation survival occurred by 25% RWC. Despite the critically low water content, the cytoplasm remained well-preserved, showing intact endoplasmic reticulum, plastoglobuli, electron-dense cytosol, and distinct (unfused) vacuoles. Small clusters of dense matter resembled polyribosomes [37]. During rehydration, NST BS cells recovered within 12 h, with chloroplasts reassembling thylakoid membranes and vacuoles coalescing. LDs previously dominating the cell walls disappeared (Figure 2F).

2.2.2. Senescent Tissue

ST BS cells largely resembled NST at full turgor, with starch-filled chloroplasts and distinguishable thylakoid membranes. Several OBs are evident, unlike in the hydrated NST (Figure 3A). At 60% RWC, ST starch appeared to increase in size, giving the appearance of amyloplasts due to the absence of thylakoid membranes (Figure 3B). This is likely due to ongoing photosynthesis; if not occurring in the BS cells, occurring in the mesophyll cells (Figure S1A). Indeed, ongoing photosynthesis was recorded at this water content, albeit at a slightly reduced capacity relative to the NST (Figure S3). Mitochondria with darkly stained cristae, indicative of active respiration, clustered around the chloroplasts. Large OBs accumulated in electron-opaque vacuoles, as seen in the NST.

By 25% RWC, ST BS cells exhibited large, irregularly sized OBs and lacked vacuolation or the small LDs seen along NST cell walls (Figure 3C). These ST OBs exhibited features typical of oleosin-deficient droplets, including fusion and enlargement [30]. ST lacked internal structures when compared to NST, with notable subcellular attrition of organelles in the air-dry state. Darkly stained cellular debris—potentially polyphenols—were evident; some wall collapse was evident and the plasmalemma was disrupted, indicating mechanical stress (Figure 3D). We propose that cell death occurs below 25% RWC. The disappearance of starch and most OBs in the desiccated state could indicate transportation to sink tissues, which is typical of nutrient salvage in age-related senescence.

During rehydration, ST cells collapsed further, with no distinguishable organelles or signs of recovery (Figure 3E). Neither BS cells nor mesophyll cells (Figure S2E) of the ST survived rehydration. For an overview of LD dynamics during drying and rehydration, see Figure S6.

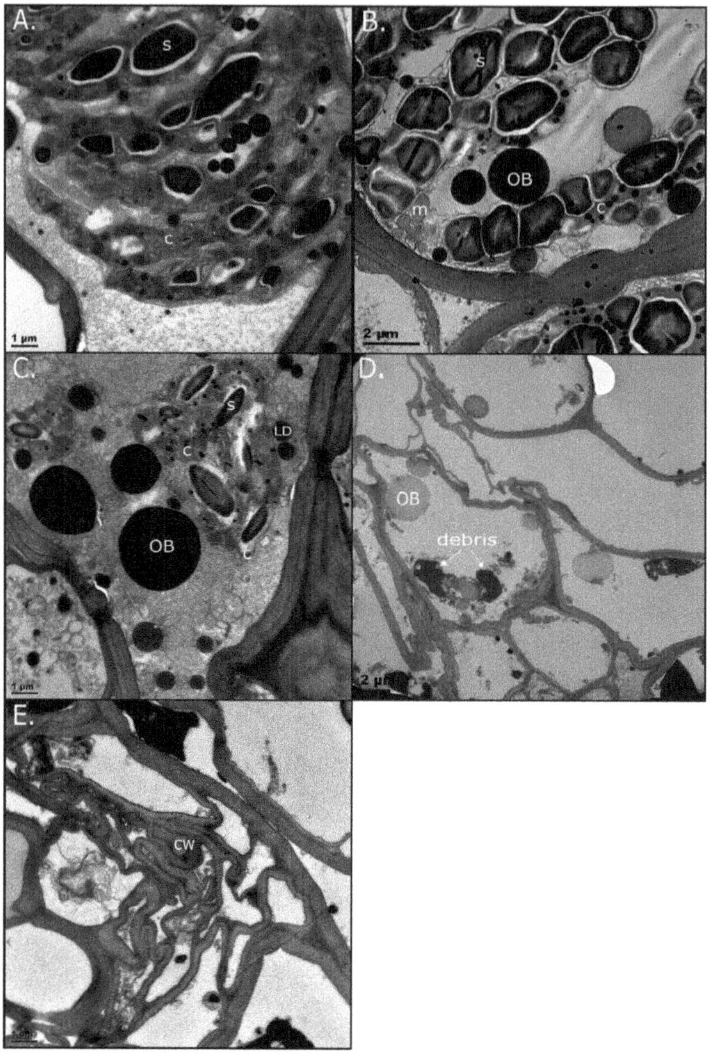

Figure 3. TEM of BS cells from ST of *Eragrostis nindensis* during drying (RWC, %) and after 12 h of rehydration. (**A**) At full hydration (100% RWC), chloroplasts (c) were intact and actively photosynthesising, with visible starch granules (s). (**B**) At 60% RWC, chloroplasts became indistinct, while large starch grains and OBs accumulated. Mitochondria (m) appeared structurally active. (**C**) At 25% RWC, chloroplasts were dismantled, starch content was reduced but still detectable, and large OBs were present alongside dispersed and small LDs. (**D**) In the desiccated state (<10% RWC), cell walls (cw) collapsed, and large OBs and cellular debris were evident. (**E**) After 12 h of rehydration, cells showed no structural recovery; cell walls remained collapsed.

2.3. *Transcriptomic Response to Drying and Rehydration*

ST was not sampled at 40% RWC, nor at 12 h rehydration. See Section 4.3 for reasoning. Thus, direct comparisons between NST and ST at these sampling points are not made.

2.3.1. Overview of DEG Patterns

A total of 8185 genes were differentially expressed (DEGs) compared to the control (100% RWC, NST), with the majority in NST (7643 genes). Of these, 4953 (60.5%)

were shared across both tissue types and all water contents, while 2690 (32.9%) were exclusive to the NST, indicating a more complex regulatory programme for desiccation tolerance (Figure 4A). Here, and at every mention below, genes were classified as differentially expressed when $\log_2$ fold change exceeded ± 2.0, with a false discovery rate (FDR; $q \leq 0.05$, [38]) applied relative to the control (100% RWC, NST).

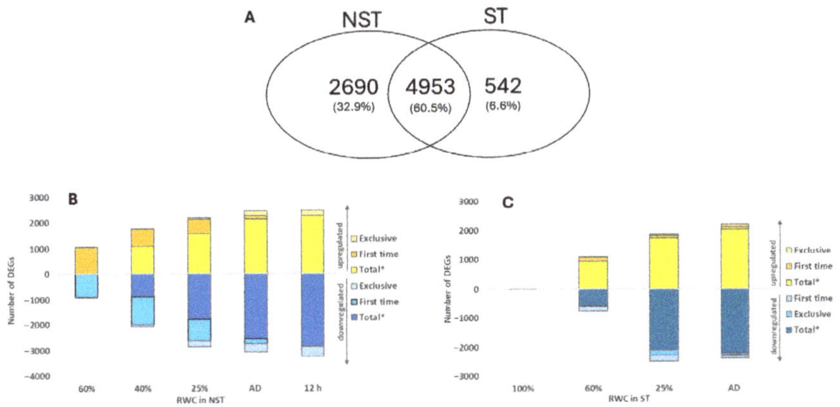

Figure 4. (**A**) Venn diagram presenting the proportion and number of differentially expressed genes (DEGs) commonly expressed across all samples, and exclusive genes per tissue type (NST, ST) during drying and rehydration. (**B**,**C**) Number of DEGs either accumulating (**B**) or diminishing (**C**) in NST or ST. Total* excludes genes that are expressed exclusively or for the first time. First time genes do not include exclusive genes.

DEGs accumulated or diminished in both tissue types throughout drying and rehydration, increasing as water contents declined (Figure 4B,C). ST had few (n = 3) DEGs at 100% RWC when compared to the control, indicating that fully hydrated leaves are transcriptionally alike (Figure 4C). Notably, distinct sets of DEGs emerged during drying, particularly at 25% RWC and at the air-dry state in both tissue types and 12 h post-rehydration in the NST. These findings suggest that *E. nindensis* maintains active transcription even at very low water contents (<25% RWC) and during early rehydration, when RWC has recovered to ca. 20%.

Overall, a greater number of DEGs diminished than accumulated in abundance, consistent with metabolic arrest, which is an established feature of desiccation tolerance [39,40]. The progressive increase in both up- and downregulated DEGs with drying further indicates dynamic transcriptional reprogramming in response to stress.

2.3.2. Core Desiccation Tolerance Mechanisms in NST

In NST, 68.5% of DEGs (2246 genes) were consistently expressed across all RWCs below 60%, reflecting a core suite of drought- and desiccation-tolerance mechanisms (Figure 5, box A). Enriched biological processes included responses to desiccation, ABA signalling, heat stress, solute metabolism and transport, photosynthetic shutdown, reproduction, and metabolite accumulation. Evidence of cell wall macromolecular catabolism signified growth cessation.

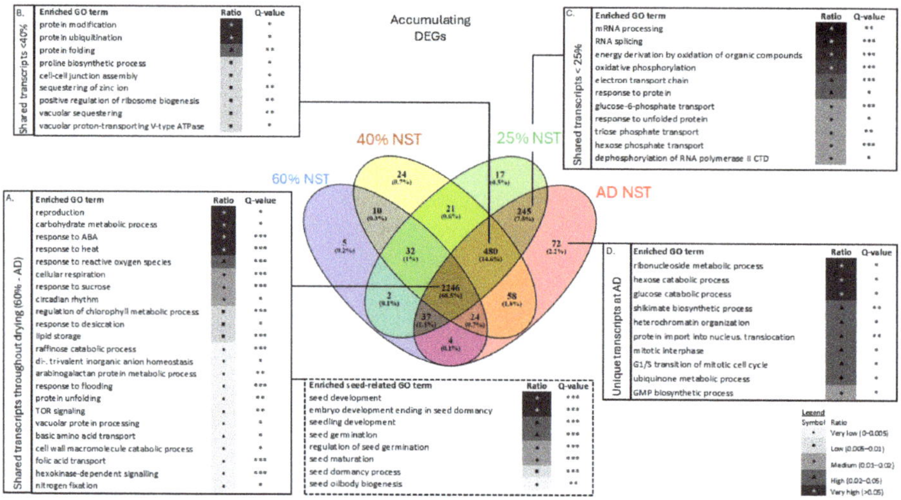

Figure 5. Drying-induced accumulating DEGs, represented by enriched Gene Ontology (GO) categories ($p < 0.05$), show distinct metabolic shifts in NST of the resurrection plant *Eragrostis nindensis* and display core traits of desiccation tolerance. The number of DEGs is relative to fully hydrated leaves (100% RWC, NST). GO categories of genes expressed throughout drying at different RWCs (shared genes) or only expressed in the desiccated state (air-dry (AD), unique genes) are depicted in boxes. The dotted box shows the seed-associated enriched GO categories that were expressed throughout drying. Significance was determined using BiNGO with FDR correction (q-value < 0.05; * q < 0.05, ** q < 0.01, *** q < 0.001). Ratio (of expected to observed number of genes) represents the degree to which each category is overrepresented as described in the legend.

Seed-related Gene Ontology (GO) terms were enriched (Figure 5, dashed box), supporting the hypothesis that resurrection plants repurpose seed maturation events for vegetative desiccation tolerance [40,41]. Indeed, seed-related protective genes such as LEAs (other than LEA2), HSPs, dehydrins and seed maturation proteins (SMPs) were also expressed during drying (Figure S4). Metabolic arrest was further supported by shifts towards vacuolation, autophagy, and stress-responsive pathways, including 'TOR signaling', 'vacuolar protein processing', and 'protein unfolding'.

During dehydration to ca. 40% RWC, transcripts related to vacuolation and autophagy were enriched (Figure 5, box B), supporting ultrastructural vacuolation observations (Figure 2B). Processes minimising macromolecular denaturation and maintaining protein integrity were dominant, reflecting shifts towards protein protection and metabolic stabilisation. The enrichment of 'proline biosynthetic process' genes align with previous findings of proline accumulation in *E. nindensis* [5].

Under severe dehydration (<25% RWC), enrichment of transcriptional regulation ('mRNA processing', 'RNA splicing', 'dephosphorylation of RNA polymerase'), solute transport ('triose phosphate transport', 'electron transport chain'), and sugar metabolism ('glucose-6- phosphate') suggest preparation for rapid recovery (Figure 5, box C). The storage and protection of RNA-encoding genes must therefore occur below 25% RWC. In the air-dry state, transcript accumulation for DNA replication, protein translation, and cell cycle regulation indicates RNA storage for immediate rehydration-driven modification or translation.

2.3.3. Rehydration and Recovery

In NST, the majority of DEGs remained differentially expressed during rehydration, including 210 DEGs that were exclusively upregulated and 393 DEGs that were uniquely downregulated (Table 1, Figure 5, box A).

Table 1. Enriched GO terms ($p < 0.05$) up- or downregulated during rehydration (12 h post-watering) in NST of *Eragrostis nindensis* indicate quick resumption of metabolism. Ratio (of expected to observed number of genes) represents the degree to which each category is overrepresented.

Expression Trend	GO Term	Description	Ratio
Accumulating DEGs	GO:0017003	protein–heme linkage	0.0136
	GO:0071258	cellular response to gravity	0.0136
	GO:0006089	lactate metabolic process	0.0204
	GO:0009438	methylglyoxal metabolic process	0.0204
	GO:0018315	molybdenum incorporation into molybdenum–molybdopterin complex	0.0136
	GO:0042040	metal incorporation into metallo–molybdopterin complex	0.0136
	GO:0046185	aldehyde catabolic process	0.0204
	GO:0019243	methylglyoxal catabolic process to D-lactate via S-lactoyl-glutathione	0.0204
Diminishing DEGs	GO:0030245	cellulose catabolic process	0.0135
	GO:0035445	borate transmembrane transport	0.0101

Functional categories associated with energy metabolism and detoxification were enriched upon rehydration, such as 'lactate metabolic process', 'molybdenum incorporation', 'metal incorporation', and 'protein-heme linkage' (Table 1), despite the absence of photosynthetic carbon. Protein complexes linked to linear electron flow ('protein-heme linkage'), aldehyde catabolism, and molybdenum cofactor enzymes were upregulated, highlighting oxidative stress mitigation. Enriched GO categories emphasised protein folding, metabolic reactivation, and RNA regulation. Growth-related genes remained suppressed (Table 1), suggesting delayed resumption of cell proliferation post-rehydration. Furthermore, 89 DEGs related to protein synthesis were upregulated during drying, the majority exclusively in NST (56%), of which 13% were only expressed at 12 h rehydration (Table S2), which may indicate potential efficacy of the translational machinery at low water contents during rehydration in *E. nindensis*.

2.3.4. Genes Suppressed During Dehydration

Fewer GO categories showed significantly downregulated DEGs during drying (Figure 6). These were primarily associated with metabolism, photosynthesis, carbon fixation, and growth, reflecting widespread metabolic suppression (Figure 6, box A). The enrichment of terms such as 'photosynthesis', 'carbon fixation', 'growth', and 'gluconeogenesis' indicates a progressive shutdown of energy-intensive processes. This metabolic arrest, including reduced expression of chlorophyll a/b binding and photosynthesis-related transcripts (Figure S5B), is characteristic of poikilochlorophyllous species and serves to limit ROS production and conserve energy [42]. Furthermore, reduced abundance of 'drought recovery' and cell wall-related genes, including multiple cellulose synthase transcripts, suggest growth inhibition. Suppressed expression of 'leaf senescence'-associated genes were also more prominent in NST than in ST (Figure S5B), aligning with irreversible degeneration in the latter.

Below 40% RWC, transcripts associated with 'chloroplast relocation', protein degradation and signalling, such as 'stomatal complex development', 'protein chromophore linkage', and 'circadian rhythm', were downregulated, indicating transcriptional suppression at lower water contents (Figure 6, box B). Categories related to protein and transcriptional upregulation at 40% RWC (Figure 5, box B) showed diminished enrichment at 25% RWC (Figure 6, box C), reflecting a metabolic shift marked by a reduced protein turnover. This coordinated downregulation of protein production genes suggests a protective strategy to maintain protein integrity by limiting molecular crowding, thereby supporting protein stabilisation and detoxification under stress [6,43,44].

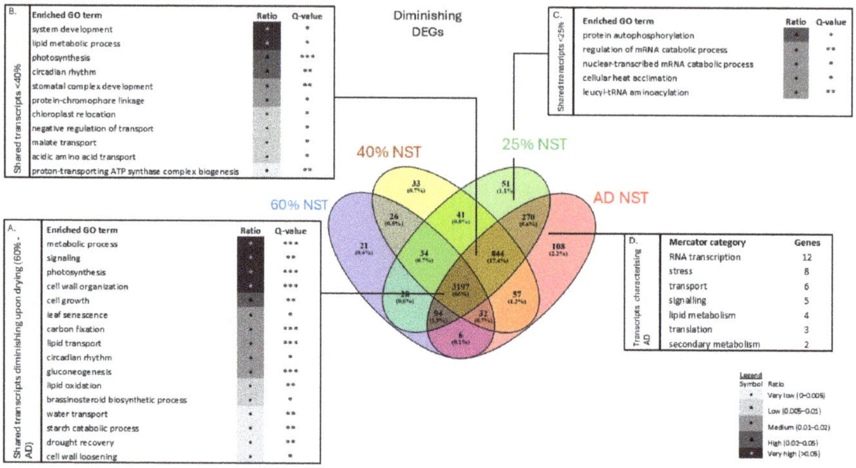

Figure 6. Drying-induced downregulated DEGs, represented by enriched GO categories ($p < 0.05$), show distinct metabolic shifts in NST of *Eragrostis nindensis*. The number of DEGs are relative to fully hydrated leaves (100% RWC, NST). Boxes summarise the most enriched GO categories of shared genes expressed during each drying stage. Mercator categories are shown for air-dry samples (AD, RWCs of <10%). Significance was determined using BiNGO with FDR correction (q-value < 0.05; * q < 0.05, ** q < 0.01, *** q < 0.001). Ratio (of expected to observed number of genes) represents the degree to which each category is overrepresented as described in the legend.

Under severe stress (<25% RWC), extensive downregulation of genes involved in translation, signalling, and metabolite transport indicated a shift towards metabolic shutdown (Figure 6, box C). In the air-dry state, the number of genes was too low to yield statistically enriched categories (Figure 6, box D). Nevertheless, the sharp decline in DEGs reflects a widespread cessation of gene expression, marking the transition to a quiescent state.

2.3.5. Exclusive Gene Expression in NST and ST

Approximately 20% of DEGs (excluding 12 h rehydration) were exclusively upregulated in NST, comprising 633 transcripts across 39 enriched GO categories (Figure 7A). Key categories include 'amino acid biosynthetic process', 'response to heat', 'chaperone-mediated protein complex assembly', 'regulation of seed germination', and 'glucose-6-phosohate transport', all supporting protein synthesis, folding, and RNA stabilisation. The exclusive enrichment of RNA processing and protein metabolism in NST suggests a mechanism for transcript stabilisation and storage during drying.

NST exhibited greater accumulation of DEGs related to post-translational regulation compared to the ST (Figure 5, boxes B, C). These genes are associated with signalling, protein degradation, post-translational modification, and transport. Furthermore, enriched categories such as 'RNA processing', 'protein folding', and 'chaperone-mediated protein complex assembly' were prominent at lower water contents (Figure 7B), suggesting active mechanisms for maintaining protein integrity. This over-representation highlights the importance of transcriptional regulation (RNA processing) and protein metabolism (translational machinery transcripts) in desiccation tolerance.

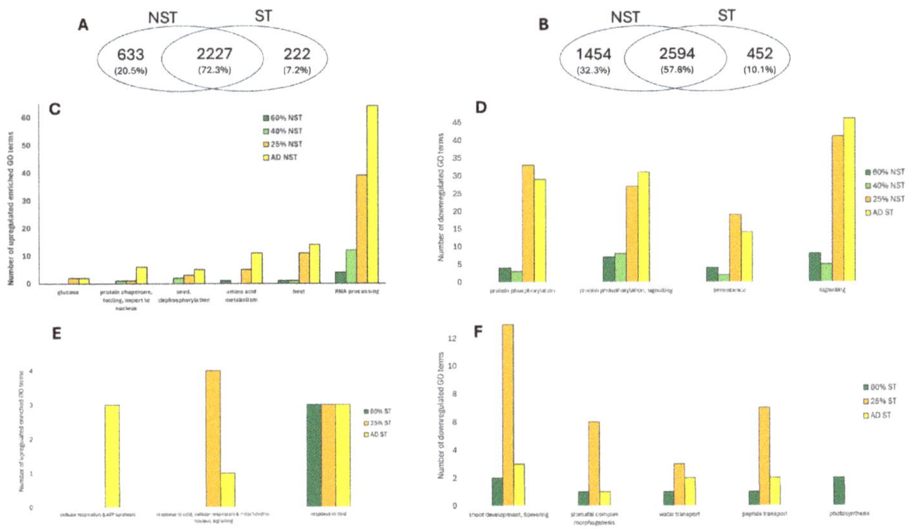

Figure 7. Differential gene expression and enriched GO categories during drying in *E. nindensis*. (**A**,**B**) Venn diagrams show shared and unique upregulated (**A**) and downregulated (**B**) DEGs between NST and ST during drying. (**C**,**E**) Enriched GO terms associated with upregulated DEGs in NST (**C**) and ST (**E**). (**D**,**F**) Enriched GO terms associated with downregulated DEGs in NST (**D**), and ST (**F**). Differential expression was defined as $\log_2$ fold change > 2 and FDR < 0.05 relative to the hydrated NST control (100% RWC).

Of the 93 DEGs belonging within the enriched 'RNA processing' and related categories, 28 were associated with transcriptional regulation, including several stress-induced transcription factors (e.g., AP2, bZIP) and genes involved in post-translational modification, such as protein synthesis and degradation (Tables S1–S3 and S5). Notably, the small nuclear ribonucleoprotein family protein (snRNP; AT3G14080), exhibited the highest fold change in NST during desiccation; 11-fold at 40% RWC and 10.7-fold at 25% RWC.

2.4. Lipidomic Shifts During Drying and Rehydration

Lipids were extracted from NST and ST samples, at 100% RWC, 25% RWC, AD, 12 h and 24 h rehydration. These were separated by reverse phase chromatography and analysed by mass spectrometry. A total of 67 lipids were annotated; TAG, DAG, MGDG, and DGDG were obtained in the positive mode, while SQDG, PE, PC, PI, and PG were obtained in the negative mode. A heatmap was generated (Figure 8), with clustering of the metabolites. In NST, drying is characterised by a relative decrease between 100% RWC and AD in MGDG 18:1/18:3, MGDG 18:3/18:3, SQDG 18:3/18:3. This is paralleled with the relative increase in PE (18:2/18:3, 18:1/18:3, 18:1/18:2), PC (16:0/18:2, 18:2/18:3, 18:1/18:3, 18:1/18:2 and 18:1/18:1), and TAG (16:0/18:2/18:3, 20:0/18:1/18:3, 16:0/18:1/18:3, 16:0/18:1/18:2, 18:1/18:1/18:2 and 18:1/18:2/18:2). For NST samples, 12 h rehydration leads to an increase in TAG (18:1/18:1/18:1, 20:1/18:3/18:3, 24:0/18:1/18:3, 20:0/18:1/18:2, 22:0/18:1/18:2). The profile after 24 h rehydration differs from that at 12 h rehydration, and from that in NST at 100% RWC, suggesting at 24 h rehydration is not sufficient to return to the 100% RWC profile. As for ST samples, dehydration is characterised by the marked decrease in DGDG (18:1/18:3, 18:2/18:3, 18:3/18:3), MGDG (18:1/18:3, 18:2/18:3, 18:3/18:3) and increase in some TAGs (18:2/18:3/18:3, 18:3/18:3/18:3 and 20:0/18:3/18:3). This clearly indicates differences between NST and ST during dehydration. Furthermore, changes during rehydration of ST are minimal. This is confirmed by the PCA (Figure 9). For legibility, we separated NST samples (Figure 9A) and ST samples (Figure 9B), however, the positions are directly comparable across panels. PC1 explains 28.7% of the variance while PC2 explains 18.2% thereof. For NST samples, drying leads to a change towards more positive values of the PC1 axis. This is also the case for ST samples; however, 25% RWC is similar to that of the AD tissues, which is not true for the NST samples. Indeed, for ST, with exception of the 100% RWC tissue, all samples are comparable, suggesting that dehydration led to irreversible changes in lipid profiles.

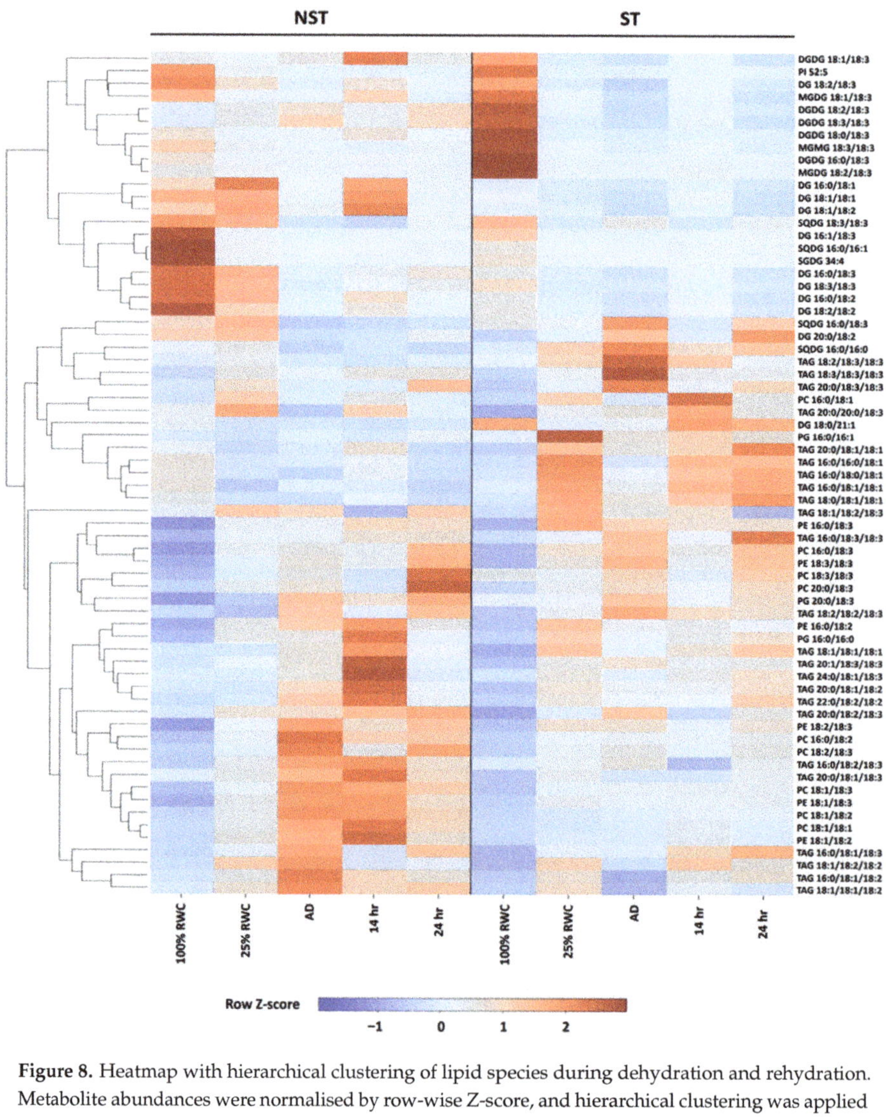

Figure 8. Heatmap with hierarchical clustering of lipid species during dehydration and rehydration. Metabolite abundances were normalised by row-wise Z-score, and hierarchical clustering was applied to group metabolites with similar response patterns. Columns represent experimental conditions in a fixed order. Warmer colours (red) indicate higher relative abundance compared to the metabolite's mean, while cooler colours (blue) indicate lower abundance. A separate horizontal colour bar indicates the Z-score scale.

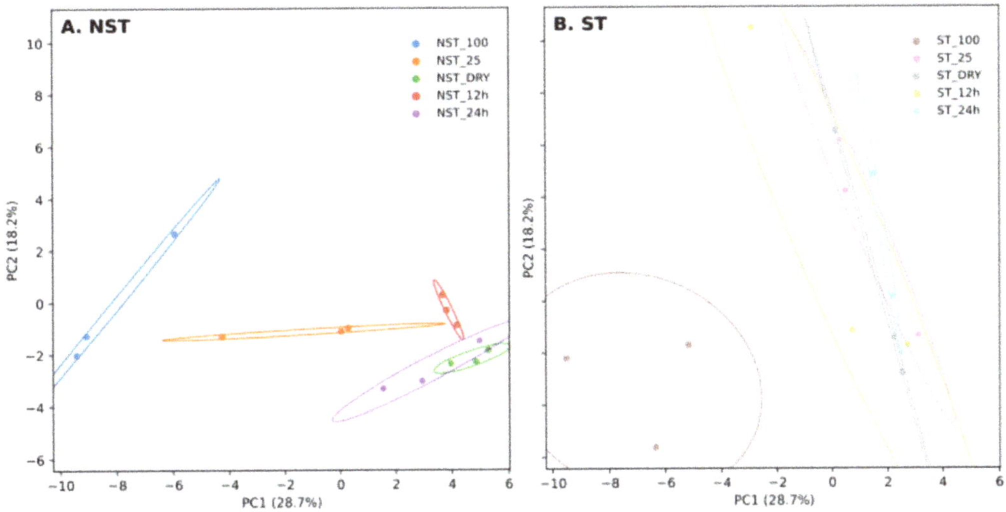

Figure 9. Principal Component Analysis (PCA) of lipid species during dehydration and rehydration. Scores for PC1 vs. PC2 are shown, with each point representing an individual sample. Panels display the same PCA projection computed once on the full dataset but visualised separately for (**A**) NST samples and (**B**) ST samples to improve legibility; positions are directly comparable across panels. Ellipses depict two-standard-deviation dispersion for each condition (bivariate SD ellipse; ca. 86% coverage under a Gaussian assumption). Axis labels report the percentage of variance explained by each principal component.

2.5. Protein Expression as a Proxy for Translational Regulation in NST

Oleosin Protein Expression

Of the eleven oleosin transcripts identified in the transcriptome, eight were differentially expressed in the NST and six in the ST during drying and rehydration (Figure 10A). The OLE1 antibody used for immunoblotting showed cross-reactivity with three *E. nindensis* isoforms (Eni_009916, Eni_003038, and Eni_007625), which share 57–67% sequence similarity with the *Arabidopsis* OLE1 protein used to generate the antibody.

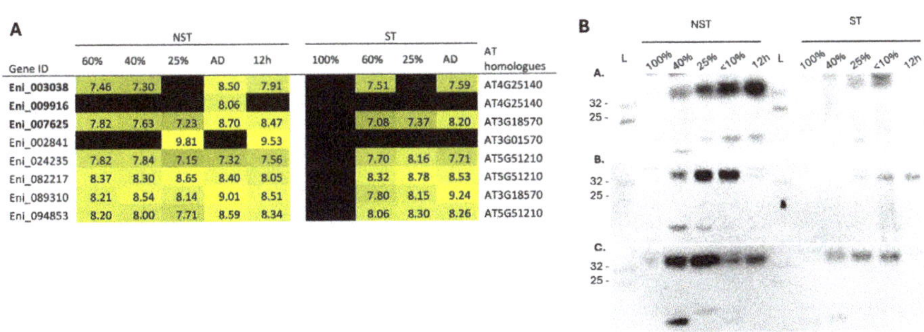

Figure 10. (**A**) Heatmap of eight differentially expressed oleosin transcripts and their corresponding *Arabidopsis thaliana* homologues identified in NST and ST during desiccation and rehydration. Differential expression is shown as $\log_2$-fold change relative to 100% NST, with lighter shades of yellow indicating higher fold change and black representing no change. (**B**) Western blot analysis of oleosin protein expression using the OLE1 antibody with three biological replicates (A.–C.). A dominant band at ca. 33 kDa was consistently observed in NST during drying, with the strongest expression at lower RWCs (25% and 10%). In contrast, ST showed faint or undetectable signals at the same position. Molecular weight was estimated using the Colour Prestained Protein Standard, Broad Range ladder

(L; 11–245 kDa, New England Biolabs). The observed molecular weight shift from the expected 19 kDa may be due to mono-ubiquitination or interaction with oleosin-like proteins such as caleosins [45].

Immunoblotting confirmed the accumulation of oleosin proteins during drying, predominantly in the NST (Figure 10A). Protein expression became detectable at 40% RWC, peaked under severe dehydration and desiccation, and generally declined during rehydration (Figure 10B). In contrast, ST samples exhibited only faint signals at the same band positions, despite comparable transcript abundances.

Consistent protein bands were detected at ca. 33 kilodaltons (kDa) across biological replicates, notably larger than the expected 19 kDa of OLE1. This may indicate binding to other oleosin isoforms, as oleosins typically range from ca. 15–30 kDa [45], or may result from mono-ubiquitination, post-translational modification, or interaction with oleosin-like proteins such as caleosin, as observed by Yu et al. [46]. Despite multiple bands, the consistent detection of the dominant band supports translational control and differential protein accumulation during drying.

3. Discussion

3.1. Cellular Ultrastructure

This study demonstrated how *E. nindensis* undergoes marked physiological and ultrastructural modifications to withstand desiccation-induced oxidative and mechanical stress, as has been reported in other resurrection plants [1,6,42,47]. These responses involve genetic reprogramming, notably photosynthetic shutdown and senescence inhibition [7,43,48,49]. While NST and ST exhibit similar ultrastructure when hydrated (Figures 2 and 3), distinct differences emerge during drying, reflecting divergent metabolic regulation.

The similar drying rates observed in NST and ST (Figure 1), indicate that senescence in ST is not triggered by differential water loss. Subcellular differences become pronounced under moderate stress (<40% RWC; Figure 2C), identifying this as a pivotal transition phase. At this point, mechanisms for biophysical protection, metabolic reorganisation, and preparation for rehydration are activated, and these are hallmarks of desiccation tolerance [3]. LDs appear in NST below 60% RWC, resembling those seen in seeds during maturation or environmental stress [36], and accumulate along the cell periphery, potentially stabilising the plasmalemma and membrane integrity [28,50]. Supporting this, BS cells in NST exhibit minimal membrane retraction, implying mechanical stabilisation. In contrast, the absence of LDs in ST likely contributes to the irreversible collapse of membrane and organelle integrity.

3.2. Desiccation Tolerance Signature

As a truly desiccation-tolerant species, *E. nindensis* shows distinct metabolic shifts across RWCs. Metabolic arrest and the accumulation of protective molecules during drying are consistent with patterns observed in other resurrection plant species [51]. At 60% RWC, photosynthesis and growth cease (Figure S4B), while protective pathways, such as unfolded protein responses, are activated at 40% RWC. Further drying (<25% RWC) enhances transcriptional control and, in the desiccated state, DNA repair pathways are enriched. Notably, NST-specific transcripts are enriched for genes related to RNA processing and translation (Figure 4). This implies that NST maintains translational control during drying, though further evidence is required to support this.

The dominant GO terms enriched at 12 h post-rehydration were 'response to gravity' and 'protein-heme linkage' (Table 1). Heme-containing proteins are essential for redox reactions, electron transport, oxygen binding, signal transduction, and micro-RNA processing [52]. However, unbound heme can trigger iron-dependent lipid peroxidation, leading to ferroptosis and cell death [52,53]. Given the abundance of polyunsaturated lipids in *E. nindensis*, particularly during drying, the upregulation of 'protein-heme linkage' may reflect protective regulation against lipid peroxidation and ROS generation during rehydration. In addition, the reduction in 'cellulose catabolism' suggests decreased cell wall remodelling, potentially limiting water loss [54].

3.3. Seed-Related Signature

Seed-related GO terms, central to dormancy and desiccation tolerance, are broadly enriched in NST across all RWCs (Figure 4) [40,55]. However, 60.5% of DEGs are shared between NST and ST (Figures 4 and S4A), indicating that senescence and cell death in ST are not caused by a failure to transcribe genes that enable desiccation tolerance. Rather, they likely result from an inability to maintain effective RNA processing (i.e., transcription and post-transcriptional regulation) or translational capacity. We propose that the preservation of these processes, beyond transcriptional activation alone, is essential for desiccation tolerance.

To demonstrate this, both HSPs and LEAs exhibit similar expression patterns in NST and ST (Figure S3), supporting their known chaperone roles during desiccation [56–63]. Similar expression profiles in the desiccation-sensitive *Eragrostis tef* further imply that transcript abundance alone does not confer tolerance [64].

3.4. Translational Regulation and Senescence

The overrepresented suppression of 'leaf senescence' in NST indicates active regulation of senescence (Figures 6 and S4A). In addition, the enrichment of post-transcriptional and translational machinery in NST suggests active protein synthesis is maintained during desiccation, a characteristic of tolerant species [65–68]. Upregulated ribosomal proteins, elongation factors, and chaperones, coupled with the appearance and organisation of cellular organelles (Figure 2), suggest that NST sustains RNA processing at low RWCs. In contrast, ST exhibits reduced RNA regulation, increased RNA degradation, and protein turnover (ubiquitination) pathways indicative of senescence (Tables S2 and S3) [48,69].

Transcript accumulation alone does not equate to functional tolerance [70]. In *Tortula ruralis*, transcripts accumulated during drying for the resumption of metabolism upon rehydration [65]. Transcripts must therefore be stabilised during desiccation, likely in ribonucleoprotein complexes [44–46]. Similar transcriptional trends observed in *E. nindensis* (Figures 4 and S4A) may allow rapid mobilisation upon rehydration [71–73], potentially through RNA sequestration in processing bodies (P-bodies), as observed in seeds [70–73]. The role of P-bodies in RNA storage under stress warrants further investigation.

RNA degradation also shapes transcript pools [74], yet mechanisms determining RNA storage, degradation or translation during drying remain unresolved [75]. Stabilising and mobilising RNA during- and post-desiccation is essential for desiccation tolerance [76,77]. In *X. humilis* PSII-related RNAs persist through desiccation, enabling recovery after rehydration through, inter alia, facilitating protein synthesis post-rehydration [78,79]. Evidence of dense polyribosomes near the endoplasmic reticulum (Figure 2E), and enriched GO terms for chaperone activity (Figure 5, box B), further supports ongoing translation and protein repair.

The exclusively upregulated snRNP (AT3G14080) in NST showed the highest fold changes among RNA processing genes (11 and 10.7 at 40% and 25% RWC, respectively; Figure 5 and Table S4). As part of the LSM1-7 decapping activator complex, it regulates RNA turnover and abiotic stress responses in *Arabidopsis* [80,81], underscoring the importance of RNA regulation in *E. nindensis* desiccation tolerance. Furthermore, elevated transcript accumulation in NST (Figure 4A) coupled with its survival post-desiccation, suggests active transcript storage during drying.

3.5. Energy Metabolism and Rehydration

TEM revealed stark differences in LD accumulation between NST and ST (Figures 2 and 3 and S5). In NST, starch was present under fully hydrated conditions, persisted during drying, and started to be metabolised (indicated by the white halo) but was still present within 12 h of rehydration. As photosynthesis does not resume this early, the starch must have been pre-stored. LDs had a different trend; they began accumulating from 60% RWC, increasing throughout desiccation. Their rapid disappearance post-rehydration suggests that stored lipids are mobilised to support membrane repair during early recovery. Recent studies have shown that LDs rapidly disappear following the cessation of stress [82–84], supporting the hypothesis that LD degradation contributes to recovery during the return to favourable conditions [28]. Transcriptomic data support this observation: starch metabolism declined during drying (Figure 6, box A), while lipid storage increased significantly (Figure 5, box A and Figure 8), indicating a metabolic shift from carbohydrate to lipid-based energy reserves.

Starch and lipid biosynthesis are competing pathways, with starch synthesis typically prevailing unless carbon supply exceeds utilisation [69]. The presence of large OBs among starch-containing chloroplasts at 40% RWC in both tissues may reflect the accumulation of excess fatty acids from thylakoid membrane breakdown. In *E. nindensis*, carbon allocation towards lipid reserves below 40% RWC likely represents an adaptive strategy for energy storage and homeostasis under extreme stress. This reliance on lipid metabolism, coupled with seed-related pathways, underscores the essential role of lipids in energy management and cellular protection during desiccation.

3.6. Desiccation-Induced Lipid Droplet Formation and Ultrastructural Remodelling

Across all NST conditions, the lipid profile remains dynamic, which is consistent with that of living, adaptable tissues. We observe a decrease in plastid lipids, notably MGDG and SQDG, together with a concomitant increase in TAG and some phospholipids, and this pattern is compatible with plastid membrane remodelling and diversion of acyl chains into OBs. TAGs, which are typically associated with developmental contexts (seeds; senescing leaves) or stress-induced regulation [83,85–87], increase under dehydrated conditions.

Plastid lipids also decline in the ST, but with a distinct signature. DGDG is more impacted by drying in ST than in NST.. TAG increases in ST, but with a different molecular species distribution than in NST. Strikingly, little additional change is observed across 25% RWC, air-dry, and rehydration, indicating that once tissues pass a severe dehydration threshold (ca. 25% RWC), the lipid state becomes largely static and persists through the air-dry state and early rehydration.

Ultrastructural observations support these findings. Uniform LDs congregating along the plasmalemma in NST at 40% and 25% RWCs contrasted with large, fused OBs in ST (Figures 2 and 3), likely resulting from the absence of oleosin expression (Figure 10). Oleosin is known to regulate LD formation and size, and its protein expression reflects translational activity at low RWCs [30]. ST failed to express oleosin adequately despite showing similar transcription trends to NST. This may imply that ST is incapable of post-transcriptional and post-translational modification, resulting in cellular disintegration.

Sequestering potentially toxic FFAs into LDs appears to enable desiccation tolerance in *E. nindensis*. LDs also provide critical mechanical stabilisation and energy-providing roles in NST. In contrast, ST cells fail to perform these protective functions and die. Together, these results indicate that desiccation tolerance in *E. nindensis* relies on preserving transcriptional and translational capacity in NST to maintain key processes such as protein expression, lipid metabolism, and appropriate metabolite compositions. Without this, stress-responsive genes seem to fail to be post-transcriptionally modified or translated into protective proteins in ST, leading to senescence.

In a recent metabolomics study on *E. nindensis*, it was shown that there were differentially abundant metabolites in NST, such as sucrose, glutamic acid, aspartic acid, proline, alpha-ketoglutaric acid, and allantoin, which are major drivers of desiccation tolerance, and furthermore aid the plant in recovering during rehydration. In contrast, metabolites which accumulated in ST—leucine, DL-isoleucine, L-valine, and cinnamic acid—indicated initiation of programmed cell death (PCD), leading to senescence [44]. However, these molecular players act downstream from the actual induction site of senescence by the environment; the upstream signalling behaviour leading up to it is still largely unknown. It is clear, however, that resurrection plants successfully suppress drought-induced senescence, which is a unique feature to maintain desiccation tolerance and successfully continue growth and development after rehydration. However, to accommodate this, there is a requirement for nutrients from senescing tissue. Senescence can be induced in leaves that are senescence-competent, i.e., those at more advanced stages of ageing. It is not yet clear how the plant regulates leaf-age timing but transcriptional networks, including those for stress-induced senescence, have been proposed [88]. It remains to be seen if the genes of these networks are also involved in the induction or repression of senescence in (drying) resurrection plants. Ultimately, such an understanding could facilitate senescence suppression in cereal crops under drought conditions.

4. Materials and Methods

4.1. Germination and Propagation

Seeds of *E. nindensis* were collected, with permission from the land manager, Mr P. Venter, from natural populations in Aggeneys, Northern Cape, South Africa (29°16′41.1″ S 19°00′25.4″ E). Surface-sterilised seeds were germinated on 4.4 g/L MS medium solidified with 0.8% agar in a growth chamber set at 24/15 °C day/night with a 16 h photoperiod (light intensity: 300 μmol m^{-2} s^{-1}) for 2–3 weeks until seedlings developed. After acclimation by removing lids, seedlings were transferred to trays with soil containing vermiculite and 0.4% Phostrogen All Purpose Plant Food (Bayer, Midrand, South Africa) supplemented with 2% calcium. Plants were watered three times a week, with a weekly light spray of Phostrogen (N16, P10, K24) nutrient solution. After one month, seedlings were transplanted into 10.5 cm pots containing a sand–vermiculite–perlite mixture (2:1:1) and grown for six months to accumulate sufficient biomass for the dehydration experiment. Two weeks before experimentation, plants were transferred into a Percival Intellus Control System (Model I-41LL, Percival Scientific, Inc., Perry, IA, USA) with a 16 h photoperiod using a light intensity of approximately 300 μmol.m-2.sec-1, with 24 °C day-time and 15 °C night-time temperatures for acclimation purposes, and they were kept well-hydrated under these controlled conditions. Plant positions were randomised throughout acclimation and experimentation.

4.2. Determination of Leaf Phenotype

Leaf phenotype was defined by position and maturity (Figure S7). NST comprised young, fully mature (unrolled) leaves positioned highest on the tiller, while ST comprised the lowest (oldest) leaves. To ensure leaf age uniformity, plants underwent a prior desiccation and rehydration cycle six weeks before the experiment. All dead leaves were pruned prior to stress treatment initiation to synchronise ST age and minimise age-dependent senescence. Senescent leaf tips were pruned. All subsequent physiological, ultrastructural, lipidomic, and transcriptomic analyses were conducted on these plants.

4.3. Desiccation and Rehydration Treatment

Desiccation was induced by withholding water until the leaves reached an air-dry desiccated state (RWC < 10%, AD). Tissues were collected when fully hydrated (100% RWC), mild (60% RWC), moderate (40% RWC), severe (<25% RWC) dehydration, desiccation (air-dry, <10%, AD) and 12 h post-rehydration (12 h) after thoroughly watering the plants. Due to financial restrictions, two sampling points had to be eliminated for the transcriptomic studies. We thus chose not to sample ST leaves at 40% RWC, which is just above the more critical threshold of below 30% RWC, and ST 12 h rehydration, as these tissues are effectively dead [12]. We also extended the period of rehydration to 24 h for the lipidomics study.

A minimum of three individual plants were used as biological replicates per RWC category. From each plant, approximately three leaves were collected from both NST and ST tissue types. Leaves were immediately weighed (fresh weight, accurate to 10 μg), then placed in microtubes filled with water and kept at 4 °C in the dark overnight. After blotting to remove surface moisture, the turgid weight was recorded. Samples were then sealed in labelled foil packets, oven-dried at 70 °C for 48 h and weighed to obtain dry weight. RWC was calculated gravimetrically according to Barr and Weatherley [89]:

$$\text{RWC (\%)} = ((\text{fresh weight} - \text{dry weight})/(\text{turgor weight} - \text{dry weight})) \times 100 \quad (1)$$

Plants were left air-dry for two weeks before rehydration. Individual plants with leaf discoloration, stress symptoms, or significantly different biomass were excluded. Sampling occurred at one-hour post-artificial dawn at each RWC stage, guided by physiological changes like leaf curling and chlorophyll loss [35]. Tissue samples from leaves with similar RWCs (±10%) were pooled into discrete RWC bins, with a minimum of three replicates per stage. Samples were either snap-frozen in liquid nitrogen and stored at −80 °C or preserved for microscopy. A two-way ANOVA was conducted to determine the difference between RWC per tissue type over time.

4.4. Transmission Electron Microscopy

Leaf segments (~4 mm^2) from three plants at each RWC stage underwent fixation and processing following Cooper and Farrant [90], with slight modifications. Tissues were fixed in 2.5% glutaraldehyde in 0.1 M phosphate buffer (pH 7.4) containing 0.5% caffeine, post-fixed in 1% osmium tetroxide, and dehydrated in an ethanol gradient. Samples were sequentially immersed overnight in 50%, 75%, and 87.5% Spurr's resin solutions (diluted with acetone) at 4 °C, then embedded in 100% Spurr's resin. Resin blocks were polymerized at 60 °C overnight, with the adaxial leaf surface oriented upwards. Cross-sections were cut using a Reichert Ultracut-S microtome (Leica Microsystems, Wetzlar, Germany), stained with 2% uranyl acetate for 10 min, followed by 1% lead citrate [91]. Ultrastructural changes were examined using an FEI Tecnai T20 transmission electron microscope (FEI Company, Hillsboro, OR, USA).

4.5. RNA Extraction and Transcriptomics

Total RNA was extracted from 50 to 100 mg of leaf tissue using TRI Reagent® (Sigma-Aldrich, Merck, Cat. No. T9424, St. Louis, MO, USA) per the manufacturer's protocol. Three replicates were used. RNA quality was assessed via Q-bit Flurometer (Thermo Fisher Scientific, Waltham, MA, USA) and Agilent 2100 Bioanalyzer (Agilent Technologies, Santa Clara, CA, USA). Libraries for RNA sequencing and RT-qPCR were prepared with barcoding for paired-end (150 nucleotide) and single-end (75 nucleotide) reads, sequenced on an Illumina HiSeq 2000 platform (Illumina, San Diego, CA, USA). Read quality was assessed using FASTQC v0.11.5, and adapter trimming was performed with Trimmomatic v0.36 [92].

Reads were aligned to the *E. nindensis* reference genome [64] using HISAT2 v2.1.0, and transcript assembly and quantification were conducted with StringTie v1.3.3 as part of the Tuxedo pipeline [93]. EdgeR normalised gene counts, and transcript abundances were $\log_2$-transformed [94]. Genes were classified as differentially expressed when $\log_2$ fold change exceeded ±2.0, with a false discovery rate (FDR; $q \leq 0.05$, [38]) applied relative to the control (100% RWC, NST).

4.6. Annotation and Quality Control

To maximise annotation confidence, multiple tools were employed. InterProScan v5 [95] identified conserved domains, Swissprot, GO, and BLAST hits, using an e-value threshold $<1 \times 10^{-10}$. Sequences were also aligned to *Arabidopsis thaliana* using BLAST+ v2.6.0 to identify homologs, with annotations assigned when more than ten *E. nindensis* reads mapped to the same *Arabidopsis* gene. Mercator v4.0 [96] was used to generate functional annotations and GO terms to analyse gene functions and biological processes [97,98]. Assembly completeness was evaluated with BUSCO v3.0 using the Embryophyta database [99].

4.7. Functional Enrichment Analysis of Co-Expressed Genes

DEGs that were common or exclusive to NST and ST were identified using Venny (v2.1; [100]). Heatmaps were generated in Genesis (v1.8.1; [101]) to visualise expression profile similarities among DEGs. Only NST rehydration (12 h) was sequenced due to financial constraints and the absence of viable tissue in ST. Consequently, NST rehydration was analysed independently.

GO enrichment analysis was performed using the Biological Networks Gene Ontology (BiNGO) tool in Cytoscape (v3.5.1) to visualise overrepresented GO categories in each sample [102]. BiNGO identifies statistically enriched GO terms based on the transcriptome-wide GO composition and applies hypergeometric tests with false discovery rate (FDR) correction (<0.05) for multiple testing.

To reduce redundancy arising from the hierarchical nature of GO terms, REViGO [103] was used to cluster semantically similar terms and eliminate redundant categories. The degree of over-representation was quantified as the ratio of observed to expected gene counts. A similarity index of 0.7 was applied, and redundant categories were filtered using a dispensability threshold of $p < 0.05$ unless otherwise specified.

4.8. Lipidomic Profiling and Analysis

Lipids were isolated using the standard Bligh and Dyer procedure [104] with a solvent mixture of methanol, chloroform, and water (2:1:0.8, $v/v/v$). This and subsequent analyses were repeated thrice. After phase separation, the organic layer was collected, evaporated under a gentle stream of nitrogen, and stored at −20 °C until further use. Prior to analysis, dried extracts were resuspended in 100 µL of isopropanol. Lipidomic analysis was performed according to [105]. Briefly, chromatographic separation was performed on a C18 Waters Acquity analytical column (100 × 2.4 mm, 1.7 µm) and a CSH C18 VanGuard precolumn (5 × 2.1 mm, 1.7 µm), maintained at 65 °C. The mobile phase consisted of solvent A (60:40 acetonitrile/water, v/v) and solvent B (90:10 isopropanol/acetonitrile, v/v). For analyses in positive-ion mode, solvents were supplemented with 10 mM ammonium formate and 0.1% formic acid, whereas in negative-ion mode they contained 10 mM ammonium acetate and 0.1% acetic acid. A constant flow of 0.6 mL/min was applied, and 10 µL of sample were injected for each run. The LC system was coupled to a quadrupole time-of-flight (QTOF) mass spectrometer (Agilent 6538, Agilent Technologies, Santa Clara, CA, USA) equipped with dual electrospray ionisation. The source was operated at 350 °C with a fragmentor voltage of 150 V and a skimmer voltage of 65 V. Data were acquired in full-scan mode across an m/z range of 100–1700 at a rate of two spectra per second. The most abundant precursor ions were selected for fragmentation at 35 eV, and MS/MS scans were performed on the six most intense ions. Continuous mass calibration was achieved using two reference ions (m/z 121.0509 and 922.0098 in positive mode; m/z 112.9856 and 1033.9881 in negative mode). Instrument parameters, data acquisition, and initial processing were controlled with MassHunter B.07 software (Agilent Technologies). Raw data files were converted from the native *.d format to *.mzML using MSConvert and processed with MS-DIAL (v5.5) [106]. Data preprocessing steps included baseline correction, peak detection, alignment, gap filling, and adduct assignment. Lipid annotation was based on accurate mass, retention time, and MS/MS spectra comparison. Potential duplicates, ghost peaks were filtered by the MSCleanR workflow implemented in MS-DIAL. Data were filtered based on a minimum blank ratio of 0.8, removed ghost peaks and inaccurate masses, and applied a maximum deviation threshold of 50. Curated lipidomics data from both positive (+) and negative (−) ionisation modes were combined into a single dataset. For Principal Component Analysis (PCA), the data were standardised using the *StandardScaler* module from *scikit-learn*, which centres each feature (metabolite) to a mean of 0 and scales it to unit variance (standard deviation of 1). The standardised data were then subjected to PCA to reduce dimensionality and visualise sample clustering. The first two principal components were plotted, with 95% confidence ellipses drawn around each condition group. Additionally, normalised (Z-scored) data were visualised as a heatmap using *seaborn*, with the order of conditions preserved as specified. The nomenclature used for lipid classes identified was according to Pauling et al. [107].

4.9. Western Blot Analysis

Total protein was extracted from the organic phase of TRI Reagent® (Sigma-Aldrich, Merck, Cat. No. T9424) treated tissue samples (5 mg DW), following the manufacturer's protocol with modifications to ensure protein purity. Precipitated protein was washed with 0.1 M ammonium acetate in methanol and acetone, air-dried, and resuspended in 2% SDS. Protein concentration was quantified using the Micro BCA™ Protein Assay Kit (Thermo Fisher Scientific, MA, USA) against a BSA standard curve.

For SDS-PAGE, 40 µg of total protein per sample was denatured at 95 °C for 5 min in Lämmli sample buffer and resolved on 15% resolving gels with a 6% stacking gel. Proteins were transferred overnight to PVDF membranes (0.22 µm pore size; Durapore®, St. Paul, MN, USA) at 40 V and 4 °C in transfer buffer following Towbin et al. [108]. Membranes were briefly stained with Ponceau S to confirm equal protein loading.

Membranes were blocked in 5% skim milk in TBS-T (0.2% Tween-20) and incubated overnight at 4 °C with a rabbit polyclonal anti-oleosin antibody (OLE1, AT4G25140; PhytoAB, #PHY0954S) at 1:1000. The expected band size is ca. 19 kDa, and cross-reactivity with *E. nindensis* homologues was confirmed by the manufacturer. After washing, membranes were incubated with HRP-conjugated goat anti-rabbit IgG (Sigma-Aldrich) at 1:2000 for 2 h at room temperature. Signal detection was performed using WesternBright ECL (Advansta, CA, USA) and imaged with a ChemiDoc system (Bio-Rad, CA, USA). Three biological replicates, each with three technical replicates, were used.

Supplementary Materials: The following supporting information can be downloaded at https://www.mdpi.com/article/10.3390/plants14213360/s1, Video S1: Time-lapse of desiccation and recovery in *Eragrostis nindensis*, https://www.youtube.com/watch?v=gV6_CJXM0sw (accessed on 26 September 2019); Figure S1: Ultrastructural changes in mesophyll cells of NST of *Eragrostis nindensis* during dehydration and rehydration; Figure S2: Ultrastructural changes in mesophyll cells of ST of *Eragrostis nindensis* during dehydration and rehydration. Figure S3: Photosynthesis as measured by Fv/Fm. Figure S4: Expression profiles of LEAs, dehydrins (DHNs), seed maturation proteins (SMPs), and heat shock proteins (HSPs) during drying and rehydration in *Eragrostis nindensis*. Figure S5: Enriched Gene Ontology (GO) categories during drying and rehydration in *Eragrostis nindensis*. Figure S6: Lipid droplet dynamics differ between NST and ST during drying and rehydration in *Eragrostis nindensis*. Figure S7: Leaf types sampled for physiological and molecular analyses in *Eragrostis nindensis*. Table S1: Transcriptome (expressed as differential gene expression) in *Eragrostis nindensis* during desiccation and rehydration; Table S2: Protein synthesis-related transcriptome changes in *Eragrostis nindensis* during desiccation and rehydration; Table S3: Protein degradation-related transcriptome changes in *Eragrostis nindensis* during drying and rehydration; Table S4: Post-transcriptional transcriptome changes in *Eragrostis nindensis* during desiccation and rehydration; Table S5. RNA processing- and translation-related DEGs exclusively accumulating in NST during drying. Table S6. TAG species.

Author Contributions: J.M.F. conceptualised the entire study. B.W. and S.M. conceptualised the transcriptomics of NST and ST for insights into senescence. E.R. conceptualised the lipidomic studies. C.F.M. did the work, with the exception of the lipidomic analysis, and wrote the paper. S.A. conducted lipidomic studies. H.W.M.H. gave input into interpretation. All authors have read and agreed to the published version of the manuscript.

Funding: This research was funded by the National Research Foundation of South Africa, via funding given to JMF, grant number 98604. SGM funded the costs of transcriptomics via 290266 QUT Research Capacity Development Account.

Institutional Review Board Statement: Not applicable.

Data Availability Statement: The original RNA-Seq data presented in the study are openly available in the Short Reads Archive at http://www.ncbi.nlm.nih.gov/bioproject/1310038 (submitted on 16 September 2025).

Acknowledgments: We thank Keren Cooper, Department of Molecular and Cell Biology at the University of Cape Town for her assistance in sectioning samples for TEM analysis.

Conflicts of Interest: The authors declare no conflicts of interest.

Abbreviations

The following abbreviations are used in this manuscript:

AD	Air-dry
AWC	Absolute water content
BiNGO	Biological networks gene ontology
BS	Bundle sheath
DEGs	Differentially expressed genes
ELIPS	Early light inducible proteins
FA	Fatty acid
FDR	False discovery rate
FFA	Free fatty acid
GO	Gene ontology
HSP	Heat shock protein
kDa	Kilodaltons
LDs	Lipid droplets
LEA	Late embryogenesis abundant
NST	Non-senescent leaf tissues
OB	Osmophilic body
P-bodies	Processing bodies
ROS	Reactive oxygen species

RWC	Relative water content
SMP	Seed maturation proteins
snRNP	Small nuclear ribonucleoprotein family protein
ST	Senescent leaf tissues
TAG	Triacylglycerol
TEM	Transmission electron micrographs/microscopy

References

1. Zia, A.; Walker, B.J.; Oung, H.M.O.; Charuvi, D.; Jahns, P.; Cousins, A.B.; Farrant, J.M.; Reich, Z.; Kirchhoff, H. Protection of the Photosynthetic Apparatus against Dehydration Stress in the Resurrection Plant *Craterostigma pumilum*. *Plant J.* **2016**, *87*, 664–680. [CrossRef]
2. Charuvi, D.; Nevo, R.; Aviv-Sharon, E.; Gal, A.; Kiss, V.; Shimoni, E.; Farrant, J.M.; Kirchhoff, H.; Reich, Z. Chloroplast Breakdown during Dehydration of a Homoiochlorophyllous Resurrection Plant Proceeds via Senescence-like Processes. *Environ. Exp. Bot.* **2019**, *157*, 100–111. [CrossRef]
3. Oliver, M.J.; Farrant, J.M.; Hilhorst, H.W.M.; Mundree, S.; Williams, B.; Bewley, J.D. Desiccation Tolerance: Avoiding Cellular Damage During Drying and Rehydration. *Annu. Rev. Plant Biol.* **2020**, *71*, 435–460. [CrossRef] [PubMed]
4. Farrant, J.M. A Comparison of Mechanisms of Desiccation Tolerance among Three Angiosperm Resurrection Plant Species. *Plant Ecol.* **2000**, *151*, 29–39. [CrossRef]
5. Willigen, C.V.; Pammenter, N.W.; Mundree, S.G.; Farrant, J.M. Mechanical Stabilization of Desiccated Vegetative Tissues of the Resurrection Grass *Eragrostis nindensis*: Does a TIP 3;1 and/or Compartmentalization of Subcellular Components and Metabolites Play a Role? *J. Exp. Bot.* **2004**, *55*, 651–661. [CrossRef]
6. Farrant, J.M.; Cooper, K.; Hilgart, A.; Abdalla, K.O.; Bentley, J.; Thomson, J.A.; Dace, H.J.W.; Peton, N.; Mundree, S.G.; Rafudeen, M.S. A Molecular Physiological Review of Vegetative Desiccation Tolerance in the Resurrection Plant *Xerophyta viscosa* (Baker). *Planta* **2015**, *242*, 407–426. [CrossRef] [PubMed]
7. Giarola, V.; Hou, Q.; Bartels, D. Angiosperm Plant Desiccation Tolerance: Hints from Transcriptomics and Genome Sequencing. *Trends Plant Sci.* **2017**, *22*, 705–717. [CrossRef]
8. Farrant, J.M.; Hilhorst, H. Crops for Dry Environments. *Curr. Opin. Biotechnol.* **2022**, *74*, 84–91. [CrossRef]
9. Dace, H.J.W. Metabolic Aspects of Vegetative Desiccation Tolerance. Ph.D. Thesis, Wageningen University, Wageningen, The Netherlands, 2024.
10. Shumayla; Alejo-Jacuinde, G.; Silva-Villatoro, P.; Nwoko, C.L.; Oliver, M.J.; Herrera-Estrella, L. The Promise of Resurrection Plants in Enhancing Crop Tolerance to Water Scarcity. *Philos. Trans. R. Soc. B Biol. Sci.* **2025**, *380*, 20240231. [CrossRef] [PubMed]
11. Tuba, Z.; Protor, M.C.F.; Csintalan, Z. Ecophysiological Responses of Homoiochlorophyllous and Poikilochlorophyllous Desiccation Tolerant Plants: A Comparison and an Ecological Perspective. *Plant Growth Regul.* **1998**, *24*, 211–217. [CrossRef]
12. Willigen, C.V.; Pammenter, N.W.; Mundree, S.; Farrant, J.M. Some Physiological Comparisons between the Resurrection Grass, Eragrostis Nindensis, and the Related Desiccation-Sensitive Species, *E. curvula*. *Plant Growth Regul.* **2001**, *35*, 121–129. [CrossRef]
13. Willigen, C.V.; Farrant, J.M.; Pammenter, N.W. Anomalous Pressure Volume Curves of Resurrection Plants Do Not Suggest Negative Turgor. *Ann. Bot.* **2001**, *88*, 537–543. [CrossRef]
14. Willigen, C.V.; Pammenter, N.W.; Jaffer, M.A.; Mundree, S.G.; Farrant, J.M. An Ultrastructural Study Using Anhydrous Fixation of *Eragrostis nindensis*, a Resurrection Grass with Both Desiccation-Tolerant and -Sensitive Tissues. *Funct. Plant Biol.* **2003**, *30*, 281–290. [CrossRef]
15. Sah, S.K.; Fan, J.; Blanford, J.; Shanklin, J.; Xu, C. Physiological Functions of Phospholipid:Diacylglycerol Acyltransferases. *Plant Cell Physiol.* **2024**, *65*, 863–871. [CrossRef]
16. Lu, J.; Xu, Y.; Wang, J.; Singer, S.D.; Chen, G. The Role of Triacylglycerol in Plant Stress Response. *Plants* **2020**, *9*, 472. [CrossRef]
17. Balogh, G.; Péter, M.; Glatz, A.; Gombos, I.; Török, Z.; Horváth, I.; Harwood, J.L.; Vígh, L. Key Role of Lipids in Heat Stress Management. *FEBS Lett.* **2013**, *587*, 1970–1980. [CrossRef] [PubMed]
18. Tshabuse, F.; Farrant, J.M.; Humbert, L.; Moura, D.; Rainteau, D.; Espinasse, C.; Idrissi, A.; Merlier, F.; Acket, S.; Rafudeen, M.S.; et al. Glycerolipid Analysis during Desiccation and Recovery of the Resurrection Plant *Xerophyta humilis* (Bak) Dur and Schinz. *Plant Cell Environ.* **2018**, *41*, 533–547. [CrossRef] [PubMed]
19. Gasulla, F.; Dorp, K.; Dombrink, I.; Za, U.; Gisch, N.; Do, P.; Bartels, D. The Role of Lipid Metabolism in the Acquisition of Desiccation Tolerance in *Craterostigma plantagineum*: A Comparative Approach. *Plant J.* **2013**, *75*, 726–741. [CrossRef]
20. Navari-Izzo, F.; Quartacci, M.F.; Pinzino, C.; Rascio, N.; Vazzana, C.; Sgherri, C.L.M. Protein Dynamics in Thylakoids of the Desiccation-Tolerant Plant *Boea hygroscopica* during Dehydration and Rehydration. *Plant Physiol.* **2000**, *124*, 1427–1436. [CrossRef] [PubMed]

21. Li, A.; Wang, D.; Yu, B.; Yu, X.; Li, W. Maintenance or Collapse: Responses of Extraplastidic Membrane Lipid Composition to Desiccation in the Resurrection Plant *Paraisometrum mileense*. *PLoS ONE* **2014**, *9*, e103430. [CrossRef]
22. Mihailova, G.; Christov, N.K.; Sárvári, É.; Solti, Á.; Hembrom, R.; Solymosi, K.; Keresztes, Á.; Velitchkova, M.; Popova, A.V.; Simova-Stoilova, L.; et al. Reactivation of the Photosynthetic Apparatus of Resurrection Plant *Haberlea rhodopensis* during the Early Phase of Recovery from Drought- and Freezing-Induced Desiccation. *Plants* **2022**, *11*, 2185. [CrossRef] [PubMed]
23. Shivaraj, Y.N.; Barbara, P.; Gugi, B.; Driouich, A.; Govind, S.R.; Devaraja, A.; Kambalagere, Y. Perspectives on Structural, Physiological, Cellular, and Molecular Responses to Desiccation in Resurrection Plants. *Scientifica* **2018**, *2018*, 9464592. [CrossRef]
24. Wu, T.; Yu, L.; Zhang, Y.; Liu, J. Characterization of Fatty Acid Desaturases Reveals Stress-Induced Synthesis of C18 Unsaturated Fatty Acids Enriched in Triacylglycerol in the Oleaginous Alga *Chromochloris zofingiensis*. *Biotechnol. Biofuels* **2021**, *14*, 184. [CrossRef]
25. Yang, Z.; Ohlrogge, J.B. Turnover of Fatty Acids during Natural Senescence of Arabidopsis, *Brachypodium*, and Switchgrass and in Arabidopsis β-Oxidation Mutants. *Plant Physiol.* **2009**, *150*, 1981–1989. [CrossRef]
26. He, M.; Ding, N.-Z. Plant Unsaturated Fatty Acids: Multiple Roles in Stress Response. *Front. Plant Sci.* **2020**, *11*, 562785. [CrossRef]
27. Hoekstra, F.A.; Golovina, E.A.; Buitink, J. Mechanisms of Plant Desiccation Tolerance. *Trends Plant Sci.* **2001**, *6*, 431–438. [CrossRef] [PubMed]
28. Bouchnak, I.; Coulon, D.; Salis, V.; D'Andréa, S.; Bréhélin, C. Lipid Droplets Are Versatile Organelles Involved in Plant Development and Plant Response to Environmental Changes. *Front. Plant Sci.* **2023**, *14*, 1193905. [CrossRef]
29. Yang, Y.; Benning, C. Functions of Triacylglycerols during Plant Development and Stress. *Curr. Opin. Biotechnol.* **2018**, *49*, 191–198. [CrossRef] [PubMed]
30. Siloto, R.M.P.; Findlay, K.; Lopez-Villalobos, A.; Yeung, E.C.; Nykiforruk, C.L.; Moloney, M.M. The Accumulation of Oleosins Determines the Size of Seed Oilbodies in Arabidopsis. *Plant Cell Online* **2006**, *18*, 1961–1974. [CrossRef]
31. Fang, Y.; Zhu, R.L.; Mishler, B.D. Evolution of Oleosin in Land Plants. *PLoS ONE* **2014**, *9*, e103806. [CrossRef]
32. Hsieh, K.; Huang, A.H. Endoplasmic Reticulum, Oleosins, and Oils in Seeds and Tapetum Cells. *Plant Physiol.* **2004**, *136*, 3427–3434. [CrossRef]
33. Xu, C.; Shanklin, J. Triacylglycerol Metabolism, Function, and Accumulation in Plant Vegetative Tissues. *Annu. Rev. Plant Biol.* **2016**, *67*, 179–206. [CrossRef]
34. Pauling, J.K.; Hermansson, M.; Hartler, J.; Christiansen, K.; Gallego, S.F.; Peng, B.; Ahrends, R.; Ejsing, C.S. Proposal for a Common Nomenclature for Fragment Ions in Mass Spectra of Lipids. *PLoS ONE* **2017**, *12*, e0188394. [CrossRef]
35. Baluku, E.; Van Der Pas, L.; Hilhorst, H.W.M.; Farrant, J.M. Metabolite Profiling of the Resurrection Grass *Eragrostis nindensis* During Desiccation and Recovery. *Plants* **2025**, *14*, 531. [CrossRef]
36. Rinne, P.L.H.; Welling, A.; Vahala, J.; Ripel, L.; Ruonala, R.; Kangasjärvi, J.; Schoot, C. van der Chilling of Dormant Buds Hyperinduces FLOWERING LOCUS T and Recruits GA-Inducible 1,3-β-Glucanases to Reopen Signal Conduits and Release Dormancy in Populus. *Plant Cell* **2011**, *23*, 130–146. [CrossRef]
37. Tymms, M.J.; Gaff, D.F.; Hallam, N.D. Protein Synthesis in the Desiccation Tolerant Angiosperm *Xerophyta villosa* during Dehydration. *J. Exp. Biol.* **1982**, *33*, 332–343.
38. Benjamini, Y.; Hochberg, Y. Controlling the False Discovery Rate: A Practical and Powerful Approach to Multiple Testing. *J. R. Stat. Soc.* **1995**, *57*, 289–300. [CrossRef]
39. Challabathula, D.; Puthur, J.T.; Bartels, D. Surviving Metabolic Arrest: Photosynthesis during Desiccation and Rehydration in Resurrection Plants. *Ann. N. Y. Acad. Sci.* **2016**, *1365*, 89–99. [CrossRef]
40. Costa, M.C.D.; Cooper, K.; Hilhorst, H.W.M.; Farrant, J.M. Orthodox Seeds and Resurrection Plants: Two of a Kind? *Plant Physiol.* **2017**, *175*, 589–599. [CrossRef]
41. Farrant, J.M.; Moore, J.P. Programming Desiccation-Tolerance: From Plants to Seeds to Resurrection Plants. *Curr. Opin. Plant Biol.* **2011**, *14*, 340–345. [CrossRef]
42. Christ, B.; Egert, A.; Süssenbacher, I.; Kräutler, B.; Bartels, D.; Peters, S.; Hörtensteiner, S. Water Deficit Induces Chlorophyll Degradation via the "PAO/Phyllobilin" Pathway in Leaves of Homoio- (*Craterostigma pumilum*) and Poikilochlorophyllous (*Xerophyta viscosa*) Resurrection Plants. *Plant Cell Environ.* **2014**, *37*, 2521–2531. [CrossRef]
43. Artur, M.A.S.; Costa, M.C.D.; Farrant, J.M.; Hilhorst, H.W.M. Genome Level Responses to the Environment: Plant Desiccation Tolerance. *Emerg. Top. Life Sci.* **2019**, *3*, 153–163. [CrossRef]
44. Mitra, R.; Wu, K.; Lee, C.; Bardwell, J.C.A. ATP-Independent Chaperones. *Annu. Rev. Biophys.* **2022**, *51*, 409–429. [CrossRef]
45. Chapman, K.D.; Dyer, J.M.; Mullen, R.T. Biogenesis and Functions of Lipid Droplets in Plants. *J. Lipid Res.* **2012**, *53*, 215–226. [CrossRef]
46. Yu, L.; Fan, J.; Zhou, C.; Xu, C. Sterols Are Required for the Coordinated Assembly of Lipid Droplets in Developing Seeds. *Nat. Commun.* **2021**, *12*, 5598. [CrossRef]
47. Mihailova, G.; Vasileva, I.; Gigova, L.; Gesheva, E.; Simova-Stoilova, L.; Georgieva, K. Antioxidant Defense during Recovery of Resurrection Plant *Haberlea rhodopensis* from Drought- and Freezing-Induced Desiccation. *Plants* **2022**, *11*, 175. [CrossRef]

48. Griffiths, C.A.; Gaff, D.F.; Neale, A.D. Drying without Senescence in Resurrection Plants. *Front. Plant Sci.* **2014**, *5*, 36. [CrossRef]
49. VanBuren, R.; Wai, C.M.; Pardo, J.; Giarola, V.; Ambrosini, S.; Song, X.; Bartels, D.; Wai, C.M.; Pardo, J.; Giarola, V.; et al. Desiccation Tolerance Evolved through Gene Duplication and Network Rewiring in *Lindernia*. *Plant Cell* **2018**, *30*, 2943–2958. [CrossRef]
50. Bewley, J.D.; Bradford, K.J.; Hilhorst, H.W.M.; Nonogaki, H. Synthesis of Storage Reserves. In *Seeds: Physiology of Development, Germination and Dormancy*, 3rd ed.; Springer New York: New York, NY, USA, 2013; pp. 85–131. ISBN 978-1-4614-4693-4.
51. Dace, H.J.W.; Reus, R.; Ricco, C.R.; Hall, R.; Farrant, J.M.; Hilhorst, H.W.M. A Horizontal View of Primary Metabolomes in Vegetative Desiccation Tolerance. *Physiol. Plant.* **2023**, *175*, e14109. [CrossRef]
52. Chiabrando, D.; Vinchi, F.; Fiorito, V.; Mercurio, S.; Tolosano, E. Heme in Pathophysiology: A Matter of Scavenging, Metabolism and Trafficking across Cell Membranes. *Front. Pharmacol.* **2014**, *5*, 61. [CrossRef]
53. Chen, X.; Li, J.; Kang, R.; Klionsky, D.J.; Tang, D. Ferroptosis: Machinery and Regulation. *Autophagy* **2020**, *17*, 2054–2081. [CrossRef]
54. Juan-Cabot, A.; Carrillo, L.; Fullana-Pericàs, M.; Galmés, J.; Medina, J.; Conesa, M.À. Mediterranean Tomato Landraces Exhibit Genotype-Specific Transcriptomic Responses to Water Stress. 2025. Available online: https://ssrn.com/abstract=5188895 (accessed on 10 July 2025).
55. VanBuren, R.; Wai, J.; Zhang, Q.; Song, X.; Edger, P.P.; Bryant, D.; Michael, T.P.; Mockler, T.C.; Bartels, D. Seed Desiccation Mechanisms Co-Opted for Vegetative Desiccation in the Resurrection Grass *Oropetium thomeaum*. *Plant Cell Environ.* **2017**, *40*, 2292–2306. [CrossRef]
56. Wang, W.; Vinocur, B.; Shoseyov, O.; Altman, A. Role of Plant Heat-Shock Proteins and Molecular Chaperones in the Abiotic Stress Response. *Trends Plant Sci.* **2004**, *9*, 244–252. [CrossRef]
57. Zhang, K.; Niu, S.; Di, D.; Shi, L.; Liu, D.; Cao, X.; Miao, H.; Wang, X.; Han, C.; Yu, J.; et al. Selection of Reference Genes for Gene Expression Studies in Virus-Infected Monocots Using Quantitative Real-Time PCR. *J. Biotechnol.* **2013**, *168*, 7–14. [CrossRef]
58. Shih, M.D.; Lin, S.C.; Hsieh, J.S.; Tsou, C.H.; Chow, T.Y.; Lin, T.P.; Hsing, Y.I.C. Gene Cloning and Characterization of a Soybean (*Glycine max* L.) LEA Protein, GmPM16. *Plant Mol. Biol.* **2004**, *56*, 689–703. [CrossRef]
59. Tunnacliffe, A.; Wise, M.J. The Continuing Conundrum of the LEA Proteins. *Naturwissenschaften* **2007**, *94*, 791–812. [CrossRef]
60. Leprince, O.; Buitink, J. Desiccation Tolerance: From Genomics to the Field. *Plant Sci.* **2010**, *179*, 554–564. [CrossRef]
61. Covarrubias, A.A.; Cuevas-Velazquez, C.L.; Romero-Pérez, P.S.; Rendón-Luna, D.F.; Chater, C.C.C. Structural Disorder in Plant Proteins: Where Plasticity Meets Sessility. *Cell. Mol. Life Sci.* **2017**, *74*, 3119–3147. [CrossRef]
62. Farrant, J.M.; Cooper, K.; Dace, H.J.W.; Bentley, J.; Hilgart, A. Desiccation Tolerance. In *Plant Stress Physiology*; CABI Publishing: Wallingford, UK, 2017; pp. 238–265. ISBN 978-1-84593-995-3.
63. Artur, M.A.S.; Zhao, T.; Ligterink, W.; Schranz, M.E.; Hilhorst, H.W.M. Dissecting the Genomic Diversification of LATE EMBRYOGENESIS ABUNDANT (LEA) Protein Gene Families in Plants. *Genome Biol. Evol.* **2018**, *11*, 459–471. [CrossRef]
64. Pardo, J.; Wai, C.M.; Chay, H.; Madden, C.F.; Hilhorst, H.W.M.; Farrant, J.M.; VanBuren, R. Intertwined Signatures of Desiccation and Drought Tolerance in Grasses. *Proc. Natl. Acad. Sci. USA* **2020**, *117*, 10079–10088. [CrossRef]
65. Oliver, M.J. Influence of Protoplasmic Water Loss on the Control of Protein Synthesis in the Desiccation-Tolerant Moss Tortula Ruralis: Ramifications for a Repair-Based Mechanism of Desiccation Tolerance. *Plant Physiol.* **1991**, *97*, 1501–1511. [CrossRef]
66. Gaff, D.F.; Bartels, D.; Gaff, J.L. Changes in Gene Expression during Drying in a Desiccation-Tolerant Grass *Sporobolus stapfianus* and a Desiccation-Sensitive Grass *Sporobolus pyramidalis*. *Funct. Plant Biol.* **1997**, *24*, 617–622. [CrossRef]
67. Blomstedt, C.K.; Griffiths, C.A.; Fredericks, D.P.; Hamill, J.D.; Gaff, D.F.; Neale, A.D. The Resurrection Plant *Sporobolus stapfianus*: An Unlikely Model for Engineering Enhanced Plant Biomass? *Plant Growth Regul.* **2010**, *62*, 217–232. [CrossRef]
68. Wang, T.; Meng, K.; Zhu, Z.; Pan, L.; Okita, T.W.; Zhang, L.; Tian, L. The Regulatory Roles of RNA-Binding Proteins in Plant Salt Stress Response. *Plants* **2025**, *14*, 1402. [CrossRef]
69. Bartels, D.; Schneider, K.; Terstappen, G.; Piatkowski, D.; Salamini, F. Molecular Cloning of Abscisic Acid-Modulated Genes Which Are Induced during Desiccation of the Resurrection Plant *Craterostigma plantagineum*. *Planta* **1990**, *181*, 27–34. [CrossRef]
70. Jang, G.-J.; Yang, J.-Y.; Hsieh, H.-L.; Wu, S.-H. Processing Bodies Control the Selective Translation for Optimal Development of Arabidopsis Young Seedlings. *Proc. Natl. Acad. Sci. USA* **2019**, *116*, 6451–6456. [CrossRef]
71. Silva, A.T.; Ribone, P.A.; Chan, R.L.; Ligterink, W.; Hilhorst, H.W.M. A Predictive Coexpression Network Identifies Novel Genes Controlling the Seed-to-Seedling Phase Transition in *Arabidopsis thaliana*. *Plant Physiol.* **2016**, *170*, 2218–2231. [CrossRef]
72. Rajjou, L.; Gallardo, K.; Debeaujon, I.; Vandekerckhove, J.; Job, C.; Job, D. The Effect of A-Amanitin on the Arabidopsis Seed Proteome Highlights the Distinct Roles of Stored and Neosynthesized mRNAs during Germination. *Plant Physiol.* **2004**, *134*, 1598–1613. [CrossRef]
73. Bai, B.; Peviani, A.; Horst, S.; van der Horst, S.; Gamm, M.; Snel, B.; Bentsink, L.; Hanson, J. Extensive Translational Regulation during Seed Germination Revealed by Polysomal Profiling. *New Phytol.* **2017**, *214*, 233–244. [CrossRef]
74. Bremer, K.; Moyes, C.D. mRNA Degradation: An Underestimated Factor in Steady-State Transcript Levels of Cytochrome c Oxidase Subunits? *J. Exp. Biol.* **2014**, *217*, 2212–2220. [CrossRef]

75. Wang, C.; Schmich, F.; Srivatsa, S.; Weidner, J.; Beerenwinkel, N.; Spang, A. Context-Dependent Deposition and Regulation of mRNAs in P-Bodies. *eLife* **2018**, *7*, e29815. [CrossRef]
76. Wood, A.J.; Oliver, M.J. Translational Control in Plant Stress: The Formation of Messenger Ribonucleoprotein Particles (mRNPs) in Response to Desiccation of *Tortula ruralis* Gametophytes. *Plant J.* **1999**, *18*, 359–370. [CrossRef]
77. Blomstedt, C.K.; Griffiths, C.A.; Gaff, D.F.; Hamill, J.; Neale, A. Plant Desiccation Tolerance and Its Regulation in the Foliage of Resurrection "Flowering-Plant" Species. *Agronomy* **2018**, *8*, 146. [CrossRef]
78. Dace, H.; Sherwin, H.W.; Illing, N.; Farrant, J.M. Use of Metabolic Inhibitors to Elucidate Mechanisms of Recovery from Desiccation Stress in the Resurrection Plant *Xerophyta humilis*. *Plant Growth Regul.* **1998**, *24*, 171–177. [CrossRef]
79. Tuba, Z.; Lichtenthaler, H.K. *Plant Desiccation Tolerance*; Springer-Verlag: Berlin/Heidelberg, Germany, 2011; ISBN 978-0-8138-1263-2.
80. Wawer, I.; Golisz, A.; Sulkowska, A.; Kawa, D.; Kulik, A.; Kufel, J. mRNA Decapping and 5′-3′ Decay Contribute to the Regulation of ABA Signaling in *Arabidopsis thaliana*. *Front. Plant Sci.* **2018**, *9*, 312. [CrossRef]
81. Chantarachot, T.; Bailey-Serres, J. Polysomes, Stress Granules, and Processing Bodies: A Dynamic Triumvirate Controlling Cytoplasmic mRNA Fate and Function. *Plant Physiol.* **2018**, *176*, 254–269. [CrossRef]
82. Zienkiewicz, K.; Zienkiewicz, A. Degradation of Lipid Droplets in Plants and Algae—Right Time, Many Paths, One Goal. *Front. Plant Sci.* **2020**, *11*, 579019. [CrossRef]
83. Mueller, S.P.; Krause, D.M.; Mueller, M.J.; Fekete, A. Accumulation of Extra-Chloroplastic Triacylglycerols in Arabidopsis Seedlings during Heat Acclimation. *J. Exp. Bot.* **2015**, *66*, 4517–4526. [CrossRef]
84. Graef, M. Lipid Droplet-Mediated Lipid and Protein Homeostasis in Budding Yeast. *FEBS Lett.* **2018**, *592*, 1291–1303. [CrossRef]
85. Higashi, Y.; Saito, K. Lipidomic Studies of Membrane Glycerolipids in Plant Leaves under Heat Stress. *Prog. Lipid Res.* **2019**, *75*, 100990. [CrossRef]
86. Gidda, S.K.; Park, S.; Pyc, M.; Yurchenko, O.; Cai, Y.; Wu, P.; Andrews, D.W.; Chapman, K.D.; Dyer, J.M.; Mullen, R.T. Lipid Droplet-Associated Proteins (LDAPs) Are Required for the Dynamic Regulation of Neutral Lipid Compartmentation in Plant Cells. *Plant Physiol.* **2016**, *170*, 2052–2071. [CrossRef]
87. Dadras, A.; Fürst-Jansen, J.M.R.; Darienko, T.; Krone, D.; Scholz, P.; Sun, S.; Herrfurth, C.; Rieseberg, T.P.; Irisarri, I.; Steinkamp, R.; et al. Environmental Gradients Reveal Stress Hubs Pre-Dating Plant Terrestrialization. *Nat. Plants* **2023**, *9*, 1419–1438. [CrossRef]
88. Schippers, J.H. Transcriptional Networks in Leaf Senescence. *Curr. Opin. Plant Biol.* **2015**, *27*, 77–83. [CrossRef]
89. Barrs, H.; Weatherley, P. A Re-Examination of the Relative Turgidity Technique for Estimating Water Deficits in Leaves. *Aust. J. Biol. Sci.* **1962**, *15*, 413–428. [CrossRef]
90. Cooper, K.; Farrant, J.M. Recovery of the Resurrection Plant *Craterostigma wilmsii* from Desiccation: Protection versus Repair. *J. Exp. Bot.* **2002**, *53*, 1805–1813. [CrossRef]
91. Reynolds, E.S. The Use of Lead Citrate at High pH as an Electron-Opaque Stain in Electron Microscopy. *J. Cell Biol.* **1963**, *17*, 208–212. [CrossRef]
92. Bolger, A.M.; Lohse, M.; Usadel, B. Trimmomatic: A Flexible Trimmer for Illumina Sequence Data. *Bioinformatics* **2014**, *30*, 2114–2120. [CrossRef]
93. Pertea, M.; Kim, D.; Pertea, G.M.; Leek, J.T.; Salzberg, S.L. Transcript-Level Expression Analysis of RNA-Seq Experiments with HISAT, StringTie and Ballgown. *Nat. Protoc.* **2016**, *11*, 1650–1667. [CrossRef]
94. McCarthy, D.J.; Chen, Y.; Smyth, G.K. Differential Expression Analysis of Multifactor RNA-Seq Experiments with Respect to Biological Variation. *Nucleic Acids Res.* **2012**, *40*, 4288–4297. [CrossRef]
95. Jones, P.; Binns, D.; Chang, H.-Y.; Fraser, M.; Li, W.; McAnulla, C.; McWilliam, H.; Maslen, J.; Mitchell, A.; Nuka, G.; et al. InterProScan 5: Genome-Scale Protein Function Classification. *Bioinformatics* **2014**, *30*, 1236–1240. [CrossRef]
96. Schwacke, R.; Ponce-Soto, G.Y.; Krause, K.; Bolger, A.M.; Arsova, B.; Hallab, A.; Gruden, K.; Stitt, M.; Bolger, M.E.; Usadel, B. MapMan4: A Refined Protein Classification and Annotation Framework Applicable to Multi-Omics Data Analysis. *Mol. Plant* **2019**, *12*, 879–892. [CrossRef]
97. Ashburner, M.; Ball, C.A.; Blake, J.A.; Botstein, D.; Butler, H.; Cherry, J.M.; Davis, A.P.; Dolinski, K.; Dwight, S.S.; Eppig, J.T.; et al. Gene Ontology: Tool for the Unification of Biology. *Nat. Genet.* **2000**, *25*, 25–29. [CrossRef]
98. The Gene Ontology Consortium; Aleksander, S.A.; Balhoff, J.; Carbon, S.; Cherry, J.M.; Drabkin, H.J.; Ebert, D.; Feuermann, M.; Gaudet, P.; Harris, N.L.; et al. The Gene Ontology Knowledgebase in 2023. *Genetics* **2023**, *224*, iyad031. [CrossRef]
99. Simão, F.A.; Waterhouse, R.M.; Ioannidis, P.; Kriventseva, E.V.; Zdobnov, E.M. BUSCO: Assessing Genome Assembly and Annotation Completeness with Single-Copy Orthologs. *Bioinformatics* **2015**, *31*, 3210–3212. [CrossRef]
100. Oliveros, J. *Venny, v2.1*; Centro Nacional de Biotecnología (CNB-CSIC): Madrid, Spain, 2007.
101. Sturn, A.; Quackenbush, J.; Trajanoski, Z. Genesis: Cluster Analysis of Microarray Data. *Bioinformatics* **2002**, *18*, 207–208. [CrossRef]
102. Maere, S.; Heymans, K.; Kuiper, M. BiNGO: A Cytoscape Plugin to Assess Overrepresentation of Gene Ontology Categories in Biological Networks. *Bioinformatics* **2005**, *21*, 3448–3449. [CrossRef]

103. Supek, F.; Bošnjak, M.; Škunca, N.; Šmuc, T. Revigo Summarizes and Visualizes Long Lists of Gene Ontology Terms. *PLoS ONE* **2011**, *6*, e21800. [CrossRef]
104. Bligh, E.G.; Dyer, W.J. A rapid method of total lipid extraction and purification. *Can. J. Biochem. Physiol.* **1959**, *37*, 911–917. [CrossRef]
105. Cajka, T.; Fiehn, O. LC–MS-Based Lipidomics and Automated Identification of Lipids Using the LipidBlast In-Silico MS/MS Library. In *Lipidomics. Methods in Molecular Biology*; Bhattacharya, S., Ed.; Humana Press: New York, NY, USA, 2017; Volume 1609. [CrossRef]
106. Takeda, H.; Matsuzawa, Y.; Takeuchi, M.; Takahashi, M.; Nishida, K.; Harayama, T.; Todoroki, Y.; Shimizu, K.; Sakamoto, N.; Oka, T.; et al. MS-DIAL 5 multimodal mass spectrometry data mining unveils lipidome complexities. *Nat. Commun.* **2024**, *15*, 9903. [CrossRef]
107. Higashi, Y.; Okazaki, Y.; Myouga, F.; Shinozaki, K.; Saito, K. Landscape of the Lipidome and Transcriptome under Heat Stress in Arabidopsis Thaliana. *Sci. Rep.* **2015**, *5*, 10533. [CrossRef]
108. Towbin, H.; Staehelint, T.; Gordon, J. Electrophoretic Transfer of Proteins from Polyacrylamide Gels to Nitrocellulose Sheets: Procedure and Some Applications. *Proc. Natl. Acad. Sci. USA* **1979**, *76*, 4350–4354. [CrossRef] [PubMed]

plants

MDPI

Article

Genome-Wide Identification, Classification, Expression Analysis, and Screening of Drought and Heat Resistance-Related Candidates of the *Rboh* Gene Family in Wheat

Miyuan Cao [1,†], Yue Zhang [1,†], Xiaoxiao Zou [1], Huangping Yin [1], Yan Yin [1], Zeqi Li [1], Wenjun Xiao [1,2], Shucan Liu [1,2], Yongliang Li [1,2,3,*] and Xinhong Guo [1,2,*]

1 College of Biology, Hunan University, Changsha 410082, China; caomiyuan@hnu.edu.cn (M.C.); zhangyue_one@hnu.edu.cn (Y.Z.); xxzou@hnu.edu.cn (X.Z.); yinhuangping@hnu.edu.cn (H.Y.); yinyan91@foxmail.com (Y.Y.); lizeqi2023@hnu.edu.cn (Z.L.); xiaowj90@hnu.edu.cn (W.X.); liushucan@hnu.edu.cn (S.L.)

2 Chongqing Research Institute, Hunan University, Chongqing 401120, China

3 Northeast Institute of Geography and Agroecology, Chinese Academy of Sciences, Changchun 130102, China

* Correspondence: yongliangli1357@163.com (Y.L.); gxh@hnu.edu.cn (X.G.)

† These authors contributed equally to this work.

Abstract: Plant respiratory burst oxidase homologs (Rbohs) are key enzymes that produce reactive oxygen species (ROS), which serve as signaling molecules regulating plant growth and stress responses. In this study, 39 *TaRboh* genes (*TaRboh01*–*TaRboh39*) were identified. These genes were distributed unevenly among the wheat genome's fourteen chromosomes, with the exception of homoeologous group 2 and 7 and chromosomes 4A, as well as one unidentified linkage group (Un). TaRbohs were classified into ten distinct clades, each sharing similar motif compositions and gene structures. The promoter regions of *TaRboh*s contained *cis*-elements related to hormones, growth and development, and stresses. Furthermore, five *TaRboh* genes (*TaRboh26*, *TaRboh27*, *TaRboh31*, *TaRboh32*, and *TaRboh34*) exhibited strong evolutionary conservation. Additionally, a Ka/Ks analysis confirmed that purifying selection was the predominant force driving the evolution of these genes. Expression profiling and qPCR results further indicated differential expression patterns of *TaRboh* genes between heat and drought stresses. *TaRboh11*, *TaRboh20*, *TaRboh22*, *TaRboh24*, *TaRboh29*, and *TaRboh34* were significantly upregulated under multiple stress conditions, whereas *TaRboh30* was only elevated in response to drought stress. Collectively, our findings provide a systematic analysis of the wheat *Rboh* gene family and establish a theoretical framework for our future research on the role of *Rboh* genes in response to heat and drought stress.

Keywords: *Triticum aestivum*; *Rbohs*; abiotic stress; reactive oxygen species; expression patterns

1. Introduction

Respiratory burst oxidase homologs (Rbohs) are known as NADPH oxidases. They are pivotal in ROS production, which is the main source of these reactive molecules in plants [1,2]. Plant Rbohs, homologous to the gp91^{phox} catalytic subunit in mammalian phagocytic cells, are plasma membrane proteins featuring six conserved transmembrane helices, two heme groups, a C-terminal domain with FAD and NADPH, and two N-terminal EF-hand domains for calcium binding [3]. These structural features enable Rbohs to catalyze the production of ROS by using NADPH or NADH as electron donors, leading to the extracellular conversion of O_2 into superoxide (O_2^-), which is then converted into hydrogen peroxide (H_2O_2) by superoxide dismutase (SOD) [3,4]. The ROS generated by *Rbohs* act as key signaling molecules, regulating plant growth, development, and stress responses, underscoring the central role of *Rbohs* in plant regulation [2,3].

The first *Rboh* family gene in plants, *OsRbohA*, was identified in *Oryza sativa* L. in 1996 [1]. Subsequently, a large number of *Rboh* genes have been identified in different plant species. They included 10 in *Arabidopsis thaliana* (*AtRbohA − AtRbohJ*) [2], 20 in *Nicotiana Tabacum* (*NtabRbohA − NtabRbohT*) [5], 9 in rice (*OsRbohA − OsRbohI*) [6], 27 in *Gossypium barbadense* (*GbRboh1 − GbRboh27*) [7], and 15 in *Zea mays* (*ZmRBOH1*∼ *ZmRBOH15*) [8]. *Rboh* genes exhibit diverse tissue-specific transcription levels, providing insights into their roles in plant biology [9]. For example, three of those in *Arabidopsis*, (*AtRbohA*, *AtRbohG*, and *AtRbohI*) were expressed in roots, two (*AtRbohH* and *AtRbohJ*) were specifically expressed in pollen, and another two (*AtRbohD* and *AtRbohF*) were expressed in all tissues [2]. For those in rice, all but two (*OsRbohD* and *OsRbohI*) of the *OsRboh* genes were expressed in roots, leaves, shoots, and calli [10]. Additionally, plant *Rboh genes* are involved in various stress responses, including drought and heat. Overexpression of the *AtRbohI* gene in *Arabidopsis* enhanced the drought stress response [11]. *AtRbohB* and *AtRbohD* mediated signal transduction under heat stress, with *AtRbohD* also responded to drought stress [12]. *OsRbohA* participated in complex signal transduction processes and contributed to plant growth and drought resistance [6]. *OsRbohB* mediated ROS and ABA signaling in response to drought stress [13]. Knocking down *ZmRBOHC* reduced the response of maize to drought stress [14]. *SlRbohB* positively regulated the drought tolerance in tomato [15].

Bread wheat (*Triticum aestivum* L.) is one of the third-largest cereal crops globally [16]. Recognizing and studying the *Rboh* gene family in wheat is essential to ensuring food security and increasing productivity, since abiotic stresses restrict plant growth and present significant challenges to the world's food supply. However, current research on the *Rboh* gene family in wheat either encounters flaws in the identification process [17] or employs improper selection criteria for *Rboh* genes [18], leading to improper identification of *Rboh* members. In this study, we used bioinformatics techniques to identify the *Rboh* gene family in wheat, determine their chromosomal locations, evolutionary relationships, conserved motifs, structures, cis-elements, collinearity, protein interaction networks, and miRNA targets. Additionally, we analyzed the expression profiles of *TaRboh* genes under different conditions using RNA-Seq data [19] and validated these profiles with quantitative real-time PCR (qRT-PCR) experiments. Overall, our findings will assist in elucidating the molecular mechanisms of *TaRboh* genes in responses to drought, heat, and combinations of these stress conditions in wheat.

2. Results

2.1. Genome-Wide Identification, Chromosome Distribution, and Physicochemical Properties of Rboh Gene Family Members in Wheat

Using HMMER 3.0 software and conserved domain identification, 39 *TaRboh* genes encoding putative proteins with conserved NADPH_Ox, Ferric_reduct, FAD_binding_8, and NAD_binding_6 structural domains were identified using a Hidden Markov Model (HMM)-based approach combined with SMART structural domain validation. Among them, TraesCS1B02G295200 and TraesCS1D02G284900 contained three and four EF-hand motifs, respectively; twenty-eight Rboh proteins had two EF-hand motifs, eight Rboh proteins had only one EF-hand motif, and one Rboh protein (TraesCS3B02G212900) lacked an EF-hand motif (Table S1). Based on their genomic localization in wheat and widely accepted standard of gene nomenclature, the 39 *TaRboh* genes were designated *TaRboh01* to *TaRboh39*. The *TaRboh* genes were unevenly distributed across chromosomes 1A, 1B, 1D, 3A, 3B, 3D, 4B, 4D, 5A, 5B, 5D, 6A, 6B, and 6D. Furthermore, most *TaRboh* genes were located in the terminal regions of the chromosomes, with few in the central regions (Figure 1). Analysis of the physicochemical properties of TaRboh proteins (Table S2) revealed that the predicted Rboh proteins comprised between 800 (TaRboh02) and 1223 (TaRboh04) amino acids, with molecular weights ranging from 90.83 kDa (TaRboh02) to 139.42 kDa (TaRboh02). Their pIs varied from 8.88 (TaRboh09) to 9.47 (TaRboh18), indicating that most of the TaRboh proteins were basic. For the instability index (II), only six Rboh proteins (TaRboh06-TaRboh07, TaRboh12, TaRboh36-TaRboh38) were predicted to be stable, with instability index values below 40. The GRAVY value was less than zero, suggesting that these proteins were hydrophilic. Moreover, subcellular localization prediction revealed that all of these TaRboh proteins were localized to the cell membrane (Table S2).

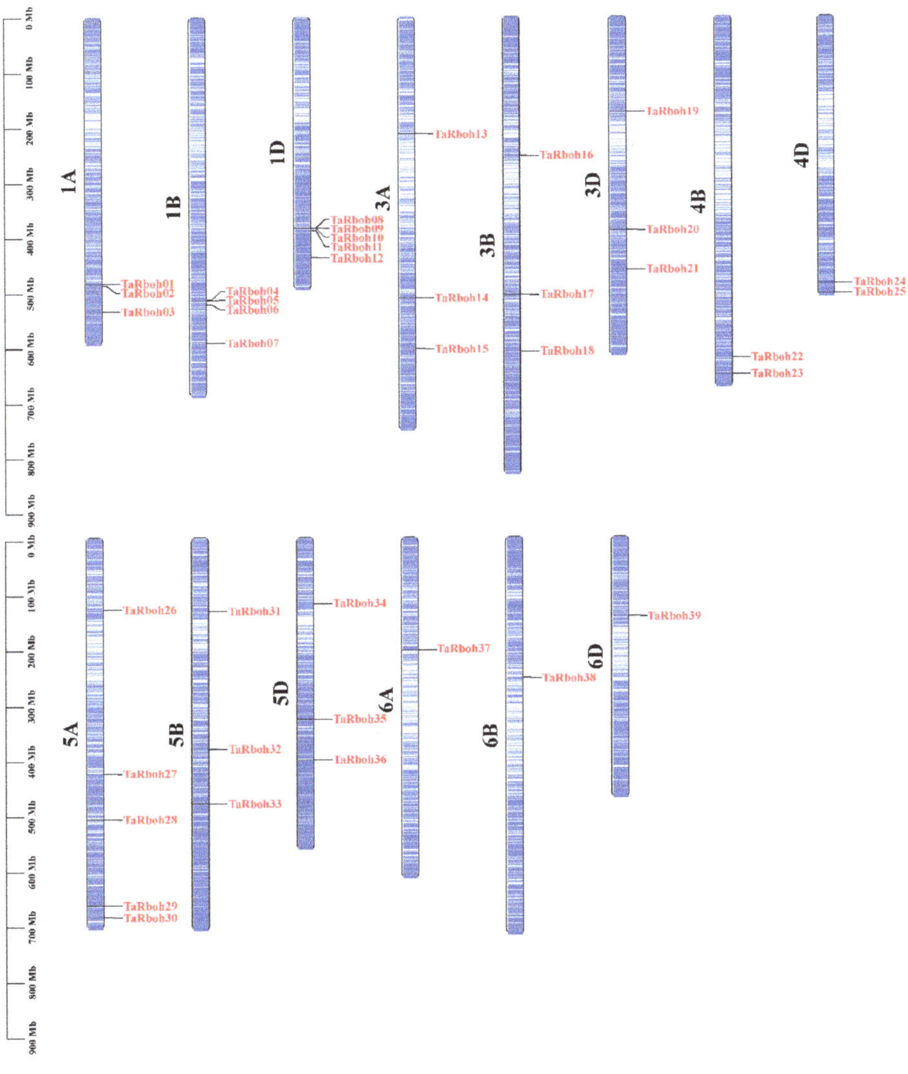

Figure 1. Locations of 39 *TaRboh* genes in the wheat genome. The color variation on the chromosomes represents gene density, ranging from low (white) to high (blue).

2.2. Sequence Analysis, Phylogenetic Analysis, and Classification of Rboh Members

Thirty-nine TaRboh protein sequences from wheat, ten AtRboh protein sequences from *Arabidopsis*, and nine OsRboh protein sequences from rice were used in a phylogenetic analysis using MEGA 11.0 software to understand the evolutionary relationships of the *TaRboh* gene family in wheat and other species. As shown in the phylogenetic tree obtained, the TaRbohs were dispersed among the five groups into which the Rbohs were clustered (Figure 2A). Twelve TaRboh proteins were in group A, nine in group B, six in group C, six in group D, and six in group E (Figure 2B). Subsequent investigation showed that the

Rboh proteins from rice and wheat were more closely grouped, suggesting a closer genetic relationship between the TaRboh family members between these two species.

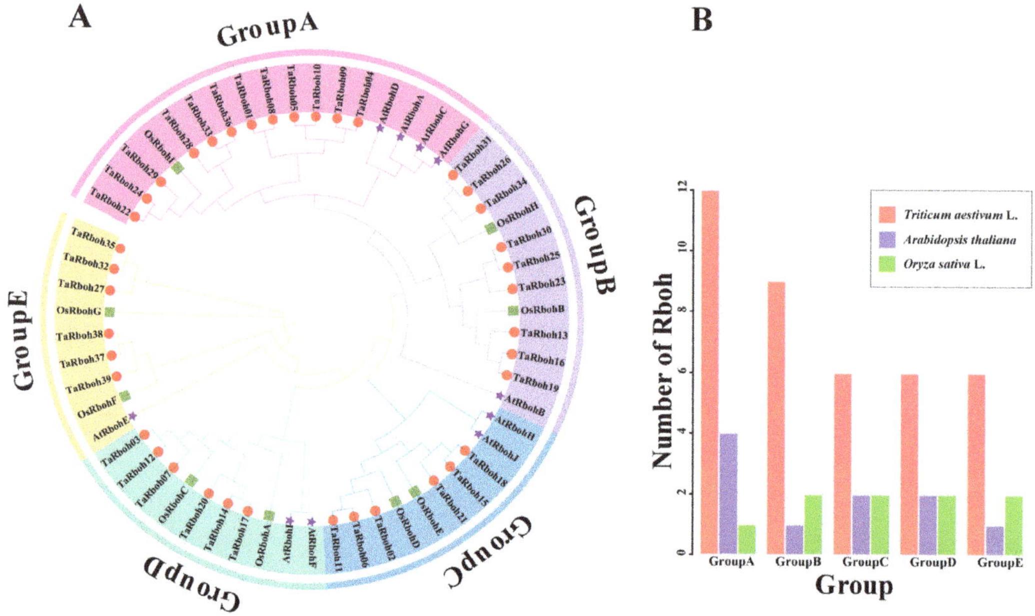

Figure 2. *Rboh* gene family in wheat, *Arabidopsis*, and rice. (**A**) The phylogenetic tree constructed for Rhoh sequences from wheat, *Arabidopsis*, and rice using MEGA-X based on the neighbor-joining (NJ) method with the p-distance substitution model (gamma = 1) and 1000 bootstrap replicates. The Rboh proteins of wheat are denoted by red circles. The Rboh proteins of rice are denoted by green squares. The Rboh proteins of *Arabidopsis* are denoted by purple pentagrams. (**B**) The number of *Rboh* genes belonging to different groups in wheat, *Arabidopsis*, and rice, respectively.

2.3. *Structure and Conserved Motifs of TaRbohs*

To explore the diversity of *TaRboh* genes, in silico phylogenetic tree construction, gene structure analysis, and motif identification were performed (Figure 3). The 39 *TaRboh* genes formed 10 groups based on the analysis using MEGA 11.0 software. Groups VI, VII, and X had the most *TaRboh* gene members (six each), while the other groups contained two or three *TaRboh* genes each (Figure 3A). Additionally, the gene structure analysis (Figure 3C) revealed that most of the *TaRboh*s had untranslated regions (UTRs). The exceptions included *TaRboh01*, *04*, *09*, *28*, and *36*. All but three (*TaRboh28*, *33*, and *36*) of these *TaRboh*s possessed introns. *TaRboh04* has the highest number of exons (16), while the numbers of exons possessed by other *TaRboh* genes varied between 1 and 14. As shown in Figure 3B, different groups of *TaRboh* genes had different motif compositions, but motifs for these genes within the same group are highly preserved. All except Group II genes had 15 motifs. Those in Group II lacked motif 14. According to domain alignment, the NADPH_Ox domain was made up of motifs 13 and 15, the NAD_binding_6 domain was made up of motifs 3, 5, 8, and 9, the Ferric_reduct domain was made up of motifs 2, 7, and 11, and the FAD_binding_8 domain was made up of motifs 1. In general, *TaRboh*s with the same types and arrangements of motifs were clustered into the same group. Genes on adjacent branches of the phylogenetic tree with similar numbers and arrangements of motifs and gene structures may have similar biological functions.

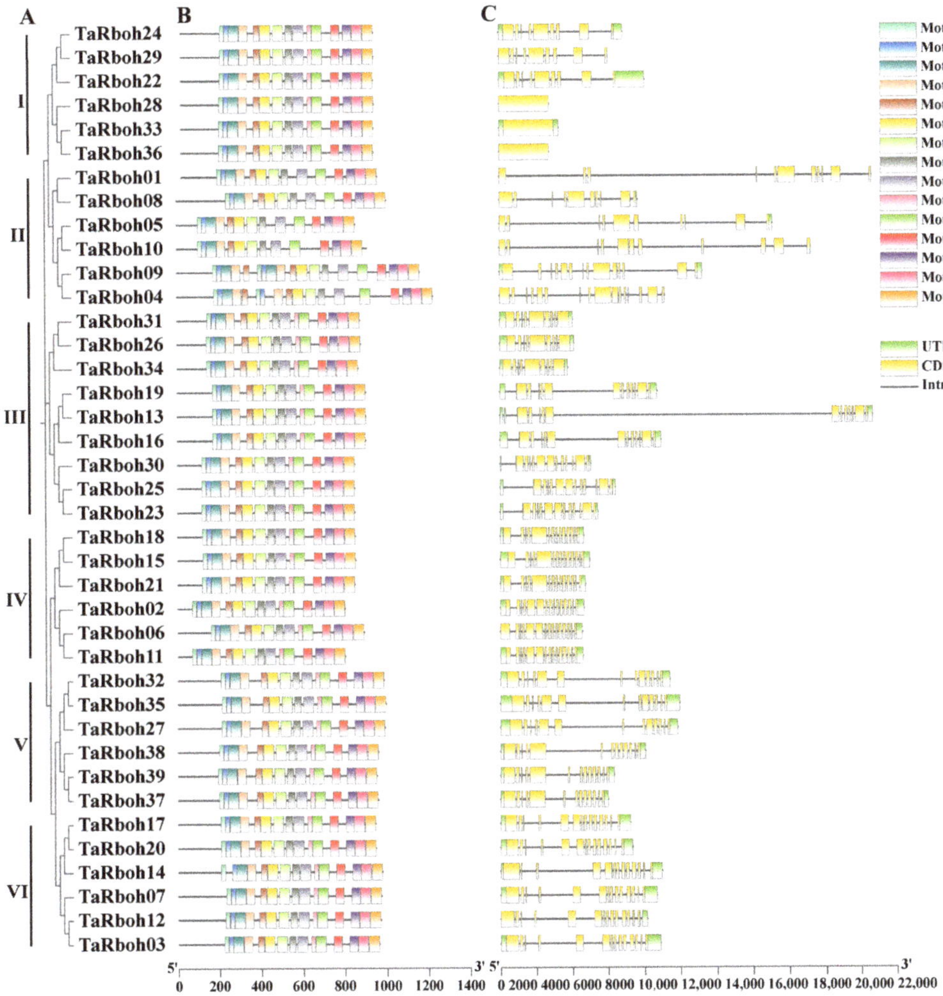

Figure 3. Evolutionary relationships, conserved motifs, and gene structures of *TaRboh* genes. (**A**) Phylogenetic tree of *TaRboh* genes constructed using MEGA-X with the Maximum Likelihood (ML) method. (**B**) Conserved motifs in TaRboh proteins, with fifteen distinct motifs identified. (**C**) Structures of *TaRboh* genes, where green rectangles represent untranslated regions (UTRs), yellow rectangles denote exons, and short lines indicate introns.

2.4. Cis-Elements in TaRboh Genes

Cis-elements in promoter regions specifically bind to transcription factors, thereby regulating the expression of downstream genes [20]. The *cis*-elements in the promoters of the *TaRboh* genes were examined using the PlantCARE website. A total of 3862 *cis*-elements were found in promoters of the 39 *TaRboh* genes. They comprised 1192 binding site elements, 197 elements connected to plant growth and development, 504 phytohormone responsive elements, 1117 elements related to abiotic and biotic stressors, and 852 elements without a functional description (Table S3). As shown in Figure 4, the majority of the biotic and abiotic stress elements are linked to wounding (CTAG-motif, TC-rich repeats, WRE3, and WUN-motif), dehydration (DRE/CRT-core and DRE1), elicitor responses (W box), low temperature (LTR), light (A-box, ACE, AE-box, ATCT-motif, Box 4, Box II-like sequence, chs-CMA2a, chs-Unit 1 m1, Gap-box, GATA-motif, G-Box, GT1-motif, GTGGC-motif, I-box, Pc-CMA2a, Pc-CMA2c, Sp1, MRE, TCCC-motif, and TCT-motif), anoxic specific inducibility (GC-motif), anaerobic induction (ARE), replication (E2Fb), stress response (STRE), and drought stress (MYB, MYB-binding site, MYB-like sequence, and MYC). Phytohormone-responsive elements included those responsive to auxin (AuxRR-core, TGA-box, and TGA-element), gibberellin (GARE-motif, P-box, and TATC-box), salicylic acid (as-1 and TCA-element), abscisic acid (ABRE, ABRE3a, and ABRE4), estrogen (ERE), glucocorticoid (GRA), and MeJA (CGTCA-motif and TGACG-motif). Plant growth and development elements are associated with endosperm expression (AAGAA-motif and GCN4_motif), xylem-specific expression (AC-I and AC-II), meristem expression (CAT-box), tissue activation (CCGTCC motif and CCGTCC-box), circadian control (circadian), meristem specific activation (dOCT), root-specific (motif I), zein metabolism (O2-site), cell cycle (re2f-1 and telo-box), seed-specific regulation (RY-element), and DNA regulation (Y-box). Abiotic and biotic stress elements, phytohormone-responsive elements, and plant growth and development elements are the three categories of *cis*-elements found in the promoters of each *TaRboh* gene; *TaRboh20*, *TaRboh12*, and *TaRboh13* had the largest numbers of these elements.

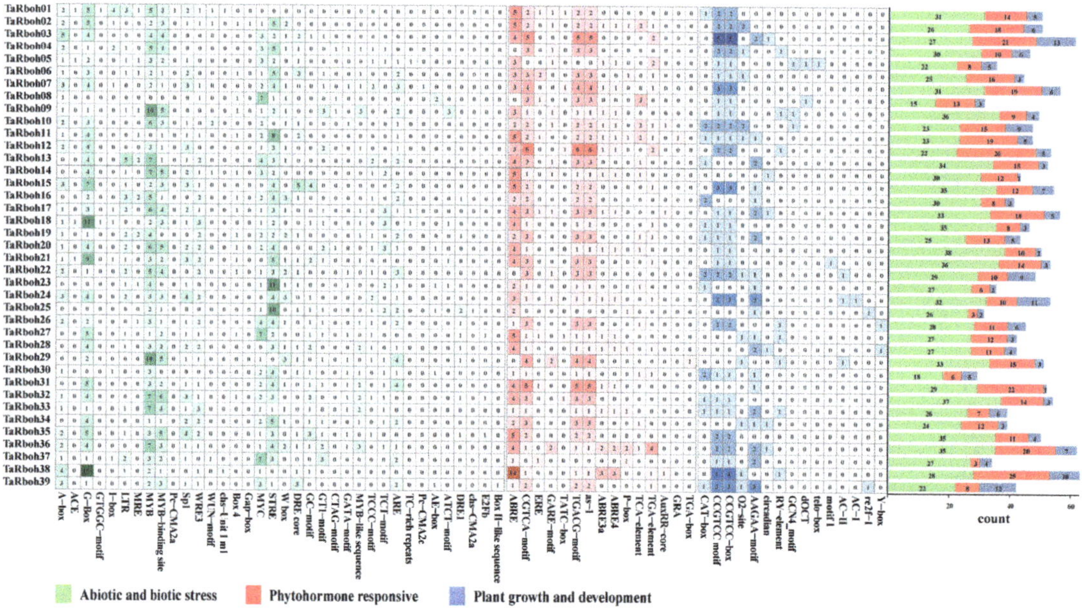

Figure 4. Distribution of *cis*-elements in the promoter regions of the *TaRboh* genes. The color and number of grids represent the quantity of *cis*-elements in each of the *TaRboh* genes.

2.5. Collinearity Relationship of the TaRboh Gene Family

Gene tandem repeats and fragment repeats are essential for gene family expansion in plants [21]. Forty-eight pairs of identical *TaRboh* genes were found using in silico collinearity analysis in wheat; these genes are unevenly distributed on homoeologous group 1, 5, and 6, as well as chromosomes 4B and 4D (Figure 5). Segmental duplications were found in all of the 48 gene pairs using MCScanX. However, tandem duplications were found in none of them. These findings imply that segmental duplications have played a major role in the evolution and diversification of the *TaRboh* genes. Values of nonsynonymous (Ka) and synonymous (Ks) ratios were computed in order to examine the underlying evolutionary selection pressure on the *TaRboh* gene family. Ka/Ks ratios for all of the 48 homologous gene pairs were smaller than 1 (Table S4), indicating that strong purifying selection might have been experienced by these genes during the evolutionary process.

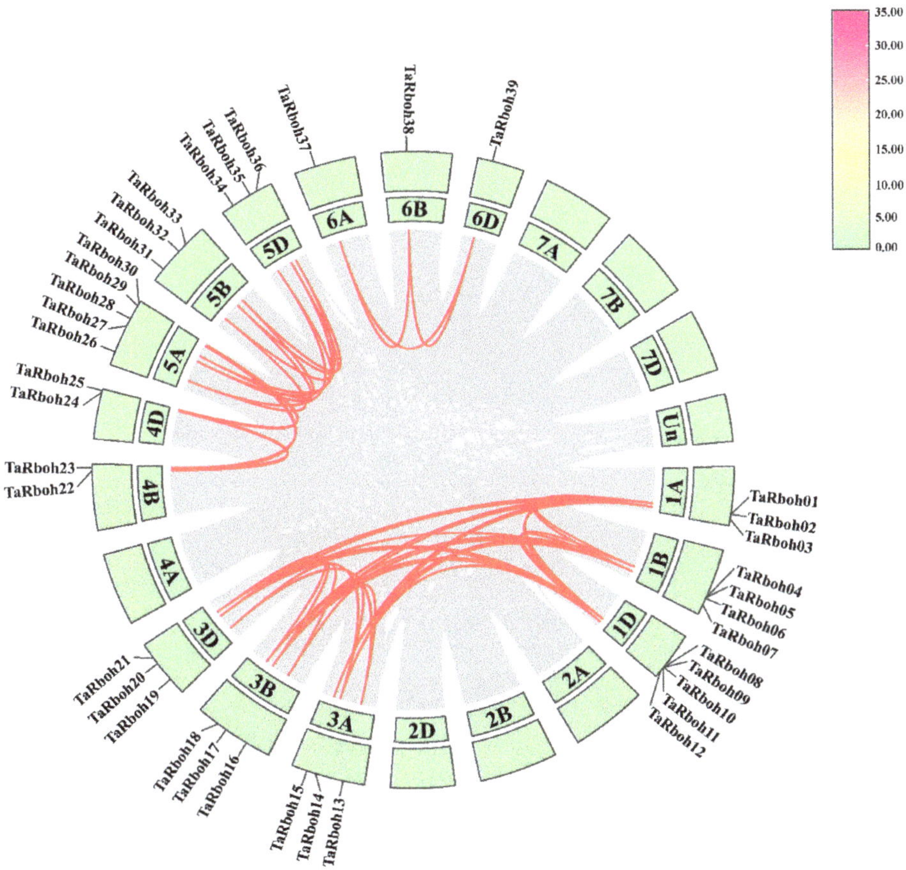

Figure 5. Collinearity analysis of *TaRboh* genes in wheat. Red lines indicate the segmentally duplicated *TaRboh* gene pairs in wheat. Gray lines indicate other gene pairs, excluding *TaRboh* genes. Chromosome lengths (Mb) are marked on the scale bar.

As shown in Figure 6A, estimated divergence times between wheat and the other six species (*Brachypodium distachyon*, rice, maize, *Glycine max*, *Arabidopsis*, and *Aegilops tauschii*) were 3.54 Mya, 33.0 Mya, 47.0 Mya, 49.1 Mya, 108.0 Mya, and 159.6 Mya, respectively. Based on the phylogenetic tree with wheat as the reference genome, 44 homologous pairs were found between wheat and *Aegilops* and they included 33 *Rboh* genes in wheat and 12 genes in *Aegilops*. From the twelve genes in *Aegilops*, eleven homologous pairs were identified between *Aegilops* and *Brachypodium* and they included six genes in *Aegilops* and seven in *Brachypodium*. Furthermore, twelve homologous pairs were identified between the seven genes in *Brachypodium* and seven genes in rice. Additionally, eleven homologous pairs were identified between seven genes in rice and six genes in maize. Three homologous pairs were identified between two genes in maize and two genes in *Glycine*. One homologous pair was identified between two genes in *Glycine* and two genes in *Arabidopsis* (Figure 6B). Among these, five *TaRboh* genes (*TaRboh26*, *TaRboh27*, *TaRboh31*, *TaRboh32*, and *TaRboh34*) were involved in twenty-four continuous collinear pairs. All of the computed *Ka*/*Ks* values for these collinear pairs (Table S4) were less than 1.0, consistent with the likelihood that purifying selection is a main drive force during evolution of these genes. Five of these *TaRboh* genes (*TaRboh26*, *TaRboh27*, *TaRboh31*, *TaRboh32*, and *TaRboh34*) also showed strong evolutionary conservation, and they seemed to be important for plant adaptation and evolution.

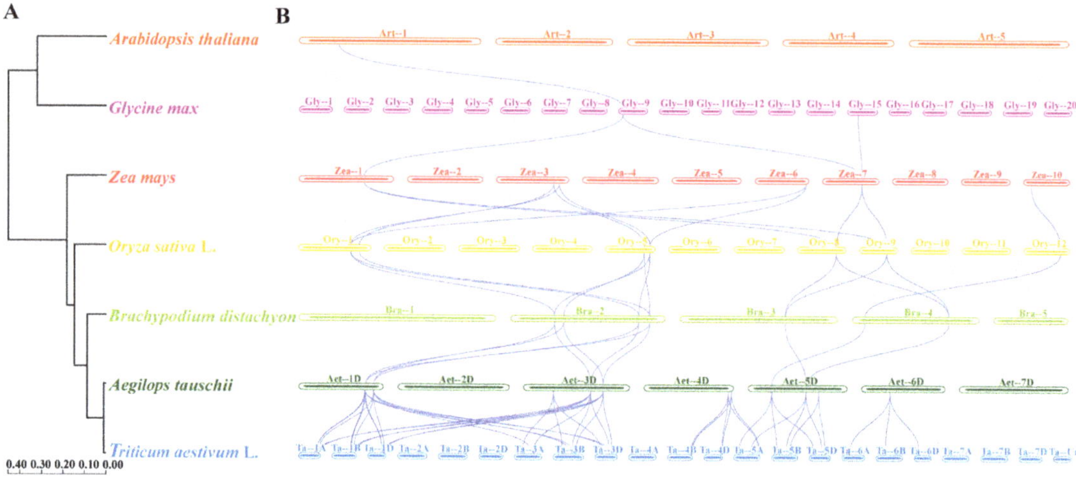

Figure 6. Synteny analysis of *TaRboh* genes between wheat and six other plant species (maize, rice, *Brachypodium*, *Glycine*, *Arabidopsis*, and *Aegilops*) based on the species' evolutionary tree on the left. (**A**) Evolutionary relationships among the seven species predicted based on their whole genomes. Node numbers represent the estimated divergence times. (**B**) Collinearity relationships among the seven plant species based on the *TaRboh* genes in wheat. Gray lines in the background indicate syntenic blocks between adjacent species, while the blue lines highlight the linked genes identified.

2.6. Protein–Protein Interactions and miRNA Targeting Events in TaRboh Genes

Protein complexes, which include a great deal of protein interaction, are necessary for the majority of life activities in an organism. To comprehend complicated biological processes, it is imperative to identify these interactions [22]. Potential interactions between TaRboh proteins were predicted using the STRING database. While TaRboh19 was found to have no interactions, the 38 TaRboh proteins were found to have 312 interaction pairs altogether. Of the TaRboh proteins, 13 (TaRboh03, TaRboh07, TaRboh12, TaRboh14, TaRboh17, TaRboh20, TaRboh22, TaRboh27, TaRboh32, TaRboh35, TaRboh37, TaRboh38, and TaRboh39) were predicted to be central nodes in the network based on the degree of connectivity within the network. Each of these proteins had 26 connections with other genes (Figure 7A).

MicroRNAs (miRNAs) are a class of endogenous, small RNAs with approximately 22 nucleotides in length. Plant miRNAs are important for growth, development, and stress responses. They also take part in post-transcriptional gene regulation [23,24]. The psRNATarget service was utilized to investigate the ways in which miRNAs control *TaRboh* gene expression. Of the *TaRboh* genes, 34 were shown to be putative targets of 31 different miRNAs. Tae-miR408 regulated the largest number of *TaRboh* genes (*TaRboh04*, *TaRboh14*, *TaRboh20*, *TaRboh22*, *TaRboh27*, *TaRboh35*, *TaRboh37*, *TaRboh38*, and *TaRboh39*). Furthermore, *TaRboh16* was the primary target (Figure 7B).

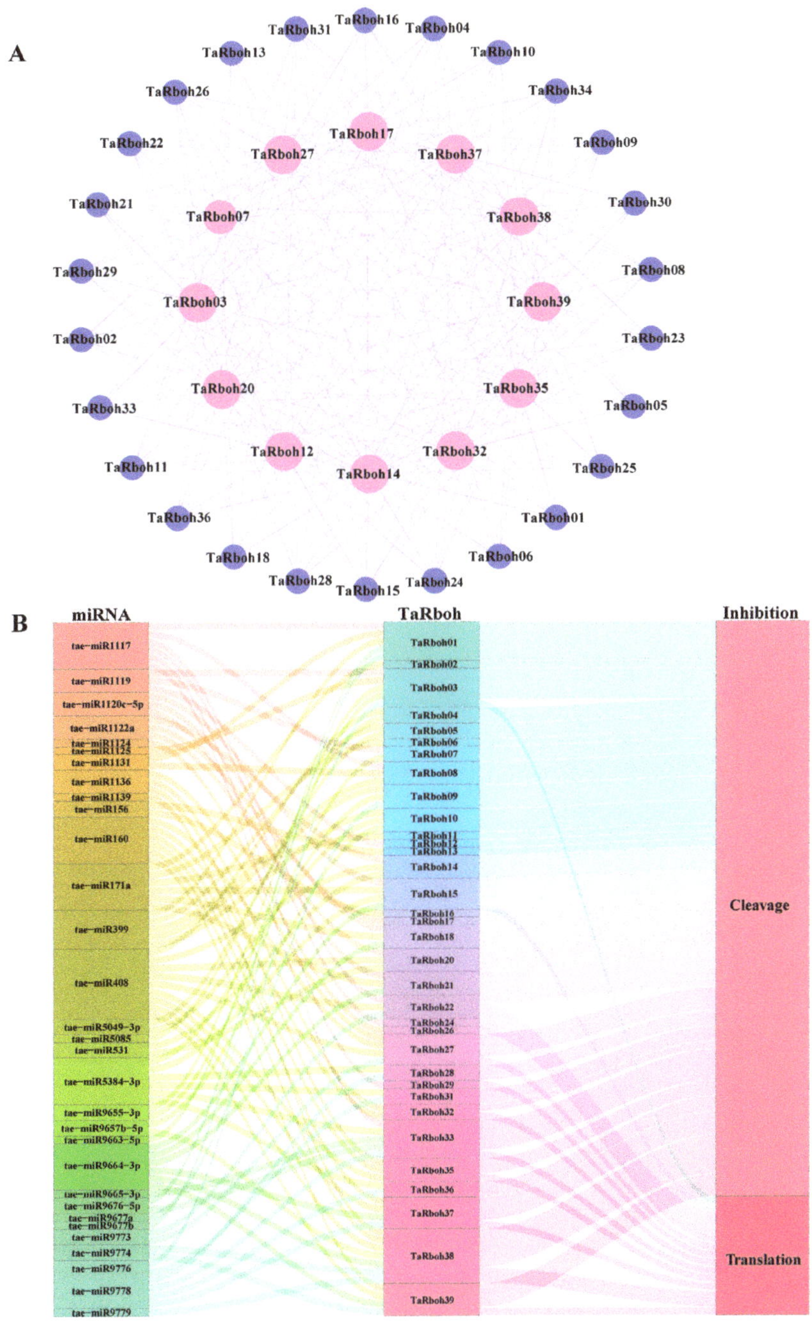

Figure 7. Protein–protein interaction network and miRNA target analysis of *TaRbohs*. (**A**) Red represents 13 TaRboh proteins with a high degree of 26. Blue represents 26 TaRboh proteins with a low degree of 12. (**B**) Sankey diagram depicting relationships of TaRboh transcripts targeted by miRNAs. The three columns represent miRNA, *TaRboh*, and the inhibition effect. Different colors are used to distinguish different mirnas, different *TaRbohs*, and different the inhibition effects.

2.7. Expression of TaRboh Genes in Different Stress Conditions

Drought and heat stress are the main factors limiting plant growth and reproduction, and they often occur simultaneously [25]. To further investigate the potential role of *TaRboh* genes, wheat was grown under a variety of stress conditions, including drought stress for one hour (DS-1), drought stress for six hours (DS-1), heat stress for one hour (HS-1), heat stress for six hours (HS-6), combined heat and drought stress for one hour (DHS-1), and combined drought and heat stress for six hours (DHS-6). An analysis of RNA-seq transcriptomic data was performed to evaluate the expression profiles of the *TaRboh* genes in various stress conditions using the Wheat Expression Browser (expVIP) (www.wheat-expression.com (accessed on 7 Dec 2023)). This analysis showed the distinct expression of these genes under different stress scenarios (Figure 8A). Compared to those in the controls, three-quarters of the *TaRboh* genes had higher expression at DS-1 and lower expression at DS-6. Although the expression of 11 *TaRboh* genes was not significantly altered at HS-1, substantial upregulation for 12 *TaRboh* genes was detected at DHS-6. Additionally, during drought stress, expressions of some genes (*TaRboh14*, *TaRboh17*, and *TaRboh20*) did not significantly change. However, the expressions of these genes were significantly increased under heat stress and combined heat and drought stress. *TaRboh28*, *TaRboh33*, and *TaRboh36* showed no significant change under other conditions, but had significantly increased expression under drought stress.

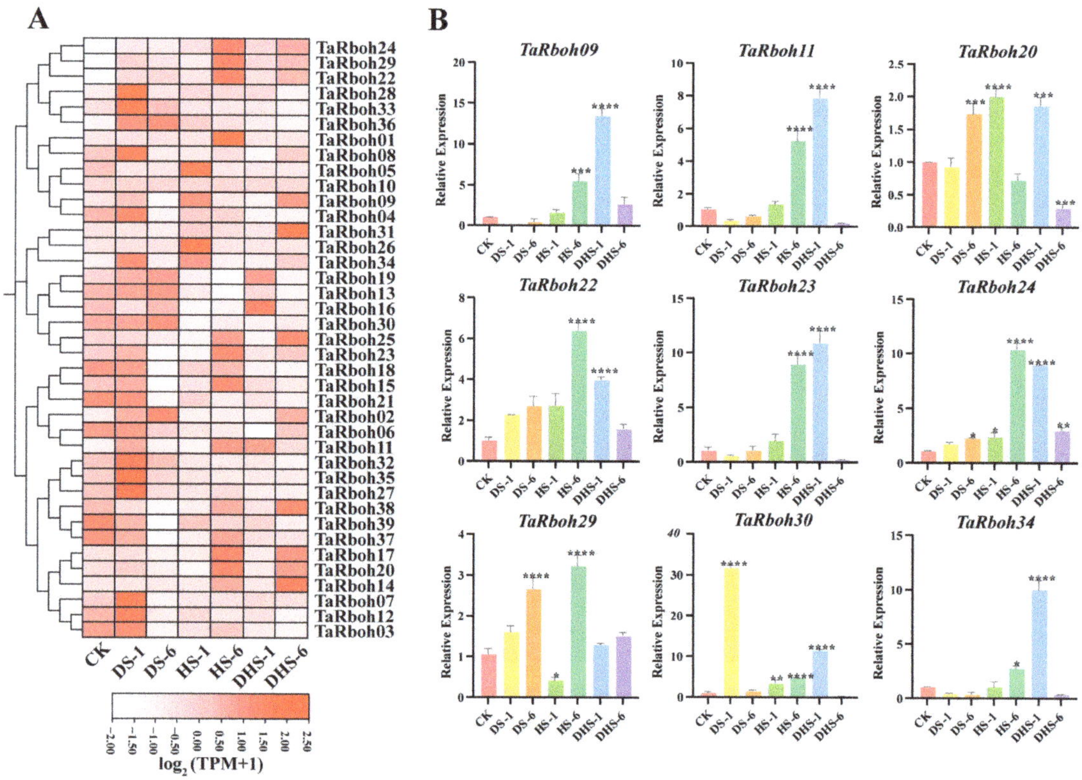

Figure 8. Expression profiles of *TaRboh* genes under different stress conditions. (**A**) Heatmap showing expression of *TaRbohs* under different treatments. Gene expression is expressed in $\log_2(TPM+1)$.

Relative expression levels under three different abiotic stress conditions (control (CK), drought stress for 1 h (DS-1), drought stress for 6 h (DS-6), heat stress for 1 h (HS-1), heat stress for 6 h (HS-6), combined drought and heat stress for 1 h (DHS-1), combined drought and heat stress for 6 h (DHS-6)). CK: control. DS-1 (6): drought stress for 1 (6) h; HS-1 (6): heat stress for 1 (6) h; DHS-1 (6): combined drought and heat stress for 1 (6) h. (**B**) Quantitative expression of nine *TaRboh* genes in response to abiotic treatments including drought, heat and drought, and heat at 1 h and 6 h following treatments. The error bars indicate the standard deviations and the values in plots corresponding to the mean ± standard deviation (SD) of three independent biological replicates (* $p < 0.05$; ** $p < 0.01$; *** $p < 0.001$; **** $p < 0.0001$).

To further verify the reliability of the transcriptome data, nine representative members from the thirty-nine *TaRboh* genes were selected for quantitative real-time PCR (qRT-PCR) experiments. As shown in Figure 8B, nine different gene expression profiles show different degrees of responsiveness to heat, drought, and a combination drought and heat conditions. Among these, six genes (*TaRboh09*, *TaRboh11*, *TaRboh22*, *TaRboh24*, *TaRboh30*, and *TaRboh34*) show expression patterns that align with the RNA-seq findings. Nevertheless, the expression profiles of three genes (*TaRboh09*, *TaRboh23*, and *TaRboh29*) differ from the RNA-seq results. The RNA-seq transcriptome data indicate that *TaRboh20* is downregulated at HS-6 compared to the control group, the expression of *TaRboh29* at HS-6 is lower than at DHS-6, and *TaRboh23* is upregulated at DHS-1 (Figure 8A). In contrast, the qRT-PCR results show that *TaRboh20* is upregulated at HS-6, *TaRboh23* exhibits a significant increase in expression at DHS-1, and *TaRboh29* shows significantly higher expression levels at HS-6 than at DHS-6. These findings suggest that distinct *TaRboh* genes have distinct patterns of expression in response to heat, drought, and combined heat and drought stress.

3. Discussion

Rboh genes encode Rboh enzymes in plants. These enzymes are essential for generating ROS and controlling plant shape, growth, development, and responses to biotic and abiotic stresses [26]. *Rboh* genes have been thoroughly investigated in a wide range of plant species. Hu et al. [17] and Sharma et al. have both attempted to identify *Rboh* family genes in wheat, but their findings are flawed. Hu's 2018 study reported 46 *Rboh* genes. However, it did not mandate the presence of all four domains (NADPH_Ox, Ferric_reduct, FAD_binding_8, and NAD_binding_6) for classification as *Rboh* genes. Furthermore, the wheat genome that was employed in their study is now only marginally relevant due to advances in sequencing technology. Sharma's recent research in wheat revealed 40 *TaRboh* genes [18]. With the exception of the extra gene *TraesCS1B02G295100* that is included in Sharma's investigation, all 39 of the *TaRboh* genes that we found through in silico analysis (Figure 1) are included in his findings. Domain analysis verified the presence of the NADPH_Ox, Ferric_reduct, FAD_binding_8, and NAD_binding_6 domains in the 39 TaRboh proteins that we discovered (Table S1). TraesCS1B02G295100, on the other hand, lacks the NADPH_Ox domain and only has the Ferric_reduct, FAD_binding_8, and NAD_binding_6 domains. The NADPH_Ox domain, located at the EF hand's N-terminus, is crucial for generating ROS via Ca2+ signaling that is mediated by Rboh proteins [27]. Considering that the NADPH_Ox domain is a crucial component of Rboh genes, TraesCS1B02G295100 most likely functions as a ferric reduction oxidase (FRO) protein as opposed to a Rboh protein [28]. Therefore, we hypothesize that *TraesCS1B02G295100* is a *FRO* gene rather than a *TaRboh* gene.

Notably, the number of *Rboh* genes in wheat is significantly higher than those in *Arabidopsis* (10) [2], rice (9) [6], maize (15) [8], *Aquilaria* (14) [29], and eggplant (8) [30], yet the number of *Rboh* genes does not seem to correlate with genome sizes [31]. Additionally, Chromosomes in homoeologous groups have an equal amount of *TaRboh* genes located in consistent positions on each chromosome. However, the distribution of *TaRboh* genes is not uniform in homoeologous groups 1 and 4, where certain genes are restricted to specific chromosomes (Figure 1). Whole-genome duplication (WGD) is the most likely explanation for these observations, as it caused the allohexaploid nature (AABBDD) of common wheat. This explains the wheat's enormous genome size and the increased number of members of the *TaRboh* gene family, which are frequently seen as three homologous alleles. But as polyploid systems get closer to states that resemble diploid states, gene loss could happen as a result of copy number reduction, pseudogenization, or silencing. Consequently, only particular chromosomes within homologous groups 1 and 4 contain *TaRboh* homologous genes [32–35]. Furthermore, collinearity analysis in wheat revealed 48 pairs of identical genes, all of which are segmental duplications (Figure 5). This finding highlights the significance of segmental duplication in the evolution of the *TaRboh* gene family. Thus, these results suggest that both segmental and whole-genome duplications may play key roles in the evolution and expansion of the *TaRboh* gene family in wheat.

Rboh protein sequences from wheat, rice, and *Arabidopsis* were used to create phylogenetic trees. Similarly to those reported earlier [29], the *Rboh* genes were clustered into five separate groups (Figure 2). Compared with those in *Arabidopsis*, *Rboh*s in wheat have closer homologous links with those in rice, indicating that these genes probably share similar structures and functions within the same group of the evolutionary tree. Group D comprises *TaRboh03*, *TaRboh07*, *TaRboh12*, *TaRboh14*, *TaRboh17*, and *TaRboh20* in wheat, and *OsRbohA* in rice, which is known for its role in drought resistance [6]. These wheat genes belong to the same group based on phylogenetic, motif, and gene structural analyses (Figure 3). They all have *cis*-regulatory elements (such as MYB and MYC) that are associated with drought tolerance (Figure 4). Results from the transcriptome analysis indicated that *TaRboh03*, *TaRboh07*, *TaRboh12*, *TaRboh14*, *TaRboh17*, and *TaRboh20* were related to drought-tolerance. Results from RT-qPCR analysis confirmed that *TaRboh20* was upregulated under drought stress (Figure 8). The other five genes also seem to be related to drought stress, although experiments are needed to validate this possibility.

In comparative genomics, homology inference is essential. However, heterozygosity, polyploidy, WGD, and repetitive sequences may limit the effectiveness of approaches based solely on sequence similarity and obscure homologous relationships [36,37]. Compared to other variables, gene content and gene order have been found to be more stable during plant evolution, making collinearity analysis especially valuable for homology inference [38]. It also aids in illuminating ancestral polyploid events [39]. We showed in this study that, although the homologous relationships of several *TaRboh* genes have gradually diminished, only five *TaRboh* genes (*TaRboh26*, *TaRboh27*, *TaRboh31*, *TaRboh32*, and *TaRboh34*) have maintained homologous relationships throughout evolution (Table S4). This may be attributed to the larger genome of wheat and the increased gene count resulting from WGD, which far exceed those of *Arabidopsis*, ultimately leading to the loss of most duplicated genes [40]. These findings indicate that *TaRboh26*, *TaRboh27*, *TaRboh31*, *TaRboh32*, and *TaRboh34* are highly conserved throughout evolution and have significant research value. However, further investigation through experiments is needed.

Transcription factors (TFs) and microRNAs (miRNAs) are important regulators of gene expression under stress that lessen the effects of adverse environmental conditions in plants [41]. The position, timing, and degree of gene expression are all determined by *cis*-elements, which are individual TF-binding sites [42]. MiRNAs attach to target mRNAs at precise locations and cleave them post-transcriptionally to control gene expression and prevent protein translation [43]. As shown in Figure 4, *cis*-element analyses of *TaRboh* gene promoters reveal a variety of *cis*-elements which could be categorized into three functional groups: stress response, hormone response, and developmental regulation. Notably, 61.4% of these elements are associated with biotic and abiotic stress responses. They included DRE core, DRE1, MYB, and MYC elements related to drought stress. Additionally, ABRE, a hormone-responsive element involved in ABA-responsive gene expression, plays a significant role in drought stress response [44]. Therefore, *TaRboh* genes are likely involved in drought stress responses. Furthermore, we identified 31 distinct miRNAs and predicted interactions between them and the *TaRboh* genes in wheat (Figure 7B). Tae-miR408 was found to have the most targets among the *Rboh* genes, and it is known to be involved in regulating heading time, phosphorus deprivation, high salinity, heavy copper stress, and stripe rust stress [45–47]. Our results indicate that Tae-miR408 may target *TaRboh04*, *TaRboh14*, *TaRboh20*, *TaRboh22*, *TaRboh27*, *TaRboh35*, *TaRboh38*, and *TaRboh39*, which requires experimental validation.

Drought and heat are major abiotic stressors that limit wheat production and often occur together. This multi-stress scenario affects plants in ways that cannot be predicted solely based on the impact of individual stresses [48,49]. Current research indicates that plant *Rboh* is capable of responding to heat, drought, and combined drought and heat stress. Through the *OsRALF45/46-OsMRLK63-OsRboh*s signaling circuit, OsMRLK63 (a kinase) controls intracellular ROS levels, hence regulating drought tolerance [50]. The FERONIA Receptor-Like Kinase in tomatoes was used by the BZR1 transcription factor to facilitate ROS signaling from *RBOH1* in response to heat stress; the *bzr1* mutant downregulated *RBOH1* expression, which reduced drought resistance, which could be partially restored by exogenous H_2O_2 [51]. *ERF74* and *ERF75* are functionally redundant in *Arabidopsis*; overexpression of *ERF74* increased ROS levels and improved resistance to heat and drought by upregulating *RbohD*, whereas the *erf74* mutant and *erf74*; *erf75* double mutant have decreased stress resistance and *RbohD* expression [52]. Additionally, hormones such as ABA (abscisic acid) and JA (jasmonic acid) also impact wheat drought resistance [53]. Expression profiling and quantitative data obtained from this study confirmed that *TaRboh* genes respond to drought, heat, and combined drought and heat stresses to varying degrees. Furthermore, several *TaRboh* genes that are elevated under single stress (heat or drought) conditions apparently have their increased expression inhibited by the dual stress (heat and drought). For instance, according to transcriptome data and qPCR results, *TaRboh22*, *TaRboh24*, and *TaRboh29* expression levels were significantly greater at HS-6 than at DHS-1 and DHS-6, and *TaRboh30* expression was significantly higher at DS-6 than at DHS-1 and DHS-6 (Figure 8). More experimental research is needed to understand this phenomenon. *TaRboh22*, a homologue of *AtRbohD*, was likely involved in responses to various abiotic stresses including drought, heat, and their combination. *TaRboh30*, with high sequence similarity to *OsRbohA*, also responds to drought stress. We found it interesting that under different stress circumstances, duplicated genes show comparable expression patterns. Examples of genes that exhibit functional redundancy among these duplicated genes include *TaRboh22*, *TaRboh24*, and *TaRboh29*. They all exhibited increased expression in response to heat, drought, and combined stress. The most notable expression increase occurred at HS-6. In summary, distinct *TaRboh*s expression patterns were noted. A possible explanation for the discrepancies between the qRT-PCR and RNA-seq results for the three *TaRboh* genes (*TaRboh09*, *TaRboh23*, and *TaRboh29*) is that we performed the treatment when the wheat reached the three-leaf stage, whereas the reference literature for RNA-seq [19] did not specify the exact timing of the treatment. In the subsequent in-depth study of the functions of these three *TaRboh* genes (*TaRboh09*, *TaRboh23*, and *TaRboh29*), we will conduct expression analysis on materials at different growth stages and under different stress treatments. The variables impacting the uneven expression patterns of *TaRboh* genes, such as the location, timing, and length of stress treatment, may be further analyzed by more research. Furthermore, a thorough investigation of the possible mechanisms behind the effects of heat and drought on wheat development is warranted.

4. Materials and Methods

4.1. Plant Materials and Stress Treatments

Wheat seeds (*Triticum aestivum* L. cv. Fielder) [54] were germinated on moist filter paper at 4 °C for 5 days, followed by 5 days at 12 °C. The seedlings were then transferred to a greenhouse with the setting of light 16 h/22 °C and dark 8 h/16 °C. The plants were grown until the three-leaf stage for stress treatments. The plants were subjected to PEG-induced dehydration mimicking drought stress (25% PEG6000 solutions), heat stress (37 °C incubator), and combined PEG-induced dehydration mimicking drought and heat stress (37 °C incubator and 25% PEG6000 solutions). Samples were collected at 0, 1, and 6 h post-treatment and were used to determine the expression levels of nine selected *TaRboh* genes under different conditions and time points.

4.2. Bioinfomatics Identification of TaRboh Family Genes in Triticum Aestivum L.

To identify *Rboh* genes in wheat, a sequence alignment was performed using HMMER 3.0 software (E-value $< 1.0 \times 10^{-15}$). The Hidden Markov Model (HMM) profile of the respiratory burst NADPH oxidase domain (PF08414; PF01794; PF08022; PF08030) was queried from the Pfam database (available online: pfam-legacy.xfam.org/ (accessed on 7 December 2023)). Wheat gene sequences, protein sequences, cDNA files, and gene annotation files were obtained from the Ensembl Plants database (https://plants.ensembl.org/index.html, accessed on 14 November 2023). Protein sequences of candidate Rbohs were further screened using SMART (http://smart.embl-heidelberg.de/, accessed on 7 Decmeber 2023) for domain prediction. Only those sequences containing the NADPH_Ox, Ferric_reduct, FAD_binding_8, and NAD_binding_6 domains were considered to be *Rboh* genes. Chromosomal localization and visualization of *TaRboh*s were performed using TBtools V2.136 software [55]. Additionally, the number of amino acids, molecular weight, theoretical pI, instability index, aliphatic index, and grand average of hydropathicity of the *TaRboh* gene family were obtained from the ExPASy website (http://web.expasy.org/protparam/, accessed on 7 Decmeber 2023) [56]. In silico Subcellular localization was predicted using the Plant-mPLoc website (version 2.0) [57] (http://www.csbio.sjtu.edu.cn/bioinf/plant-multi/, accessed on 7 Decmeber 2023).

4.3. Multiple Alignments and Phylogenetic Tree Analysis

Protein sequences of TaRbohs were aligned using the Clustal X 2.1 tool [58], and the alignments were edited using the ESPrit 2.2-ENDscript 1.0 tool [59]. *Arabidopsis* and rice sequence data used in this study were obtained from the National Center for Biotechnology Information (NCBI) (http://www.ncbi.nlm.nih.gov). A phylogenetic tree was constructed using the MEGA 11.0 software with the neighbor-joining method, and reliability was confirmed using 1000 replicates of the bootstrap analysis [60]. Additionally, iTOL 6.9 (https://itol.embl.de) [61] was used to visualize the phylogenetic tree.

4.4. Conserved Motif and Gene Structure Analysis of TaRbohs

The MEME server v 5.5.5 (http://meme-suite.org/tools/meme, accessed on 7 Decmeber 2023) [62] and the Gene Structure Display Server (GSDS) website (http://gsds.gao-lab.org/) were used for the analysis of conserved motifs and gene structure of *TaRbohs*, respectively. For the MEME analysis, the maximum number of motifs was set to 15, with all other parameters set to default.

4.5. Cis-Elements Analysis of TaRboh Genes

To analyze the *cis*-elements of *TaRboh* genes, 1.5 kb upstream regions from the transcription start sites of each *TaRboh* gene were extracted from the wheat gene files. *Cis*-elements were predicted using PlantCARE website (http://bioinformatics.psb.ugent.be/webtools/plantcare/html/, accessed on 7 Decmeber 2023) [63], and results were visualized using R language.

4.6. Gene Duplication and Collinerity Analysis

Gene duplication events within the wheat genome were analyzed using the Multiple Collinearity Scan toolbox (MCScanX) program [64] with a BLASTp search (e-value $< 10^{-5}$). Apart from wheat, we have chosen three monocots (*Brachypodium*, rice, and maize) and three dicots (*Glycine*, *Arabidopsis*, and *Aegilops*) to examine the evolutionary traits of *Rboh* genes in seven different species. Evolutionary relationships among wheat, *Aegilops*, *Brachypodium*, rice, maize, *Arabidopsis*, and *Glycine* were investigated using the TimeTree website [65]. Using the wheat genome as the reference, TBtools V2.136 software was employed to identify collinear gene pairs among these species. Gene duplication within species and interspecies gene collinearity relationships were visualized using TBtools V2.136 software. The non-synonymous (Ka) and synonymous (Ks) substitution rates for gene pairs were calculated with KaKs_Calculator 2.0 software [66]. Positive selection is indicated by a Ka/Ks ratio greater than 1, negative selection is shown by a Ka/Ks ratio less than 1, and neutral selection is indicated by a Ka/Ks ratio equal to 1. Divergence times of collinear gene pairs were estimated using the formula $T = Ks/(2\lambda \times 10^{-6})$ Mya, with λ set to 6.5×10^{-9}.

4.7. Prediction of Protein–Protein Interactions Network and MicroRNA Targets

The STRING database (version 12.0) [67] (http://string-db.org/, accessed on 26 July 2023) was used to analyze protein–protein interactions among all TaRboh proteins with medium confidence (0.4). The interaction network was then visualized and enhanced using Cytoscape software (version 3.10.1) (http://www.cytoscape.org/) [68]. To predict potential miRNA binding sites for *TaRboh* genes, *TaRboh* cDNA sequences were submitted to the psRNATarget website with published miRNAs in wheat using default settings [69].

4.8. Expression Analysis of TaRboh Genes

RNA-seq data for wheat *TaRboh* genes under different stress conditions (drought, heat, and combined drought and heat) were obtained from the expVIP database [19]. The raw data were processed using $\log_2(\text{TPM} + 1)$ transformation and visualized using the 'HeatMap' function in TBtools V2.136 software.

4.9. RNA Isolation and Quantitative Real-Time PCR (qRT-PCR)

Total RNA was extracted from plants subjected to drought, heat, and combined drought and heat stress at 1 h and 6 h using the Spin Column Plant Total RNA Purification Kit from Sangon Biotech. The extracted RNA was qualitatively assessed using electrophoresis and a NanoDrop 1000. First-strand cDNA synthesis was performed using NovoScript® Plus All-in-One 1st Strand cDNA Synthesis SuperMix (gDNA Purge). Quantitative real-time PCR (qRT-PCR) was conducted using Hieff® qPCR SYBR Green Master Mix (No Rox), with *TaRP15* as the internal reference gene. Nine representative *TaRboh* genes that strongly responded to heat and drought were validated using qRT-PCR investigation. The primers used for qRT-PCR are listed in Table S5.

4.10. Statistical Analysis

The relative expression levels were calculated from triplicate biological and technical measurements using the $2^{-\triangle\triangle Ct}$ method, where $\triangle\triangle Ct = CT_{target/Cd} - CT_{actin/Cd} - \left(CT_{target/control} - CT_{actin/control}\right)$. The information was displayed as mean ± standard error (S.E.). Statistical comparisons were conducted using one-way ANOVA, with differences considered significant at $p < 0.05$. GraphPad Prism v9.5.0 was employed to visualize the obtained results.

5. Conclusions

Using bioinformatics analysis, 39 *TaRboh* genes were found in this study based on the most recent wheat genome assembly. The results from phylogenetic study show that *TaRbohs* were closely related to *OsRbohs*. *TaRbohs* were divided into ten groups based on their phylogenetic relationships, structures, and motifs. According to a collinearity analysis, *TaRbohs* had severe purifying selection during evolution, which led to the loss of certain wheat genes, in contrast to *Arabidopsis*. An investigation of protein–protein interactions revealed connections between the TaRboh proteins. According to miRNA target predictions, *TaRboh* gene regulation might be influenced by a number of miRNAs. Furthermore, *TaRboh* promoter predictions suggested that *TaRbohs* are important in plant growth and development, responses to biotic and abiotic stressors, and phytohormone signaling. As shown in the expression analysis, *TaRbohs* responded to heat, drought, and a combination heat and drought conditions, and these results were confirmed in the qPCR analysis. Results presented in this study form a solid foundation for future research into *TaRboh*-related stress resistance mechanisms in wheat.

Supplementary Materials: The following supporting information can be downloaded at: https://www.mdpi.com/article/10.3390/plants13233377/s1, Table S1: Domains of TaRboh proteins predicted by SMART analysis. Table S2: Physicochemical properties and subcellular localization predictions of *TaRboh* proteins. Table S3: Function and classification of *cis*-elements in *TaRboh* genes. Table S4: Homologous pairs and Ka/Ks value from collinearity analysis. Table S5: The primers of *TaRbohs* and refrerence gene for the qRT-PCR analysis.

Author Contributions: Conceptualization, X.G. and Y.L.; methodology, M.C.; validation, M.C. and Y.Z.; investigation, X.Z. and H.Y.; visualization, Y.Y. and Z.L.; resources, W.X.; writing—original draft, M.C.; writing—review and editing, Y.Z., S.L., and Y.L.; funding acquisition, X.G., S.L., Y.L., and W.X. All authors have read and agreed to the published version of the manuscript.

Funding: This work was supported by the National Natural Science Foundation of China (32372124, 32300456, 82304652), the China Postdoctoral Science Foundation (2022M721101, 2021M701160, 2023M731065), the Hunan Provincial Department of Science and Technology (2023JJ40132, 2022JJ40051, 2023JJ40199), the Changsha Natural Science Foundation (kq2202149), and the Natural Science Foundation of Chongqing, China (CSTB2023NSCQ-MSX0542, CSTB2023NSCQ-MSX1031, CSTB2022NSCQ-MSX0517, CSTB2022NSCQ-MSX1138).

Data Availability Statement: Data are contained within the article and Supplementary Materials.

Conflicts of Interest: The authors declare no conflicts of interest.

References

1. Torres, M.A.; Dangl, J.L. Functions of the Respiratory Burst Oxidase in Biotic Interactions, Abiotic Stress and Development. *Curr. Opin. Plant Biol.* **2005**, *8*, 397–403. [CrossRef] [PubMed]
2. Sagi, M.; Fluhr, R. Production of Reactive Oxygen Species by Plant NADPH Oxidases. *Plant Physiol.* **2006**, *141*, 336–340. [CrossRef] [PubMed]
3. Marino, D.; Dunand, C.; Puppo, A.; Pauly, N. A Burst of Plant NADPH Oxidases. *Trends Plant Sci.* **2012**, *17*, 9–15. [CrossRef] [PubMed]
4. Suzuki, N.; Koussevitzky, S.; Mittler, R.; Miller, G. ROS and Redox Signalling in the Response of Plants to Abiotic Stress. *Plant Cell Environ.* **2012**, *35*, 259–270. [CrossRef]
5. Yu, S.; Kakar, K.U.; Yang, Z.; Nawaz, Z.; Lin, S.; Guo, Y.; Ren, X.-L.; Baloch, A.A.; Han, D. Systematic Study of the Stress-Responsive Rboh Gene Family in Nicotiana Tabacum: Genome-Wide Identification, Evolution and Role in Disease Resistance. *Genomics* **2020**, *112*, 1404–1418. [CrossRef]
6. Wang, G.-F.; Li, W.-Q.; Li, W.-Y.; Wu, G.-L.; Zhou, C.-Y.; Chen, K.-M. Characterization of Rice NADPH Oxidase Genes and Their Expression under Various Environmental Conditions. *Int. J. Mol. Sci.* **2013**, *14*, 9440–9458. [CrossRef]
7. Chang, Y.; Li, B.; Shi, Q.; Geng, R.; Geng, S.; Liu, J.; Zhang, Y.; Cai, Y. Comprehensive Analysis of Respiratory Burst Oxidase Homologs (Rboh) Gene Family and Function of GbRboh5/18 on Verticillium Wilt Resistance in Gossypium Barbadense. *Front. Genet.* **2020**, *11*, 788. [CrossRef]
8. Zhang, H.; Wang, X.; Yan, A.; Deng, J.; Xie, Y.; Liu, S.; Liu, D.; He, L.; Weng, J.; Xu, J. Evolutionary Analysis of Respiratory Burst Oxidase Homolog (RBOH) Genes in Plants and Characterization of ZmRBOHs. *Int. J. Mol. Sci.* **2023**, *24*, 3858. [CrossRef]
9. Chapman, J.M.; Muhlemann, J.K.; Gayomba, S.R.; Muday, G.K. RBOH-Dependent ROS Synthesis and ROS Scavenging by Plant Specialized Metabolites To Modulate Plant Development and Stress Responses. *Chem. Res. Toxicol.* **2019**, *32*, 370–396. [CrossRef]
10. Wong, H.L.; Pinontoan, R.; Hayashi, K.; Tabata, R.; Yaeno, T.; Hasegawa, K.; Kojima, C.; Yoshioka, H.; Iba, K.; Kawasaki, T.; et al. Regulation of Rice NADPH Oxidase by Binding of Rac GTPase to Its N-Terminal Extension. *Plant Cell* **2008**, *19*, 4022–4034. [CrossRef]
11. He, H.; Yan, J.; Yu, X.; Liang, Y.; Fang, L.; Scheller, H.V.; Zhang, A. The NADPH-Oxidase AtRbohI Plays a Positive Role in Drought-Stress Response in Arabidopsis Thaliana. *Biochem. Biophys. Res. Commun.* **2017**, *491*, 834–839. [CrossRef] [PubMed]
12. Rivas, F.J.M.; Fernie, A.R.; Aarabi, F. Roles and Regulation of the RBOHD Enzyme in Initiating ROS-Mediated Systemic Signaling during Biotic and Abiotic Stress. *Plant Stress* **2024**, *11*, 100327. [CrossRef]
13. Shi, Y.; Chang, Y.-L.; Wu, H.-T.; Shalmani, A.; Liu, W.-T.; Li, W.-Q.; Xu, J.-W.; Chen, K.-M. OsRbohB-Mediated ROS Production Plays a Crucial Role in Drought Stress Tolerance of Rice. *Plant Cell Rep.* **2020**, *39*, 1767–1784. [CrossRef] [PubMed]
14. Gao, H.; Cui, J.; Liu, S.; Wang, S.; Lian, Y.; Bai, Y.; Zhu, T.; Wu, H.; Wang, Y.; Yang, S.; et al. Natural Variations of ZmSRO1d Modulate the Trade-off between Drought Resistance and Yield by Affecting ZmRBOHC-Mediated Stomatal ROS Production in Maize. *Mol. Plant* **2022**, *15*, 1558–1574. [CrossRef] [PubMed]

15. Li, X.; Zhang, H.; Tian, L.; Huang, L.; Liu, S.; Li, D.; Song, F. Tomato SlRbohB, a Member of the NADPH Oxidase Family, Is Required for Disease Resistance Against Botrytis Cinerea and Tolerance to Drought Stress. *Front. Plant Sci.* **2015**, *6*, 463. [CrossRef]
16. Guo, S.; Song, J.; Zhang, B.; Jiang, H.; Ma, R.; Yu, M. Genome-Wide Identification and Expression Analysis of Beta-Galactosidase Family Members During Fruit Softening of Peach [*Prunus persica* (L.) Batsch]. *Postharvest Biol. Technol.* **2018**, *136*, 111–123. [CrossRef]
17. Hu, C.-H.; Wei, X.-Y.; Yuan, B.; Yao, L.-B.; Ma, T.-T.; Zhang, P.-P.; Wang, X.; Wang, P.-Q.; Liu, W.-T.; Li, W.-Q.; et al. Genome-Wide Identification and Functional Analysis of NADPH Oxidase Family Genes in Wheat During Development and Environmental Stress Responses. *Front. Plant Sci.* **2018**, *9*, 906. [CrossRef]
18. Sharma, Y.; Ishu; Shumayla; Dixit, S.; Singh, K.; Upadhyay, S.K. Decoding the Features and Potential Roles of Respiratory Burst Oxidase Homologs in Bread Wheat. *Curr. Plant Biol.* **2024**, *37*, 100315. [CrossRef]
19. Liu, Z.; Xin, M.; Qin, J.; Peng, H.; Ni, Z.; Yao, Y.; Sun, Q. Temporal Transcriptome Profiling Reveals Expression Partitioning of Homeologous Genes Contributing to Heat and Drought Acclimation in Wheat (*Triticum aestivum* L.). *BMC Plant Biol.* **2015**, *15*, 152. [CrossRef]
20. Priest, H.D.; Filichkin, S.A.; Mockler, T.C. Cis-Regulatory Elements in Plant Cell Signaling. *Curr. Opin. Plant Biol.* **2009**, *12*, 643–649. [CrossRef]
21. Leister, D. Tandem and Segmental Gene Duplication and Recombination in the Evolution of Plant Disease Resistance Genes. *Trends Genet.* **2004**, *20*, 116–122. [CrossRef] [PubMed]
22. De Las Rivas, J.; Fontanillo, C. Protein-Protein Interaction Networks: Unraveling the Wiring of Molecular Machines within the Cell. *Brief. Funct. Genom.* **2012**, *11*, 489–496. [CrossRef] [PubMed]
23. Bartel, D.P. MicroRNAs. *Cell* **2004**, *116*, 281–297. [CrossRef] [PubMed]
24. Chen, X. Small RNAs and Their Roles in Plant Development. *Annu. Rev. Cell Dev. Biol.* **2009**, *25*, 21–44. [CrossRef] [PubMed]
25. Sato, H.; Mizoi, J.; Shinozaki, K.; Yamaguchi-Shinozaki, K. Complex Plant Responses to Drought and Heat Stress Under Climate Change. *Plant J.* **2024**, *117*, 1873–1892. [CrossRef]
26. Kimura, S.; Hunter, K.; Vaahtera, L.; Tran, H.C.; Citterico, M.; Vaattovaara, A.; Rokka, A.; Stolze, S.C.; Harzen, A.; Meißner, L.; et al. CRK2 and C-Terminal Phosphorylation of NADPH Oxidase RBOHD Regulate Reactive Oxygen Species Production in Arabidopsis. *Plant Cell* **2020**, *32*, 1063–1080. [CrossRef]
27. Torres, M.A.; Onouchi, H.; Hamada, S.; Machida, C.; Hammond-Kosack, K.E.; Jones, J.D. Six Arabidopsis Thaliana Homologues of the Human Respiratory Burst Oxidase (Gp91phox). *Plant J.* **1998**, *14*, 365–370. [CrossRef]
28. Chang, Y.-L.; Li, W.-Y.; Miao, H.; Yang, S.-Q.; Li, R.; Wang, X.; Li, W.-Q.; Chen, K.-M. Comprehensive Genomic Analysis and Expression Profiling of the NOX Gene Families Under Abiotic Stresses and Hormones in Plants. *Genome Biol. Evol.* **2016**, *8*, 791–810. [CrossRef]
29. Begum, K.; Das, A.; Ahmed, R.; Akhtar, S.; Kulkarni, R.; Banu, S. Genome-Wide Analysis of Respiratory Burst Oxidase Homolog (Rboh) Genes in Aquilaria Species and Its Association with Agarwood Formation. *Front. Plant Sci.* **2024**, *14*, 1326080. [CrossRef]
30. Du, L.; Jiang, Z.; Zhou, Y.; Shen, L.; He, J.; Xia, X.; Zhang, L.; Yang, X. Genome-Wide Identification and Expression Analysis of Respiratory Burst Oxidase Homolog (RBOH) Gene Family in Eggplant (*Solanum melongena* L.) under Abiotic and Biotic Stress. *Genes* **2023**, *14*, 1665. [CrossRef]
31. Cheng, C.; Che, Q.; Su, S.; Liu, Y.; Wang, Y.; Xu, X. Genome-Wide Identification and Characterization of Respiratory Burst Oxidase Homolog Genes in Six Rosaceae Species and an Analysis of Their Effects on Adventitious Rooting in Apple. *PLoS ONE* **2020**, *15*, e0239705. [CrossRef] [PubMed]
32. Li, T.; Kong, C.; Deng, P.; Li, C.; Zhao, G.; Li, H.; Gao, L.; Cui, D.; Jia, J. Intra-Varietal Diversity and Its Contribution to Wheat Evolution, Domestication, and Improvement in Wheat. *Int. J. Mol. Sci.* **2023**, *24*, 10217. [CrossRef] [PubMed]
33. Zimin, A.V.; Puiu, D.; Hall, R.; Kingan, S.; Clavijo, B.J.; Salzberg, S.L. The First Near-Complete Assembly of the Hexaploid Bread Wheat Genome, *Triticum aestivum*. *GigaScience* **2017**, *6*, gix097. [CrossRef]
34. Bottley, A.; Xia, G.M.; Koebner, R.M.D. Homoeologous Gene Silencing in Hexaploid Wheat. *Plant J.* **2006**, *47*, 897–906. [CrossRef] [PubMed]
35. Juery, C.; Concia, L.; De Oliveira, R.; Papon, N.; Ramírez-González, R.; Benhamed, M.; Uauy, C.; Choulet, F.; Paux, E. New Insights into Homoeologous Copy Number Variations in the Hexaploid Wheat Genome. *Plant Genome* **2021**, *14*, e20069. [CrossRef] [PubMed]
36. Fleury, D.; Baumann, U.; Langridge, P. Plant Genome Sequencing. In *Plant Biotechnology and Agriculture*; Elsevier: Amsterdam, The Netherlands, 2012; pp. 83–97. ISBN 978-0-12-381466-1.
37. Jung, S.; Jiwan, D.; Cho, I.; Lee, T.; Abbott, A.; Sosinski, B.; Main, D. Synteny of Prunus and Other Model Plant Species. *BMC Genom.* **2009**, *10*, 76. [CrossRef]
38. Bennetzen, J.L. Comparative Sequence Analysis of Plant Nuclear Genomes: Microcolinearity and Its Many Exceptions. *Plant Cell* **2000**, *12*, 1021–1029. [CrossRef]
39. Proost, S.; Fostier, J.; De Witte, D.; Dhoedt, B.; Demeester, P.; Van De Peer, Y.; Vandepoele, K. I-ADHoRe 3.0—Fast and Sensitive Detection of Genomic Homology in Extremely Large Data Sets. *Nucleic Acids Res.* **2012**, *40*, e11. [CrossRef]
40. Panchy, N.; Lehti-Shiu, M.; Shiu, S.-H. Evolution of Gene Duplication in Plants. *Plant Physiol.* **2016**, *171*, 2294–2316. [CrossRef]

41. Rao, S.; Balyan, S.; Jha, S.; Bansal, C.; Das, J.R.; Gupta, A.; Mathur, S. Orchestration of MicroRNAs and Transcription Factors in the Regulation of Plant Abiotic Stress Response. In *Plant Stress Biology*; Giri, B., Sharma, M.P., Eds.; Springer: Singapore, 2020; pp. 251–277. ISBN 9789811593796.
42. Vandepoele, K.; Quimbaya, M.; Casneuf, T.; De Veylder, L.; Van De Peer, Y. Unraveling Transcriptional Control in Arabidopsis Using Cis-Regulatory Elements and Coexpression Networks. *Plant Physiol.* **2009**, *150*, 535–546. [CrossRef]
43. Chen, K.; Rajewsky, N. The Evolution of Gene Regulation by Transcription Factors and microRNAs. *Nat. Rev. Genet.* **2007**, *8*, 93–103. [CrossRef] [PubMed]
44. Fujita, Y.; Fujita, M.; Satoh, R.; Maruyama, K.; Parvez, M.M.; Seki, M.; Hiratsu, K.; Ohme-Takagi, M.; Shinozaki, K.; Yamaguchi-Shinozaki, K. AREB1 Is a Transcription Activator of Novel ABRE-Dependent ABA Signaling That Enhances Drought Stress Tolerance in Arabidopsis. *Plant Cell* **2005**, *17*, 3470–3488. [CrossRef] [PubMed]
45. Zhao, X.Y.; Hong, P.; Wu, J.Y.; Chen, X.B.; Ye, X.G.; Pan, Y.Y.; Wang, J.; Zhang, X.S. The Tae-miR408-Mediated Control of TaTOC1 Genes Transcription Is Required for the Regulation of Heading Time in Wheat. *Plant Physiol.* **2016**, *170*, 1578–1594. [CrossRef] [PubMed]
46. Bai, Q.; Wang, X.; Chen, X.; Shi, G.; Liu, Z.; Guo, C.; Xiao, K. Wheat miRNA TaemiR408 Acts as an Essential Mediator in Plant Tolerance to Pi Deprivation and Salt Stress via Modulating Stress-Associated Physiological Processes. *Front. Plant Sci.* **2018**, *9*, 499. [CrossRef] [PubMed]
47. Feng, H.; Zhang, Q.; Wang, Q.; Wang, X.; Liu, J.; Li, M.; Huang, L.; Kang, Z. Target of Tae-miR408, a Chemocyanin-like Protein Gene (TaCLP1), Plays Positive Roles in Wheat Response to High-Salinity, Heavy Cupric Stress and Stripe Rust. *Plant Mol. Biol.* **2013**, *83*, 433–443. [CrossRef]
48. Li, L.; Mao, X.; Wang, J.; Chang, X.; Reynolds, M.; Jing, R. Genetic Dissection of Drought and Heat-responsive Agronomic Traits in Wheat. *Plant Cell Environ.* **2019**, *42*, 2540–2553. [CrossRef]
49. Suzuki, N.; Rivero, R.M.; Shulaev, V.; Blumwald, E.; Mittler, R. Abiotic and Biotic Stress Combinations. *New Phytol.* **2014**, *203*, 32–43. [CrossRef]
50. Jing, X.-Q.; Shi, P.-T.; Zhang, R.; Zhou, M.-R.; Shalmani, A.; Wang, G.-F.; Liu, W.-T.; Li, W.-Q.; Chen, K.-M. Rice Kinase OsMRLK63 Contributes to Drought Tolerance by Regulating Reactive Oxygen Species Production. *Plant Physiol.* **2024**, *194*, 2679–2696. [CrossRef]
51. Yin, Y.; Qin, K.; Song, X.; Zhang, Q.; Zhou, Y.; Xia, X.; Yu, J. BZR1 Transcription Factor Regulates Heat Stress Tolerance Through FERONIA Receptor-Like Kinase-Mediated Reactive Oxygen Species Signaling in Tomato. *Plant Cell Physiol.* **2018**, *59*, 2239–2254. [CrossRef]
52. Yao, Y.; He, R.J.; Xie, Q.L.; Zhao, X.H.; Deng, X.M.; He, J.B.; Song, L.; He, J.; Marchant, A.; Chen, X.; et al. ETHYLENE RESPONSE FACTOR 74 (ERF74) Plays an Essential Role in Controlling a Respiratory Burst Oxidase Homolog D (RbohD)-dependent Mechanism in Response to Different Stresses in Arabidopsis. *New Phytol.* **2017**, *213*, 1667–1681. [CrossRef]
53. Wang, X.; Li, Q.; Xie, J.; Huang, M.; Cai, J.; Zhou, Q.; Dai, T.; Jiang, D. Abscisic Acid and Jasmonic Acid Are Involved in Drought Priming-Induced Tolerance to Drought in Wheat. *Crop J.* **2021**, *9*, 120–132. [CrossRef]
54. Li, Y.; Li, Y.; Zou, X.; Jiang, S.; Cao, M.; Chen, F.; Yin, Y.; Xiao, W.; Liu, S.; Guo, X. Bioinformatic Identification and Expression Analyses of the MAPK–MAP4K Gene Family Reveal a Putative Functional MAP4K10-MAP3K7/8-MAP2K1/11-MAPK3/6 Cascade in Wheat (*Triticum aestivum* L.). *Plants* **2024**, *13*, 941. [CrossRef] [PubMed]
55. Chen, C.; Chen, H.; Zhang, Y.; Thomas, H.R.; Frank, M.H.; He, Y.; Xia, R. TBtools: An Integrative Toolkit Developed for Interactive Analyses of Big Biological Data. *Mol. Plant* **2020**, *13*, 1194–1202. [CrossRef] [PubMed]
56. Artimo, P.; Jonnalagedda, M.; Arnold, K.; Baratin, D.; Csardi, G.; De Castro, E.; Duvaud, S.; Flegel, V.; Fortier, A.; Gasteiger, E.; et al. ExPASy: SIB Bioinformatics Resource Portal. *Nucleic Acids Res.* **2012**, *40*, W597–W603. [CrossRef]
57. Chou, K.-C.; Shen, H.-B. Plant-mPLoc: A Top-Down Strategy to Augment the Power for Predicting Plant Protein Subcellular Localization. *PLoS ONE* **2010**, *5*, e11335. [CrossRef]
58. Thompson, J.D.; Gibson, T.J.; Higgins, D.G. Multiple Sequence Alignment Using ClustalW and ClustalX. *Curr. Protoc. Bioinform.* **2003**, 002. [CrossRef]
59. Robert, X.; Gouet, P. Deciphering Key Features in Protein Structures with the New ENDscript Server. *Nucleic Acids Res.* **2014**, *42*, W320–W324. [CrossRef]
60. Kumar, S.; Stecher, G.; Tamura, K. MEGA7: Molecular Evolutionary Genetics Analysis Version 7.0 for Bigger Datasets. *Mol. Biol. Evol.* **2016**, *33*, 1870–1874. [CrossRef]
61. Ciccarelli, F.D.; Doerks, T.; Von Mering, C.; Creevey, C.J.; Snel, B.; Bork, P. Toward Automatic Reconstruction of a Highly Resolved Tree of Life. *Science* **2006**, *311*, 1283–1287. [CrossRef]
62. Bailey, T.L.; Williams, N.; Misleh, C.; Li, W.W. MEME: Discovering and Analyzing DNA and Protein Sequence Motifs. *Nucleic Acids Res.* **2006**, *34*, W369–W373. [CrossRef]
63. Lescot, M. PlantCARE, a Database of Plant Cis-Acting Regulatory Elements and a Portal to Tools for in Silico Analysis of Promoter Sequences. *Nucleic Acids Res.* **2002**, *30*, 325–327. [CrossRef] [PubMed]
64. Wang, Y.; Tang, H.; DeBarry, J.D.; Tan, X.; Li, J.; Wang, X.; Lee, T.-h.; Jin, H.; Marler, B.; Guo, H.; et al. MCScanX: A Toolkit for Detection and Evolutionary Analysis of Gene Synteny and Collinearity. *Nucleic Acids Res.* **2012**, *40*, e49. [CrossRef] [PubMed]
65. Kumar, S.; Suleski, M.; Craig, J.M.; Kasprowicz, A.E.; Sanderford, M.; Li, M.; Stecher, G.; Hedges, S.B. TimeTree 5: An Expanded Resource for Species Divergence Times. *Mol. Biol. Evol.* **2022**, *39*, msac174. [CrossRef] [PubMed]

66. Wang, D.; Zhang, Y.; Zhang, Z.; Zhu, J.; Yu, J. KaKs_Calculator 2.0: A Toolkit Incorporating Gamma-Series Methods and Sliding Window Strategies. *Genom. Proteom. Bioinform.* **2010**, *8*, 77–80. [CrossRef] [PubMed]
67. Szklarczyk, D.; Gable, A.L.; Nastou, K.C.; Lyon, D.; Kirsch, R.; Pyysalo, S.; Doncheva, N.T.; Legeay, M.; Fang, T.; Bork, P.; et al. The STRING Database in 2021: Customizable Protein–Protein Networks, and Functional Characterization of User-Uploaded Gene/Measurement Sets. *Nucleic Acids Res.* **2021**, *49*, D605–D612. [CrossRef]
68. Shannon, P.; Markiel, A.; Ozier, O.; Baliga, N.S.; Wang, J.T.; Ramage, D.; Amin, N.; Schwikowski, B.; Ideker, T. Cytoscape: A Software Environment for Integrated Models of Biomolecular Interaction Networks. *Genome Res.* **2003**, *13*, 2498–2504. [CrossRef]
69. Dai, X.; Zhuang, Z.; Zhao, P.X. psRNATarget: A Plant Small RNA Target Analysis Server (2017 Release). *Nucleic Acids Res.* **2018**, *46*, W49–W54. [CrossRef]

plants

MDPI

Article

Phosphoproteomic Analysis Reveals Impairment of Rice Germination by Chloramphenicol

Rui Li [1,2], Narumon Phaonakrop [3], Sittiruk Roytrakul [3], Karan Lohmaneeratana [2] and Arinthip Thamchaipenet [2,4,*]

1 Interdisciplinary Graduate Program in Bioscience, Faculty of Science, Kasetsart University, Bangkok 10900, Thailand; rui.l@ku.th
2 Department of Genetics, Faculty of Science, Kasetsart University, Bangkok 10900, Thailand; karan.l@ku.th
3 National Center for Genetic Engineering and Biotechnology, National Science and Technology Development Agency, Pathum Thani 12120, Thailand; narumon.pha@biotec.or.th (N.P.); sittiruk@biotec.or.th (S.R.)
4 Omics Center for Agriculture, Bioresources, Food, and Health, Kasetsart University (OmiKU), Bangkok 10900, Thailand
* Correspondence: arinthip.t@ku.ac.th

Abstract: Seed germination is a critical phase in rice production and is highly sensitive to environmental and chemical stresses. Chloramphenicol (CAM), a known phytotoxic antibiotic, has been reported to suppress rice seedling establishment, yet its underlying molecular mechanisms remain poorly understood. In this study, we investigated the effects of varying CAM concentrations on rice germination and early seedling establishment. While CAM significantly retarded germination speed and seedling growth, the final germination rates remained largely unaffected, even at high concentrations. To uncover the molecular basis of CAM phytotoxicity, we conducted time-resolved phosphoproteomic profiling during both the germination and early seedling stages. Our analyses revealed dynamic, stage-specific phosphorylation changes: moderate alterations affecting metabolic and cytokinesis-related processes during germination, and extensive disruptions in metabolic pathways, stress response mechanisms, DNA replication, and hormone signaling during early seedling establishment. Collectively, these findings demonstrate that CAM disrupts rice development by remodeling phosphorylation networks and modulating key physiological and signaling pathways. This study provides novel insights into the molecular mechanisms underlying antibiotic-induced growth inhibition and advances our understanding of plant stress responses during early development.

Keywords: chloramphenicol; rice (*Oryza sativa* L.) germination; phosphoproteomics; signaling pathways

1. Introduction

Rice is a staple food for a significant proportion of the global population and plays a vital role in ensuring food security and economic stability worldwide [1]. With rising demand [2], improving rice yield remains a critical objective for production. Thailand is a major contributor to the global rice trade [3], with Khao Dawk Mali 105 (*Oryza sativa* L. cv. KDML 105), commonly known as Thai Jasmine rice, being one of its key export cultivars [4]. Although KDML 105 is favored in the market for its exceptional quality and distinctive aroma, its yield per hectare is typically lower than that of other white rice varieties [5]. Therefore, enhancing the yield of KDML 105 is crucial for bolstering Thailand's agricultural economy and supporting the country's contribution to the global rice trade.

Germination is the initial step in rice production, crucial for subsequent vegetative growth and ultimately influencing optimal yield. From an omics perspective, this process can be divided into two main stages: (i) the germination stage, occurring from 0 to 72 h, and (ii) the early seedling establishment stage, occurring from 72 to 168 h [6,7]. During the germination stage, seeds undergo three distinct phases: Phase I involves water uptake; Phase II entails activation of seed storage enzymes; and Phase III marks the initiation of key metabolic pathways [8]. These events culminate in the mobilization of storage reserves to support seedling establishment. Disruption of this process can significantly compromise plant vigor and productivity, making it a key target for crop improvement efforts.

Chloramphenicol (CAM) is a broad-spectrum antibiotic that inhibits protein synthesis by binding to the 50S ribosomal subunit, thereby preventing peptide bond formation [9]. Although primarily effective against prokaryotes, CAM persists in agricultural environments through wastewater contamination [10] and soil microbial activity [11], raising concerns about its ecological impact. Previous studies have revealed that CAM exerts phytotoxic effects across different plant species [12] by disrupting key physiological processes such as nutrient and water uptake [13,14], photosynthesis [15], reactive oxygen species (ROS) homeostasis [15]. These disruptions underscore CAM's role as a potential environmental stressor in agroecosystems. In the context of seed biology, CAM has been proposed to inhibit germination by suppressing α-amylase activity [16] and compromising membrane integrity through hydrophobic interactions [17]. In tomato, CAM was found to inhibit early seedling establishment more strongly than germination initiation [18], suggesting stage-specific sensitivity.

Plant hormones regulate seed germination and early seedling growth in response to environmental cues. Early work by Noodén and Thimann (1965) demonstrated that CAM significantly disrupts auxin-mediated plant growth during germination. They further showed that exogenous application of indole-3-acetic acid (IAA) partially alleviates CAM-induced growth inhibition [19], suggesting that CAM interferes with the auxin signaling pathway. Although it was proposed that this interference results from the inhibition of protein biosynthesis required for auxin signaling [19], the specific protein targets affected by CAM remain unidentified. Given CAM's established role as a protein biosynthesis inhibitor, investigating CAM-induced proteomic alterations is critical for elucidating its underlying phytotoxic mechanisms.

Protein phosphorylation is a critical post-translational modification that regulates protein structure and function, playing essential roles in enzyme activity, signal transduction, and stress responses [20]. Phosphoproteomic analysis enables dynamic, system-level profiling of protein activation states and underlying signaling networks. Prior studies have demonstrated that phosphorylation patterns in rice seeds change dynamically during germination. Notably, proteins involved in hormonal signaling are predominantly phosphorylated during early germination, while those linked to storage mobilization and stress adaptation dominate later stages [21]. Additionally, spatial phosphoproteomic analyses have revealed phosphorylation of nuclear zinc finger proteins as regulators of germination at the subcellular level [22]. These findings highlight phosphoproteomics as a promising approach to elucidate how rice responds to CAM-induced stress during germination.

To address current knowledge gaps, the present study employed time-resolved phosphoproteomic analysis using LC-MS/MS to characterize dynamic changes in protein phosphorylation across distinct germination stages under CAM treatment. By identifying key regulatory proteins and signaling pathways affected by CAM, this work provides novel insights into the molecular basis of antibiotic-induced phytotoxicity and lays the groundwork for developing strategies to mitigate its adverse effects on rice productivity.

2. Results

2.1. The Effect of Chloramphenicol (CAM) on Seed Germination

The investigation of the effects of CAM (FUJIFILM Wako Pure Chemical Corporation, Osaka, Japan) at various concentrations suggests that CAM minimally affects overall seed germinability. No significant differences in germination rates were observed between the CAM-treated seeds and the water control, with all groups achieving germination rates above 98% (Figure 1a). Furthermore, treatments with 500 μg/mL, 1500 μg/mL, and 2500 μg/mL CAM significantly delayed the median germination time from 3 to 4.5 days, whereas no significant differences were observed between the control and the lower concentration treatments (Figure 1b). In contrast to its minimal effect on germination, CAM treatment significantly inhibited seedling establishment. Exposure to CAM concentrations above 15 μg/mL led to a noticeable reduction in shoot length at 3-day (germination stage) (Figure 1c). This inhibitory effect became more pronounced at 6-day (early seedling establishment stage), resulting in substantial suppression of shoot growth (Figure 1d). A similar result was observed in root development. Root growth was strikingly restricted when exposed to CAM concentrations above 5 μg/mL during the germination stage and was markedly reduced during the early seedling establishment stage (Figure 1e,f). The seedling vigor index (SVI), which integrates germination percentage and seedling growth, was used to assess overall seedling performance under CAM treatment. A significant reduction in SVI was observed even at the lowest CAM concentration of 5 μg/mL. However, increasing the CAM concentration beyond 50 μg/mL did not result in further statistically significant changes in SVI (Figure 1g).

Moreover, CAM exposure caused marked alterations in rice seedling morphology (Figure 2a). Crown root formation was completely inhibited at all tested CAM concentrations (Figure 2b). No normal seedlings, with both root and first leaf, developed when seeds were treated with CAM concentrations exceeding 15 μg/mL (Figure 2c). While no statistically significant differences in SVI were observed between the 15 and 50 μg/mL CAM treatments (Figure 1g), 15 μg/mL was the highest concentration that allowed for the successful establishment of normal seedlings. Based on this, 15 μg/mL CAM was selected for sample preparation in the subsequent phosphoproteomic analysis.

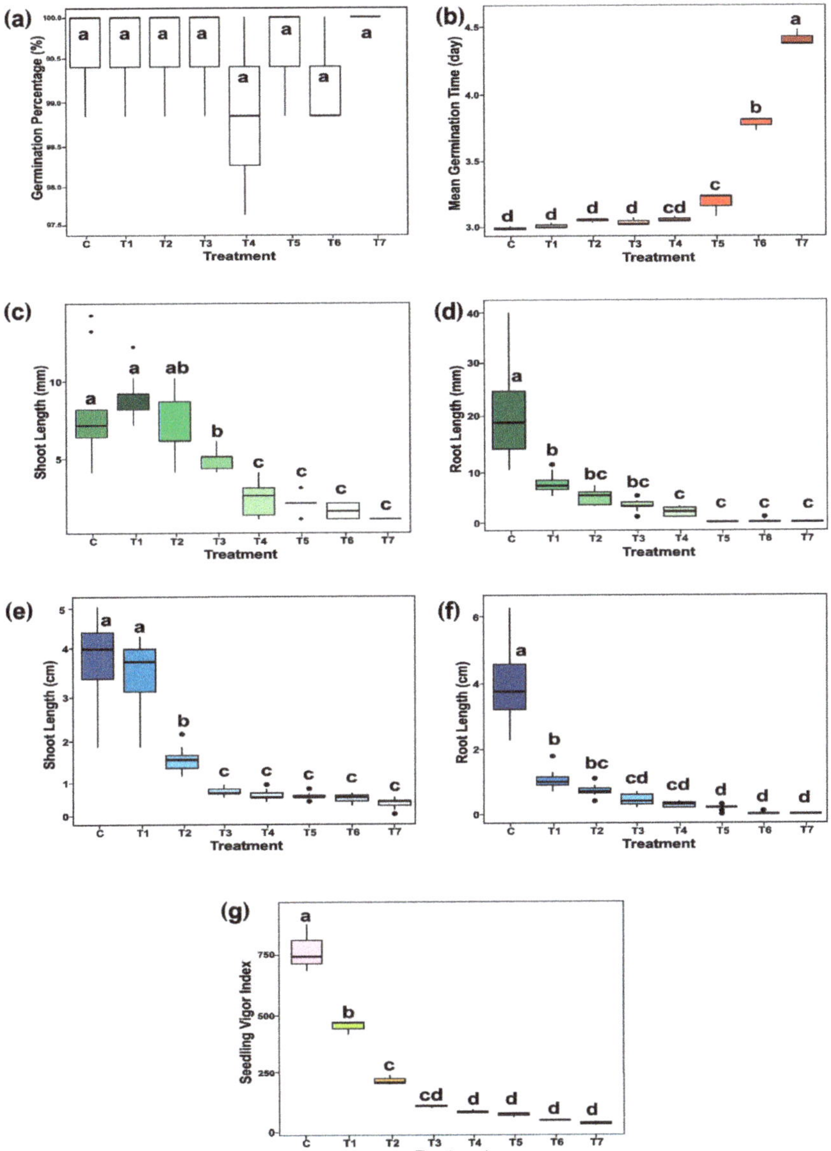

Figure 1. Effects of chloramphenicol (CAM) in rice (*Oryza sativa* L. cv. KDML 105) at 3-day (germination stage) and 6-day (early seedling establishment stage). (**a**) Germination percentage, (**b**) median germination time, (**c**) shoot length at 3-day, (**d**) shoot length at 6-day, (**e**) root length at 3-day, (**f**) root length at 6-day, and (**g**) seedling vigor index (SVI). Different lowercase letters above the bars indicate significant differences between treatments based on Tukey's HSD test ($p < 0.05$). C, water control; T1, T2, T3, T4, T5, T6, and T7 represent treatments of 5, 15, 50, 150, 500, 1500, and 2500 µg/mL CAM, respectively.

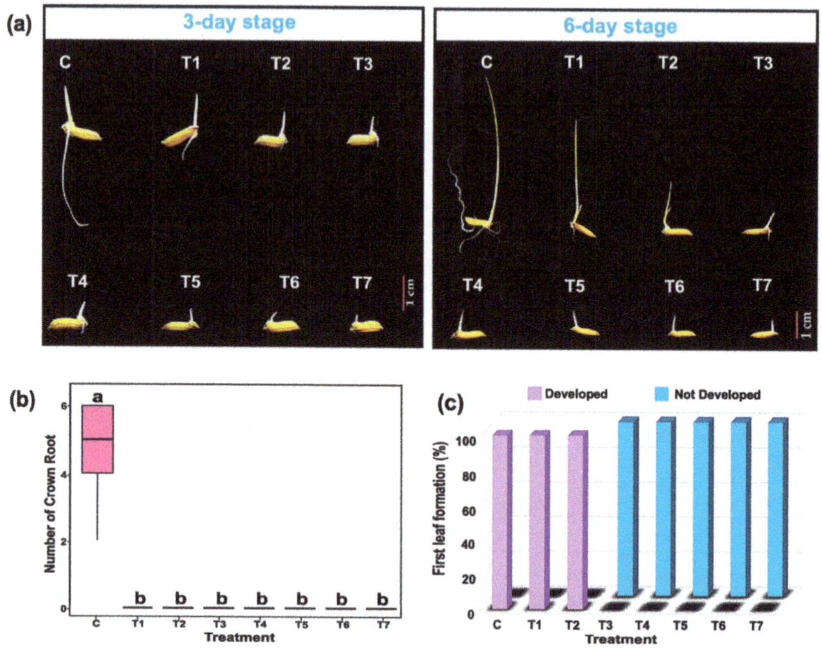

Figure 2. Effects of chloramphenicol (CAM) on the morphology of rice (*Oryza sativa* L. cv. KDML 105) seedlings. (**a**) Representative seedling phenotypes at 3-day (germination stage) and 6-day (early seedling establishment stage), (**b**) crown root number, and (**c**) percentage of first leaf formation. Different lowercase letters above the bars indicate significant differences between treatments based on Tukey's HSD test ($p < 0.05$). C, water control; T1, T2, T3, T4, T5, T6, and T7 represent treatments of 5, 15, 50, 150, 500, 1500, and 2500 µg/mL CAM, respectively.

2.2. The Effect of CAM on the Phosphoprotein Profiles of Germinating Seeds

Phosphoprotein samples derived from germinated seeds treated with sterile distilled water (control) and 15 µg/mL CAM (treatment) were analyzed using LC-MS/MS. A total of 8509 phosphoproteins were identified (Table S1). This analysis revealed that CAM treatment altered the global phosphoprotein profiles of germinating seeds (Figure 3a). Partial least squares discriminant analysis (PLS-DA) further resolved the data into four distinct clusters, demonstrating statistically significant variations between control and CAM treatment across the two germination stages (Figure 3b). Stage-specific PLS-DA analyses produced comparable results (Figure 3c,d). Subsequently, the differentially expressed phosphoprotein (DEPP) analysis revealed that at the germination stage (3-day), 16 phosphoproteins were upregulated, while 24 were downregulated (Figure 3e, Table S2). At the early seedling establishment stage (6-day), 18 phosphoproteins were upregulated, and 29 were downregulated (Figure 3f, Table S2). These DEPPs were identified as CAM-responsive phosphoproteins for further bioinformatics analysis.

Gene ontology (GO) annotation indicated that CAM-response phosphoproteins across a broad spectrum of biological processes (Figure 4). Several biological processes, including cellular component organization, nucleic acid metabolism, response to stimulus, development, cell cycle, signaling, and lipid metabolism, were influenced by CAM independent of developmental stage. Biological processes related to tetrahydrobiopterin, hormone metabolism, and cell differentiation were specifically activated at the germination stage (3-day), whereas biological processes associated with carbohydrate derivative metabolism, protein metabolism, cell division, and the innate immune response became more prominent at the early seedling establishment stage (6-day). Similarly, CAM treatment affected a variety of cellular components, with an observable shift from membrane, plastid, and mitochondrion localization at the germination stage to nucleus, cytoplasm, Golgi apparatus, endosome, and Cajal body localization at 6-day. Catalytic activity, metal ion binding, and nucleic acid binding were the top three molecular functions associated with the identified phosphoproteins at both stages (Figure 4).

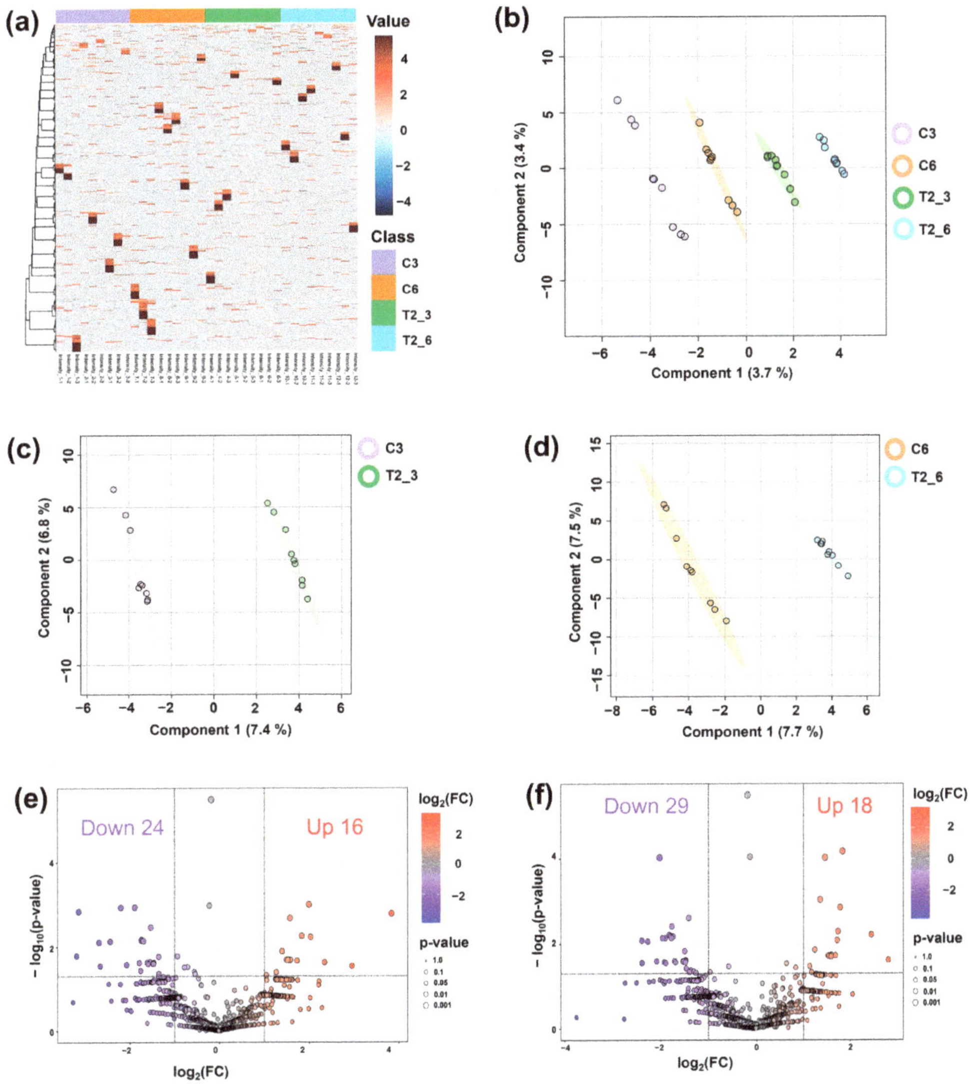

Figure 3. Phosphoprotein profiles of germinated rice seeds (*Oryza sativa* L. cv. KDML 105) in response to CAM at 3-day (germination stage) and 6-day (early seedling establishment stage). (**a**) Heatmap depicting the phosphoprotein profiles of CAM-treated rice at 3-day and 6-day. Two-dimensional PLS-DA plots of phosphoprotein profiles of (**b**) controls and CAM-treated rice at 3-day and 6-day, (**c**) control and CAM-treated rice at 3-day, (**d**) control and CAM-treated rice at 6-day. Volcano plots of CAM-responsive phosphoproteins at (**e**) 3-day and (**f**) 6-day. C3, 3-day control; C6, 6-day control; T2_3, 3-day CAM treatment; and T2_6, 6-day CAM treatment.

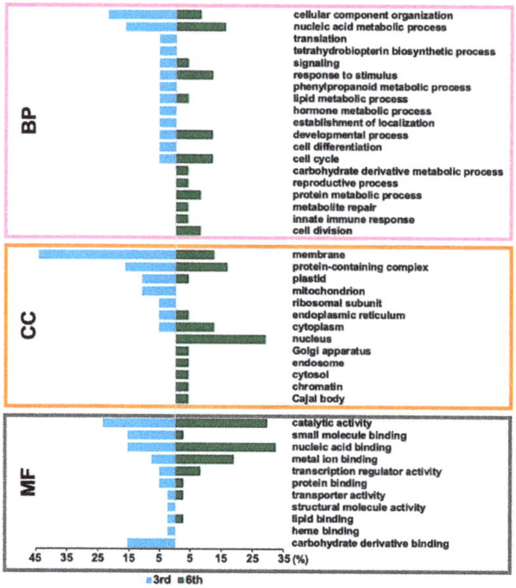

Figure 4. Gene Ontology (GO) annotation of identified CAM-responsive phosphoproteins at 3-day (germination stage) and 6-day (early seedling establishment stage). MF, molecular function; BP, biological process; CC, cellular component; blue bar, 3-day; and green bars, 6-day annotations.

2.3. Enrichment Analysis of Identified CAM-Responsive Phosphoproteins

Pathway enrichment analysis revealed that CAM's impact on germinating rice seeds intensified over time (Figure 5). At 3-day (germination stage), CAM upregulated phosphoproteins involved in macromolecule modification, GPI-anchor biosynthesis, cytokinin biosynthesis, and cytoskeleton-dependent cytokinesis, while downregulating those associated with tetrahydrobiopterin biosynthesis and anatomical structure development. By 6-day (early seedling establishment stage), CAM-responsive phosphoproteins were predominantly enriched in pathways related to macromolecule metabolism, regulation of the mitotic cell cycle, anatomical structure development, response to stimuli, protein monoubiquitination, the innate immune response, and ethylene-activated signaling. In contrast, proteins involved in metabolite repair, intracellular cholesterol transport, DNA replication, and DNA repair were downregulated. Notably, phosphoproteins related to anatomical structure development responded to CAM at both stages, with greater enrichment observed at the later stage (Table 1).

Table 1. Pathway enrichment analysis results of CAM-responsive phosphoproteins of germinated rice seeds (*Oryza sativa* L. cv. KDML 105) at 3-day (germination stage) and 6-day (early seedling establishment stage).

Term Name	Term ID	$-\log_{10}$ (p-Value)	Count	Uniprot ID	Subcellular Localization
Upregulation at 3-day stage					
Assembly of actomyosin apparatus involved in cytokinesis	GO:0000912	2.408	1	Q5NAV7	Cytoplasm, Nucleus
GPI anchor biosynthetic process	GO:0006506	1.355	1	Q84MV9	Endoplasmic reticulum
Cytokinin biosynthetic process	GO:0009691	1.355	1	Q33CD3	Mitochondrion
Macromolecule modification	GO:0043412	1.302	3	Q5NAV7	Nucleus
				Q84MV9	Endoplasmic reticulum
				Q33CD3	Mitochondrion

Table 1. *Cont.*

Term Name	Term ID	$-\log_{10}$ (*p*-Value)	Count	Uniprot ID	Subcellular Localization
Downregulation at 3-day stage					
Tetrahydrobiopterin biosynthetic process	GO:0006729	1.904	1	Q5SN39	Mitochondrion
Anatomical structure development	GO:0032989	1.558	1	Q42982	Cytoplasm
Upregulation at 6-day stage					
Regulation of mitotic cell cycle	GO:0007346	2.161	2	Q8H2X8	Nucleus
				Q7XU27	Nucleus
Anatomical structure development	GO:0048366	2.148	2	Q8H2X8	Nucleus
				Q7XU27	Nucleus
Protein monoubiquitination	GO:0006513	1.792	1	Q7XU27	Nucleus
Macromolecule metabolic process	GO:0043170	1.696	5	Q8H2X8	Nucleus
				Q7XU27	Nucleus
				Q651A5	Nucleus
				Q6EUK2	Nucleus
				A0A0P0WTE8	Cytoplasm
Innate immune response	GO:0045087	1.532	1	Q7XU27	Nucleus
Ethylene-activated signaling pathway	GO:0009873	1.497	1	Q651A5	Nucleus
Response to stimulus	GO:0050896	1.337	1	Q7XU27	Nucleus
Downregulation at 6-day stage					
Intracellular cholesterol transport	GO:0032367	1.679	1	Q5QMW8	Cytoplasm
Metabolite repair	GO:0110051	1.554	1	Q2QQ94	Cytoplasm
Nuclear cell cycle DNA replication initiation	GO:1902315	1.525	1	Q5JKB0	Nucleus
DNA repair	GO:0006281	1.379	1	Q5JKB0	Nucleus

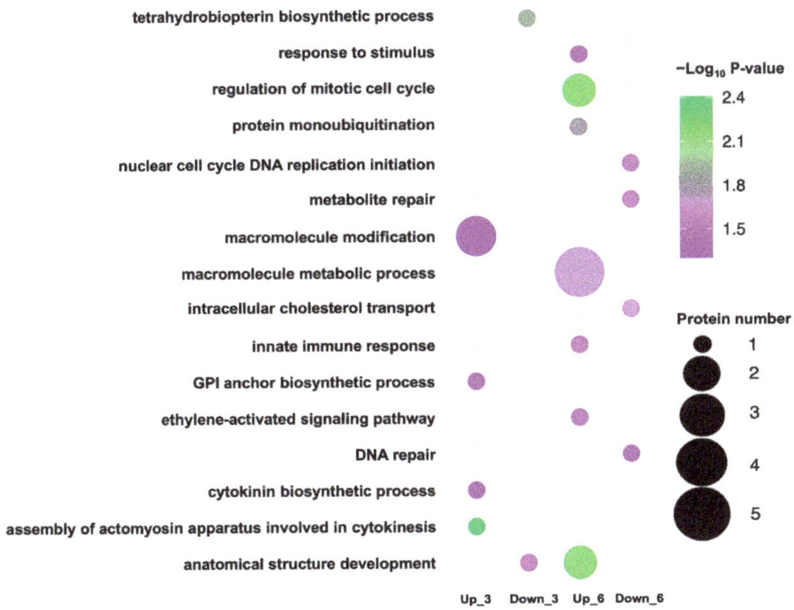

Figure 5. Bubble plot of pathway enrichment in CAM-response phosphoproteins at 3-day (germination stage) and 6-day (early seedling establishment stage). Bubble corresponds to a biological pathway, bubble size indicates the number of enriched proteins, and bubble color reflects the direction of regulation. The x-axis shows different conditions of upregulated CAM-response phosphoproteins at 3-day (Up_3) and 6-day (Up_6) and downregulated CAM-response phosphoproteins at 3-day (Down_3) and 6-day (Down_6). The y-axis categorizes pathways by their annotations.

3. Discussion

3.1. CAM-Induced Physiological and Phosphoproteomic Changes During Germination

The physiological results demonstrated that although a high concentration of chloramphenicol (CAM) (2500 µg/mL, approaching its solubility limit in water) significantly delayed median germination time from 3 to 4.5 days, the germination rate was still reasonably high (98%, Figure 1a,b). This observation aligns with previous research in lettuce, which reported that 3000 µg/mL CAM had minimal effects on seed germinability [23].

In contrast, all tested CAM concentrations significantly inhibited seedling growth, with the degree of inhibition varying by organ and developmental stage. Roots—highly responsive to environmental stimuli during rice germination [24]—were more sensitive to CAM than shoots. Primary root growth was markedly suppressed beginning at 5 µg/mL CAM during both germination and early seedling establishment, whereas shoot growth remained unaffected at this concentration (Figure 1e,f). This organ-specific sensitivity concurs with previous studies in *Poaceae* and *Brassicaceae* reporting similar root inhibition at comparable CAM levels [12], likely due to preferential CAM accumulation in root tissues as observed in broad bean plants [25]. Additionally, CAM suppressed crown root formation during germination (Figure 2a,b). These results suggest that the reduction in seedling vigor is not solely a consequence of delayed germination but also reflects broader CAM-induced toxicity affecting seedling physiology.

At the molecular level, CAM-induced alterations in phosphorylation profiles exhibited clear stage specificity. Among the identified differentially expressed phosphoproteins (DEPPs), only two—A0A0P0X308 and Q2QQ28—were consistently downregulated across both germination stages (Figure S1 and Table S6). A0A0P0X308 remains uncharacterized, whereas Q2QQ28 encodes a zinc finger protein of the Cys-Cys-His-Cys (CCHC) type, which is predicted to localize to the nucleus (Figure S2). Notably, previous studies have shown that altering the stage-specific phosphorylation of nucleus-localized zinc finger proteins can significantly delay rice germination [22]. Therefore, our findings suggest that CAM may slow rice germination by modifying the phosphorylation status of this CCHC-type zinc finger protein.

Further enrichment analysis revealed clear stage-specific differences (Figure 5) that support our physiological observations. At the early seedling establishment stage, CAM-responsive phosphoproteins participated in a broader spectrum of biological processes compared to those observed at the germination stage. This finding aligns with previous reports describing dynamic phosphoproteomic changes during early rice germination [26] and supports observations of the accelerated inhibitory effects of CAM on plant growth [19]. Notably, a significant increase in the number of nucleus-localized CAM-responsive phosphoproteins was detected at the later germination stage (Figure 4, Table 1), diverging from earlier studies that documented a general decline in nuclear phosphoproteins during germination [22]. Nevertheless, CAM-responsive phosphoproteins associated with morphological development were identified at both stages, underscoring the impact of CAM on developmental regulation during rice germination.

Collectively, these physiological and phosphoproteomic results reveal that sensitivity to CAM during germination is both organ-specific and dependent on the developmental stage. Furthermore, the effects of CAM become more pronounced over time, progressively affecting a broader and more complex array of biological processes as the duration of exposure extends.

3.2. The Impact of CAM Exposure on the 3-Day Germination Stage in Rice

3.2.1. CAM Treatment Disrupts Phragmoplast-Based Cytokinesis

RUNKEL (RUK) (Q5NAV7), a member of the FUSED serine/threonine kinase family, regulates phragmoplast-based cytokinesis by directly binding and modulating the localization of microtubule-associated proteins (MAPs) [27], including the kinesin-like protein KIN-7A [28], a key promoter of phragmoplast expansion [29]. Loss of RUK function causes cytokinesis defects due to KIN-7A mislocalization [28]. In this study, exposure to CAM for 72 h significantly increased phosphorylated RUK levels in germinating rice (Figure 6). Although RUK is classified as a pseudokinase and likely lacks catalytic activity [30], its phosphorylation may interfere with its ability to bind MAP targets. Given that mislocalization of KIN-7A leads to growth inhibition [28], we propose that CAM-induced phosphorylation diminishes RUK's binding affinity, resulting in MAP mislocalization and the subsequent disruption of phragmoplast-based cytokinesis. This hypothesis is further supported by protein network analysis, which reveals predicted interactions between RUK and KIN-7A (Figure 6).

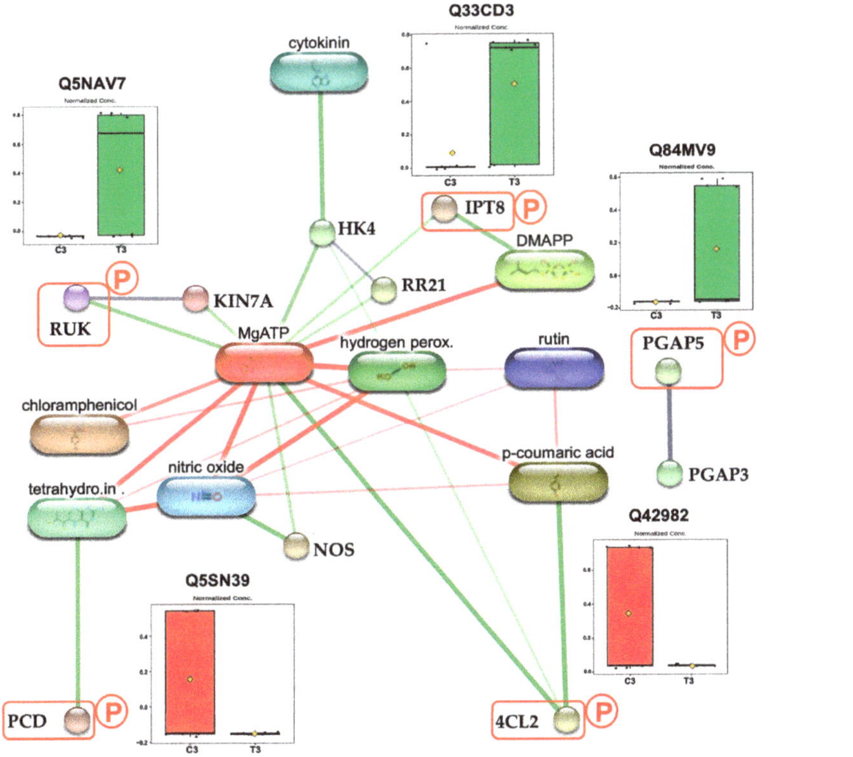

Figure 6. Protein–protein interaction network of CAM-responsive phosphoproteins and their potential functional partners identified at the germination stage (3-day). Phosphoproteins identified in this study are highlighted with a red square marked 'P', and the abundance of each protein under control (C) and treatment (T) conditions is displayed in a box plot with statistical comparisons using Fisher's LSD test ($p < 0.05$). Red and green boxes represent phosphoproteins abundance in control and treatment samples, respectively.

3.2.2. CAM Disrupts GPI Anchor Protein Maturation

The Os03g0397300 protein (Q84MV9) is a rice homolog of post-GPI attachment to proteins factor 5 (PGAP5). In eukaryotes, PGAP5 functions primarily as a metallophosphoesterase that regulates glycan core remodeling during glycosylphosphatidylinositol (GPI) anchor biosynthesis [31]. This remodeling occurs in the endoplasmic reticulum (ER) and is essential for the efficient transport of GPI-APs from the ER to the Golgi apparatus [31,32], where they undergo further maturation via PGAP3-mediated lipid remodeling before being delivered to the plasma membrane and cell wall [33,34]. In plants, GPI anchor–mediated protein localization is important for cell wall formation and seed germination [35]. Disruption of lipid remodeling during GPI anchor biosynthesis reduces cellulose production, consequently inhibits growth in rice [36]. In this study, we found that three days of CAM exposure increases PGAP5 phosphorylation in rice (Figure 6). Although direct evidence for phosphorylation's role in PGAP5 regulation is lacking, the growth inhibition observed under CAM treatment—consistent with that seen in *PGAP5* mutants [37]—suggests that CAM-induced phosphorylation may interfere with PGAP5 activity. We propose that CAM suppresses seedling growth by promoting PGAP5 phosphorylation, which disrupts glycan core remodeling, reduces transport of GPI-anchored proteins from the ER to the Golgi, and ultimately weakens cell wall formation.

3.2.3. CAM Treatment Enhances Mitochondrial CK Biosynthesis

Adenylate isopentenyltransferase 8 (IPT8, Q33CD3) regulates cytokinin (CK) biosynthesis primarily by catalyzing the transfer of an isopentenyl group from dimethylallyl pyrophosphate (DMAPP) to adenine nucleotides, mainly AMP [38,39]. IPT8 expression is predominantly localized in roots and positively correlates with endogenous trans-zeatin levels, the major active CK form in plants [40,41]. Overexpression of AtIPT8 has been shown to impair seedling growth, particularly root development, by increasing CK levels, which in turn elevate reactive oxygen species (ROS) such as hydrogen peroxide (H_2O_2) [42]. Although CK signaling in rice involves a well-characterized phosphorelay cascade—including the histidine kinase receptor OsHK4 and the type-B response regulator OsRR21 [43,44]—the functional consequences of OsIPT8 phosphorylation on its enzymatic activity remain unknown. In this study, we observed increased phosphorylation of OsIPT8 following 72 h of CAM treatment (Figure 6). Given that the morphological changes induced by CAM resemble the AtIPT8 overexpression phenotype [42], we hypothesize that CAM-induced phosphorylation activates OsIPT8, thereby enhancing cytokinin biosynthesis and contributing to growth inhibition through elevated oxidative stress.

3.2.4. CAM Exposure Disrupts Nitric Oxide Biosynthesis

Pterin-4-alpha-carbinolamine dehydratase (PCD) is a key enzyme in tetrahydrobiopterin (BH_4) metabolism, catalyzing the conversion of 4a-hydroxytetrahydrobiopterin back into its active form [45,46]. Given that BH_4 serves as an essential cofactor for nitric oxide synthase (NOS) [47], PCD indirectly regulates nitric oxide (NO) production [48]. In rice, inhibition of NOS-mediated NO synthesis has been shown to compromise oxidative stress tolerance, resulting in impaired germination and root development [49,50]. In our study, we detected a decrease in phosphorylated PCD (Q5SN39) levels after 72 h of CAM treatment (Figure 6). While direct evidence linking phosphorylation to PCD activation remains to be established, the well-documented role of NO in rice stress tolerance [51] supports the hypothesis that CAM-induced reduction in PCD phosphorylation disrupts BH_4 metabolism and NO signaling, thereby contributing to the observed germination and growth deficits.

3.2.5. CAM Treatment Reduces Flavonoid Metabolic Pathway

4-Coumarate-CoA ligase 2 (4CL2) catalyzes the conversion of 4-coumaric acid into 4-coumaroyl-CoA, a key intermediate in the flavonoid biosynthesis pathway [52]. Unlike other members of the 4CL family, 4CL2 is primarily associated with flavonoid production rather than lignin formation [53]. In *Camellia sinensis*, the expression level of Cs4CL2 has been shown to correlate with the endogenous accumulation of quercetin-3-O-rutinoside (rutin) [54], a major flavonoid compound. Rutin contributes to plant development through its role in anatomical structure formation, and exogenous application of rutin has been reported to promote plant growth [55]. Moreover, rutin functions as a potent antioxidant [56], enhancing stress tolerance by facilitating the scavenging of ROS [57], including hydrogen peroxide (H_2O_2) [58]. In rice, Os4CL2 displays high enzymatic activity in roots [53], and increased accumulation of rutin under salinity stress has been identified as part of the adaptive response to abiotic stress conditions [59]. Although serine residues have been identified as potential phosphorylation sites in Bn4CL3 [60], the regulatory role of phosphorylation in 4CL enzymes remains largely uncharacterized. In this study, we observed that CAM treatment significantly reduced the phosphorylation level of Os4CL2 (Q42982) (Figure 6) as well as root development during early germination (3-day). Given the established role of rutin in promoting rice root growth [61], our findings suggest that phosphorylation may be critical for Os4CL2 activation and subsequent rutin biosynthesis. We propose that CAM-mediated dephosphorylation diminishes Os4CL2 activity, leading to reduced rutin production and impaired ROS scavenging capacity, thereby negatively affecting root development.

3.3. *The Impact of CAM Exposure on the 6-Day Early Seedling Establishment Stage in Rice*

3.3.1. CAM Enhances Nuclear Export of SHR

SHORT-ROOT (SHR) is a pivotal transcription factor specifically expressed in stele cells that regulates plant growth, particularly root architecture, by coordinating cell division and tissue patterning [62,63]. For SHR to function as a non-cell-autonomous regulator of root cell differentiation, it must be exported from the nucleus into the cytoplasm of stele cells, allowing its movement via plasmodesmata into adjacent endodermal cells [64]. Within endodermal cells, SHR interacts with SCARECROW (SCR) and accumulates in the nucleus to regulate gene expression [65]. In rice, OsSHR1 similarly regulated root development through interactions with SCR in the endodermis [66], and loss-of-function mutants of OsSHR1 inhibited seedling growth and impaired root development [67]. In *Arabidopsis*, SHR phosphorylation at specific threonine residues modulates its nuclear export and intercellular trafficking [64], suggesting post-translational modification as a key regulatory mechanism. Computational predictions from our study indicate a high potential for the serine phosphorylation of OsSHR1 in rice under CAM treatment (Figure S3). Consistent with this, we observed a significant upregulation of phosphorylated OsSHR1(Q8H2X8) levels after 144 h of CAM treatment (Figure 7). Intriguingly, the phenotypic effects of CAM treatment resemble those observed in OsSHR1 loss-of-function mutants, characterized by reduced root development [67]. These findings suggest that CAM-induced phosphorylation of OsSHR1 may impair its mobility between cells, consequently disrupting its regulatory functions and reducing cell division during rice germination.

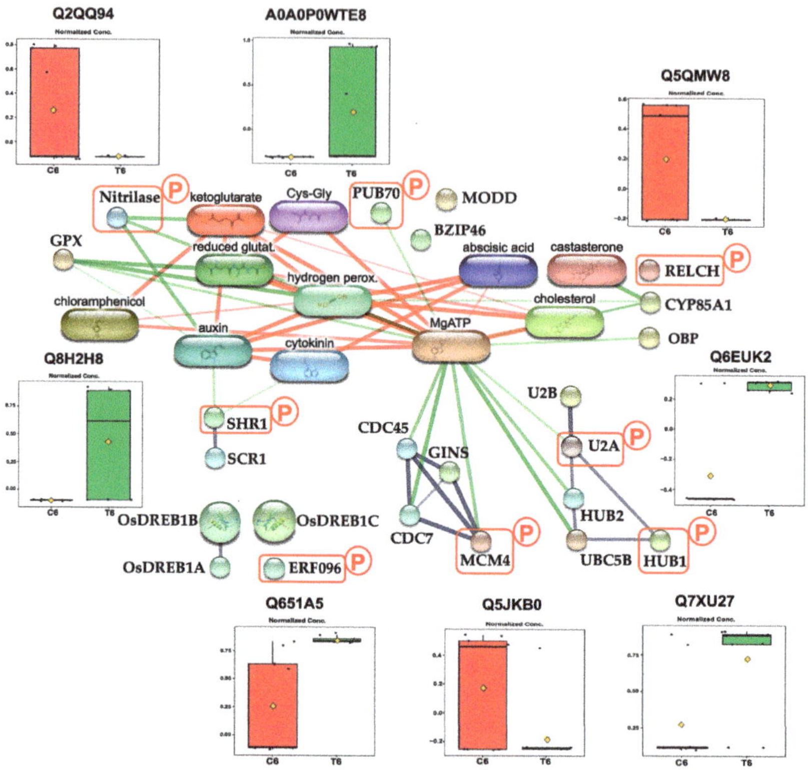

Figure 7. Protein–protein interaction network of CAM-responsive phosphoproteins and their predicted functional partners at the early seedling establishment stage (6-day). Phosphoproteins identified in this study are highlighted with a red square marked 'P', and the abundance of each protein under control (C) and treatment (T) conditions is displayed in a box plot with statistical comparisons using Fisher's LSD test ($p < 0.05$). Red and green boxes represent phosphoproteins abundance in control and treatment samples, respectively.

3.3.2. CAM Exposure Distributs Histone Modification

Histone monoubiquitination 1 (HUB1) functions as an E3 ubiquitin ligase, collaborating with HUB2 and E2 ubiquitin-conjugating enzymes (UBCs) to catalyze the monoubiquitination of histone H2B (H2Bub) by attaching a single ubiquitin molecule [68]. This histone modification is a critical regulatory mechanism essential for rice development and stress adaptation [69,70]. Disruption of HUB1 expression impairs rice growth [70]. In the present study, we observed that phosphorylated OsHUB1 (Q7XU27) levels were significantly elevated after 144 h of CAM treatment (Figure 7). Notably, this increase in OsHUB1 phosphorylation was accompanied by reduced seedling growth, which contrasts with earlier reports indicating that HUB1-mediated H2Bub promotes plant growth [71,72]. This inconsistency may be due to the regulatory effects of phosphorylation on E3 ubiquitin ligase activity. For instance, in *Arabidopsis*, phosphorylation by SnRK2 kinases suppresses the E3 ligase function of HUB2 without disrupting the HUB1/HUB2 complex [73]. Based on these findings together with our results, we propose that CAM treatment induces phosphorylation that inactivates HUB1's E3 ligase activity, thereby reducing H2Bub levels and inhibiting seedling growth.

3.3.3. CAM Treatment Suppresses AP2/ERF-Mediated Stress Tolerance

The Os09g0571700 protein (Q651A5), an ethylene-responsive transcription factor 96 (ERF096), is a DNA-binding protein that specifically targets the GCC-box and dehydration-responsive element (DRE) cis-elements [74]. In rice, OsERF096 functions as a negative regulator of stress tolerance by repressing dehydration-responsive element-binding (DREB) factors [74], which are key transcription factors that activate stress-responsive genes through an ABA-independent signaling pathway [75,76]. We found that CAM treatment significantly increased the levels of phosphorylated OsERF096 during the early seedling establishment stage (6-day) (Figure 7). Although phosphorylation is known to be essential for OsERF activation [77–79], direct evidence of its role in regulating OsERF096 remains lacking. Given the consistent growth inhibition observed in OsERF096 overexpression mutants [74], our results suggest that CAM-induced phosphorylation may be necessary for OsERF096 activation. Taken together, the CAM-induced increase in phosphorylated OsERF096 appears to enhance its repressive effect on the expression of OsDREB1 genes, thereby reducing rice growth under CAM treatment conditions.

3.3.4. CAM Exposure Influence Pre-mRNA Splicing

U2 small nuclear ribonucleoprotein A (U2A) is a core component of the U2 small nuclear ribonucleoprotein (U2 snRNP), which is essential for the function of the major spliceosome [80]. U2A contributes to pre-mRNA splicing by forming a heterodimer with U2B [81]. Disruption of this complex results in severe splicing defects [81], which can compromise cell wall integrity and reduce stress tolerance [82]. In plants, U2A is both evolutionarily conserved and stress-responsive [80], with its expression upregulated under stress conditions [83]. Although phosphorylation is a well-established regulatory mechanism in U2 snRNP-mediated splicing [84], the specific functional consequences of U2A phosphorylation remain unclear. Our proteomic analysis revealed elevated levels of U2A (Q6EUK2) phosphorylation after 144 h CAM treatment (Figure 7). Given the observed phenotypes, which are consistent with splicing defects [85], we propose that CAM-induced phosphorylation of U2A disrupts its heterodimerization with U2B, thereby impairing U2 snRNP-mediated splicing. This disruption likely compromises cell wall integrity and ultimately inhibits seedling growth.

3.3.5. CAM Exposure Reduce ABA-Mediated Stress Response

Os06g0163000 protein (A0A0P0WTE8), a plant U-box protein 70 (PUB70), functions as a U-box E3 ubiquitin ligase. It negatively regulates abscisic acid (ABA) signaling by mediating the ubiquitination and subsequent degradation of OsbZIP46 through its interaction with the mediator of OsbZIP46 deactivation and degradation (MODD) [86]. The E3 ubiquitin ligase activity of OsPUB70 is essential for OsbZIP46 ubiquitination and degradation [86]. Loss of OsPUB70 function impairs the degradation of OsbZIP46, which in turn leads to increased ABA sensitivity [86]. Our proteomic analysis shows that phosphorylated OsPUB70 levels increase during the early seedling establishment stage (Figure 7) in response to CAM treatment. Given the demonstrated disruption of protein phosphorylation on the activation of E3 ubiquitin ligases [87], we propose that increased phosphorylation induced by CAM may inactivate the E3 ubiquitin ligase activity of OsPUB70, thereby reducing the degradation of OsbZIP46. This, in turn, could ultimately inhibit growth by enhancing ABA sensitivity in rice.

3.3.6. CAM Exposure Impairs Castasterone Biosynthesis

The HEAT repeat-containing protein Os01g0514300 (Q5QMW8) belongs to the RELCH family, which facilitates cholesterol non-vesicular transport between recycling endosomes and the trans-Golgi network through interactions with the small GTPase RAB11 and oxysterol-binding protein (OBP) [88,89]. In plants, non-vesicular transport is the primary mechanism for cholesterol trafficking and is essential for maintaining cholesterol homeostasis [90]. This pathway efficiently delivers cholesterol to biosynthetic enzymes, such as cytochrome P450 85A1 (CYP85A1), supporting the biosynthesis of castasterone (CAS) [91]. Disruption of cholesterol-derived CAS biosynthesis has been shown to impair seedling development [91]. In rice, CAS is the predominant biologically active brassinosteroid (BR), with a more significant role than brassinolide [92]. Applying CAS externally has been shown to promote seedling growth in rice when exposed to low-temperature stress [93]. In the present study, we observed that CAM treatment reduces the phosphorylation levels of RELCH (Figure 7). Although direct evidence linking phosphorylation to RELCH activity remains to be established, the phenotypic similarity between our observation and those seen in cholesterol-driven CAS production inhibition mutants in *Arabidopsis* [91], as well as BR biosynthesis inhibition phenotypes in rice [94], suggests that phosphorylation is essential for RELCH activation in cholesterol-driven CAS production. Therefore, we propose that CAM treatment reduces RELCH phosphorylation, disrupts cholesterol non-vesicular transport, and consequently suppresses seedling growth under stress by diminishing CAS production.

3.3.7. CAM Exposure Increases Deaminated Glutathione

Nitrilase 1 (NIT1) functions as a metabolite repair enzyme by catalyzing the hydrolysis of deaminated glutathione (dGSH)—a toxic byproduct of glutathione metabolism—into α-ketoglutarate and cysteinylglycine [95]. This reaction prevents the accumulation of dGSH and contributes to the maintenance of the cellular antioxidant GSH pool [96]. Loss of NIT1 function causes accumulation of dGSH, which potentially inhibits glutathione peroxidase (GPX) activity [96]. This inhibition compromises the detoxification of hydrogen peroxide and lipid hydroperoxides, leading to elevated lipid peroxidation and oxidative damage to cellular membranes [97]. Our proteomic analysis revealed a significant decrease in the phosphorylation level of NIT1 homolog Q2QQ94 at the early seedling establishment stage after CAM treatment (Figure 7). Although the role of phosphorylation in regulating NIT1 remains unclear, the observed root growth inhibition—resembling the phenotype of NIT1 knockout mutants of rice? [98]—suggests that phosphorylation is necessary for the amidase activation of NIT1. We propose that CAM-induced phosphorylated NIT1 reduction leads to the accumulation of dGSH, which increases oxidative damage to cellular membranes and ultimately inhibits seedling growth.

3.3.8. CAM Exposure Impairs DNA Replication

Mini-chromosome maintenance complex protein 4 (MCM4, Q5JKB0) is a vital subunit of the MCM2–7 helicase complex, which is essential for DNA replication [99]. During the G1 phase of the cell cycle, the MCM2–7 complex is loaded onto replication origins as an inactive double hexamer and is subsequently activated in the S phase [100]. This activation depends on the phosphorylation of MCM4 by cell division cycle 7 (CDC7) kinase, which facilitates the recruitment of CDC45 and the GINS complex to form the active CDC45–MCM–GINS (CMG) helicase [100]. Failure to phosphorylate MCM4 disrupts CMG assembly, impairs helicase activity, and ultimately inhibit cell proliferation by compromises DNA replication [101]. In this study, we observed a reduction in phosphorylated MCM4 levels on the sixth day of CAM treatment (Figure 7). The observed growth inhibition is consistent with the phenotype of MCM4 knockout mutants of xxx? [102], suggests that phosphorylation is essential for MCM4 activation. We propose that the CAM-induced phosphorylated MCM4 reduction disrupts CMG complex formation, thereby impairing

DNA replication and inhibiting seedling establishment in rice during the early seedling establishment stage.

3.4. CAM Inhibits Root System Development During Germination

3.4.1. CAM Inhibits Primary Root Formation at Germination Stage

During the 3-day germination stage, CAM treatment significantly decreased the phosphorylation level of 4CL2 (Q42982, Figure 6), a key enzyme involved in flavonoid biosynthesis, particularly affecting rutin production (Section 3.2.5). Enrichment analysis further revealed a strong association between 4CL2 and anatomical structure development (Table 1), highlighting its potential role in seedling establishment. Considering that rutin has been reported to promote root growth in rice seedlings [61], we propose that CAM suppresses primary root elongation during germination by inhibiting rutin biosynthesis. This hypothesis is supported by evidence that CK signaling negatively regulates rutin biosynthesis in *Morus alba* L. [103]. Notably, CAM treatment was correlated with enhanced CK biosynthesis, as indicated by increased phosphorylation of the CK biosynthesis enzyme IPT8 (Section 3.2.3). Taken together, these findings suggest that CAM inhibits primary root growth during the germination stage by attenuating rutin biosynthesis via elevated CK signaling (Figure 8a).

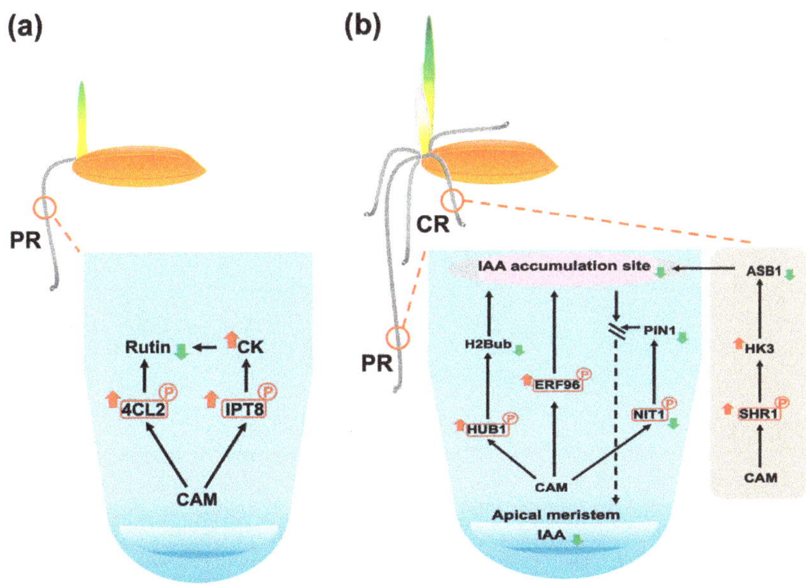

Figure 8. Identified phosphoproteins and potential pathways associated with root system development in rice during CAM treatment at (**a**) germination stage (3-day) and (**b**) early seedling establishment stage (6-day). PR, primary root; CR, crown root; red arrows, upregulation; green arrows, downregulation; black arrows, downstream consequences; double lines and dashed arrows, blocked IAA transport processes; and red square marked 'P', phosphorylated proteins.

3.4.2. CAM Inhibits Primary Root Formation at Early Seedling Establishment Stage

During the early seedling establishment stage (6-day), CAM exposure significantly increased phosphorylation of HUB1 (Q7XU27, Figure 7), which may lead to reduced H2Bub levels (Section 3.3.2). Considering the well-established role of H2Bub in promoting auxin-mediated primary root growth [104] and the strong correlation between HUB1 and anatomical structure development identified by enrichment analysis (Table 1), we propose that CAM impairs primary root elongation by attenuating auxin (AUX) signaling.

Moreover, CAM treatment resulted in elevated levels of active OsERF096 (Section 3.3.3), decreased expression of OsNIT1 (Section 3.3.7), and reduced mobility of SHORT-ROOT1 (SHR1) (Section 3.3.1). Given that OsERF096 facilitates the conjugation and inactivation of IAA [105], OsNIT1 is critical for maintaining proper AUX distribution in roots [98], and OsSHR1 dysfunction represses root system development by inhibiting the PIN1-mediated auxin signaling pathway [67], these molecular changes further support the hypothesis that CAM inhibits primary root elongation through disruption of AUX homeostasis and signaling (Figure 8b).

3.4.3. CAM Inhibits Crown Root Formation at Early Seedling Establishment Stage

In rice, root system formation during germination is a tightly regulated process that begins with primary root emergence, followed by crown root initiation and outgrowth [106]. This developmental stage relies on the finely tuned antagonistic regulatory roles of AUX and CK signaling pathways [107,108], where a lowered AUX-to-CK ratio inhibits crown root formation [108]. In *Arabidopsis*, the SHR-SCR complex modulates HISTIDINE KINASE 3 (AHK3) activity, thereby regulating CK signaling that influences AUX biosynthesis via ANTHRANILATE SYNTHASE BETA 1 (ASB1) [109]. This regulatory network enables the SHR-SCR complex to partially govern the AUX–CK balance that is essential for proper cell differentiation.

Our proteomics results revealed that CAM treatment at the early seedling establishment stage significantly increased phosphorylation of SHORT-ROOT1 (SHR1, Q8H2X8) (Figure 7), which may inhibit the formation of the SHR-SCR complex (Section 3.3.1). The phenotypic similarity in crown root development between CAM-treated rice seedlings and CK signaling enhancement mutants [108] suggests that CAM disrupts crown root development by lower the AUX-to-CK ratio. We propose that this effect may involve reduced SHR-SCR–mediated repression of the rice AHK3 homolog, OsHK3, leading to decreased OsASB1-dependent AUX biosynthesis [43]. Taken together, these findings suggest that CAM disrupts hormone homeostasis critical for crown root formation through the SHR-SCR–OsHK3–OsASB1 regulatory pathway (Figure 8b).

4. Materials and Methods

4.1. Plant Material Preparation

Healthy Thai jasmine rice (*Oryza sativa* L. cv. KDML 105) seeds obtained from the Rice Seed Division, Rice Department, Ministry of Agriculture and Cooperatives, Thailand were surface-sterilized following a modified protocol based on Md-Zali et al. [110]. Briefly, seeds were initially soaked in sterile distilled water for 40 min at room temperature with agitation at 120 rpm, followed by immersion in 1% sodium hypochlorite for 30 min and then 75% ethanol for 2 min. Sterilization was concluded with ten rinses using sterile distilled water. To assess the impact of CAM, 80 seeds were germinated in solutions containing chloramphenicol (CAM; FUJIFILM Wako Pure Chemical Corporation, Osaka, Japan) at concentrations of 5, 15, 50, 150, 500, 1500, and 2500 μg/mL in 90 × 15 mm polystyrene Petri dishes. Sterile distilled water served as the control. The highest concentration that allowed normal seedling development was identified and used for subsequent analyses. For phosphoproteomic analysis, each replicate consisted of 50 seeds germinated in 5 mL of the selected CAM concentration in 220 mL tissue culture glass jars. Control samples were prepared under the same conditions, using sterile distilled water instead of CAM. Both germination experiments were conducted in triplicate under dark conditions at 30 ± 3 °C. Samples were collected at 3 days, representing the germination stage, and at 6 days, corresponding to the early seedling establishment stage.

4.2. Physiological Analysis of Rice Germination

The germination status of rice seeds was assessed by counting the number of germinated seeds at 3 and 6 days after sowing. Shoot and root lengths were measured at the same time points using a ruler. Measurements were taken from ten seeds (n = 10) at 72 h and fifteen seeds (n = 15) at 144 h. The seedling vigor index (SVI) was calculated to provide a more comprehensive assessment of seedling development, using the previously described formula [111,112]: SVI = (Germination percentage × (shoot length + root length)).

4.3. Total Protein Extraction

At the 3-day and 6-day germination stages, seeds were harvested and stored at −80 °C. Total protein extraction was performed by grinding the entire germinated seed (including the primary leaf, coleoptile, seed coat, endosperm, embryo, and roots) to a fine powder under liquid nitrogen using a porcelain mortar. A 150 mg aliquot of the homogenized material was transferred to a 1.5 mL centrifuge tube and vortexed with 1% SDS solution (pH 7.0) for 30 min at room temperature [113]. Following centrifugation at 12,000× *g* for 7 min at 4 °C, the supernatant was mixed with a 1:1 (*v*/*v*) solution of 20% TCA/acetone to precipitate proteins and subsequently incubated at −20 °C for 1 h [114]. The resulting protein pellets were air-dried for 3 min and then resuspended in ultrapure water. Protein concentration was determined using the Lowry assay [115].

4.4. Label-Free Quantitative Phosphoproteomics Analysis Using LC-MS/MS

Prior to phosphoprotein enrichment (Pierce™ Phosphoprotein Enrichment Kit, Rockford, IL, USA), protein concentrations for each sample were adjusted to 10 mg/mL, and Halt™ Phosphatase Inhibitor Cocktail (Thermo Scientific, Rockford, IL, USA) was added at 1× concentration to inhibit phosphatase activity and preserve phosphorylation. The enriched phosphoproteins were then concentrated and desalted using a 9 kDa molecular weight cut-off membrane column and gel filtration, respectively (Thermo Scientific, Rockford, IL, USA). For in-solution digestion, 5 μg of the protein sample was completely dissolved in 10 mM ammonium bicarbonate (AMBIC). Disulfide bonds were reduced with 5 mM dithiothreitol (DTT) in 10 mM AMBIC at 60 °C for 1 h, followed by alkylation of sulfhydryl groups with 15 mM iodoacetamide (IAA) in 10 mM AMBIC for 45 min at room temperature in the dark. Trypsin digestion was performed overnight at 37 °C using sequencing grade trypsin (Promega, Madison, WI, USA) at a 1:20 enzyme-to-protein ratio (50 ng/μL trypsin concentration). Prior to LC-MS/MS analysis, digested samples were dried and reconstituted in 0.1% formic acid for protonation.

Phosphopeptide separation was performed using a nano LC-MS/MS system consisting of an Ultimate 3000 LC system (Dionex, Thermo Fisher Scientific, Waltham, MA, USA) coupled to a ZenoTOF 7600 system (SCIEX, Framingham, MA, USA). For each treatment at each time point, three biological replicates were analyzed, each with three technical replicates, resulting in a total of nine LC-MS/MS injections. A 75 μm I.D. × 15 cm Acclaim PepMap RSLC C18 column (2 μm, 100 Å, nanoViper; Thermo Fisher Scientific) was used, with the column temperature maintained at 60 °C. Mobile phases consisted of 0.1% formic acid in water (solvent A) and 0.1% formic acid in 80% acetonitrile (solvent B). Phosphopeptides were eluted using a linear gradient of 5–55% solvent B over 30 min at a constant flow rate of 0.30 μL/min. The ZenoTOF 7600 system was operated in positive ion mode with the following source parameters: ion source gas 1 at 8 psi, curtain gas at 35 psi, CAD gas at 7 psi, source temperature at 200 °C, and spray voltage at 3300 V. Data-dependent acquisition (DDA) was employed, selecting the top 50 most abundant precursor ions from each survey MS1 scan for subsequent MS/MS analysis. Dynamic exclusion was enabled, with precursors excluded for 12 s after two MS/MS sampling events (dynamic collision energy enabled). MS2 spectra were acquired in the 100–1800 *m/z* range with a 50 ms accumulation time and Zeno trap activation. Collision energy parameters were as follows: 80 V declustering potential, no DP spread, and a 0 V CE spread. Time bins were summed using all channels with a 150,000 cps Zeno trap threshold. The Top 60 DDA method cycle time was set to 3.0 s.

4.5. Quantification and Identification of Phosphoproteins

Protein quantitation in individual samples was carried out using MaxQuant 2.4.9.0 with the Andromeda search engine [116]. The search correlated MS/MS spectra to the Uniprot *Oryza sativa* database (downloaded on 26 April 2023). The label-free quantitation was performed using MaxQuant's standard settings, including a maximum of three missed cleavages, a mass tolerance of 0.6 dalton, trypsin as the digesting enzyme, carbamidomethylation of cysteine as a fixed modification, phosphorylation of STY, and oxidation of methionine as a variable modification. Raw MS/MS data and analysis files have been deposited at the ProteomeXchange Consortium under the dataset PXD050963 (https://proteomecentral.proteomexchange.org/cgi/GetDataset?ID=PXD050963, accessed on 5 August 2025) and jPOST dataset JPST003006.

Visualization and statistical analyses of the phosphoproteome data, including partial least squares discriminant analysis (PLS-DA) and differential expression analysis, were conducted using MetaboAnalyst version 6.0 [117]. Peak intensities from control and treatment groups were compared separately for each sampling time point. Data quality was improved by applying a relative standard deviation (RSD) reliability filter and a median absolute deviation (MAD) variance filter. Differentially expressed protein (DEP) analysis was then performed, and a volcano plot was generated using a fold change (FC) threshold of 2.0 and a raw p-value threshold of 0.05. Potential phosphorylation sites within the identified DEPs were predicted using NetPhos version 3.1 (https://services.healthtech.dtu.dk/services/NetPhos-3.1/ (accessed on 19 July 2025)) [118]. Pathway enrichment analysis of DEPs was conducted using g:Profiler (stable version available at https://biit.cs.ut.ee/gprofiler/gost (accessed on 22 March 2025)) [119], restricted to the *Oryza sativa* Japonica group. A false discovery rate (FDR) of 0.05 was applied using the Benjamini–Hochberg procedure as the significance threshold. Subcellular localization of DEPs associated with biological pathways identified by enrichment analysis was predicted using DeepLoc version 2.1 (https://services.healthtech.dtu.dk/services/DeepLoc-2.1/ (accessed on 19 July 2025)) [120]. Protein–protein interaction analysis was performed using STITCH version 5.0 (http://stitch.embl.de) [121], restricting the search to the *Oryza sativa* japonica group. A minimum interaction score of 0.4 was applied, with restrictions on first-shell and second-shell interactors.

4.6. Statistical Analysis

Statistical analyses and graphical representations were generated using R version 4.4.1 with the RStudio interface (version 2023.06.1.524) [122]. Analysis of variance (ANOVA) was performed to assess variance within the physiological dataset, employing a significance level of $\alpha = 0.05$. Post hoc multiple comparisons of means were conducted using Tukey's Honestly Significant Difference (TukeyHSD) test at a 95% family-wise confidence level. The 'multcompView' and 'ggplot2' R packages were utilized for data visualization [123].

5. Conclusions

In conclusion, this study demonstrates that chloramphenicol (CAM) effectively slows rice germination and seedling establishment without markedly affecting overall germinability, exhibiting organ-specific sensitivity. Our phosphoproteomic analysis further reveals that CAM induces stage-specific phosphorylation changes disrupting key physiological and molecular pathways. Notably, during the germination stage, CAM interferes with cytokinesis, cell wall formation, and oxidative damage resistance. During early seedling establishment, CAM additionally impacts DNA replication, cell division, stress responses, and critical hormonal signaling pathways, including auxin, abscisic acid, and brassinosteroids. These findings provide important insights into the molecular mechanisms underlying the CAM-induced inhibition of seed germination and lay a foundation for developing strategies to mitigate growth suppression caused by antibiotic contamination. Future studies comparing tissue-specific responses and variations among different rice cultivars could further elucidate differential sensitivity to CAM and help identify more tolerant genotypes or targeted mitigation strategies.

Supplementary Materials: The following supporting information can be downloaded at https://www.mdpi.com/article/10.3390/plants14182845/s1: Table S1. Raw data of all identified phosphoproteins from germinated rice, Table S2. CAM-response phosphoproteins identified from rice seeds at day 3 (germination stage) and day 6 (early seedling establishment stage). Table S3. Enrichment analysis results of CAM-response phosphoproteins from 3-day and 6-day rice seeds, used for bubble plot visualization. Table S4. Raw data of germination. Table S5. Raw data of root/shoot lengths (3-day and 6-day), crown root number, SVI, and first leaf formation (6-day). Table S6. Venn diagram of differentially expressed phosphoproteins (DEPPs) identified at 3-day (germination stage) and 6-day (early seedling establishment stage) in response to chloramphenicol (CAM). Figure S1. Venn diagram of differentially expressed phosphoproteins (DEPPs) identified at 3-day (germination stage) and 6-day (early seedling establishment stage) in response to chloramphenicol (CAM). Figure S2. Subcellular localization prediction of the CCHC-type zinc finger protein (Q2QQ28), generated using DeepLoc version 2.1. Figure S3. Predicted phosphorylation sites of OsSHR1 (Q8H2X8), generated using NetPhos version 3.1.

Author Contributions: Conceptualization R.L., S.R. and A.T.; data curation, R.L., N.P., K.L. and S.R.; methodology, R.L. and A.T.; writing—original draft preparation, R.L. and A.T.; writing—review and editing, R.L., S.R. and A.T.; visualization, R.L.; supervision, A.T.; project administration, A.T.; funding acquisition, A.T. All authors have read and agreed to the published version of the manuscript.

Funding: This research was funded by grants from Kasetsart University Research and Development Institute [FF(KU)5.64] and the NSRF via the Program Management Unit for Human Resources & Institutional Development, Research and Innovation [B13F670076].

Data Availability Statement: The MS/MS raw data and analysis are available in the ProteomeXchange Consortium under the dataset PXD050963 (https://proteomecentral.proteomexchange.org/cgi/GetDataset?ID=PXD050963, accessed on 5 August 2025) and jPOST dataset JPST003006.

Acknowledgments: This research was supported by the Interdisciplinary Graduate Program in Bioscience, Faculty of Science, Kasetsart University; and Bioinformatics Academic Association of Thailand (BAT). We are very grateful to the Rice Seed Division, Rice Department, Ministry of Agriculture and Cooperatives, Thailand, for providing seeds of the Thai jasmine rice cultivar Khao Dawk Mali 105 (*Oryza sativa* L. cv. KDML 105).

Conflicts of Interest: The authors declare no conflicts of interest.

References

1. Dominic, N.; Cenggoro, T.W.; Pardamean, B. Systematic literature review: Accelerate the rice production for global food security. *AIP Conf. Proc.* **2023**, *2594*, 80001. [CrossRef]
2. United States Department of Agriculture. Available online: https://downloads.usda.library.cornell.edu/usda-esmis/files/3t945q76s/tx31sh59w/np195824g/wasde0725.pdf (accessed on 24 July 2025).
3. Co, H.C.; Boosarawongse, R. Forecasting Thailand's rice export: Statistical techniques vs. artificial neural networks. *Comput. Ind. Eng.* **2007**, *53*, 610–627. [CrossRef]

4. Thai Rice Exporters Association. Available online: http://www.thairiceexporters.or.th/statistic_2025.html (accessed on 24 July 2025).
5. Wangcharoen, W.; Phanchaisri, C.; Daengpok, W.; Phuttawong, R.; Hangsoongnern, T.; Phanchaisri, B. Consumer acceptance test and some related properties of selected KDML 105 rice mutants. *J. Food Sci. Technol.* **2016**, *53*, 3550–3556. [CrossRef]
6. He, D.; Yang, P. Proteomics of rice seed germination. *Front. Plant Sci.* **2013**, *4*, 246. [CrossRef]
7. Hazra, A.; Das, S. The molecular and metabolic events behind different germination stages of rice seeds: A metabolomics perspective. *JSFA Rep.* **2024**, *4*, 118–134. [CrossRef]
8. Yang, P.; Li, X.; Wang, X.; Chen, H.; Chen, F.; Shen, S. Proteomic analysis of rice (*Oryza sativa*) seeds during germination. *Proteomics* **2007**, *7*, 3358–3368. [CrossRef]
9. Weber, M.J.; DeMoss, J.A. Inhibition of the peptide bond synthesizing cycle by chloramphenicol. *J. Bacteriol.* **1969**, *97*, 1099–1105. [CrossRef]
10. Pan, M.; Wong, C.K.C.; Chu, L.M. Distribution of antibiotics in wastewater-irrigated soils and their accumulation in vegetable crops in the Pearl River Delta, Southern China. *J. Agric. Food Chem.* **2014**, *62*, 11062–11069. [CrossRef]
11. Nguyen, L.M.; Nguyen, N.T.T.; Nguyen, T.T.T.; Nguyen, T.T.; Nguyen, D.T.C.; Tran, T.V. Occurrence, toxicity and adsorptive removal of the chloramphenicol antibiotic in water: A review. *Environ. Chem. Lett.* **2022**, *20*, 1929–1963. [CrossRef] [PubMed]
12. Chen, Q.Y.; Wu, Z.H.; Liu, J.L. Ecotoxicity of chloramphenicol and Hg acting on the root elongation of crops in North China. *Int. J. Environ. Res.* **2011**, *5*, 909–916. [CrossRef]
13. Sutcliffe, J.F. New evidence for a relationship between ion absorption and protein turnover in plant cells. *Nature* **1960**, *188*, 294–297. [CrossRef]
14. Önder, N. Comparative studies on water uptake and water permeability of potato tissue under the effect of indoleacetic acid, chloramphenicol and actinomycin D. *Physiol. Plant.* **1974**, *31*, 1–3. [CrossRef]
15. Rehman, A.U.; Kodru, S.; Vass, I. Chloramphenicol mediates superoxide production in photosystem II and enhances its photodamage in isolated membrane particles. *Front. Plant Sci.* **2016**, *7*, 479. [CrossRef]
16. Srivastava, B.I.S.; Meredith, W.O.S. Mechanism of action of gibberellic acid: Inhibition of α-amylase development during germination of barley by chloramphenicol and its reversal by gibberellic acid. *Can. J. Bot.* **1962**, *40*, 1257–1265. [CrossRef]
17. Pan, M.; Chu, L.M. Phytotoxicity of veterinary antibiotics to seed germination and root elongation of crops. *Ecotoxicol. Environ. Saf.* **2016**, *126*, 228–237. [CrossRef]
18. Bellino, A.; Lofrano, G.; Carotenuto, M.; Libralato, G.; Baldantoni, D. Antibiotic effects on seed germination and root development of tomato (*Solanum lycopersicum* L.). *Ecotoxicol. Environ. Saf.* **2018**, *148*, 135–141. [CrossRef]
19. Noodén, L.D.; Thimann, K.V. Inhibition of protein synthesis and of auxin-induced growth by chloramphenicol. *Plant Physiol.* **1965**, *40*, 193–201. [CrossRef]
20. Zhang, W.J.; Zhou, Y.; Zhang, Y.; Su, Y.H.; Xu, T. Protein phosphorylation: A molecular switch in plant signaling. *Cell Rep.* **2023**, *42*, 112729. [CrossRef]
21. Han, C.; Wang, K.; Yang, P. Gel-based comparative phosphoproteomic analysis on rice embryo during germination. *Plant Cell Physiol.* **2014**, *55*, 1376–1394. [CrossRef]
22. Li, M.; Yin, X.; Sakata, K.; Yang, P.; Komatsu, S. Proteomic analysis of phosphoproteins in the rice nucleus during the early stage of seed germination. *J. Proteome Res.* **2015**, *14*, 2884–2896. [CrossRef]
23. Frankland, B.; Smith, H. Temperature and other factors affecting chloramphenicol stimulation of the germination of light-sensitive lettuce seeds. *Planta* **1967**, *77*, 354–366. [CrossRef]
24. Li, R.; Phaonakrop, N.; Lohmaneeratana, K.; Roytrakul, S.; Thamchaipenet, A. Phosphoproteomic insights into the regulation of root length in rice (*Oryza sativa* L. cv. KDML 105): Uncovering key events and pathways involving phosphorylated proteins. *PeerJ* **2025**, *13*, e19361. [CrossRef] [PubMed]
25. Pramer, D. The movement of chloramphenicol and streptomycin in broad bean and tomato plants. *Ann. Bot.* **1954**, *18*, 463–470. [CrossRef]
26. Han, C.; Yang, P.; Sakata, K.; Komatsu, S. Quantitative proteomics reveals the role of protein phosphorylation in rice embryos during early stages of germination. *J. Proteome Res.* **2014**, *13*, 1766–1782. [CrossRef]
27. Krupnova, T.; Sasabe, M.; Ghebreghiorghis, L.; Gruber, C.W.; Hamada, T.; Dehmel, V.; Strompen, G.; Stierhof, Y.-D.; Lukowitz, W.; Kemmerling, B.; et al. Microtubule-associated kinase-like protein RUNKEL needed for cell plate expansion in *Arabidopsis* cytokinesis. *Curr. Biol.* **2009**, *19*, 518–523. [CrossRef]
28. Krupnova, T.; Stierhof, Y.D.; Hiller, U.; Strompen, G.; Müller, S. The microtubule-associated kinase-like protein RUNKEL functions in somatic and syncytial cytokinesis. *Plant J.* **2013**, *74*, 781–791. [CrossRef]
29. Takahashi, Y.; Soyano, T.; Kosetsu, K.; Sasabe, M.; Machida, Y. HINKEL kinesin, ANP MAPKKKs and MKK6/ANQ MAPKK, which phosphorylates and activates MPK4 MAPK, constitute a pathway that is required for cytokinesis in *Arabidopsis thaliana*. *Plant Cell Physiol.* **2010**, *51*, 1766–1776. [CrossRef]

30. Preuss, F.; Chatterjee, D.; Mathea, S.; Shrestha, S.; St-Germain, J.; Saha, M.; Kannan, N.; Raught, B.; Rottapel, R.; Knapp, S. Nucleotide binding, evolutionary insights, and interaction partners of the pseudokinase Unc-51-like kinase 4. *Structure* **2020**, *28*, 1184–1196.e1186. [CrossRef]
31. Fujita, M.; Maeda, Y.; Ra, M.; Yamaguchi, Y.; Taguchi, R.; Kinoshita, T. GPI glycan remodeling by PGAP5 regulates transport of GPI-anchored proteins from the ER to the Golgi. *Cell* **2009**, *139*, 352–365. [CrossRef]
32. Komath, S.S.; Fujita, M.; Hart, G.W.; Ferguson, M.A.; Kinoshita, T. Glycosylphosphatidylinositol Anchors. In *Essentials of Glycobiology*, 4th ed.; Varki, A., Cummings, R.D., Esko, J.D., Stanley, P., Hart, G.W., Aebi, M., Mohnen, D., Kinoshita, T., Packer, N.H., Prestegard, J.H., Schnaar, R.L., Seeberger P.H., Eds.; Cold Spring Harbor Laboratory Press: New York, NY, USA, 2022; pp. 195–210.
33. Fujita, M.; Umemura, M.; Yoko-o, T.; Jigami, Y. PER1 Is required for GPI-phospholipase A2 activity and involved in lipid remodeling of GPI-anchored proteins. *Mol. Biol. Cell.* **2006**, *17*, 5253–5264. [CrossRef]
34. Bernat-Silvestre, C.; Ma, Y.; Johnson, K.; Ferrando, A.; Aniento, F.; Marcote, M.J. Characterization of *Arabidopsis* post-glycosylphosphatidylinositol attachment to proteins phospholipase 3 like genes. *Front. Plant Sci.* **2022**, *13*, 817915. [CrossRef]
35. Gillmor, C.S.; Lukowitz, W.; Brininstool, G.; Sedbrook, J.C.; Hamann, T.; Poindexter, P.; Somerville, C. Glycosylphosphatidylinositol-anchored proteins are required for cell wall synthesis and morphogenesis in *Arabidopsis*. *Plant Cell* **2005**, *17*, 1128–1140. [CrossRef]
36. Xu, Z.; Gao, Y.; Gao, C.; Mei, J.; Wang, S.; Ma, J.; Yang, H.; Cao, S.; Wang, Y.; Zhang, F.; et al. Glycosylphosphatidylinositol anchor lipid remodeling directs proteins to the plasma membrane and governs cell wall mechanics. *Plant Cell* **2022**, *34*, 4778–4794. [CrossRef]
37. Vazquez, H.M.; Vionnet, C.; Roubaty, C.; Conzelmann, A. Cdc1 removes the ethanolamine phosphate of the first mannose of GPI anchors and thereby facilitates the integration of GPI proteins into the yeast cell wall. *Mol. Biol. Cell* **2014**, *25*, 3375–3388. [CrossRef]
38. Yadav, P.; Yadav, S.K.; Singh, M.; Singh, M.P.; Chinnusamy, V. Genome wide identification and characterization of Isopentenyl transferase (IPT) gene family associated with cytokinin synthesis in rice. *Plant Physiol. Rep.* **2024**, *29*, 207–225. [CrossRef]
39. Sun, J.; Niu, Q.-W.; Tarkowski, P.; Zheng, B.; Tarkowska, D.; Sandberg, G.r.; Chua, N.-H.; Zuo, J. The *Arabidopsis* AtIPT8/PGA22 gene encodes an isopentenyl transferase that Is involved in de novo dytokinin biosynthesis. *Plant Physiol.* **2003**, *131*, 167–176. [CrossRef]
40. Sakano, Y.; Okada, Y.; Matsunaga, A.; Suwama, T.; Kaneko, T.; Ito, K.; Noguchi, H.; Abe, I. Molecular cloning, expression, and characterization of adenylate isopentenyltransferase from hop (*Humulus lupulus* L.). *Phytochemistry* **2004**, *65*, 2439–2446. [CrossRef]
41. Takei, K.; Sakakibara, H.; Sugiyama, T. Identification of genes encoding adenylate isopentenyltransferase, a cytokinin biosynthesis enzyme, in *Arabidopsis thaliana*. *J. Biol. Chem.* **2001**, *276*, 26405–26410. [CrossRef]
42. Wang, Y.; Shen, W.; Chan, Z.; Wu, Y. Endogenous cytokinin overproduction modulates ROS homeostasis and decreases salt stress resistance in *Arabidopsis thaliana*. *Front. Plant Sci.* **2015**, *6*, 1004. [CrossRef]
43. Tsai, Y.C.; Weir, N.R.; Hill, K.; Zhang, W.; Kim, H.J.; Shiu, S.H.; Schaller, G.E.; Kieber, J.J. Characterization of genes involved in cytokinin signaling and metabolism from rice. *Plant Physiol.* **2012**, *158*, 1666–1684. [CrossRef]
44. Chun, Y.; Fang, J.; Savelieva, E.M.; Lomin, S.N.; Shang, J.; Sun, Y.; Zhao, J.; Kumar, A.; Yuan, S.; Yao, X.; et al. The cytokinin receptor OHK4/OsHK4 regulates inflorescence architecture in rice via an IDEAL PLANT ARCHITECTURE1/WEALTHY FARMER'S PANICLE-mediated positive feedback circuit. *Plant Cell* **2023**, *36*, 40–64. [CrossRef]
45. Eichwald, T.; da Silva, L.B.; Staats Pires, A.C.; Niero, L.; Schnorrenberger, E.; Filho, C.C.; Espíndola, G.; Huang, W.L.; Guillemin, G.J.; Abdenur, J.E.; et al. Tetrahydrobiopterin: Beyond its traditional role as a cofactor. *Antioxidants* **2023**, *12*, 1037. [CrossRef]
46. Naponelli, V.; Noiriel, A.; Ziemak, M.J.; Beverley, S.M.; Lye, L.-F.; Plume, A.M.; Botella, J.R.; Loizeau, K.; Ravanel, S.p.; Rébeillé, F.; et al. Phylogenomic and runctional analysis of pterin-4a-carbinolamine dehydratase family (COG2154) proteins in plants and microorganisms. *Plant Physiol.* **2008**, *146*, 1515–1527. [CrossRef]
47. Berka, V.; Yeh, H.-C.; Gao, D.; Kiran, F.; Tsai, A.-L. Redox function of tetrahydrobiopterin and effect of L-arginine on oxygen binding in endothelial nitric oxide synthase. *Biochemistry* **2004**, *43*, 13137–13148. [CrossRef]
48. Kim, H.K.; Ha, S.H.; Han, J. Potential therapeutic applications of tetrahydrobiopterin: From inherited hyperphenylalaninemia to mitochondrial diseases. *Ann. N. Y. Acad. Sci.* **2010**, *1201*, 177–182. [CrossRef]
49. Peres da Rocha Oliveiros Marciano, D.; Toledo Ramos, F.; Neiva Alvim, M.; Ronaldo Magalhaes, J.; Giovanni Costa França, M. Nitric oxide reduces the stress effects of aluminum on the process of germination and early root growth of rice. *J. Plant Nutr. Soil Sci.* **2010**, *173*, 885–891. [CrossRef]
50. Cao, X.; Zhu, C.; Zhong, C.; Zhang, J.; Wu, L.; Jin, Q.; Ma, Q. Nitric oxide synthase-mediated early nitric oxide burst alleviates water stress-induced oxidative damage in ammonium-supplied rice roots. *BMC Plant Biol.* **2019**, *19*, 108. [CrossRef]
51. Teng, Z.; Zheng, Q.; Peng, Y.; Li, Y.; Meng, S.; Liu, B.; Peng, Y.; Duan, M.; Yuan, D.; Zhang, J.; et al. Nitrate reductase–dependent nitric oxide production mediates nitrate-conferred salt tolerance in rice seedlings. *Plant Physiol.* **2025**, *197*, kiaf080. [CrossRef]

52. Li, Y.; Kim, J.I.; Pysh, L.; Chapple, C. Four isoforms of *Arabidopsis* 4-coumarate:CoA ligase have overlapping yet distinct roles in phenylpropanoid metabolism. *Plant Physiol.* **2015**, *169*, 2409–2421. [CrossRef]
53. Sun, H.; Li, Y.; Feng, S.; Zou, W.; Guo, K.; Fan, C.; Si, S.; Peng, L. Analysis of five rice 4-coumarate:coenzyme A ligase enzyme activity and stress response for potential roles in lignin and flavonoid biosynthesis in rice. *Biochem. Biophys. Res. Commun.* **2013**, *430*, 1151–1156. [CrossRef]
54. Li, M.; Guo, L.; Wang, Y.; Li, Y.; Jiang, X.; Liu, Y.; Xie, D.-Y.; Gao, L.; Xia, T. Molecular and biochemical characterization of two 4-coumarate: CoA ligase genes in tea plant (*Camellia sinensis*). *Plant Mol. Biol.* **2022**, *109*, 579–593. [CrossRef]
55. Singh, A.; Gupta, R.; Pandey, R. Rice seed priming with picomolar rutin enhances rhizospheric *Bacillus subtilis* CIM colonization and plant growth. *PLoS ONE* **2016**, *11*, e0146013. [CrossRef]
56. Shomali, A.; Das, S.; Arif, N.; Sarraf, M.; Zahra, N.; Yadav, V.; Aliniaeifard, S.; Chauhan, D.K.; Hasanuzzaman, M. Diverse physiological roles of flavonoids in plant environmental stress responses and tolerance. *Plants* **2022**, *11*, 3158. [CrossRef]
57. Ismail, H.; Dragišic Maksimovic, J.; Maksimovic, V.; Shabala, L.; Živanovic, B.D.; Tian, Y.; Jacobsen, S.E.; Shabala, S. Rutin, a flavonoid with antioxidant activity, improves plant salinity tolerance by regulating K^+ retention and Na^+ exclusion from leaf mesophyll in quinoa and broad beans. *Funct. Plant Biol.* **2015**, *43*, 75–86. [CrossRef]
58. Yildiztugay, E.; Ozfidan-Konakci, C. Profiling of rutin-mediated alleviation of cadmium-induced oxidative stress in *Zygophyllum fabago*. *Environ. Toxicol.* **2015**, *30*, 816–835. [CrossRef]
59. Quoc, C.D.; Lan, A.B.; Ngoc, T.N.; Le Hong, T.; Thanh, T.T.; Tran, G.-B.; Hoa, S.P.; Hung, T.N.; Ngoc, T.N.H.; Cong, H.N.; et al. Variations in some metabolic compounds in the roots of rice varieties (*Oryza sativa* L.) with different salinity tolerance under salinity stress during the seedling stage. *Plant Physiol. Rep.* **2024**, *29*, 660–677. [CrossRef]
60. Tang, Y.H.; Liu, F.; Mao, K.Q.; Xing, H.C.; Chen, J.R.; Guo, Q.Q. Cloning and characterization of the key 4-coumarate CoA ligase genes in *Boehmeria nivea*. *S. Afr. J. Bot.* **2018**, *116*, 123–130. [CrossRef]
61. Singh, A.; Gupta, R.; Pandey, R. Exogenous application of rutin and gallic acid regulate antioxidants and alleviate reactive oxygen generation in *Oryza sativa* L. *Physiol. Mol. Biol. Plants* **2017**, *23*, 301–309. [CrossRef]
62. Lucas, M.; Swarup, R.; Paponov, I.A.; Swarup, K.; Casimiro, I.; Lake, D.; Peret, B.; Zappala, S.; Mairhofer, S.; Whitworth, M.; et al. SHORT-ROOT regulates primary, lateral, and adventitious root development in *Arabidopsis*. *Plant Physiol.* **2010**, *155*, 384–398. [CrossRef]
63. Montiel, G.g.; Gantet, P.; Jay-Allemand, C.; Breton, C. Transcription factor networks. pathways to the knowledge of root development. *Plant Physiol.* **2004**, *136*, 3478–3485. [CrossRef]
64. Gallagher, K.L.; Paquette, A.J.; Nakajima, K.; Benfey, P.N. Mechanisms regulating SHORT-ROOT intercellular movement. *Curr. Biol.* **2004**, *14*, 1847–1851. [CrossRef] [PubMed]
65. Cui, H.; Levesque, M.P.; Vernoux, T.; Jung, J.W.; Paquette, A.J.; Gallagher, K.L.; Wang, J.Y.; Blilou, I.; Scheres, B.; Benfey, P.N. An evolutionarily conserved mechanism delimiting SHR movement defines a single layer of endodermis in plants. *Science* **2007**, *316*, 421–425. [CrossRef] [PubMed]
66. Wu, S.; Lee, C.M.; Hayashi, T.; Price, S.; Divol, F.; Henry, S.; Pauluzzi, G.; Perin, C.; Gallagher, K.L. A plausible mechanism, based upon Short-Root movement, for regulating the number of cortex cell layers in roots. *Proc. Natl. Acad. Sci. USA* **2014**, *111*, 16184–16189. [CrossRef]
67. Xing, Y.; Wang, N.; Zhang, T.; Zhang, Q.; Du, D.; Chen, X.; Lu, X.; Zhang, Y.; Zhu, M.; Liu, M.; et al. SHORT-ROOT 1 is critical to cell division and tracheary element development in rice roots. *Plant J.* **2021**, *105*, 1179–1191. [CrossRef]
68. Hu, M.; Pei, B.L.; Zhang, L.F.; Li, Y.Z. Histone H2B monoubiquitination is involved in regulating the dynamics of microtubules during the defense response to *Verticillium dahliae* toxins in *Arabidopsis*. *Plant Physiol.* **2014**, *164*, 1857–1865. [CrossRef]
69. Xu, Z.; Yang, Y.; Zhang, F.; Li, H.; Ma, H.; Wu, W.; Ding, Y. OsbZIP27 coordinates with OsHUB1 and OsHUB2 to modulate drought tolerance in rice. *J. Genet. Genom.* **2025**, *52*, 168–178. [CrossRef] [PubMed]
70. Xu, Z.; Li, E.; Xue, G.; Zhang, C.; Yang, Y.; Ding, Y. OsHUB2 inhibits function of OsTrx1 in heading date in rice. *Plant J.* **2022**, *110*, 1670–1680. [CrossRef] [PubMed]
71. Fleury, D.; Himanen, K.; Cnops, G.; Nelissen, H.; Boccardi, T.M.; Maere, S.; Beemster, G.T.S.; Neyt, P.; Anami, S.; Robles, P.; et al. The *Arabidopsis thaliana* homolog of Yeast BRE1 has a function in cell cycle regulation during early leaf and root growth. *Plant Cell* **2007**, *19*, 417–432. [CrossRef] [PubMed]
72. Cao, H.; Li, X.; Wang, Z.; Ding, M.; Sun, Y.; Dong, F.; Chen, F.; Liu, L.; Doughty, J.; Li, Y.; et al. Histone H2B monoubiquitination mediated by HISTONE MONOUBIQUITINATION1 and HISTONE MONOUBIQUITINATION2 is involved in anther development by regulating tapetum degradation-related genes in rice. *Plant Physiol.* **2015**, *168*, 1389–1405. [CrossRef]
73. Li, Q.; Wang, L.; Grubb, L.E.; Talasila, M.; Rodriguez Gallo, M.C.; Mehta, D.; Scandola, S.; Uhrig, R.G. B4 Raf-like MAPKKK RAF24 regulates *Arabidopsis thaliana* flowering time through HISTONE MONO-UBIQUITINATION 2. *New Phytol.* **2025**, *247*, 897–915. [CrossRef] [PubMed]
74. Sun, M.; Shen, Y.; Chen, Y.; Wang, Y.; Cai, X.; Yang, J.; Jia, B.; Dong, W.; Chen, X.; Sun, X. Osa-miR1320 targets the ERF transcription factor OsERF096 to regulate cold tolerance via JA-mediated signaling. *Plant Physiol.* **2022**, *189*, 2500–2516. [CrossRef]

75. Dubouzet, J.G.; Sakuma, Y.; Ito, Y.; Kasuga, M.; Dubouzet, E.G.; Miura, S.; Seki, M.; Shinozaki, K.; Yamaguchi-Shinozaki, K. OsDREB genes in rice, *Oryza sativa* L., encode transcription activators that function in drought-, high-salt- and cold-responsive gene expression. *Plant J.* **2003**, *33*, 751–763. [CrossRef]
76. Wani, A.B.; Noor, W.; Pandit, A.; Husaini, A.M. Upregulated expression of MYB4, DREB1 and AP37 transcription factors modulates cold stress response in high-altitude Himalayan rice via time-dependent ROS regulation. *Mol. Biol. Rep.* **2025**, *52*, 417. [CrossRef] [PubMed]
77. Xu, L.; Yang, L.; Li, A.; Guo, J.; Wang, H.; Qi, H.; Li, M.; Yang, P.; Song, S. An AP2/ERF transcription factor confers chilling tolerance in rice. *Sci. Adv.* **2024**, *10*, eado4788. [CrossRef]
78. Cheong, Y.H.; Moon, B.C.; Kim, J.K.; Kim, C.Y.; Kim, M.C.; Kim, I.H.; Park, C.Y.; Kim, J.C.; Park, B.O.; Koo, S.C.; et al. BWMK1, a rice mitogen-activated protein kinase, locates in the nucleus and mediates pathogenesis-related gene expression by activation of a transcription factor. *Plant Physiol.* **2003**, *132*, 1961–1972. [CrossRef] [PubMed]
79. Lu, J.; Ju, H.; Zhou, G.; Zhu, C.; Erb, M.; Wang, X.; Wang, P.; Lou, Y. An EAR-motif-containing ERF transcription factor affects herbivore-induced signaling, defense and resistance in rice. *Plant J.* **2011**, *68*, 583–596. [CrossRef] [PubMed]
80. Liu, Y.; Tian, Y.; Wang, L.-X.; Fan, T.; Zhang, J.; Chen, M.-X.; Liu, Y.-G. Phylogeny and conservation of plant U2A/U2A′, a core splicing component in U2 spliceosomal complex. *Planta* **2021**, *255*, 25. [CrossRef]
81. Caspary, F.; Séraphin, B. The yeast U2A′/U2B″ complex is required for pre-spliceosome formation. *EMBO J.* **1998**, *17*, 6348–6358. [CrossRef]
82. Ma, L.; Zhang, X.; Li, C.; Ma, X.; Zhao, X.; Zhao, X.; Zhang, P.; Zhu, X. A U2 snRNP-specific protein, U2A′, is involved in stress response and drug resistance in Cryptococcus deneoformans. *Biochimie* **2024**, *220*, 179–187. [CrossRef]
83. Bae, M.S.; Cho, E.J.; Choi, E.-Y.; Park, O.K. Analysis of the *Arabidopsis* nuclear proteome and its response to cold stress. *Plant J.* **2003**, *36*, 652–663. [CrossRef]
84. Chen, L.; Weinmeister, R.; Kralovicova, J.; Eperon, L.P.; Vorechovsky, I.; Hudson, A.J.; Eperon, I.C. Stoichiometries of U2AF35, U2AF65 and U2 snRNP reveal new early spliceosome assembly pathways. *Nucleic Acids Res.* **2016**, *45*, 2051–2067. [CrossRef]
85. Wang, Z.; Hong, Y.; Yao, J.; Huang, H.; Qian, B.; Liu, X.; Chen, Y.; Pang, J.; Zhan, X.; Zhu, J.-K.; et al. Modulation of plant development and chilling stress responses by alternative splicing events under control of the spliceosome protein SmEb in *Arabidopsis*. *Plant Cell Environ.* **2022**, *45*, 2762–2779. [CrossRef]
86. Tang, N.; Ma, S.; Zong, W.; Yang, N.; Lv, Y.; Yan, C.; Guo, Z.; Li, J.; Li, X.; Xiang, Y.; et al. MODD mediates deactivation and degradation of OsbZIP46 to negatively regulate ABA signaling and drought resistance in rice. *Plant Cell* **2016**, *28*, 2161–2177. [CrossRef]
87. Perez, J.M.; Chen, Y.; Xiao, T.S.; Abbott, D.W. Phosphorylation of the E3 ubiquitin protein ligase ITCH diminishes binding to its cognate E2 ubiquitin ligase. *J. Biol. Chem.* **2018**, *293*, 1100–1105. [CrossRef]
88. Kanerva, K.; Uronen, R.-L.; Blom, T.; Li, S.; Bittman, R.; Lappalainen, P.; Peränen, J.; Raposo, G.; Ikonen, E. LDL cholesterol recycles to the plasma membrane via a Rab8a-Myosin5b-Actin-Dependent membrane transport route. *Dev. Cell* **2013**, *27*, 249–262. [CrossRef] [PubMed]
89. Sobajima, T.; Yoshimura, S.I.; Maeda, T.; Miyata, H.; Miyoshi, E.; Harada, A. The Rab11-binding protein RELCH/KIAA1468 controls intracellular cholesterol distribution. *J. Cell Biol.* **2018**, *217*, 1777–1796. [CrossRef] [PubMed]
90. Kumar, K.; Gibbs, H.C.; Yeh, A.T.; Griffing, L.R. The sterol trafficking pathway in *Arabidopsis thaliana*. *Front. Plant Sci.* **2021**, *12*, 616631. [CrossRef]
91. Joo, S.-H.; Kim, T.-W.; Son, S.-H.; Lee, W.S.; Yokota, T.; Kim, S.-K. Biosynthesis of a cholesterol-derived brassinosteroid, 28-norcastasterone, in *Arabidopsis thaliana*. *J. Exp. Bot.* **2011**, *63*, 1823–1833. [CrossRef]
92. Kim, B.K.; Fujioka, S.; Takatsuto, S.; Tsujimoto, M.; Choe, S. Castasterone is a likely end product of brassinosteroid biosynthetic pathway in rice. *Biochem. Biophys. Res. Commun.* **2008**, *374*, 614–619. [CrossRef] [PubMed]
93. Fujii, S.; Saka, H. The promotive effect of brassinolide on lamina joint-cell elongation, germination and seedling growth under low-temperature stress in rice (*Oryza sativa* L.). *Plant Prod. Sci.* **2001**, *4*, 210–214. [CrossRef]
94. Jantapo, K.; Wimonchaijit, W.; Wang, W.; Chaiwanon, J. Supraoptimal brassinosteroid levels inhibit root growth by reducing root meristem and cell elongation in rice. *Plants* **2021**, *10*, 1962. [CrossRef]
95. Peracchi, A.; Veiga-da-Cunha, M.; Kuhara, T.; Ellens, K.W.; Paczia, N.; Stroobant, V.; Seliga, A.K.; Marlaire, S.; Jaisson, S.; Bommer, G.T.; et al. Nit1 is a metabolite repair enzyme that hydrolyzes deaminated glutathione. *Proc. Natl. Acad. Sci. USA* **2017**, *114*, e3233–e3242. [CrossRef]
96. Niehaus, T.D.; Patterson, J.A.; Alexander, D.C.; Folz, J.S.; Pyc, M.; MacTavish, B.S.; Bruner, S.D.; Mullen, R.T.; Fiehn, O.; Hanson, A.D. The metabolite repair enzyme Nit1 is a dual-targeted amidase that disposes of damaged glutathione in *Arabidopsis*. *Biochem. J.* **2019**, *476*, 683–697. [CrossRef]

97. Zhang, L.; Wu, M.; Teng, Y.; Jia, S.; Yu, D.; Wei, T.; Chen, C.; Song, W. Overexpression of the glutathione peroxidase 5 (RcGPX5) gene from *Rhodiola crenulata* increases drought tolerance in *Salvia miltiorrhiza*. *Front. Plant Sci.* **2018**, *9*, 1950. [CrossRef]
98. Song, M.; Fan, X.; Chen, J.; Qu, H.; Luo, L.; Xu, G. OsNAR2.1 interaction with OsNIT1 and OsNIT2 functions in root-growth responses to nitrate and ammonium. *Plant Physiol.* **2020**, *183*, 289–303. [CrossRef]
99. Forsburg, S.L. Eukaryotic MCM proteins: Beyond replication initiation. *Microbiol. Mol. Biol. Rev.* **2004**, *68*, 109–131. [CrossRef]
100. van Deursen, F.; Sengupta, S.; De Piccoli, G.; Sanchez-Diaz, A.; Labib, K. Mcm10 associates with the loaded DNA helicase at replication origins and defines a novel step in its activation. *EMBO J.* **2012**, *31*, 2195–2206. [CrossRef]
101. Masai, H.; Taniyama, C.; Ogino, K.; Matsui, E.; Kakusho, N.; Matsumoto, S.; Kim, J.M.; Ishii, A.; Tanaka, T.; Kobayashi, T.; et al. Phosphorylation of MCM4 by Cdc7 kinase facilitates its interaction with Cdc45 on the chromatin. *J. Biol. Chem.* **2006**, *281*, 39249–39261. [CrossRef] [PubMed]
102. Herridge, R.P.; Day, R.C.; Macknight, R.C. The role of the MCM2-7 helicase complex during *Arabidopsis* seed development. *Plant Mol. Biol.* **2014**, *86*, 69–84. [CrossRef] [PubMed]
103. Lee, Y.; Lee, D.-E.; Lee, H.-S.; Kim, S.-K.; Lee, W.S.; Kim, S.-H.; Kim, M.-W. Influence of auxins, cytokinins, and nitrogen on production of rutin from callus and adventitious roots of the white mulberry tree (*Morusalba* L.). *Plant Cell Tiss. Organ Cult.* **2011**, *105*, 9–19. [CrossRef]
104. Zhang, L.; Luo, P.; Bai, J.; Wu, L.; Di, D.-W.; Liu, H.-Q.; Li, J.-J.; Liu, Y.-L.; Khaskheli, A.J.; Zhao, C.-M.; et al. Function of histone H2B monoubiquitination in transcriptional regulation of auxin biosynthesis in *Arabidopsis*. *Commun. Biol.* **2021**, *4*, 206. [CrossRef] [PubMed]
105. Cai, X.; Chen, Y.; Wang, Y.; Shen, Y.; Yang, J.; Jia, B.; Sun, X.; Sun, M. A comprehensive investigation of the regulatory roles of OsERF096, an AP2/ERF transcription factor, in rice cold stress response. *Plant Cell Rep.* **2023**, *42*, 2011–2022. [CrossRef]
106. Rebouillat, J.; Dievart, A.; Verdeil, J.L.; Escoute, J.; Giese, G.; Breitler, J.C.; Gantet, P.; Espeout, S.; Guiderdoni, E.; Périn, C. Molecular genetics of rice root development. *Rice* **2009**, *2*, 15–34. [CrossRef]
107. Mao, C.; He, J.; Liu, L.; Deng, Q.; Yao, X.; Liu, C.; Qiao, Y.; Li, P.; Ming, F. OsNAC2 integrates auxin and cytokinin pathways to modulate rice root development. *Plant Biotechnol. J.* **2020**, *18*, 429–442. [CrossRef]
108. Gao, S.; Fang, J.; Xu, F.; Wang, W.; Sun, X.; Chu, J.; Cai, B.; Feng, Y.; Chu, C. CYTOKININ OXIDASE/DEHYDROGENASE4 integrates cytokinin and auxin signaling to control rice crown root formation. *Plant Physiol.* **2014**, *165*, 1035–1046. [CrossRef]
109. Salvi, E.; Di Mambro, R.; Pacifici, E.; Dello Ioio, R.; Costantino, P.; Moubayidin, L.; Sabatini, S. SCARECROW and SHORTROOT control the auxin/cytokinin balance necessary for embryonic stem cell niche specification. *Plant Signal Behav.* **2018**, *13*, e1507402.
110. Md Zali, A.Z.; Ja'afar, Y.; Paramisparan, K.; Ismail, S.I.; Saad, N.; Mohd Hata, E.; Md Hatta, M.A.; Ismail, M.R.; Yusof, M.T.; Zulperi, D. First report of *Burkholderia gladioli* causing bacterial panicle blight of rice in Malaysia. *Plant Dis.* **2023**, *107*, 551. [CrossRef]
111. Debta, H.; Kunhamu, T.K.; Petrík, P.; Fleischer, P.; Jisha, K.C. Effect of hydropriming and osmopriming on the germination and seedling vigor of the East Indian Sandalwood (*Santalum album* L.). *Forests* **2023**, *14*, 1076. [CrossRef]
112. Ros, C.; Bell, R.W.; White, P.F. Seedling vigour and the early growth of transplanted rice (*Oryza sativa*). *Plant Soil* **2003**, *252*, 325–337. [CrossRef]
113. Chen, Y.; Gin, J.W.; Wang, Y.; de Raad, M.; Tan, S.; Hillson, N.J.; Northen, T.R.; Adams, P.D.; Petzold, C.J. Alkaline-SDS cell lysis of microbes with acetone protein precipitation for proteomic sample preparation in 96-well plate format. *PLoS ONE* **2023**, *18*, e0288102. [CrossRef] [PubMed]
114. Niu, L.; Zhang, H.; Wu, Z.; Wang, Y.; Liu, H.; Wu, X.; Wang, W. Modified TCA/acetone precipitation of plant proteins for proteomic analysis. *PLoS ONE* **2018**, *13*, e0202238. [CrossRef]
115. Waterborg, J.H.; Matthews, H.R. *Basic Protein and Peptide Protocols*; Humana Press: Berlin, Germany, 1994; pp. 1–4.
116. Tyanova, S.; Temu, T.; Cox, J. The MaxQuant computational platform for mass spectrometry-based shotgun proteomics. *Nat. Protoc.* **2016**, *11*, 2301–2319. [CrossRef] [PubMed]
117. Pang, Z.; Lu, Y.; Zhou, G.; Hui, F.; Xu, L.; Viau, C.; Spigelman, A.F.; MacDonald, P.E.; Wishart, D.S.; Li, S. MetaboAnalyst 6.0: Towards a unified platform for metabolomics data processing, analysis and interpretation. *Nucleic Acids Res.* **2024**, *52*, W398–W406. [CrossRef]
118. Blom, N.; Sicheritz-Pontén, T.; Gupta, R.; Gammeltoft, S.; Brunak, S. Prediction of post-translational glycosylation and phosphorylation of proteins from the amino acid sequence. *PROTEOMICS* **2004**, *4*, 1633–1649. [CrossRef]
119. Kolberg, L.; Raudvere, U.; Kuzmin, I.; Adler, P.; Vilo, J.; Peterson, H. g:Profiler—Interoperable web service for functional enrichment analysis and gene identifier mapping (2023 update). *Nucleic Acids Res.* **2023**, *51*, W207–W212. [CrossRef]
120. Ødum, M.T.; Teufel, F.; Thumuluri, V.; Almagro Armenteros, J.J.; Johansen, A.R.; Winther, O.; Nielsen, H. DeepLoc 2.1: Multi-label membrane protein type prediction using protein language models. *Nucleic Acids Res.* **2024**, *52*, W215–W220. [CrossRef]
121. Szklarczyk, D.; Santos, A.; von Mering, C.; Jensen, L.J.; Bork, P.; Kuhn, M. STITCH 5: Augmenting protein-chemical interaction networks with tissue and affinity data. *Nucleic Acids Res.* **2016**, *44*, D380–D384. [CrossRef]

122. R Core Team. *R: A Language and Environment for Statistical Computing*; R Foundation for Statistical Computing: Vienna, Austria, 2013.
123. Villanueva, R.A.M.; Chen, Z.J. ggplot2: Elegant graphics for data analysis. *Meas. Interdiscip. Res. Perspect.* **2019**, *17*, 160–167. [CrossRef]

plants

MDPI

Review

Advances in Functional Genomics for Exploring Abiotic Stress Tolerance Mechanisms in Cereals

Tatenda Goche [1,*], Peter Mavindidze [2] and Tinashe Zenda [3,*]

[1] Crop Science Department, Faculty of Agriculture and Environmental Sciences, Bindura University of Science Education, Bindura P.O. Box 1020, Zimbabwe
[2] Department of Biosciences, Durham University, Durham DH1 3LE, UK; peter.mavindidze@durham.ac.uk
[3] Crop Science Department, Faculty of Plant and Animal Sciences and Technology, Marondera University of Agricultural Sciences and Technology, Marondera P.O. Box 35, Zimbabwe
* Correspondence: tgoche@buse.ac.zw (T.G.); tzenda@muast.ac.zw (T.Z.)

Abstract: Climate change, population growth and the increasing demand for food and nutritional security necessitate the development of climate-resilient cereal crops. This requires first gaining mechanistic insights into the molecular mechanisms underpinning plant abiotic and biotic stress tolerance. Although this is challenging, recent conceptual and technological advances in functional genomics, coupled with computational biology, high-throughput plant phenotyping and artificial intelligence, are now aiding our uncovering of the molecular mechanisms underlying plant stress tolerance. Integrating other innovative approaches such as genome editing, modern plant breeding and synthetic biology facilitates the development of climate-smart cereal crops. Here, we discuss major recent advances in plant functional genomic approaches and techniques such as third-generation sequencing, transcriptomics, pangenomes, genome-wide association studies and epigenomics, which have advanced our understanding of the molecular basis of stress tolerance and development of stress-resilient cereals. Further, we highlight how these genomics approaches are successfully integrated into new plant breeding methods for effective development of stress-tolerant crops. Overall, harnessing these advances and improved knowledge of crop stress tolerance could accelerate development of climate-resilient cereals for global food and nutrition security.

Keywords: cereal crops; functional genomics; abiotic stress tolerance; third-generation sequencing; transcriptomics; genome-editing techniques; metagenomics

1. Introduction

Crop plants often endure serious abiotic (drought, heat, salinity, anoxia, nutritional imbalances, etc.) and biotic (pathogenic, pest, etc.) stresses, which hinder their growth and development [1–3], consequently reducing crop productivity, quality and quantity [4,5]. The rapidly changing global climate renders these stresses increasingly important by escalating their incidence and severity [6–8]. At the same time, the global human population has increased to approximately 8 billion people [9], amplifying global food demand [1,10,11], against the backdrop of a shrinking natural agricultural resource base [12]. Consequently, agriculture faces a mammoth task to sufficiently and nutritiously feed the growing population, as scientists grapple with a climate adaptation–resource conservation–food security dichotomy [6,13]. Therefore, adaptation of cereals to climate change is critical in ensuring future food and nutrition security [14].

Cereal crops such as wheat (*Triticum aestivum*), rice (*Oryza sativa*), maize (*Zea mays*), barley (*Hordeum vulgarie*) and sorghum (*Sorghum bicolor*) constitute a major portion of global human diets, with over 40% of calories directly derived from their grains, and another percentage from the use of cereals as animal feed and raw materials in beverages manufacture [15,16]. Consequently, the effects of climate change-exacerbated abiotic and biotic stresses on cereal crops pose huge global food and nutrition security challenges [16–18]. For instance, complex drought patterns have been shown to robustly explain global yield loss for major crops, with wheat, soybean (*Glycine max*) and maize yields across southern and eastern Europe, the Americas and sub-Saharan Africa exhibiting high and multi-crop susceptibility to more complex droughts than previously assessed [19]. Current yield loss estimates for soybean, wheat and maize for mid-century are 16%, 7.7% and 8.3%, respectively [20]. Meanwhile, a meta-analysis has revealed that combined drought and heat stress significantly reduce crop yields by diminishing harvest index, shortening crop life cycles and altering seed number, size and composition, with their impact being more severe during the reproductive stage of plants [21]. Both abiotic and biotic stresses disrupt several cellular and whole-plant functions, disturb cellular homeostasis and hinder physiological and metabolic processes, which negatively impact overall plant growth and development [22–24] (Figure 1). Therefore, gaining mechanistic insights into the fundamental aspects of crop stress tolerance and developing climate-resilient cereals could aid us in adapting agriculture to climate change and achieving global food and nutrition security [3,4,25,26].

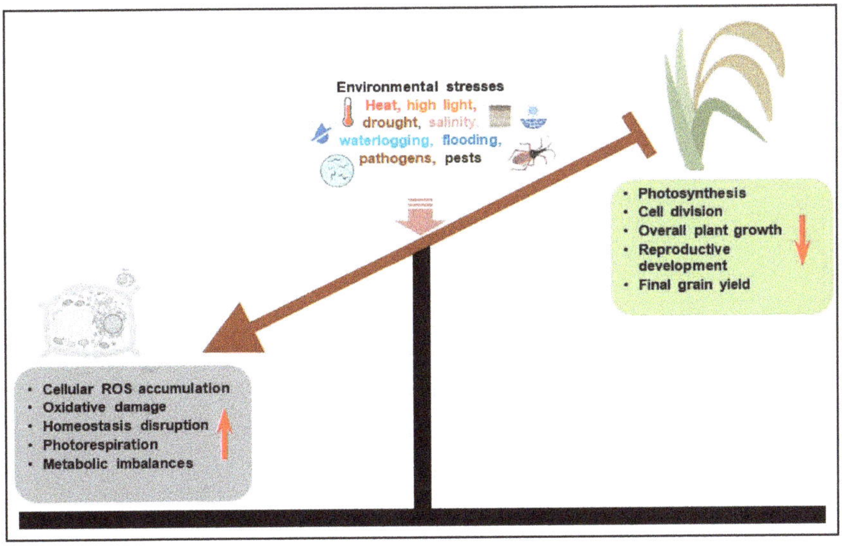

Figure 1. The impact of environmental stress on cereal crop plants. Environmental stresses repress plant physiological processes, overall growth and productivity, by increasing cellular reactive oxygen species accumulation, oxidative stress damage, homeostasis disruption and metabolic burden. Note: red arrows signify direction of change in parameters. Brown arrow and blunt ends signify promotive effect and repression, respectively.

Meanwhile, plants have evolved several mechanisms to tolerate abiotic and biotic stresses, including stress signal transduction, osmotic adjustment, antioxidant systems activation, transcriptional factor activation, phytohormonal regulation, defence-response-related or stress-responsive genes activation and epigenetic regulation [27–30].

Moreover, facilitated by epigenetic mechanisms, plants develop stress memory within their system, allowing them to 'remember' the initial stress when it recurs and institute apt responses [31,32]. Exploring the genetic and molecular mechanisms underlying cereal crops' responses and tolerance to different abiotic and biotic stresses facilitates cereal crop improvement for climate resilience through targeted breeding [33,34]. In this regard, omics-driven elucidation of mechanisms underlying crop plants tolerance to environmental stress has significantly streamlined the identification of key candidate genes, proteins, metabolites and/or metabolic pathways linked to specific quantitative traits that have been successfully analysed (for instance, through quantitative trait mapping), validated, manipulated and/or incorporated into crop variety improvement programs much "earlier" than would the conventional approaches [35].

In the past, limited genetic information and resources hindered efforts to uncover the molecular basis of complex traits and identify stress-responsive genes in major cereals [34]. Although technologies like short-read Illumina sequencing have played a critical role in advancing genome sequencing projects [36,37], their inherent limitations, such as large assembly gaps and inability to span long genomic regions, have resulted in fragmented and error-prone assemblies. These shortcomings have, in turn, affected the accuracy of downstream analyses in many plant genomes [38,39]. Fortunately, rapid advances in nanopore technologies for single-molecule and long-read sequencing have facilitated the closing of gaps, improved accuracy, read length and throughput, yielded haplotype-resolved genome assemblies and simplified parsing of differences between plant genomes [40–42]. Evidently, therefore, the latest advances in functional genomics have greatly expanded our capacity to explore the genetic and molecular basis of plant stress tolerance and engineer climate resilience in cereals [25,43,44]. For instance, advanced genome sequencing techniques have facilitated the decoding of whole genomes, trait discovery and mapping of genomic regions controlling those quantitative traits in several cereal species [44]. RNA-seq transcriptomics analysis has successfully revealed key drought-responsive genes in maize and wheat [45,46], isobaric tags for relative and absolute quantitation (iTRAQ) proteomics has revealed key waterlogging-responsive proteins in wheat [47] and liquid chromatography combined with mass spectrometry (LC-MS) metabolomics analysis has successfully revealed key low-temperature stress-responsive metabolites in rice [48], etc. (summarised herein in Table 1).

Meanwhile, the now-available high-quality reference genomes [49,50], pangenomes [51–54], gene model annotations, expression atlases, single nucleotide polymorphisms (SNPs), genome-wide association studies (GWAS), gene networks, sequenced mutation populations, etc., constitute a repertoire of genetic and genomic resources underpinning functional genomics analysis of stress tolerance and breeding of crops [18,34,55]. These are supported by improved experimental and transformation protocols [56–59], high-throughput plant phenotyping platforms [60,61], multi-omics and single-cell omics [35,62] data analytics [63] and machine learning (ML) [64,65] approaches. Integrating these innovations with technological advances in biotechnology [66,67], genome editing [68–70] and modern breeding [18,71,72] approaches aids the exploration of stress tolerance and engineering of climate-resilience in cereals.

In this review, we discuss the various genomic approaches and tools that are driving the exploration of abiotic stress tolerance mechanisms in cereal crops and facilitating the development of climate-resilient varieties. We begin by briefly outlining the key mechanisms underlying plant responses to abiotic stresses, followed by an overview of genomics-based strategies used to investigate these mechanisms. Finally, we highlight how functional genomics is being integrated with modern plant breeding techniques to accelerate the development of stress-tolerant cereal crops.

Table 1. Selected comparative omics studies revealing abiotic stress-responsive genes/proteins/metabolites and tolerance mechanisms in major cereal crop species.

Abiotic Stress	Crop Species	Genotypes	Environmental Condition/Context	Outcome/Key Findings	Reference
Transcriptomics					
Drought and salinity	Wheat	Drought/salt tolerant JM22 and salt sensitive YM20	Genome-wide transcriptome analysis of wheat cultivars' response to drought (10% soil moisture) and salinity (100 mM NaCl) stresses.	10 DEGs, annotated to cellular process, metabolic process, osmotic regulation, and MAPK signalling pathway, were co-identified as drought and salinity tolerance-associated DEGs contributing to better stress tolerance in JM22.	[45]
Drought	Maize	Drought-tolerant CML69 and susceptible LX9801 inbred lines	Comparative transcriptomic and physiological analyses of maize seedling leaf tissues response to 3–5 days of drought treatment.	Among other key results, the tolerant line showed significantly higher leaf RWC, and lower electrolyte leakage and MDA levels than the susceptible line, which possibly contributed to its better drought tolerance.	[73]
Waterlogging	Maize	Tropical line Suwan-2 and temperate line Cim-3	Comparative physiological and transcriptomic analysis of maize inbred line tolerance to waterlogging stress.	Crucial waterlogging-responsive DEGs in Suwan-2 were related to TF modulation, cellular redox homeostasis maintenance, and plant hormone biosynthesis regulation.	[74]
Salinity	Rice	Salt-tolerant HH11 and salt-sensitive IR29 cultivars	Transcriptome analysis of rice response to 200 mM NaCl salt for 0 h, 6 h, 24 h and 48 h at the 3 leaf stage.	HH11 showed more favourable antioxidant and osmotic adjustments than IR29 upon salt stress exposure, thus, better salt tolerance.	[75]
Heat stress (HS)	Wheat	Heat-tolerant genotype WH-730	Transcriptomic analysis of a heat-tolerant wheat genotype response to control and heat treatment conditions.	5610 heat-responsive DEGs were identified, and participate in HS response pathways, e.g., HSPs, antioxidant defence, and metabolic adjustments. Among them, peroxidase was dominant, enhancing HS tolerance, possibly via regulation of lignin biosynthesis.	[76]
Cold	Rice	Cold-tolerant cultivar Huaidao5 and Cold-sensitive cultivar Huaidao9	Differential expression and co-expression network analyses of rice panicle and flag leaf transcriptomes under reproductive-stage cold stress.	Huaidao5 showed better panicle tolerance to cold stress due to higher expression levels of cold-responsive genes in related pathways, e.g., MAPK signalling pathway, glutathione metabolism, plant hormone signal transduction, etc.	[77]

Table 1. *Cont.*

Abiotic Stress	Crop Species	Genotypes	Environmental Condition/Context	Outcome/Key Findings	Reference
Cadmium	Rice	ZZ143 (low grain Cd) and YX409 (high grain Cd)	Genotypes subjected to 100 µmol/L Cd stress for 10 days.	ZZ143 showed higher root Cd tolerance than the susceptible genotype, possibly due to its greater root sulphur assimilation, and higher number of Cd-responsive DEGs and pathways, e.g., secondary metabolites biosynthesis, MAPK signalling, etc.	[78]
Low nitrogen	Sorghum	N-efficient (398B) and the N-inefficient (CS3541) inbred lines	Comparative phenotypic and transcriptome analysis of sorghum genotypes under low N hydroponic and field conditions	398B exhibited superior low N tolerance than CS3541 under both field and hydroponic conditions, due to its higher photosynthetic performance and sustenance of N metabolism-related enzyme activities.	[79]
Proteomics					
Drought	Wheat	Tolerant BW35695 and drought-sensitive BW4074	Physiological, biochemical, and iTRAQ leaf proteome analyses of wheat responses to drought.	Tolerant variety showed greater osmotic adjustment, antioxidant capacity, and high upregulation of protein synthesis-related proteins, contributing to better stress tolerance.	[80]
Drought	Maize	Drought-tolerant YE8112 and drought-sensitive MO17	Physiological and iTRAQ leaf proteome analyses of maize responses to drought.	A total of 721 DAPs were identified. Most DAPs in YE8112 were associated with photosynthesis antenna proteins pathway, and YE8112 had better tolerance due to its activation of photosynthesis proteins related to balancing light capture and utilisation.	[81]
Heat	Rice	Heat-tolerant variety 9311 and sensitive variety Guangluai4 (GLA4)	Phosphoproteomic analysis of high temp (30–38 °C for 1 to 9 days)-induced changes in indica rice developing grains.	A total of 9994 phosphosites from 3216 phosphoproteins were identified in all endosperm samples. Several HS-induced consensus phosphorylation motifs were identified, and revealed a core set of HS-responsive protein kinases, splicing factors, and regulatory factors, especially those involved in starch metabolism.	[82]
DS and elevated temp (ET)	Barley	7 spring barley RILs (hybrids of European and Syrian accessions)	LC-MS based proteomic analysis of barley flag leaf response to drought and ET (20/30 °C night/day).	Several protein accumulation changes under DS, ET and combined stresses were identified, including for photosynthetic apparatus-related proteins. Dehydrins were found among universally stress-responsive proteins.	[83]

Table 1. *Cont.*

Abiotic Stress	Crop Species	Genotypes	Environmental Condition/Context	Outcome/Key Findings	Reference
Waterlogging	Wheat	Tolerant XM 55 and sensitive genotypes YM 158	iTRAQ proteomic analysis of wheat responses to waterlogging stress.	Of the 7710 DAPs identified, 16 were distinct between the 2 cultivars under stress; 11 DAPs were upregulated and 5 were downregulated. 9 DAPs, including DEAD-box ATP-dependent RNA helicase 3, responded to waterlogging with non-cultivar specificity.	[47]
Salinity	Pearl millet (*Pennisetum glaucum*)	Tolerant (Tol) and sensitive (Sen) accessions	2DE-based whole proteome analysis of pearl millet response to 150 mm NaCl treatment	295 and 315 protein spots were identified in tolerant and sensitive accessions, respectively. Salinity tolerance of the tolerant accession was attributed to its higher upregulation of stress-responsive proteins.	[84]
Salinity	Wheat	Kharchia-65 salt-tolerant) and PBW-373 (salt-sensitive)	LC–MS/MS based proteomic analysis of wheat responses to 0 and 300 mM NaCl treatment for 48 h.	21,863 proteins and 5133 protein groups were identified. There was higher upregulation of stress-responsive proteins, e.g., auxin-responsive, peroxidase, etc., in tolerant genotype and comparative downregulation in susceptible genotype.	[85]
Low temperature (LT)	Maize	LT-tolerant Gurez local and LT-sensitive GM6	2D-PAGE based proteomic analysis of maize leaf responses to low temp (6 °C) exposure for 12 h at 3-leaf stage.	19 and 10 proteins were identified in Gurez local and GM6, respectively, including 3 novel abiotic stress- and LT-responsive proteins (e.g., nodulin-like protein) identified from Gurez local.	[86]
Aluminium (Al)	Barley	Al-sensitive barley cultivar ZU9	TMT-based quantitative proteomic analysis of barley response to aluminium stress under phosphorus-*Piriformospora indica* interaction	DEPs were mostly enriched in the phenylpropanoid biosynthesis pathway, among which peroxidases were prominent. *P. indica* in combination with P helped barley plants to endure Al-induced stress by modulating antioxidative defence system.	[87]
Low inorganic phosphorus (Pi)	Wheat	Higher PUE genotype TM98 and a lower PUE genotype H4399	Label-free quantitative proteomic analysis of wheat leaf response to low Pi.	2110 high-confidence proteins were identified, among them 244 and 133 DAPs under Pi deficiency in H4399 and TM98, respectively. Abundance of energy metabolism-related proteins was decreased by Pi deficiency in H4399 shoots, but not in TM98.	[88]

Table 1. *Cont.*

Abiotic Stress	Crop Species	Genotypes	Environmental Condition/Context	Outcome/Key Findings	Reference
Drought	Sorghum	Drought-sensitive S4 and S4-1, and drought-resistant T33 and T14	nano-LC-MS/MS-based leaf proteome analysis of sorghum response to drought.	A total of 3927 proteins were quantified, with 46, 36, 35, and 102 DAPs identified in S4, S4-1, T14, and T33 varieties, respectively. Tolerant genotypes showed enhanced TCA cycle and influenced aminoacyl-tRNA biosynthesis.	[89]
Metabolomics					
Low temp (LT)	Rice	Varieties 02428 (*japonica*) and YZX (*indica*)	LC–MS/MS-based metabolomics analysis of rice response to LT (15 °C for 4 days) at germination.	A total of 730 metabolites were detected by LC-MS/MS method. 7 key LT-responsive metabolites were identified, and these metabolites were observed to participate in biosynthesis of amino acids and phenylpropanoids, as well as metabolism of glutathione and inositol phosphate.	[48]
Low nitrogen	Wheat	Zheng Mai 366 and Ai Kang58, dominant species in Henan.	UPLC-QTOF-based analysis of wheat flag leaf response to low N stress.	Chemical analyses identified 11 secondary metabolites, considered biomarkers of low N stress. Most of these secondary metabolites were flavonoids and their related derivatives, such as iso-vitexin, iso-orientin, etc.	[90]
Low nitrogen	Sorghum	10 diverse entries (including inbreds and hybrids)	UPLC-MS/MS-based analysis of sorghum roots' response to low N.	Roots from plants with N stress contained reduced phenylalanine, a precursor for salicylic acid, providing evidence for compromised metabolic capacity for defence response under low N conditions.	[91]
Drought	Barley	German variety Maresi and Syrian breeding line Cam/B1//CI08887/CI05761	Untargeted GC-MS based metabolomics profiling of barley leaf and root responses to drought.	Compatible solutes and osmolytes were the major group of compounds accumulated under drought, and revealed changes in accumulation of some metabolites, e.g., proline and other amino acids, CHOs or carboxylic acids were considered a basic plant strategy for acquiring drought stress tolerance.	[92]
Drought	Wheat	Drought-tolerant T13 and drought-susceptible T2	Integrated transcriptome and metabolomics analyses of wheat responses to drought.	Flavonoids and phenolic acids metabolism were associated with wheat seedlings' drought tolerance, with their biosynthesis-related DEMs and genes possibly being key factors underlining the difference in drought tolerance.	[93]

Table 1. *Cont.*

Abiotic Stress	Crop Species	Genotypes	Environmental Condition/Context	Outcome/Key Findings	Reference
Low phosphorus (LP)	Wheat	G28 (LP-tolerant) and L143 (LP-sensitive) varieties	Metabolomics and transcriptomics analysis of wheat response to 72 h of LP stress.	A total of 181 and 163 DAMs were detected in G28LP and L143LP under LP stress, respectively. Additionally, joint metabolomics and transcriptomic analysis revealed that wheat LP tolerance was closely related to 15 metabolites and 18 key genes in the sugar and amino acid metabolism pathways.	[94]
Salt and heat	Wheat	Warm-adapted Fahng60 and heat-sensitive Samerng2 cultivars	Physiological and metabolomics analysis of seedlings' response to salt (150 mM NaCl) and HS (42 °C for 4 h) treatments.	Amino acids, sugars, and sugar derivatives were the major responsive metabolites in leaves under the stress. Additionally, in both genotypes, the ABC transporters, glucosinolate metabolism, aminoacyl-tRNA biosynthesis, etc., were the key overrepresented pathways under the stress combination.	[95]
Saline-alkaline	Rice	Saline–alkali-tolerant cultivar Tongxi926	Integrated transcriptome and metabolomics analysis of rice response to high saline–alkali stress (pH > 9.5).	9347 DEGs and 693 DAMs were identified. Among the DAMs, lipid and amino acid accumulation were greatly enhanced, and pathways related to ABC transporter, amino acid biosynthesis, glutathione metabolism, TCA cycle, etc., were significantly enriched.	[96]
Drought	Maize	Drought-tolerant line si287 and a drought-sensitive line X178	Transcriptomic and metabolomics analysis of maize response to a 7-day drought at the 3-leaf stage	DEGs and DEMs were significantly enriched in flavonoid biosynthesis, starch and sucrose metabolism, and amino acids biosynthesis-related pathways. Joint analysis identified proline, tryptophan and phenylalanine as key stress-responsive amino acids.	[97]
Salinity	Barley	GN2 (salt-tolerant) and GN18 (salt-sensitive)	Proteomic and metabolomics analysis of barley response to salt stress at germination stage.	Besides the stress-responsive DAPs, a total of 187 salt-regulated metabolites were identified, which were mainly related to ABC transporters, amino acid metabolism, CHO metabolism and lipid metabolism.	[98]

Note: DEGs, differentially expressed genes; DAPs, differentially abundant proteins; DEPs, differentially expressed proteins; DEMs, differentially expressed metabolites; DAMs, differentially accumulated metabolites; PUE, phosphorus utilization efficiency; HSPs, heat shock proteins; CHO, carbohydrates; Cd, cadmium; iTRAQ, isobaric tags for relative and absolute quantitation; LC-MS/MS, liquid chromatography combined with mass spectrometry; nano-LC-MS/MS, nano-scale liquid chromatography mass spectrometry; 2D-PAGE, two dimensional gel electrophoresis; LFQ-based, label-free quantification based; UPLC-QTOF, ultra-performance liquid chromatography-quadrupole time-of-flight mass spectrometry.

2. Overview of Abiotic Stress Tolerance Mechanisms in Cereals

Creating stress-tolerant crops requires first gaining a mechanistic understanding of the plant biological processes and fundamental mechanisms underlying responses to those different stresses [1,99]. Cereals have evolved to thrive in environments that routinely expose them to a variety of stressors, such as high temperatures, drought, salt, mineral toxicity and water scarcity. To cope with these pressures, cereals have developed a variety of stress tolerance mechanisms. These include the accumulation of compatible solutes such as proline, betaine and sugars to maintain cell turgor and prevent water loss [100–104]. In addition, cereals can cope with drought stress by reducing transpiration rates and increasing water-use efficiency [105,106]. They can also produce deeper roots to access deeper soil water reserves [107,108]. Furthermore, cereals possess the capacity to withstand high temperatures due to their ability to release heat shock proteins, which protect proteins from denaturation and oxidative damage [109]. To withstand cold stress, some cereals produce antifreeze proteins, which inhibit ice from forming and shield membranes from damage [110,111]. Cereals also possess an inherent ability to tolerate high salt concentrations by excluding salt from the roots and accumulating it in the vacuoles, as well as balancing the expression of reactive oxygen species (ROS)-scavenging genes [104,112–117].

Understanding root development and architecture holds the promise of exploiting and manipulating root traits to boost food plant output and optimise agricultural land usage. Cereals alter their root architecture to increase nutrient uptake under stress. The plants accomplish this via the production of enzymes that aid in nutrient mobilisation and assimilation. Siddiqui et al. [118] and Halder et al. [119] detail the genomics of root system variation and root system architecture under abiotic stress, respectively.

Under field conditions, different environmental stresses usually occur concurrently. The simultaneous occurrence of several environmental stressors and biotic variables triggers complex plant responses due to crosstalk mechanisms. Combined stresses are usually more damaging to plant systems in comparison to individual stress [120]. Combined heat and drought stress on winter wheat inflicted more damage compared to individual stress and the recovery rates of the physiological traits affected by the combined stress [121]. Another research team has also highlighted the complexity of combined stress on sorghum [122]. Understanding these mechanisms is essential for creating stress-tolerant cereal cultivars that can ensure food security in the face of climate change. Rigorous approaches are required in the development of multi-stress-tolerant cereal crops. Schimt et al. [123] identified novel alleles for combined heat and drought stress in wheat, whilst Yang et al. [124] showed that the maize primary root response to combined heat and drought stress is regulated via the salicylic acid pathway. Further combined stress studies have been performed in maize [124,125], wheat [126–128], rice [129,130], sorghum [131] and barley [132].

3. Recent Advances in Crop Functional Genomics

In this section, we briefly discuss various genomic approaches underpinning the exploration of stress tolerance mechanisms in crop plants. These include genome sequencing and annotation, transcriptomics, proteomics and metabolomics, among others.

3.1. Third Generation Sequencing, Long Reads and Pangenomes

The evolvement of genomics technologies over the past decade has revolutionised genome sequencing, annotation and analysis, which has advanced our understanding of plant genetic architecture of environmental adaptation [44,50,133]. Cost-effective high-throughput next-generation sequencing (NGS) approaches (Illumina/Solexa, Ion Torrent PGM, etc.) have enabled whole-genome sequencing of most crop species [44,50,134–136]; however, NGS approaches produced fragmented assemblies due to short-read length, which have low accuracy, hence are not able to detect low-abundance mutations and SNPs [25]. Fortunately, the current third-generation sequencing (TGS) technologies, especially Oxford Nanopore and Pacific Bioscience platforms [137,138], anchored on long reads [49,133], have facilitated the assemblage of high-quality chromosomal-scale plant genomes [25] and pangenomes for most major crops, some minor species and crop wild relatives (CWRs) [35,139,140]. This has offered us a holistic 360-degree view of the plant genomes, viz., the structure, organisation and regulation [41], as well as facilitating the mining of novel agronomic and climate-responsive alleles from CWRs through de novo domestication [25,141–144].

Despite short-read sequencing being cost-effective, accurate and amenable to diverse analysis tools and pipelines, it cannot span lengthy nucleic acid polymers and complicates reconstruction and counting of the original molecules [40]. On the other hand, long-read sequencing (LRS) allows for gap closing and improved accuracy. TGS approaches now offer possibilities to resolve transposon repeats through generation of long reads that span over transposon regions, including unknown locations, thereby facilitating the production of high-quality assemblies for complex genomes [10,145]. These LRS-based methods combine nanopore and single-molecule real-time sequencing to yield long reads (of ~60–200 Kb length), which can span complex and repetitive regions [10,133,146]. They are usually coupled to optical mapping and confirmation capture to generate draft genomes of unmatched contiguity [25,147]. Thus, long reads can significantly enhance de novo assembly, mapping certainty, transcript isoform detection and structural variant calling. Furthermore, LRS of native DNA and/or RNA molecules removes amplification bias while preserving base modifications [40,148]. Taken together, the era of LRS approaches is offering a window for de novo assembling of genomes, resolving sequence assembly ambiguities and gap filling. Moreover, this has facilitated comparisons between related species and identification of subtle genetic variations that may be key in improving elite crops [44].

Molecular markers, especially genome-scale SNPs, now permit quick dissection of genome-wide variations and key agronomic and climate-adaptive traits, from both elite and CWRs, which is essential for traits introgression, marker-assisted selection (MAS), marker-assisted breeding (MAB) and other genomic applications [43,149,150]. The greater abundance of SNPs permits greater coverage of loci across the genome [151]. Moreover, SNPs are amenable to high-throughput (HTP) and automated profiling, thereby facilitating rapid and HTP high-density SNP-marker-based genotyping [152,153]. Thus, SNPs are now widely deployed (more than other genetic markers such as single-sequence repeats) in marker-assisted breeding, GWAS, genomic selection (GS) and map-based cloning [151,154,155].

Feasibly, high-resolution GWAS has been conveniently applied to identify key genetic factors (QTLs, genes' genomic regions and haplotypes) governing several key agronomic, climate-adaptive and nutritional traits [35,149,156–161]. For instance, recent advancements in functional genomics approaches and resources have unprecedentedly facilitated functional characterisation of genes in crops with large genomes, such as polyploidy wheat (AABBDD), which has aided crop improvement [34]. Integrating these novel genetic and genomic resources (including high-quality draft genomes, pangenomes, gene model annotations, expression atlases, gene networks, sequence indexed mutant population, structured natural population, omics databases, etc. [18,34,149,162]) offers rapid and comprehensive methods for gene functional analysis, which enhances our understanding of the molecular mechanisms underpinning crop climate adaptation and accelerates the designing of climate-resilient crops [18].

3.2. *Transcriptomics*

The comprehensive examination of all expressed RNA transcripts a given organism has produced is known as transcriptomics. This covers not only the conceptual changes in the genome but also the mechanism through which the information stored in the genome is utilised by the cell, as well as the flow of biological information from the genome to the cell. Advances in transcriptomics include RNA sequencing (RNA-seq), gene co-expression network analysis, alternative splicing analysis, long non-coding RNA (lncRNA) analysis and functional annotation and pathway analysis. RNA-seq is a technique that uses NGS to investigate the quantity and sequences of RNA in a sample. It examines the transcriptome to determine which genes encoded in DNA are activated or suppressed and to what degree [163]. RNA-Seq, as the technology of choice for whole-transcriptome studies, has been successfully employed in cereal crop stress response studies, offering accurate gene expression profiling analysis and revealing key stress-responsive genes, metabolic pathways and stress tolerance mechanisms [164–167] (Table 1; Supplementary Table S1). Its applications are of great value in cereal crop gene functional validation. For example, RNA-seq and qRT-PCR were utilised to confirm that the overexpression of the heat shock factor gene (*TaHsfA6bT*) from wheat provides thermotolerance in barley [168].

Gene co-expression network analysis is a computational tool for identifying gene modules that are co-regulated under stress situations. This technique can assist in the identification of essential regulatory mechanisms involved in cereal stress tolerance, for example, drought stress tolerance in maize [169]. Critical modules and genes for drought tolerance in wheat have been identified using this technique [170]. In addition, Chopra et al. [171] identified complex gene interactions for cold and heat stresses in sorghum after evaluating their co-expression networks. Furthermore, tandem maize paralogues were shown to be consistently co-expressed under cadmium (cd) stress in a study that elucidated the genetic basis of plant response to cd stress. Alternative splicing involves the production of several mRNA isoforms from a single gene, and alternative splicing variation can be exploited in crop improvement programs for stress tolerance [172]. Lin et al. [173] performed alternative splicing of *ZmrbohB* cDNA to obtain two transcript isoforms in maize. Transcriptomics can be used to investigate alternative splicing events in response to stress [174]. This method can aid in the discovery of new transcripts and isoforms implicated in stress-response pathways. Mastrangelo et al. [175] comprehensively detailed transcriptomic plasticity through alternative splicing.

Long non-coding RNAs (lncRNAs) are a type of non-coding RNA that plays a role in gene expression regulation [176]. They have been defined as transcripts longer than 200 nucleotides with limited phylogenetic conservation, expressed at low levels and characterised by tissue/organ-specific expression profiles [177]. These transcripts have been characterised and shown to be responsive to drought, salt, cold, boron and heat stress in studies on maize [178–183]. Novel lncRNAs engaged in stress-response pathways have been identified concurrently. Further research has been carried out in sorghum, wheat, barley and the closely related *Brachypodium distachyon* [184]. Overall, improvements in transcriptomics have significantly increased our understanding of the molecular pathways underlying cereal stress tolerance. This knowledge can be utilised to create stress-tolerant cultivars that are better suited to changing environmental conditions.

3.3. *Proteomics*

The proteome offers a unique avenue for the study of intricate biological functions involving vast numbers and networks of proteins. Protein identification makes use of a variety of technologies. These are categorised into gel-based and non-gel-based techniques, coupled with mass spectrometry and bioinformatics for putative functional annotation.

Recent developments in mass spectrometry (MS)-based tools and methodologies have led to a rise in the use of a variety of unique proteomics approaches on cereals. Traditionally, two-dimensional polyacrylamide gel electrophoresis (2-D PAGE) or two-dimensional electrophoresis (2-DE) [185] has been used as a standard procedure for proteomics research. However, gel-based technologies are labour-intensive and time-consuming, making them unsuitable for processing large quantities of samples. MS-based methods such as the iTRAQ are the most widely used technology for identifying, characterising and quantifying proteins and their proteoforms, mainly because they offer high throughput and applicability to large sample numbers [186]. Genomics and transcriptomics are concerned with genes present in a genome and their expression, whereas proteomics is concerned with the qualitative and quantitative components of expressed proteins, as well as how the expressed levels fluctuate under different environmental situations. Aside from protein abundance, the existence of protein modifications can help in understanding biological processes or chemical adjustments such as signalling and food processing, respectively [187,188].

Proteomics has emerged as a powerful tool for exploring stress tolerance in cereals as it allows researchers to identify and quantify differentially expressed proteins in response to different stresses (Table 1; Supplementary Table S2). The availability of cereal crop complete genome sequences is a significant step forward in proteomics studies since it allows for the identification of genes linked with essential agronomic features. Advances in proteomics for exploring stress tolerance in cereals include the identification of stress-responsive proteins [89,107,189], the mapping of protein–protein interactions [190,191], the quantification of protein expression levels [89,192] and the development of stress-tolerant cultivars [193–195]. Some of the milestones achieved in cereals include the identification of heat shock proteins, dehydrins and late embryogenesis abundant (LEA) proteins that are upregulated in response to various stresses in maize and sorghum [107,189], protein complexes involved in the response to salt stress in rice [190] and changes in the expression levels of enzymes involved in the synthesis of compatible solutes in response to salt stress in barley [192] (Table 1; Supplementary Table S2). In addition, proteins involved in stress response, with potential to be targeted for crop genetic improvement, have been identified in most cereals, as recently reviewed [196].

3.4. Metabolomics

Metabolomics technologies speed up crop development research through the discovery of different metabolic pathways for abiotic/biotic stress tolerance. Metabolites have been regarded as "the ultimate response of biological systems to genetic or environmental changes" since they constitute one quantifiable endpoint of all cellular regulatory functions [197]. Modern metabolomics methods, such as improved mass spectrometry (MS), nuclear magnetic resonance (NMR) and cutting-edge imaging technologies like nanoscale-secondary ion mass spectrometry (NanoSIMS), have created new opportunities for in-depth study of the cereal metabolome. The full spectrum of techniques that can be used to study the plant metabolome has been extensively reviewed [198].

Metabolomic analyses can be classified as directed/targeted or undirected/non-targeted [199]. Directed metabolomics entails focused analyses that aim to identify and quantify as many metabolites within a given chemical group as possible, whilst the goal of undirected metabolomics is to detect as many metabolite groups as possible in order to create patterns or fingerprints, with or without identifying the identified molecules or compounds. Stress-related amino acids have been identified in tropical sorghum through non-targeted metabolomics [200]. Through directed metabolomics, Sheflin et al. [91] identified and quantified phenylalanine, a precursor for salicylic acid, providing evidence for compromised metabolic capacity for defence response under low-nitrogen conditions.

In contrast to the transcriptome and proteome, the metabolome is not necessarily linked to the plant genome [201]. It is, however, closely linked to the phenotype [202], providing a perfect platform to probe stress-responsive metabolites and stress-induced regulatory pathways that govern abiotic stress tolerance (Table 1). However, there is a gap between metabolite expression and the sophisticated stress tolerance process for identifying desired metabolites that operate effectively under stress conditions. Filling this gap could aid in the development of an effective breeding plan and facilitate the gathering of data for the identification of cultivars that perform well under abiotic stress conditions. In an effort to address this, Zhang et al. [203] successfully used a multi-omics approach to reveal the roles and mechanisms of two hub genes, *Bx12* and *ZmGLK44*, in regulating maize metabolite biosynthesis and drought tolerance.

4. Genome Editing Technologies

Over the last decade, genome editing has emerged as a novel approach that allows researchers to directly control literally every gene in a wide variety of cell types and organisms. Genome-editing technologies are tools that enable the modification of individual genes to understand their roles [204]. Technologies developed for genome editing include the Clustered Regularly Interspaced Short Palindromic Repeats-CRISPR-associated protein 9 (CRISPR-Cas9), CRISPR-Prevotella and Francisella 1 (CRISPR-Cpf1), Transcription Activator-Like Effector Nucleases (TALENs) and Zinc Finger Nucleases (ZFNs). The CRISPR-Cas9 enables precise and efficient gene editing by utilising a guide RNA (gRNA) to target a specific DNA sequence and the Cas9 enzyme to introduce changes, such as gene knockouts or precise gene modifications. CRISPR-Cas9 has been widely applied in cereal crop research to better understand the function of certain genes implicated in stress response, including drought in maize [205], heat in several species [206], chilling stress in rice [207] and salinity in rice and wheat [208]. This is due to its simplicity, accessibility, adaptability, flexibility and wide applicability. CRISPR-Cpf1 is similar to CRISPR-Cas9, except instead of Cas9, the Cpf1 (Cas12a) enzyme is used. CRISPR-Cpf1, like CRISPR-Cas9, can be utilised for precise gene editing, such as gene knockouts and alterations. It has various advantages over Cas9, such as a reduced gRNA size and different target site preferences.

ZFNs and TALENs are chimeric nucleases made up of sequence-specific DNA-binding modules coupled to a non-specific DNA cleavage domain [209,210]. These nucleases enable a wide range of genetic alterations by causing DNA double-strand breaks, which drive error-prone non-homologous end joining or homology-directed repair at specified genomic loci [211]. The programmability of the DNA-binding domains that are derived from zinc-finger and transcription activator-like effector proteins has previously established these technologies as the most preferred in genetic engineering. CRISPR-Cas9, ZFNs and TALENs, however, faced challenges that CRISPR-Cpf1/Cas12a managed to fill [212]. These include re-engineering at the same target DNA region. This was because of the loss of the target site after the initial engineering with CRISPR-Cas9, ZFNs or TALENs. Cpf1 (Cas12a) functions as both an endoribonuclease and an endodeoxyribonuclease, cleaving target sequences and generating double-stranded breaks. Cpf1 also enables multiplexed genome editing because a single crRNA array transcript can target several loci in the genome. Although many successes were reported on model cereal crops such as rice [213–216], initial CRISPR-Cpf1 research in other cereals required optimisation to improve its applicability [217].

Overall, these genome editing technologies can be used to explore the function of genes involved in cereal crop stress responses. The involvement of genes of interest in stress tolerance systems can be ascertained, and plants with greater stress resistance can be designed by precise alterations. However, it is critical to evaluate the ethical and regulatory elements of genome editing technologies in order to ensure their responsible and safe use.

5. Epigenomics

Environmental stresses induce epigenetic changes such as cytosine DNA methylation and histone post-translational modifications, which alter the chromatin structure and spatiotemporal gene expression. Stable epigenetic alterations are inheritable across filial generations, and this facilitates plant adaptation to environmental change [218]. Thus, epigenomics, which involves the study of chromatin states, chromatin alterations, their impact on gene regulation and aspects of epigenetic inheritance [219,220], has emerged as a new avenue for exploring molecular mechanisms of tolerance and improving crop adaptation to environmental stresses [221–224]. Epigenomics leverages chromatin information to better annotate and decode plant genomes and offers complementary strategies to identify and select heritable epialleles that govern crop traits independent of underlying genotype [220]. Considering that epigenetic mechanisms play crucial roles in other biological processes and stress responses [225,226], gaining mechanistic insights into the stress-induced epigenetic changes in plants helps us better understand the molecular adaptation of plants to environmental stresses [218,219]. Additionally, elucidation of the epigenetics-mediated cryptic molecular switches that drive the capacity of plants to memorise stress and reproduce stress-resistant progenies will help us in breeding climate-resilient crop species [221]. Leveraging recent advances in genome analysis tools, several epigenetic resources, including epialleles, epigenetic quantitative trait loci (epiQTLs), epigenome-wide association (epiGWAS), etc., can be utilised in epibreeding for improved stress tolerance in crops [72,218,227,228].

Meanwhile, plant microRNAs (miRNAs) [229,230] and long non-coding RNAs (lncRNAs) [231] have emerged as key post-transcriptional and translational regulators of gene-expression related to growth, development and stress response modulation [44,232]. To date, several stress-responsive miRNAs have been identified in different crops (including miR156, miR159, miR168, miR398, etc.) and their key roles in epigenetic regulation of stress responses established [233–236]. Meanwhile, miRNA-driven RNA-interference (RNAi) is rapidly becoming the preferred technology for improving crop traits and plasticity to environmental stress [237]. Thus, epigenetics can be integrated into the analyses of non-coding RNA, cis-regulatory elements and other non-genic variations controlling plant stress adaptation, thereby facilitating epigenetics-assisted breeding of crops [35,44,224,238]. Furthermore, supported by new epigenome editing tools, epigenetic engineering has already been initiated [218]. For instance, target-specific epigenetic engineering has been attempted by exogenous RNAi, facilitated by VIGS (virus-induced gene silencing) and grafting [225]. Going forward, new innovative techniques such as CRISPR-Cas9 and synthetic biology are anticipated to anchor epigenetic engineering of plant genomes, making epibreeding of crops for stress tolerance possible [225,239]. Further, induced epigenetic modifications can be a source of phenotypic variations in natural populations that can be passed on to multiple filial generations [240]. In light of the decreased genetic variation in conventional crop breeding due to selection pressure and climate change [67,225], epigenetic modifications may contribute to crop breeding by providing epigenome diversity, useful markers and eustressors for use in epibreeding of climate-resilient crops [222,223,225,226].

6. Integrating Novel Breeding Methods for Quick Trait Fixation and Optimization

Despite being characterised with comparably longer periods (~7–10 years), contemporary plant breeding approaches significantly anchored the identification and introgression of several key agronomic and stress-responsive traits/genes in several crops in the past [33,66,158,241–244]. However, the fast-changing climate dictates that rapid breeding cycles should be achieved, and, thus, calls for new breeding innovations [8,245]. Building on the recent advances in functional genomics, we can now innovate with several new breeding methods for improved selection and breeding efficiency, quick fixation of key agronomic and climate-adaptive traits and optimisation of plant performance [71,246]. Top among these novel approaches are genomic selection [247–249], speed breeding [250], haplotype-based breeding [251], fast-forward breeding [252] and smart breeding [253].

Genomic selection (GS) now offers unprecedented opportunities for improving selection accuracy, minimising phenotyping, reducing cycle time and increasing speed and genetic gains of the breeding programs [71,254–256]. GS can facilitate selection for complex or polygenic traits (governed by multiple additive effects genes) such as drought and/or heat stress tolerance [10], by means of the net genetic merit of an individual obtained by making use of several genome-wide markers [257]. Unlike MAS, which relies on the association between the trait of interest and other genes/traits [258], GS is practically viable even when the causal association is not known [254]. In GS, the individual effects of genome-wide HTP markers are estimated through prediction models [249]. Then, genomic-estimated breeding values (GEBVs) of these individuals are computed to select the elite lines; this does not require phenotyping [247,259]. Two populations are used, i.e., a training or reference population comprising a cohort of individuals with known genotypic and phenotypic information, and a testing or breeding population comprising candidate breeding lines with genotypic data only [247,254]. In the present climate change scenario, GS holds much possibility for improving the genetic gain of breeding programs and ensuring rapid development of climate-tolerant crops, including cereals [154,255,256].

NGS and TGS-based approaches and SNP arrays permit and expedite genotyping by sequencing (GBS) of large germplasm pools for GWAS and GS [153,155], as well as facilitating haplotype-based breeding [251,260]. These innovations enhance the capturing of species-wide genetic variation and increase the efficiency of breeding programs, as ML can be used to analyse genotypic data [261]. Meanwhile, speed breeding and GETs considerably accelerate the development of climate-resilient crops [7,143]. Speed breeding involves nurturing thousands to billions of plants under fully enclosed glasshouses and artificial light-controlled conditions [250] and considerably shortens the plant growth cycles, thereby fast-tracking crop improvement programs [262]. For instance, up to seven generation cycles per year can be attained in chickpea [263], which considerably saves breeding and variety release time by ~50%. Moreover, by shortening the breeding cycles, speed breeding can be used to investigate a range of physiological indices and optimise the transgenic pipeline, gene stacking, GS and genome editing for crop improvement [264].

Haplotype-based breeding (HBB) involves the identification of superior haplotypes and their deployment in breeding programs [265]. HBB circumvents the challenges faced by SNPs in predicting complex traits, increases the resolution of candidate genomic regions [251] and enhances exploitation of natural genetic variation within genetic populations, consequently improving breeding efficiency and precision [266]. Fast-forward breeding integrates TGS, high-throughput plant-phenotyping platforms (HT3Ps), systems biology, QTL mapping, GS and genomic prediction, machine learning (ML), artificial intelligence (AI) and other novel breeding approaches to significantly enhance the genetic base of breeding programs and accelerate genetic gains, thereby quickening the creation of stress-tolerant crop cultivars [252]. Smart breeding, on the other hand, is driven by big data (gathered spatiotemporally over multiple-environmental trial sites), AI, optimised prediction models and integrated genomic-enviromic prediction (iGEP) [253] and provides a platform for coupling multi-omics information to ML and AI for targeted designing of crop breeding pipelines [253]. Furthermore, HT3Ps, including novel sensors and high-resolution imagery, can quantify plant performance in specific environments with the generated phenomics data optimised for genetic gains [61,253,267]. Taken together, integrating novel plant breeding methods with functional genomics approaches considerably accelerates the identification and introgression of novel and key stress-responsive traits into elite crops, consequently fast-tracking the creation of new stress-tolerant crop cultivars [25,44,71,246,262] (Figure 2).

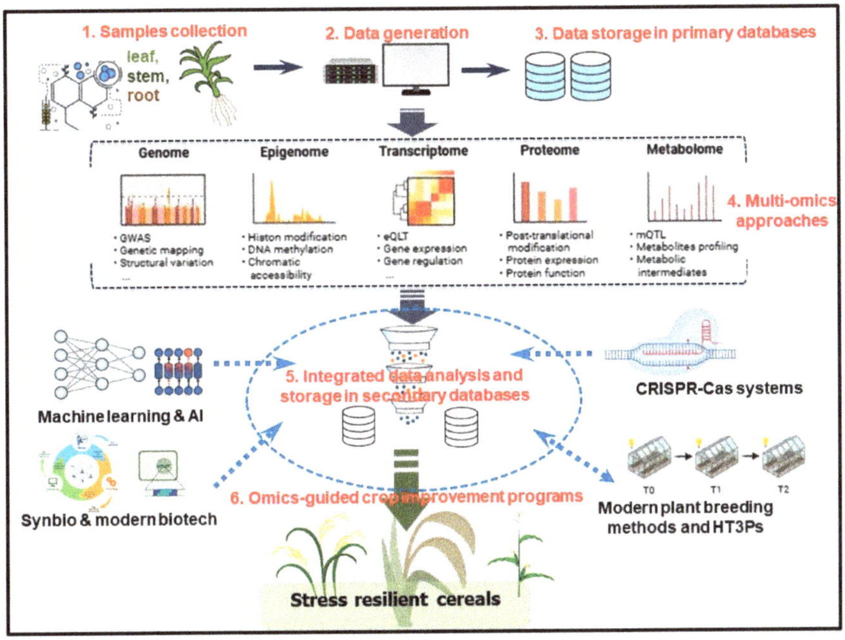

Figure 2. Integration of multi-omics approaches to modern plant breeding, biotechnology, machine learning, artificial intelligence (AI) and other technologies could considerably accelerate the identification and introgression of novel and key stress-responsive traits into elite crops, consequently fast-tracking the development of new stress-tolerant cereals. Note: HT3Ps, high-throughput plant phenotyping platforms; Synbio, synthetic biology; light blue arrows and circles signify integration.

7. Metagenomics

A cocktail of mutualistic microorganisms (bacteria, fungi, etc.) inhabits plant surfaces, interacting with each other and with their plant hosts intimately [268]. These microbial assemblies constitute the dynamic plant microbiome and are critical for plants' survival [269,270]. Plant microbiomes confer health and fitness benefits to their plant hosts, such as providing novel nutritional and biosynthetic capacities that invigorate plant growth and improve resistance to different abiotic and biotic stresses [270,271]. Recently, plant microbiomes have been reported to enhance plant capacities for disease resistance [272] and tolerance to abiotic stresses such as drought [273], heat [274] and salinity [29,271,274–276]. Furthermore, they influence plant metabolism and phytohormonal pathways aimed at instituting enhanced plant fitness and survival [270]. Thus, at the moment, plant microbiomes are a hot topic for research for improved plant stress tolerance and crop climate resilience [29,276–279].

A global analysis of the heterogeneity and exact functions of the plant microbiomes will provide new insights into the mechanisms of plant stress response and tolerance to biotic and abiotic stresses [270,280]. Moreover, beneficial interactions, such as nitrogen fixation, can be harnessed to enhance crop productivity and environmental functioning [281,282]. Despite the associated conceptual and technical bottlenecks, metagenomics is a powerful tool that can facilitate rapid analysis of microbial heterogeneity, thereby aiding us in deciphering the associations among microorganisms and their environment, as well as the overall functioning of the microbiome [275,280]. Buoyed by the recent conceptual and technological advancements in metagenomics and metatranscriptomics, computational biology [283], ML and HT3P platforms [272,284], as well as the availability of huge microbiome datasets generated from genome sequencing [285,286], identification and analysis of microbiomes is increasingly becoming feasible and meaningful. In particular, the recent progress in metabolomics methods and data analysis tools is anticipated to play a significant role in this regard [280,287].

Among other research targets, the focus may be on characterisation of crop-specific microbiomes, splitting cultivar-related and environment-related microbiomes, partitioning microbiomes' exact contributions to plant fitness, stress response and yield and analysis of metagenomes and microbiome metabolic pathways [281]. Insights gained from such investigations are key in enabling the custom design of microbiome-based techniques and methods for improving crop growth and stress tolerances [277,281]. Moreover, in view of climate change and the apparent geographical variations in extreme weather events and environmental stress consequences, metagenomics potentially assists in identifying the most beneficial crop–microbiomes combinations and effective mapping of suitable crops to specific locations [25].

8. Challenges and Perspectives

Developing stress-tolerant cereal crops is daunting. Despite undeniably revolutionising the dissection of molecular mechanisms and identification of key candidate traits underpinning crop plants' adaptation to various environmental stresses [35], the use of omics techniques has its own limitations. Firstly, several of the key candidate stress-responsive genes and proteins remain largely of unknown function due to limitations in their functional characterisation [288]. Owing to limited functional annotation capabilities of some bioinformatics tools, voluminous multiomics data remain just unanalysed "big data" in databases, with little biological relevance [289].

Secondly, the heterogeneity of multiomics data (endowed with different inherent biases, noises and scaling issues) from various platforms makes its unification and normalisation difficult. Moreover, its interpretation is complicated by the intricate nature of biological networks and omics levels that demand efficient analysis tools [290]. Fortunately, advances in computational biology and data analysis approaches, coupled with machine learning (ML), could help disentangle the complex "big" data, providing the much-needed new information of biological significance, including identification of novel genes and proteins, and revealing their functions in crop stress tolerance and yield [291–293].

Thirdly, most of the key candidate stress-responsive genes identified are yet to be functionally verified. One of the reasons is that cereal crop species such as sorghum are not amenable or have low transformation efficiencies due to recalcitrance [294]. To address the problem of transformation incompatibilities in such cereal crops, modern gene-editing technologies, such as nanoparticle-based CRISPR-Cas delivery systems, can be used [68,295]. However, since abiotic stress tolerance is a complex multigenic trait, the use of gene editing, which often targets single or a few genes, in the development of new stress-tolerant cereal crop cultivars may be limited. Moreover, gene editing depends on the mechanistic understanding of the specific genes underpinning plant tolerance to stress; yet, these causal genes have yet to be pinpointed and validated, especially under field conditions [296]. This is further convoluted by the complex crosstalk among several gene regulatory networks and stress signalling pathways. Nonetheless, we anticipate the amalgamation of multiomics, GWAS, genomic selection, phenomics and genomic prediction approaches to ease the identification of the major genetic factors regulating stress tolerance, whilst computational biology, Synbio and ML will facilitate deconstruction and elucidation of complex stress response networks, thereby aiding stress tolerance engineering in cereal crops [291,292,297,298].

Moreover, several studies have functionally verified candidate genes and networks, mainly in the model species Arabidopsis and not in cereal crops [299]. However, cross-species validation may not fully reflect gene or protein function in cereals due to differences in plant/crop physiology and regulatory context [300]. Thus, future studies involving gene knockouts or overexpression directly in cereals would better elucidate their native functional roles. Further, most of the identified stress-responsive genes are yet to be validated under field conditions. Whereas most stress-response experiments are performed under controlled lab environments, where usually single stresses are studied, crop plants growing in the field face unstable, harsh environmental conditions, often involving complex multiple stresses [301,302]. Therefore, translation of "impressive" lab results to the field (outside lab resource-constrained) settings has not yielded expected outcomes with regard to crop stress tolerance. Going forward, carrying out stress-response studies, especially examination of the physiological responses or the causal inferences between phenotypes and candidate genes, under actual field conditions, may provide reliable results and help in translating foundational discoveries to field crops.

9. Conclusions and Future Outlook

Cereal crops, like any other plants, are routinely exposed to a variety of abiotic and biotic stresses, against which they must develop appropriate responses. Therefore, creating stress-tolerant crops requires first gaining a mechanistic understanding of the plant biological processes and fundamental mechanisms underlying responses to those stresses. Here, we have discussed how the recent progress in genomics technologies and approaches, including third-generation sequencing platforms, long-read sequences, pangenomes, GWAS and multi-omics, has considerably aided the exploration of mechanisms underlying crops' tolerance to abiotic and biotic stresses. Some of the identified candidate genes and metabolic pathways have been deployed in genomics-assisted or marker-assisted breeding programs via molecular breeding approaches or genetic-engineering methodologies. Integrating other new and innovative approaches such as genome-editing techniques, HT3Ps, modern plant breeding and synthetic biology could facilitate rapid identification and introgression of novel and key stress-responsive traits into elite crops and create climate-smart cereal cultivars (Figure 2). Going forward, we anticipate single-cell omics approaches to dominate the landscape of identifying, monitoring and analysing spatiotemporal expressions of key stress-responsive genes and metabolites [29,303,304]. This will be supported by advances in HT3Ps, machine learning and artificial intelligence [292]. Resolving biological questions at the single-cell level will reveal novel insights into the spatiotemporal interplay among transcriptional, metabolic, phytohormonal and epigenetic regulatory networks, thereby aiding the design of climate-resilient cereals [305]. Moreover, synthetic biology is expected to become popular in metabolic pathway engineering for stress tolerance in crops, aided by novel genome-editing technologies [292,306]. Overall, the advances in plant functional genomics approaches, in integration with recent plant breeding, genome-editing and computational biology techniques, will simplify the untangling of complex mechanisms of plant stress tolerance and facilitate the design of stress-resilient cereals for a food-secure future.

Supplementary Materials: The following supporting information can be downloaded at: https://www.mdpi.com/article/10.3390/plants14162459/s1, Supplementary Table S1. Selected comparative transcriptomic studies revealing abiotic stress-responsive genes and tolerance mechanisms in major cereal crops. Supplementary Table S2. Selected comparative proteomics studies revealing abiotic stress-responsive proteins and tolerance mechanisms in major cereal crops. Refs. [307–332] are cited in the Supplementary Materials.

Author Contributions: Conceptualisation, T.Z. and T.G.; writing—original draft preparation, T.Z. and T.G.; writing—review and editing, T.Z., T.G. and P.M.; visualisation—T.Z. All authors have read and agreed to the published version of the manuscript.

Funding: This research received no external funding.

Data Availability Statement: No new data was generated in this research.

Acknowledgments: We acknowledge all the authors whose work we cited in this review.

Conflicts of Interest: The authors declare no conflict of interest.

References

1. Bailey-Serres, J.; Parker, J.E.; Ainsworth, E.A.; Oldroyd, G.E.D.; Schroeder, J.I. Genetic Strategies for Improving Crop Yields. *Nature* **2019**, *575*, 109–118. [CrossRef] [PubMed]
2. Raza, A.; Razzaq, A.; Mehmood, S.S.; Zou, X.; Zhang, X.; Lv, Y.; Xu, J. Impact of Climate Change on Crops Adaptation and Strategies to Tackle Its Outcome: A Review. *Plants* **2019**, *8*, 34. [CrossRef]
3. Rivero, R.M.; Mittler, R.; Blumwald, E.; Zandalinas, S.I. Developing Climate-Resilient Crops: Improving Plant Tolerance to Stress Combination. *Plant J. Cell Mol. Biol.* **2022**, *109*, 373–389. [CrossRef]
4. Dhankher, O.P.; Foyer, C.H. Climate Resilient Crops for Improving Global Food Security and Safety. *Plant Cell Environ.* **2018**, *41*, 877–884. [CrossRef]
5. Mariem, S.B.; Soba, D.; Zhou, B.; Loladze, I.; Morales, F.; Aranjuelo, I. Climate Change, Crop Yields, and Grain Quality of C_3 Cereals: A Meta-analysis of [CO_2], Temperature, and Drought Effects. *Plants* **2021**, *10*, 1052. [CrossRef]
6. Zandalinas, S.I.; Fritschi, F.B.; Mittler, R. Global Warming, Climate Change, and Environmental Pollution: Recipe for a Multifactorial Stress Combination Disaster. *Trends Plant Sci.* **2021**, *26*, 588–599. [CrossRef]
7. Eckardt, N.A.; Ainsworth, E.A.; Bahuguna, R.N.; Broadley, M.R.; Busch, W.; Carpita, N.C.; Castrillo, G.; Chory, J.; DeHaan, L.R.; Duarte, C.M.; et al. Climate Change Challenges, Plant Science Solutions. *Plant Cell* **2022**, *35*, 24–66. [CrossRef]
8. Xiong, W.; Reynolds, M.; Xu, Y. Climate Change Challenges Plant Breeding. *Curr. Opin. Plant Biol.* **2022**, *70*, 102308. [CrossRef] [PubMed]
9. United Nations Department of Economic and Social Affairs, Population Division. *World Population Prospects 2022: Summary of Results*; United Nations: New York, NY, USA, 2022.
10. Scheben, A.; Yuan, Y.; Edwards, D. Advances in Genomics for Adapting Crops to Climate Change. *Curr. Plant Biol.* **2016**, *6*, 2–10. [CrossRef]
11. Hemathilake, D.M.K.S.; Gunathilake, D.M.C.C. Chapter 31—Agricultural Productivity and Food Supply to Meet Increased Demands. In *Future Foods*; Bhat, R., Ed.; Academic Press: Cambridge, MA, USA, 2022; pp. 539–553, ISBN 978-0-323-91001-9.
12. Umesha, S.; Manukumar, H.M.G.; Chandrasekhar, B. Chapter 3—Sustainable Agriculture and Food Security. In *Biotechnology for Sustainable Agriculture*; Singh, R.L., Mondal, S., Eds.; Woodhead Publishing: Cambridge, UK, 2018; pp. 67–92, ISBN 978-0-12-812160-3.
13. Shahzad, A.; Ullah, S.; Dar, A.A.; Sardar, M.F.; Mehmood, T.; Tufail, M.A.; Shakoor, A.; Haris, M. Nexus on Climate Change: Agriculture and Possible Solution to Cope Future Climate Change Stresses. *Environ. Sci. Pollut. Res. Int.* **2021**, *28*, 14211–14232. [CrossRef]
14. Anderson, J.; Song, B.-H. Plant Adaptation to Climate Change—Where Are We? *J. Syst. Evol.* **2020**, *58*, 533–545. [CrossRef]
15. Ray, D.K.; Mueller, N.D.; West, P.C.; Foley, J.A. Yield Trends Are Insufficient to Double Global Crop Production by 2050. *PLoS ONE* **2013**, *8*, e66428. [CrossRef]
16. Farooq, A.; Farooq, N.; Akbar, H.; Hassan, Z.U.; Gheewala, S.H. A Critical Review of Climate Change Impact at a Global Scale on Cereal Crop Production. *Agronomy* **2023**, *13*, 162. [CrossRef]
17. Fatima, Z.; Ahmed, M.; Hussain, M.; Abbas, G.; Ul-Allah, S.; Ahmad, S.; Ahmed, N.; Ali, M.A.; Sarwar, G.; Haque, E.U.; et al. The Fingerprints of Climate Warming on Cereal Crops Phenology and Adaptation Options. *Sci. Rep.* **2020**, *10*, 18013. [CrossRef] [PubMed]

18. Thudi, M.; Palakurthi, R.; Schnable, J.C.; Chitikineni, A.; Dreisigacker, S.; Mace, E.; Srivastava, R.K.; Satyavathi, C.T.; Odeny, D.; Tiwari, V.K.; et al. Genomic Resources in Plant Breeding for Sustainable Agriculture. *J. Plant Physiol.* **2021**, *257*, 153351. [CrossRef]
19. Santini, M.; Noce, S.; Antonelli, M.; Caporaso, L. Complex Drought Patterns Robustly Explain Global Yield Loss for Major Crops. *Sci. Rep.* **2022**, *12*, 5792. [CrossRef]
20. Hendrawan, V.S.A.; Komori, D.; Kim, W. Possible Factors Determining Global-Scale Patterns of Crop Yield Sensitivity to Drought. *PLoS ONE* **2023**, *18*, e0281287. [CrossRef] [PubMed]
21. Cohen, I.; Zandalinas, S.I.; Huck, C.; Fritschi, F.B.; Mittler, R. Meta-Analysis of Drought and Heat Stress Combination Impact on Crop Yield and Yield Components. *Physiol. Plant.* **2021**, *171*, 66–76. [CrossRef]
22. Pandey, P.; Irulappan, V.; Bagavathiannan, M.V.; Senthil-Kumar, M. Impact of Combined Abiotic and Biotic Stresses on Plant Growth and Avenues for Crop Improvement by Exploiting Physio-Morphological Traits. *Front. Plant Sci.* **2017**, *8*, 537. [CrossRef] [PubMed]
23. Anzano, A.; Bonanomi, G.; Mazzoleni, S.; Lanzotti, V. Plant Metabolomics in Biotic and Abiotic Stress: A Critical Overview. *Phytochem. Rev.* **2022**, *21*, 503–524. [CrossRef]
24. Zhang, H.; Zhu, J.; Gong, Z.; Zhu, J.-K. Abiotic Stress Responses in Plants. *Nat. Rev. Genet.* **2022**, *23*, 104–119. [CrossRef]
25. Pourkheirandish, M.; Golicz, A.A.; Bhalla, P.L.; Singh, M.B. Global Role of Crop Genomics in the Face of Climate Change. *Front. Plant Sci.* **2020**, *11*, 922. [CrossRef]
26. Zhang, H.; Li, Y.; Zhu, J.-K. Developing Naturally Stress-Resistant Crops for a Sustainable Agriculture. *Nat. Plants* **2018**, *4*, 989–996. [CrossRef]
27. Ndlovu, E.; van Staden, J.; Maphosa, M. Morpho-Physiological Effects of Moisture, Heat and Combined Stresses on *Sorghum bicolor* [Moench (L.)] and Its Acclimation Mechanisms. *Plant Stress* **2021**, *2*, 100018. [CrossRef]
28. Saharan, B.S.; Brar, B.; Duhan, J.S.; Kumar, R.; Marwaha, S.; Rajput, V.D.; Minkina, T. Molecular and Physiological Mechanisms to Mitigate Abiotic Stress Conditions in Plants. *Life* **2022**, *12*, 1634. [CrossRef]
29. Shelake, R.M.; Kadam, U.S.; Kumar, R.; Pramanik, D.; Singh, A.K.; Kim, J.-Y. Engineering Drought and Salinity Tolerance Traits in Crops through CRISPR-Mediated Genome Editing: Targets, Tools, Challenges, and Perspectives. *Plant Commun.* **2022**, *3*, 100417. [CrossRef] [PubMed]
30. Waadt, R.; Seller, C.A.; Hsu, P.-K.; Takahashi, Y.; Munemasa, S.; Schroeder, J.I. Plant Hormone Regulation of Abiotic Stress Responses. *Nat. Rev. Mol. Cell Biol.* **2022**, *23*, 680–694. [CrossRef]
31. Chang, Y.-N.; Zhu, C.; Jiang, J.; Zhang, H.; Zhu, J.-K.; Duan, C.-G. Epigenetic Regulation in Plant Abiotic Stress Responses. *J. Integr. Plant Biol.* **2020**, *62*, 563–580. [CrossRef] [PubMed]
32. Jacques, C.; Salon, C.; Barnard, R.L.; Vernoud, V.; Prudent, M. Drought Stress Memory at the Plant Cycle Level: A Review. *Plants* **2021**, *10*, 1873. [CrossRef] [PubMed]
33. Kole, C.; Muthamilarasan, M.; Henry, R.; Edwards, D.; Sharma, R.; Abberton, M.; Batley, J.; Bentley, A.; Blakeney, M.; Bryant, J.; et al. Application of Genomics-Assisted Breeding for Generation of Climate Resilient Crops: Progress and Prospects. *Front. Plant Sci.* **2015**, *6*, 563. [CrossRef]
34. Adamski, N.M.; Borrill, P.; Brinton, J.; Harrington, S.A.; Marchal, C.; Bentley, A.R.; Bovill, W.D.; Cattivelli, L.; Cockram, J.; Contreras-Moreira, B.; et al. A Roadmap for Gene Functional Characterisation in Crops with Large Genomes: Lessons from Polyploid Wheat. *eLife* **2020**, *9*, e55646. [CrossRef]
35. Zenda, T.; Liu, S.; Dong, A.; Li, J.; Wang, Y.; Liu, X.; Wang, N.; Duan, H. Omics-Facilitated Crop Improvement for Climate Resilience and Superior Nutritive Value. *Front. Plant Sci.* **2021**, *12*, 774994. [CrossRef]
36. Goodwin, S.; McPherson, J.D.; McCombie, W.R. Coming of Age: Ten Years of next-Generation Sequencing Technologies. *Nat. Rev. Genet.* **2016**, *17*, 333–351. [CrossRef]
37. Bansal, V.; Boucher, C. Sequencing Technologies and Analyses: Where Have We Been and Where Are We Going? *iScience* **2019**, *18*, 37–41. [CrossRef]
38. Chaney, L.; Sharp, A.R.; Evans, C.R.; Udall, J.A. Genome Mapping in Plant Comparative Genomics. *Trends Plant Sci.* **2016**, *21*, 770–780. [CrossRef] [PubMed]
39. Wang, P.; Meng, F.; Moore, B.M.; Shiu, S.-H. Impact of Short-Read Sequencing on the Misassembly of a Plant Genome. *BMC Genom.* **2021**, *22*, 99. [CrossRef] [PubMed]
40. Amarasinghe, S.L.; Su, S.; Dong, X.; Zappia, L.; Ritchie, M.E.; Gouil, Q. Opportunities and Challenges in Long-Read Sequencing Data Analysis. *Genome Biol.* **2020**, *21*, 30. [CrossRef]
41. Michael, T.P.; VanBuren, R. Building Near-Complete Plant Genomes. *Curr. Opin. Plant Biol.* **2020**, *54*, 26–33. [CrossRef]
42. Wang, Y.; Zhao, Y.; Bollas, A.; Wang, Y.; Au, K.F. Nanopore Sequencing Technology, Bioinformatics and Applications. *Nat. Biotechnol.* **2021**, *39*, 1348–1365. [CrossRef] [PubMed]
43. Henry, R.J. Innovations in Plant Genetics Adapting Agriculture to Climate Change. *Curr. Opin. Plant Biol.* **2020**, *56*, 168–173. [CrossRef]

44. Zenda, T.; Liu, S.; Dong, A.; Duan, H. Advances in Cereal Crop Genomics for Resilience under Climate Change. *Life* **2021**, *11*, 502. [CrossRef]
45. Dugasa, M.T.; Feng, X.; Wang, N.-H.; Wang, J.; Wu, F. Comparative Transcriptome and Tolerance Mechanism Analysis in the Two Contrasting Wheat (*Triticum aestivum* L.) Cultivars in Response to Drought and Salinity Stresses. *Plant Growth Regul.* **2021**, *94*, 101–114. [CrossRef]
46. Zenda, T.; Liu, S.; Wang, X.; Liu, G.; Jin, H.; Dong, A.; Yang, Y.; Duan, H. Key Maize Drought-Responsive Genes and Pathways Revealed by Comparative Transcriptome and Physiological Analyses of Contrasting Inbred Lines. *Int. J. Mol. Sci.* **2019**, *20*, 1268. [CrossRef]
47. Yang, R.; Li, M.; Harrison, M.T.; Fahad, S.; Wei, M.; Li, X.; Yin, L.; Sha, A.; Zhou, M.; Liu, K. iTRAQ Proteomic Analysis of Wheat (*Triticum aestivum* L.) Genotypes Differing in Waterlogging Tolerance. *Front. Plant Sci.* **2022**, *13*, 890083. [CrossRef]
48. Yang, M.; Yang, J.; Su, L.; Sun, K.; Li, D.; Liu, Y.; Wang, H.; Chen, Z.; Guo, T. Metabolic Profile Analysis and Identification of Key Metabolites during Rice Seed Germination under Low-Temperature Stress. *Plant Sci.* **2019**, *289*, 110282. [CrossRef]
49. Pucker, B.; Irisarri, I.; de Vries, J.; Xu, B. Plant Genome Sequence Assembly in the Era of Long Reads: Progress, Challenges and Future Directions. *Quant. Plant Biol.* **2022**, *3*, e5. [CrossRef] [PubMed]
50. Sun, Y.; Shang, L.; Zhu, Q.-H.; Fan, L.; Guo, L. Twenty Years of Plant Genome Sequencing: Achievements and Challenges. *Trends Plant Sci.* **2022**, *27*, 391–401. [CrossRef] [PubMed]
51. Ni, L.; Liu, Y.; Ma, X.; Liu, T.; Yang, X.; Wang, Z.; Liang, Q.; Liu, S.; Zhang, M.; Wang, Z.; et al. Pan-3D Genome Analysis Reveals Structural and Functional Differentiation of Soybean Genomes. *Genome Biol.* **2023**, *24*, 12. [CrossRef] [PubMed]
52. Raza, A.; Bohra, A.; Varshney, R.K. Pan-Genome for Pearl Millet That Beats the Heat. *Trends Plant Sci.* **2023**, *28*, 857–860. [CrossRef] [PubMed]
53. Shi, J.; Tian, Z.; Lai, J.; Huang, X. Plant Pan-Genomics and Its Applications. *Mol. Plant* **2023**, *16*, 168–186. [CrossRef]
54. Yan, H.; Sun, M.; Zhang, Z.; Jin, Y.; Zhang, A.; Lin, C.; Wu, B.; He, M.; Xu, B.; Wang, J.; et al. Pangenomic Analysis Identifies Structural Variation Associated with Heat Tolerance in Pearl Millet. *Nat. Genet.* **2023**, *55*, 507–518. [CrossRef]
55. Shivhare, R.; Lata, C. Exploration of Genetic and Genomic Resources for Abiotic and Biotic Stress Tolerance in Pearl Millet. *Front. Plant Sci.* **2017**, *7*, 2069. [CrossRef]
56. Masters, A.; Kang, M.; McCaw, M.; Zobrist, J.D.; Gordon-Kamm, W.; Jones, T.; Wang, K. Agrobacterium-Mediated Immature Embryo Transformation of Recalcitrant Maize Inbred Lines Using Morphogenic Genes. *J. Vis. Exp. JoVE* **2020**, e60782. [CrossRef]
57. Chen, Z.; Debernardi, J.M.; Dubcovsky, J.; Gallavotti, A. Recent Advances in Crop Transformation Technologies. *Nat. Plants* **2022**, *8*, 1343–1351. [CrossRef] [PubMed]
58. Kang, M.; Lee, K.; Finley, T.; Chappell, H.; Veena, V.; Wang, K. An Improved Agrobacterium-Mediated Transformation and Genome-Editing Method for Maize Inbred B104 Using a Ternary Vector System and Immature Embryos. *Front. Plant Sci.* **2022**, *13*, 860971. [CrossRef] [PubMed]
59. Lee, K.; Wang, K. Strategies for Genotype-Flexible Plant Transformation. *Curr. Opin. Biotechnol.* **2023**, *79*, 102848. [CrossRef]
60. Smith, D.T.; Potgieter, A.B.; Chapman, S.C. Scaling up High-Throughput Phenotyping for Abiotic Stress Selection in the Field. *TAG Theor. Appl. Genet. Theor. Angew. Genet.* **2021**, *134*, 1845–1866. [CrossRef] [PubMed]
61. Hall, R.D.; D'Auria, J.C.; Ferreira, A.C.S.; Gibon, Y.; Kruszka, D.; Mishra, P.; Zedde, R. van de High-Throughput Plant Phenotyping: A Role for Metabolomics? *Trends Plant Sci.* **2022**, *27*, 549–563. [CrossRef]
62. Scossa, F.; Alseekh, S.; Fernie, A.R. Integrating Multi-Omics Data for Crop Improvement. *J. Plant Physiol.* **2021**, *257*, 153352. [CrossRef]
63. Dwivedi, S.L.; Goldman, I.; Ceccarelli, S.; Ortiz, R. Chapter Three—Advanced Analytics, Phenomics and Biotechnology Approaches to Enhance Genetic Gains in Plant Breeding. In *Advances in Agronomy*; Sparks, D.L., Ed.; Academic Press: Cambridge, MA, USA, 2020; Volume 162, pp. 89–142.
64. Mohd Saad, N.S.; Neik, T.X.; Thomas, W.J.W.; Amas, J.C.; Cantila, A.Y.; Craig, R.J.; Edwards, D.; Batley, J. Advancing Designer Crops for Climate Resilience through an Integrated Genomics Approach. *Curr. Opin. Plant Biol.* **2022**, *67*, 102220. [CrossRef]
65. Sidak, D.; Schwarzerová, J.; Weckwerth, W.; Waldherr, S. Interpretable Machine Learning Methods for Predictions in Systems Biology from Omics Data. *Front. Mol. Biosci.* **2022**, *9*, 926623. [CrossRef] [PubMed]
66. Munaweera, T.I.K.; Jayawardana, N.U.; Rajaratnam, R.; Dissanayake, N. Modern Plant Biotechnology as a Strategy in Addressing Climate Change and Attaining Food Security. *Agric. Food Secur.* **2022**, *11*, 26. [CrossRef]
67. Steinwand, M.A.; Ronald, P.C. Crop Biotechnology and the Future of Food. *Nat. Food* **2020**, *1*, 273–283. [CrossRef]
68. Demirer, G.S.; Silva, T.N.; Jackson, C.T.; Thomas, J.B.; Ehrhardt, D.W.; Rhee, S.Y.; Mortimer, J.C.; Landry, M.P. Nanotechnology to Advance CRISPR-Cas Genetic Engineering of Plants. *Nat. Nanotechnol.* **2021**, *16*, 243–250. [CrossRef] [PubMed]
69. Gao, C. Genome Engineering for Crop Improvement and Future Agriculture. *Cell* **2021**, *184*, 1621–1635. [CrossRef] [PubMed]
70. Pixley, K.V.; Falck-Zepeda, J.B.; Paarlberg, R.L.; Phillips, P.W.B.; Slamet-Loedin, I.H.; Dhugga, K.S.; Campos, H.; Gutterson, N. Genome-Edited Crops for Improved Food Security of Smallholder Farmers. *Nat. Genet.* **2022**, *54*, 364–367. [CrossRef] [PubMed]

71. Sinha, P.; Singh, V.K.; Bohra, A.; Kumar, A.; Reif, J.C.; Varshney, R.K. Genomics and Breeding Innovations for Enhancing Genetic Gain for Climate Resilience and Nutrition Traits. *TAG Theor. Appl. Genet.* **2021**, *134*, 1829–1843. [CrossRef]
72. Varshney, R.K.; Bohra, A.; Yu, J.; Graner, A.; Zhang, Q.; Sorrells, M.E. Designing Future Crops: Genomics-Assisted Breeding Comes of Age. *Trends Plant Sci.* **2021**, *26*, 631–649. [CrossRef]
73. Waititu, J.K.; Zhang, X.; Chen, T.; Zhang, C.; Zhao, Y.; Wang, H. Transcriptome Analysis of Tolerant and Susceptible Maize Genotypes Reveals Novel Insights about the Molecular Mechanisms Underlying Drought Responses in Leaves. *Int. J. Mol. Sci.* **2021**, *22*, 6980. [CrossRef]
74. Yao, Q. Crucial Waterlogging-Responsive Genes and Pathways Revealed by Comparative Physiology and Transcriptome in Tropical and Temperate Maize (*Zea mays* L.) Inbred Lines. *J. Plant Biol.* **2021**, *64*, 313–325. [CrossRef]
75. Fang, X.; Mo, J.; Zhou, H.; Shen, X.; Xie, Y.; Xu, J.; Yang, S. Comparative Transcriptome Analysis of Gene Responses of Salt-Tolerant and Salt-Sensitive Rice Cultivars to Salt Stress. *Sci. Rep.* **2023**, *13*, 19065. [CrossRef] [PubMed]
76. Lamba, K.; Kumar, M.; Singh, V.; Chaudhary, L.; Gupta, V. Transcriptome Analysis for Heat Stress Related Genes in Wheat Genotype WH-730. *Cereal Res. Commun.* **2025**, *53*, 793–805. [CrossRef]
77. Niu, Y.; Fan, S.; Cheng, B.; Li, H.; Wu, J.; Zhao, H.; Huang, Z.; Yan, F.; Qi, B.; Zhang, L. Comparative Transcriptomics and Co-Expression Networks Reveal Cultivar-Specific Molecular Signatures Associated with Reproductive-Stage Cold Stress in Rice. *Plant Cell Rep.* **2023**, *42*, 707–722. [CrossRef]
78. Zhao, S.; Zhang, Q.; Xiao, W.; Chen, D.; Hu, J.; Gao, N.; Huang, M.; Ye, X. Comparative Transcriptomic Analysis Reveals the Important Process in Two Rice Cultivars with Differences in Cadmium Accumulation. *Ecotoxicol. Environ. Saf.* **2023**, *252*, 114629. [CrossRef] [PubMed]
79. Liu, C.; Gu, W.; Liu, C.; Shi, X.; Li, B.; Zhou, Y. Comparative Phenotypic and Transcriptomic Analysis Reveals Genotypic Differences in Nitrogen Use Efficiency in Sorghum. *Plant Physiol. Biochem.* **2024**, *215*, 109028. [CrossRef]
80. Moloi, S.J.; Alqarni, A.O.; Brown, A.P.; Goche, T.; Shargie, N.G.; Moloi, M.J.; Gokul, A.; Chivasa, S.; Ngara, R. Comparative Physiological, Biochemical, and Leaf Proteome Responses of Contrasting Wheat Varieties to Drought Stress. *Plants* **2024**, *13*, 2797. [CrossRef]
81. Zenda, T.; Liu, S.; Wang, X.; Jin, H.; Liu, G.; Duan, H. Comparative Proteomic and Physiological Analyses of Two Divergent Maize Inbred Lines Provide More Insights into Drought-Stress Tolerance Mechanisms. *Int. J. Mol. Sci.* **2018**, *19*, 3225. [CrossRef]
82. Pang, Y.; Hu, Y.; Bao, J. Comparative Phosphoproteomic Analysis Reveals the Response of Starch Metabolism to High-Temperature Stress in Rice Endosperm. *Int. J. Mol. Sci.* **2021**, *22*, 10546. [CrossRef]
83. Mikołajczak, K.; Kuczyńska, A.; Krajewski, P.; Kempa, M.; Witaszak, N. Global Proteome Profiling Revealed the Adaptive Reprogramming of Barley Flag Leaf to Drought and Elevated Temperature. *Cells* **2023**, *12*, 1685. [CrossRef] [PubMed]
84. Jha, S. Proteome Responses of Pearl Millet Genotypes under Salinity. *Plant Gene* **2022**, *29*, 100347. [CrossRef]
85. Yadav, R.; Santal, A.R.; Singh, N.P. Comparative Root Proteome Analysis of Two Contrasting Wheat Genotypes Kharchia-65 (Highly Salt-Tolerant) and PBW-373 (Salt-Sensitive) for Salinity Tolerance Using LC–MS/MS Approach. *Vegetos* **2022**, *35*, 133–139. [CrossRef]
86. Ramazan, S.; Jan, N.; John, R. Comparative Protein Analysis of Two Maize Genotypes with Contrasting Tolerance to Low Temperature. *BMC Plant Biol.* **2023**, *23*, 183. [CrossRef]
87. Feng, Q.; Sehar, S.; Zhou, F.; Wei, D.; Askri, S.M.H.; Ma, Z.; Adil, M.F.; Shamsi, I.H. Physiological and TMT-Based Quantitative Proteomic Responses of Barley to Aluminium Stress under Phosphorus-Piriformospora Indica Interaction. *Plant Physiol. Biochem.* **2023**, *196*, 634–646. [CrossRef] [PubMed]
88. Zheng, L.; Wang, R.; Zhou, P.; Pan, Y.; Shen, R.; Lan, P. Comparative Physiological and Proteomic Response to Phosphate Deficiency between Two Wheat Genotypes Differing in Phosphorus Utilization Efficiency. *J. Proteom.* **2023**, *280*, 104894. [CrossRef]
89. Li, Y.; Tan, B.; Wang, D.; Mu, Y.; Li, G.; Zhang, Z.; Pan, Y.; Zhu, L. Proteomic Analysis Revealed Different Molecular Mechanisms of Response to PEG Stress in Drought-Sensitive and Drought-Resistant Sorghums. *Int. J. Mol. Sci.* **2022**, *23*, 13297. [CrossRef]
90. Zhang, Y.; Ma, X.; Wang, X.; Liu, J.; Huang, B.; Guo, X.; Xiong, S.; La, G.-X. UPLC-QTOF Analysis Reveals Metabolomic Changes in the Flag Leaf of Wheat (*Triticum aestivum* L.) under Low-Nitrogen Stress. *Plant Physiol. Biochem.* **2017**, *111*, 30–38. [CrossRef] [PubMed]
91. Sheflin, A.M.; Chiniquy, D.; Yuan, C.; Goren, E.; Kumar, I.; Braud, M.; Brutnell, T.; Eveland, A.L.; Tringe, S.; Liu, P.; et al. Metabolomics of Sorghum Roots during Nitrogen Stress Reveals Compromised Metabolic Capacity for Salicylic Acid Biosynthesis. *Plant Direct* **2019**, *3*, e00122. [CrossRef] [PubMed]
92. Swarcewicz, B.; Sawikowska, A.; Marczak, Ł.; Łuczak, M.; Ciesiołka, D.; Krystkowiak, K.; Kuczyńska, A.; Piślewska-Bednarek, M.; Krajewski, P.; Stobiecki, M. Effect of Drought Stress on Metabolite Contents in Barley Recombinant Inbred Line Population Revealed by Untargeted GC–MS Profiling. *Acta Physiol. Plant.* **2017**, *39*, 158. [CrossRef]
93. Guo, X.; Lv, L.; Zhao, A.; Zhao, W.; Liu, Y.; Li, Z.; Li, H.; Chen, X. Integrated Transcriptome and Metabolome Analysis Revealed Differential Drought Stress Response Mechanisms of Wheat Seedlings with Varying Drought Tolerance. *BMC Plant Biol.* **2025**, *25*, 571. [CrossRef]

94. Li, P.; Ma, X.; Wang, J.; Yao, L.; Li, B.; Meng, Y.; Si, E.; Yang, K.; Shang, X.; Zhang, X. Integrated Analysis of Metabolome and Transcriptome Reveals Insights for Low Phosphorus Tolerance in Wheat Seedling. *Int. J. Mol. Sci.* **2023**, *24*, 14840. [CrossRef]
95. Shunkao, S.; Theerakulpisut, P.; Wanichthanarak, K.; Pongdontri, P.; Thitisaksakul, M. Integrative Physiological and Metabolomics Study Reveals Adaptive Strategies of Wheat Seedlings to Salt and Heat Stress Combination. *Plant Growth Regul.* **2023**, *100*, 181–196. [CrossRef]
96. Qian, G.; Wang, M.; Wang, X.; Liu, K.; Li, Y.; Bu, Y.; Li, L. Integrated Transcriptome and Metabolome Analysis of Rice Leaves Response to High Saline–Alkali Stress. *Int. J. Mol. Sci.* **2023**, *24*, 4062. [CrossRef]
97. Li, Y.; Su, Z.; Lin, Y.; Xu, Z.; Bao, H.; Wang, F.; Liu, J.; Hu, S.; Wang, Z.; Yu, X. Utilizing Transcriptomics and Metabolomics to Unravel Key Genes and Metabolites of Maize Seedlings in Response to Drought Stress. *BMC Plant Biol.* **2024**, *24*, 34. [CrossRef]
98. Chen, Y.; Wang, J.; Yao, L.; Li, B.; Ma, X.; Si, E.; Yang, K.; Li, C.; Shang, X.; Meng, Y. Combined Proteomic and Metabolomic Analysis of the Molecular Mechanism Underlying the Response to Salt Stress during Seed Germination in Barley. *Int. J. Mol. Sci.* **2022**, *23*, 10515. [CrossRef]
99. Snowdon, R.J.; Wittkop, B.; Chen, T.-W.; Stahl, A. Crop Adaptation to Climate Change as a Consequence of Long-Term Breeding. *TAG Theor. Appl. Genet.* **2021**, *134*, 1613–1623. [CrossRef] [PubMed]
100. Gong, Z.; Chen, W.; Bao, G.; Sun, J.; Ding, X.; Fan, C. Physiological Response of *Secale cereale* L. Seedlings under Freezing-Thawing and Alkaline Salt Stress. *Environ. Sci. Pollut. Res. Int.* **2020**, *27*, 1499–1507. [CrossRef] [PubMed]
101. Aycan, M.; Baslam, M.; Asiloglu, R.; Mitsui, T.; Yildiz, M. Development of New High-Salt Tolerant Bread Wheat (*Triticum aestivum* L.) Genotypes and Insight into the Tolerance Mechanisms. *Plant Physiol. Biochem. PPB* **2021**, *166*, 314–327. [CrossRef]
102. Frimpong, F.; Windt, C.W.; van Dusschoten, D.; Naz, A.A.; Frei, M.; Fiorani, F. A Wild Allele of Pyrroline-5-Carboxylate Synthase1 Leads to Proline Accumulation in Spikes and Leaves of Barley Contributing to Improved Performance Under Reduced Water Availability. *Front. Plant Sci.* **2021**, *12*, 633448. [CrossRef]
103. Nefissi Ouertani, R.; Abid, G.; Karmous, C.; Ben Chikha, M.; Boudaya, O.; Mahmoudi, H.; Mejri, S.; Jansen, R.K.; Ghorbel, A. Evaluating the Contribution of Osmotic and Oxidative Stress Components on Barley Growth under Salt Stress. *AoB Plants* **2021**, *13*, plab034. [CrossRef]
104. Zhang, M.-X.; Bai, R.; Nan, M.; Ren, W.; Wang, C.-M.; Shabala, S.; Zhang, J.-L. Evaluation of Salt Tolerance of Oat Cultivars and the Mechanism of Adaptation to Salinity. *J. Plant Physiol.* **2022**, *273*, 153708. [CrossRef]
105. Reddy, P.S.; Dhaware, M.G.; Sivasakthi, K.; Divya, K.; Nagaraju, M.; Sri Cindhuri, K.; Kavi Kishor, P.B.; Bhatnagar-Mathur, P.; Vadez, V.; Sharma, K.K. Pearl Millet Aquaporin Gene PgPIP2;6 Improves Abiotic Stress Tolerance in Transgenic Tobacco. *Front. Plant Sci.* **2022**, *13*, 820996. [CrossRef]
106. Zhang, X.; Liu, W.; Lv, Y.; Li, T.; Tang, J.; Yang, X.; Bai, J.; Jin, X.; Zhou, H. Effects of Drought Stress during Critical Periods on the Photosynthetic Characteristics and Production Performance of Naked Oat (*Avena nuda* L.). *Sci. Rep.* **2022**, *12*, 11199. [CrossRef]
107. Goche, T.; Shargie, N.G.; Cummins, I.; Brown, A.P.; Chivasa, S.; Ngara, R. Comparative Physiological and Root Proteome Analyses of Two Sorghum Varieties Responding to Water Limitation. *Sci. Rep.* **2020**, *10*, 11835. [CrossRef]
108. Liao, Q.; Chebotarov, D.; Islam, M.S.; Quintana, M.R.; Natividad, M.A.; De Ocampo, M.; Beredo, J.C.; Torres, R.O.; Zhang, Z.; Song, H.; et al. Aus Rice Root Architecture Variation Contributing to Grain Yield under Drought Suggests a Key Role of Nodal Root Diameter Class. *Plant Cell Environ.* **2022**, *45*, 854–870. [CrossRef]
109. Ngcala, M.G.; Goche, T.; Brown, A.P.; Chivasa, S.; Ngara, R. Heat Stress Triggers Differential Protein Accumulation in the Extracellular Matrix of Sorghum Cell Suspension Cultures. *Proteomes* **2020**, *8*, 29. [CrossRef]
110. Griffith, M.; Ala, P.; Yang, D.S.; Hon, W.C.; Moffatt, B.A. Antifreeze Protein Produced Endogenously in Winter Rye Leaves. *Plant Physiol.* **1992**, *100*, 593–596. [CrossRef]
111. Yu, X.-M.; Griffith, M. Winter Rye Antifreeze Activity Increases in Response to Cold and Drought, but Not Abscisic Acid. *Physiol. Plant.* **2001**, *112*, 78–86. [CrossRef]
112. Cheng, X.; He, Q.; Tang, S.; Wang, H.; Zhang, X.; Lv, M.; Liu, H.; Gao, Q.; Zhou, Y.; Wang, Q.; et al. The miR172/IDS1 Signaling Module Confers Salt Tolerance through Maintaining ROS Homeostasis in Cereal Crops. *New Phytol.* **2021**, *230*, 1017–1033. [CrossRef]
113. Punia, H.; Tokas, J.; Malik, A.; Bajguz, A.; El-Sheikh, M.A.; Ahmad, P. Ascorbate-Glutathione Oxidant Scavengers, Metabolome Analysis and Adaptation Mechanisms of Ion Exclusion in Sorghum under Salt Stress. *Int. J. Mol. Sci.* **2021**, *22*, 13249. [CrossRef]
114. Tian, Q.; Shen, L.; Luan, J.; Zhou, Z.; Guo, D.; Shen, Y.; Jing, W.; Zhang, B.; Zhang, Q.; Zhang, W. Rice Shaker Potassium Channel OsAKT2 Positively Regulates Salt Tolerance and Grain Yield by Mediating K(+) Redistribution. *Plant Cell Environ.* **2021**, *44*, 2951–2965. [CrossRef]
115. Alam, M.S.; Yang, Z.-K.; Li, C.; Yan, Y.; Liu, Z.; Nazir, M.M.; Xu, J.-H. Loss-of-Function Mutations of OsbHLH044 Transcription Factor Lead to Salinity Sensitivity and a Greater Chalkiness in Rice (*Oryza sativa* L.). *Plant Physiol. Biochem. PPB* **2022**, *193*, 110–123. [CrossRef]
116. Sun, S.; Li, X.; Gao, S.; Nie, N.; Zhang, H.; Yang, Y.; He, S.; Liu, Q.; Zhai, H. A Novel WRKY Transcription Factor from Ipomoea Trifida, ItfWRKY70, Confers Drought Tolerance in Sweet Potato. *Int. J. Mol. Sci.* **2022**, *23*, 686. [CrossRef] [PubMed]

117. Wen, D.; Bao, L.; Huang, X.; Qian, X.; Chen, E.; Shen, B. OsABT Is Involved in Abscisic Acid Signaling Pathway and Salt Tolerance of Roots at the Rice Seedling Stage. *Int. J. Mol. Sci.* **2022**, *23*, 10656. [CrossRef]
118. Siddiqui, M.N.; Léon, J.; Naz, A.A.; Ballvora, A. Genetics and Genomics of Root System Variation in Adaptation to Drought Stress in Cereal Crops. *J. Exp. Bot.* **2021**, *72*, 1007–1019. [CrossRef]
119. Halder, T.; Choudhary, M.; Liu, H.; Chen, Y.; Yan, G.; Siddique, K.H.M. Wheat Proteomics for Abiotic Stress Tolerance and Root System Architecture: Current Status and Future Prospects. *Proteomes* **2022**, *10*, 17. [CrossRef]
120. Ramegowda, V.; Senthil, A.; Senthil-Kumar, M. Stress Combinations and Their Interactions in Crop Plants. *Plant Physiol. Rep.* **2024**, *29*, 1–5. [CrossRef]
121. Ru, C.; Hu, X.; Chen, D.; Wang, W.; Zhen, J.; Song, T. Individual and Combined Effects of Heat and Drought and Subsequent Recovery on Winter Wheat (*Triticum aestivum* L.) Photosynthesis, Nitrogen Metabolism, Cell Osmoregulation, and Yield Formation. *Plant Physiol. Biochem.* **2023**, *196*, 222–235. [CrossRef] [PubMed]
122. Ngara, R.; Ndimba, B.K. Understanding the Complex Nature of Salinity and Drought-Stress Response in Cereals Using Proteomics Technologies. *Proteomics* **2014**, *14*, 611–621. [CrossRef]
123. Schmidt, J.; Tricker, P.J.; Eckermann, P.; Kalambettu, P.; Garcia, M.; Fleury, D. Novel Alleles for Combined Drought and Heat Stress Tolerance in Wheat. *Front. Plant Sci.* **2019**, *10*, 1800. [CrossRef]
124. Yang, X.; Zhu, X.; Wei, J.; Li, W.; Wang, H.; Xu, Y.; Yang, Z.; Xu, C.; Li, P. Primary Root Response to Combined Drought and Heat Stress Is Regulated via Salicylic Acid Metabolism in Maize. *BMC Plant Biol.* **2022**, *22*, 417. [CrossRef]
125. Barua, D.; Mishra, A.; Kirti, P.B.; Barah, P. Identifying Signal-Crosstalk Mechanism in Maize Plants during Combined Salinity and Boron Stress Using Integrative Systems Biology Approaches. *BioMed Res. Int.* **2022**, *2022*, 1027288. [CrossRef]
126. Zhang, X.; Shi, Z.; Jiang, D.; Högy, P.; Fangmeier, A. Independent and Combined Effects of Elevated CO_2 and Post-Anthesis Heat Stress on Protein Quantity and Quality in Spring Wheat Grains. *Food Chem.* **2019**, *277*, 524–530. [CrossRef]
127. Adrees, M.; Khan, Z.S.; Rehman, M.Z.U.; Rizwan, M.; Ali, S. Foliar Spray of Silicon Nanoparticles Improved the Growth and Minimized Cadmium (Cd) in Wheat under Combined Cd and Water-Limited Stress. *Environ. Sci. Pollut. Res. Int.* **2022**, *29*, 77321–77332. [CrossRef]
128. Itam, M.O.; Mega, R.; Gorafi, Y.S.A.; Yamasaki, Y.; Tahir, I.S.A.; Akashi, K.; Tsujimoto, H. Genomic Analysis for Heat and Combined Heat-Drought Resilience in Bread Wheat under Field Conditions. *TAG Theor. Appl. Genet.* **2022**, *135*, 337–350. [CrossRef]
129. Da Costa, M.V.J.; Ramegowda, V.; Sreeman, S.; Nataraja, K.N. Targeted Phytohormone Profiling Identifies Potential Regulators of Spikelet Sterility in Rice under Combined Drought and Heat Stress. *Int. J. Mol. Sci.* **2021**, *22*, 11690. [CrossRef]
130. Da Costa, M.V.J.; Ramegowda, V.; Ramakrishnan, P.; Nataraja, K.N.; Sheshshayee, M.S. Comparative Metabolite Profiling of Rice Contrasts Reveal Combined Drought and Heat Stress Signatures in Flag Leaf and Spikelets. *Plant Sci. Int. J. Exp. Plant Biol.* **2022**, *320*, 111262. [CrossRef]
131. Johnson, S.M.; Lim, F. L.; Finkler, A.; Fromm, H.; Slabas, A.R.; Knight, M.R. Transcriptomic Analysis of Sorghum Bicolor Responding to Combined Heat and Drought Stress. *BMC Genom.* **2014**, *15*, 456. [CrossRef]
132. Mikołajczak, K.; Kuczyńska, A.; Krajewski, P.; Kempa, M.; Nuc, M. Transcriptome Profiling Disclosed the Effect of Single and Combined Drought and Heat Stress on Reprogramming of Genes Expression in Barley Flag Leaf. *Front. Plant Sci.* **2022**, *13*, 1096685. [CrossRef]
133. Li, C.; Lin, F.; An, D.; Wang, W.; Huang, R. Genome Sequencing and Assembly by Long Reads in Plants. *Genes* **2017**, *9*, 6. [CrossRef]
134. Sharma, T.R.; Devanna, B.N.; Kiran, K.; Singh, P.K.; Arora, K.; Jain, P.; Tiwari, I.M.; Dubey, H.; Saklani, B.; Kumari, M.; et al. Status and Prospects of Next Generation Sequencing Technologies in Crop Plants. *Curr. Issues Mol. Biol.* **2018**, *27*, 1–36. [CrossRef]
135. Jung, H.; Winefield, C.; Bombarely, A.; Prentis, P.; Waterhouse, P. Tools and Strategies for Long-Read Sequencing and De Novo Assembly of Plant Genomes. *Trends Plant Sci.* **2019**, *24*, 700–724. [CrossRef]
136. Michael, T.P.; VanBuren, R. Progress, Challenges and the Future of Crop Genomes. *Curr. Opin. Plant Biol.* **2015**, *24*, 71–81. [CrossRef] [PubMed]
137. Lu, H.; Giordano, F.; Ning, Z. Oxford Nanopore MinION Sequencing and Genome Assembly. *Genom. Proteom. Bioinform.* **2016**, *14*, 265–279. [CrossRef] [PubMed]
138. Rhoads, A.; Au, K.F. PacBio Sequencing and Its Applications. *Genom. Proteom. Bioinform.* **2015**, *13*, 278–289. [CrossRef] [PubMed]
139. Tao, Y.; Zhao, X.; Mace, E.; Henry, R.; Jordan, D. Exploring and Exploiting Pan-Genomics for Crop Improvement. *Mol. Plant* **2019**, *12*, 156–169. [CrossRef]
140. Zanini, S.F.; Bayer, P.E.; Wells, R.; Snowdon, R.J.; Batley, J.; Varshney, R.K.; Nguyen, H.T.; Edwards, D.; Golicz, A.A. Pangenomics in Crop Improvement-from Coding Structural Variations to Finding Regulatory Variants with Pangenome Graphs. *Plant Genome* **2022**, *15*, e20177. [CrossRef]
141. Khan, A.W.; Garg, V.; Roorkiwal, M.; Golicz, A.A.; Edwards, D.; Varshney, R.K. Super-Pangenome by Integrating the Wild Side of a Species for Accelerated Crop Improvement. *Trends Plant Sci.* **2020**, *25*, 148–158. [CrossRef]

142. Mammadov, J.; Buyyarapu, R.; Guttikonda, S.K.; Parliament, K.; Abdurakhmonov, I.Y.; Kumpatla, S.P. Wild Relatives of Maize, Rice, Cotton, and Soybean: Treasure Troves for Tolerance to Biotic and Abiotic Stresses. *Front. Plant Sci.* **2018**, *9*, 886. [CrossRef]
143. Tian, Z.; Wang, J.-W.; Li, J.; Han, B. Designing Future Crops: Challenges and Strategies for Sustainable Agriculture. *Plant J. Cell Mol. Biol.* **2021**, *105*, 1165–1178. [CrossRef]
144. Bohra, A.; Tiwari, A.; Kaur, P.; Ganie, S.A.; Raza, A.; Roorkiwal, M.; Mir, R.R.; Fernie, A.R.; Smýkal, P.; Varshney, R.K. The Key to the Future Lies in the Past: Insights from Grain Legume Domestication and Improvement Should Inform Future Breeding Strategies. *Plant Cell Physiol.* **2022**, *63*, 1554–1572. [CrossRef]
145. Yasir, M.; Turner, A.K.; Lott, M.; Rudder, S.; Baker, D.; Bastkowski, S.; Page, A.J.; Webber, M.A.; Charles, I.G. Long-Read Sequencing for Identification of Insertion Sites in Large Transposon Mutant Libraries. *Sci. Rep.* **2022**, *12*, 3546. [CrossRef]
146. Method of the Year 2022: Long-Read Sequencing. *Nat. Methods* **2023**, *20*, 1. [CrossRef]
147. Shi, J.; Ma, X.; Zhang, J.; Zhou, Y.; Liu, M.; Huang, L.; Sun, S.; Zhang, X.; Gao, X.; Zhan, W.; et al. Chromosome Conformation Capture Resolved near Complete Genome Assembly of Broomcorn Millet. *Nat. Commun.* **2019**, *10*, 464. [CrossRef]
148. Depledge, D.P.; Srinivas, K.P.; Sadaoka, T.; Bready, D.; Mori, Y.; Placantonakis, D.G.; Mohr, I.; Wilson, A.C. Direct RNA Sequencing on Nanopore Arrays Redefines the Transcriptional Complexity of a Viral Pathogen. *Nat. Commun.* **2019**, *10*, 754. [CrossRef]
149. Bohra, A.; Chand Jha, U.; Godwin, I.D.; Kumar Varshney, R. Genomic Interventions for Sustainable Agriculture. *Plant Biotechnol. J.* **2020**, *18*, 2388–2405. [CrossRef]
150. Hasan, N.; Choudhary, S.; Naaz, N.; Sharma, N.; Laskar, R.A. Recent Advancements in Molecular Marker-Assisted Selection and Applications in Plant Breeding Programmes. *J. Genet. Eng. Biotechnol.* **2021**, *19*, 128. [CrossRef]
151. Mohanta, T.K.; Bashir, T.; Hashem, A.; Abd Allah, E.F. Systems Biology Approach in Plant Abiotic Stresses. *Plant Physiol. Biochem. PPB* **2017**, *121*, 58–73. [CrossRef]
152. Perkel, J. SNP Genotyping: Six Technologies That Keyed a Revolution. *Nat. Methods* **2008**, *5*, 447–453. [CrossRef]
153. Zhang, J.; Yang, J.; Zhang, L.; Luo, J.; Zhao, H.; Zhang, J.; Wen, C. A New SNP Genotyping Technology Target SNP-Seq and Its Application in Genetic Analysis of Cucumber Varieties. *Sci. Rep.* **2020**, *10*, 5623. [CrossRef]
154. Nepolean, T.; Kaul, J.; Mukri, G.; Mittal, S. Genomics-Enabled Next-Generation Breeding Approaches for Developing System-Specific Drought Tolerant Hybrids in Maize. *Front. Plant Sci.* **2018**, *9*, 361. [CrossRef]
155. Wang, N.; Yuan, Y.; Wang, H.; Yu, D.; Liu, Y.; Zhang, A.; Gowda, M.; Nair, S.K.; Hao, Z.; Lu, Y.; et al. Applications of Genotyping-by-Sequencing (GBS) in Maize Genetics and Breeding. *Sci. Rep.* **2020**, *10*, 16308. [CrossRef]
156. Du, H.; Qin, R.; Li, H.; Du, Q.; Li, X.; Yang, H.; Kong, F.; Liu, B.; Yu, D.; Wang, H. Genome-Wide Association Studies Reveal Novel Loci for Herbivore Resistance in Wild Soybean (*Glycine soja*). *Int. J. Mol. Sci.* **2022**, *23*, 8016. [CrossRef] [PubMed]
157. Abdelraheem, A.; Thyssen, G.N.; Fang, D.D.; Jenkins, J.N.; McCarty, J.C.; Wedegaertner, T.; Zhang, J. GWAS Reveals Consistent QTL for Drought and Salt Tolerance in a MAGIC Population of 550 Lines Derived from Intermating of 11 Upland Cotton (*Gossypium hirsutum*) Parents. *Mol. Genet. Genom. MGG* **2021**, *296*, 119–129. [CrossRef]
158. Puranik, S.; Sahu, P.P.; Beynon, S.; Srivastava, R.K.; Sehgal, D.; Ojulong, H.; Yadav, R. Genome-Wide Association Mapping and Comparative Genomics Identifies Genomic Regions Governing Grain Nutritional Traits in Finger Millet (*Eleusine coracana* L. Gaertn.). *Plants People Planet* **2020**, *2*, 649–662. [CrossRef]
159. Li, L.; Mao, X.; Wang, J.; Chang, X.; Reynolds, M.; Jing, R. Genetic Dissection of Drought and Heat-Responsive Agronomic Traits in Wheat. *Plant Cell Environ.* **2019**, *42*, 2540–2553. [CrossRef]
160. Yang, L.; Lei, L.; Liu, H.; Wang, J.; Zheng, H.; Zou, D. Whole-Genome Mining of Abiotic Stress Gene Loci in Rice. *Planta* **2020**, *252*, 85. [CrossRef]
161. Raza, A.; Mubarik, M.S.; Sharif, R.; Habib, M.; Jabeen, W.; Zhang, C.; Chen, H.; Chen, Z.-H.; Siddique, K.H.M.; Zhuang, W.; et al. Developing Drought-Smart, Ready-to-Grow Future Crops. *Plant Genome* **2022**, *16*, e20279. [CrossRef]
162. Lu, X.; Liu, J.; Ren, W.; Yang, Q.; Chai, Z.; Chen, R.; Wang, L.; Zhao, J.; Lang, Z.; Wang, H.; et al. Gene-Indexed Mutations in Maize. *Mol. Plant* **2018**, *11*, 496–504. [CrossRef]
163. Wang, Z.; Gerstein, M.; Snyder, M. RNA-Seq: A Revolutionary Tool for Transcriptomics. *Nat. Rev. Genet.* **2009**, *10*, 57–63. [CrossRef] [PubMed]
164. Gelli, M.; Duo, Y.; Konda, A.R.; Zhang, C.; Holding, D.; Dweikat, I. Identification of Differentially Expressed Genes between Sorghum Genotypes with Contrasting Nitrogen Stress Tolerance by Genome-Wide Transcriptional Profiling. *BMC Genom.* **2014**, *15*, 179. [CrossRef] [PubMed]
165. Zhang, M.; Kong, X.; Xu, X.; Li, C.; Tian, H.; Ding, Z. Comparative Transcriptome Profiling of the Maize Primary, Crown and Seminal Root in Response to Salinity Stress. *PLoS ONE* **2015**, *10*, e0121222. [CrossRef]
166. He, X.; Ma, H.; Zhao, X.; Nie, S.; Li, Y.; Zhang, Z.; Shen, Y.; Chen, Q.; Lu, Y.; Lan, H.; et al. Comparative RNA-Seq Analysis Reveals That Regulatory Network of Maize Root Development Controls the Expression of Genes in Response to N Stress. *PLoS ONE* **2016**, *11*, e0151697. [CrossRef]
167. Tian, T.; Chen, L.; Ai, Y.; He, H. Selection of Candidate Genes Conferring Blast Resistance and Heat Tolerance in Rice through Integration of Meta-QTLs and RNA-Seq. *Genes* **2022**, *13*, 224. [CrossRef] [PubMed]

168. Poonia, A.K.; Mishra, S.K.; Sirohi, P.; Chaudhary, R.; Kanwar, M.; Germain, H.; Chauhan, H. Overexpression of Wheat Transcription Factor (TaHsfA6b) Provides Thermotolerance in Barley. *Planta* **2020**, *252*, 53. [CrossRef] [PubMed]
169. Liu, S.; Zenda, T.; Dong, A.; Yang, Y.; Wang, N.; Duan, H. Global Transcriptome and Weighted Gene Co-Expression Network Analyses of Growth-Stage-Specific Drought Stress Responses in Maize. *Front. Genet.* **2021**, *12*, 645443. [CrossRef] [PubMed]
170. Lv, L.; Zhang, W.; Sun, L.; Zhao, A.; Zhang, Y.; Wang, L.; Liu, Y.; Li, Z.; Li, H.; Chen, X. Gene Co-Expression Network Analysis to Identify Critical Modules and Candidate Genes of Drought-Resistance in Wheat. *PLoS ONE* **2020**, *15*, e0236186. [CrossRef]
171. Chopra, R.; Burow, G.; Burke, J.J.; Gladman, N.; Xin, Z. Genome-Wide Association Analysis of Seedling Traits in Diverse Sorghum Germplasm under Thermal Stress. *BMC Plant Biol.* **2017**, *17*, 12. [CrossRef]
172. Dwivedi, S.L.; Siddique, K.H.M.; Farooq, M.; Thornton, P.K.; Ortiz, R. Using Biotechnology-Led Approaches to Uplift Cereal and Food Legume Yields in Dryland Environments. *Front. Plant Sci.* **2018**, *9*, 1249. [CrossRef]
173. Lin, F.; Zhang, Y.; Jiang, M.-Y. Alternative Splicing and Differential Expression of Two Transcripts of Nicotine Adenine Dinucleotide Phosphate Oxidase B Gene from *Zea mays*. *J. Integr. Plant Biol.* **2009**, *51*, 287–298. [CrossRef]
174. Laloum, T.; Martín, G.; Duque, P. Alternative Splicing Control of Abiotic Stress Responses. *Trends Plant Sci.* **2018**, *23*, 140–150. [CrossRef]
175. Mastrangelo, A.M.; Marone, D.; Laidò, G.; De Leonardis, A.M.; De Vita, P. Alternative Splicing: Enhancing Ability to Cope with Stress via Transcriptome Plasticity. *Plant Sci. Int. J. Exp. Plant Biol.* **2012**, *185–186*, 40–49. [CrossRef]
176. Mirdar Mansuri, R.; Azizi, A.-H.; Sadri, A.-H.; Shobbar, Z.-S. Long Non-Coding RNAs as the Regulatory Hubs in Rice Response to Salt Stress. *Sci. Rep.* **2022**, *12*, 21696. [CrossRef]
177. De Quattro, C.; Pè, M.E.; Bertolini, E. Long Noncoding RNAs in the Model Species *Brachypodium distachyon*. *Sci. Rep.* **2017**, *7*, 11252. [CrossRef]
178. Zhang, W.; Han, Z.; Guo, Q.; Liu, Y.; Zheng, Y.; Wu, F.; Jin, W. Identification of Maize Long Non-Coding RNAs Responsive to Drought Stress. *PLoS ONE* **2014**, *9*, e98958. [CrossRef] [PubMed]
179. Huanca-Mamani, W.; Arias-Carrasco, R.; Cárdenas-Ninasivincha, S.; Rojas-Herrera, M.; Sepúlveda-Hermosilla, G.; Caris-Maldonado, J.C.; Bastías, E.; Maracaja-Coutinho, V. Long Non-Coding RNAs Responsive to Salt and Boron Stress in the Hyper-Arid Lluteño Maize from Atacama Desert. *Genes* **2018**, *9*, 170. [CrossRef] [PubMed]
180. Pang, J.; Zhang, X.; Ma, X.; Zhao, J. Spatio-Temporal Transcriptional Dynamics of Maize Long Non-Coding RNAs Responsive to Drought Stress. *Genes* **2019**, *10*, 138. [CrossRef]
181. Hu, X.; Wei, Q.; Wu, H.; Huang, Y.; Peng, X.; Han, G.; Ma, Q.; Zhao, Y. Identification and Characterization of Heat-Responsive lncRNAs in Maize Inbred Line CM1. *BMC Genom.* **2022**, *23*, 208. [CrossRef]
182. Liu, P.; Zhang, Y.; Zou, C.; Yang, C.; Pan, G.; Ma, L.; Shen, Y. Integrated Analysis of Long Non-Coding RNAs and mRNAs Reveals the Regulatory Network of Maize Seedling Root Responding to Salt Stress. *BMC Genom.* **2022**, *23*, 50. [CrossRef]
183. Xuhui, L.; Weiwei, C.; Siqi, L.; Junteng, F.; Hang, Z.; Xiangbo, Z.; Yongwen, Q. Full-Length Transcriptome Analysis of Maize Root Tips Reveals the Molecular Mechanism of Cold Stress during the Seedling Stage. *BMC Plant Biol.* **2022**, *22*, 398. [CrossRef] [PubMed]
184. De Quattro, C.; Mica, E.; Pè, M.E.; Bertolini, E. *Brachypodium distachyon* Long Noncoding RNAs: Genome-Wide Identification and Expression Analysis. *Methods Mol. Biol.* **2018**, *1667*, 31–42. [CrossRef]
185. O'Farrell, P.H. High Resolution Two-Dimensional Electrophoresis of Proteins. *J. Biol. Chem.* **1975**, *250*, 4007–4021. [CrossRef] [PubMed]
186. Zhang, Z.; Wu, S.; Stenoien, D.L.; Paša-Tolić, L. High-Throughput Proteomics. *Annu. Rev. Anal. Chem.* **2014**, *7*, 427–454. [CrossRef]
187. Andjelković, U.; Josić, D. Mass Spectrometry Based Proteomics as Foodomics Tool in Research and Assurance of Food Quality and Safety. *Trends Food Sci. Technol.* **2018**, *77*, 100–119. [CrossRef]
188. Chawade, A.; van Ham, J.; Blomquist, H.; Bagge, O.; Alexandersson, E.; Ortiz, R. High-Throughput Field-Phenotyping Tools for Plant Breeding and Precision Agriculture. *Agronomy* **2019**, *9*, 258. [CrossRef]
189. Ali, A.E.; Husselmann, L.H.; Tabb, D.L.; Ludidi, N. Comparative Proteomics Analysis between Maize and Sorghum Uncovers Important Proteins and Metabolic Pathways Mediating Drought Tolerance. *Life* **2023**, *13*, 170. [CrossRef]
190. López-Cristoffanini, C.; Bundó, M.; Serrat, X.; San Segundo, B.; López-Carbonell, M.; Nogués, S. A Comprehensive Study of the Proteins Involved in Salinity Stress Response in Roots and Shoots of the FL478 Genotype of Rice (*Oryza sativa* L. ssp. Indica). *Crop J.* **2021**, *9*, 1154–1168. [CrossRef]
191. Yadav, R.; Chakraborty, S.; Ramakrishna, W. Wheat Grain Proteomic and Protein–Metabolite Interactions Analyses Provide Insights into Plant Growth Promoting Bacteria–Arbuscular Mycorrhizal Fungi–Wheat Interactions. *Plant Cell Rep.* **2022**, *41*, 1417–1437. [CrossRef] [PubMed]
192. Widodo; Patterson, J.H.; Newbigin, E.; Tester, M.; Bacic, A.; Roessner, U. Metabolic Responses to Salt Stress of Barley (*Hordeum vulgare* L.) Cultivars, Sahara and Clipper, Which Differ in Salinity Tolerance. *J. Exp. Bot.* **2009**, *60*, 4089–4103. [CrossRef]
193. Paul, S.; Gayen, D.; Datta, S.K.; Datta, K. Dissecting Root Proteome of Transgenic Rice Cultivars Unravels Metabolic Alterations and Accumulation of Novel Stress Responsive Proteins under Drought Stress. *Plant Sci.* **2015**, *234*, 133–143. [CrossRef]

194. Cheng, L.; Wang, Y.; He, Q.; Li, H.; Zhang, X.; Zhang, F. Comparative Proteomics Illustrates the Complexity of Drought Resistance Mechanisms in Two Wheat (*Triticum aestivum* L.) Cultivars under Dehydration and Rehydration. *BMC Plant Biol.* **2016**, *16*, 188. [CrossRef]
195. Yousuf, P.Y.; Abd Allah, E.F.; Nauman, M.; Asif, A.; Hashem, A.; Alqarawi, A.A.; Ahmad, A. Responsive Proteins in Wheat Cultivars with Contrasting Nitrogen Efficiencies under the Combined Stress of High Temperature and Low Nitrogen. *Genes* **2017**, *8*, 356. [CrossRef] [PubMed]
196. Kosová, K.; Prášil, I.T.; Klíma, M.; Nesvadba, Z.; Vítámvás, P.; Ovesná, J. Proteomics of Wheat and Barley Cereals in Response to Environmental Stresses: Current State and Future Challenges. *J. Proteomics* **2023**, *282*, 104923. [CrossRef]
197. Fiehn, O. Metabolomics—The Link between Genotypes and Phenotypes. *Plant Mol. Biol.* **2002**, *48*, 155–171. [CrossRef]
198. Carrera, F.P.; Noceda, C.; Maridueña-Zavala, M.G.; Cevallos-Cevallos, J.M. Metabolomics, a Powerful Tool for Understanding Plant Abiotic Stress. *Agronomy* **2021**, *11*, 824. [CrossRef]
199. De Vos, R.C.; Moco, S.; Lommen, A.; Keurentjes, J.J.; Bino, R.J.; Hall, R.D. Untargeted Large-Scale Plant Metabolomics Using Liquid Chromatography Coupled to Mass Spectrometry. *Nat. Protoc.* **2007**, *2*, 778–791. [CrossRef]
200. Ramalingam, A.P.; Mohanavel, W.; Premnath, A.; Muthurajan, R.; Prasad, P.V.V.; Perumal, R. Large-Scale Non-Targeted Metabolomics Reveals Antioxidant, Nutraceutical and Therapeutic Potentials of Sorghum. *Antioxidants* **2021**, *10*, 1511. [CrossRef]
201. Alseekh, S.; Bermudez, L.; de Haro, L.A.; Fernie, A.R.; Carrari, F. Crop Metabolomics: From Diagnostics to Assisted Breeding. *Metabolomics* **2018**, *14*, 148. [CrossRef]
202. Jendoubi, T. Approaches to Integrating Metabolomics and Multi-Omics Data: A Primer. *Metabolites* **2021**, *11*, 184. [CrossRef] [PubMed]
203. Zhang, F.; Wu, J.; Sade, N.; Wu, S.; Egbaria, A.; Fernie, A.R.; Yan, J.; Qin, F.; Chen, W.; Brotman, Y.; et al. Genomic Basis Underlying the Metabolome-Mediated Drought Adaptation of Maize. *Genome Biol.* **2021**, *22*, 260. [CrossRef] [PubMed]
204. Hwarari, D.; Radani, Y.; Ke, Y.; Chen, J.; Yang, L. CRISPR/Cas Genome Editing in Plants: Mechanisms, Applications, and Overcoming Bottlenecks. *Funct. Integr. Genom.* **2024**, *24*, 50. [CrossRef]
205. Shi, J.; Gao, H.; Wang, H.; Lafitte, H.R.; Archibald, R.L.; Yang, M.; Hakimi, S.M.; Mo, H.; Habben, J.E. ARGOS8 Variants Generated by CRISPR-Cas9 Improve Maize Grain Yield under Field Drought Stress Conditions. *Plant Biotechnol. J.* **2017**, *15*, 207–216. [CrossRef]
206. Pandey, S.; Divakar, S.; Singh, A. Genome Editing Prospects for Heat Stress Tolerance in Cereal Crops. *Plant Physiol. Biochem.* **2024**, *215*, 108989. [CrossRef]
207. Zhang, L.; Xiang, Z.; Li, J.; Wang, S.; Chen, Y.; Liu, Y.; Mao, D.; Luan, S.; Chen, L. bHLH57 Confers Chilling Tolerance and Grain Yield Improvement in Rice. *Plant Cell Environ.* **2023**, *46*, 1402–1418. [CrossRef]
208. Nazir, R.; Mandal, S.; Mitra, S.; Ghorai, M.; Das, N.; Jha, N.K.; Majumder, M.; Pandey, D.K.; Dey, A. Clustered Regularly Interspaced Short Palindromic Repeats (CRISPR)/CRISPR-Associated Genome-Editing Toolkit to Enhance Salt Stress Tolerance in Rice and Wheat. *Physiol. Plant.* **2022**, *174*, e13642. [CrossRef]
209. Urnov, F.D.; Rebar, E.J.; Holmes, M.C.; Zhang, H.S.; Gregory, P.D. Genome Editing with Engineered Zinc Finger Nucleases. *Nat. Rev. Genet.* **2010**, *11*, 636–646. [CrossRef] [PubMed]
210. Carroll, D. Genome Engineering with Zinc-Finger Nucleases. *Genetics* **2011**, *188*, 773–782. [CrossRef] [PubMed]
211. Wyman, C.; Kanaar, R. DNA Double-Strand Break Repair: All's Well That Ends Well. *Annu. Rev. Genet.* **2006**, *40*, 363–383. [CrossRef]
212. Bin Moon, S.; Lee, J.M.; Kang, J.G.; Lee, N.-E.; Ha, D.-I.; Kim, D.Y.; Kim, S.H.; Yoo, K.; Kim, D.; Ko, J.-H.; et al. Highly Efficient Genome Editing by CRISPR-Cpf1 Using CRISPR RNA with a Uridinylate-Rich 3′-Overhang. *Nat. Commun.* **2018**, *9*, 3651. [CrossRef]
213. Endo, A.; Masafumi, M.; Kaya, H.; Toki, S. Efficient Targeted Mutagenesis of Rice and Tobacco Genomes Using Cpf1 from Francisella Novicida. *Sci. Rep.* **2016**, *6*, 38169. [CrossRef]
214. Wang, M.; Mao, Y.; Lu, Y.; Tao, X.; Zhu, J.-K. Multiplex Gene Editing in Rice Using the CRISPR-Cpf1 System. *Mol. Plant* **2017**, *10*, 1011–1013. [CrossRef] [PubMed]
215. Xu, R.; Qin, R.; Li, H.; Li, D.; Li, L.; Wei, P.; Yang, J. Generation of Targeted Mutant Rice Using a CRISPR-Cpf1 System. *Plant Biotechnol. J.* **2017**, *15*, 713–717. [CrossRef] [PubMed]
216. Yin, X.; Biswal, A.K.; Dionora, J.; Perdigon, K.M.; Balahadia, C.P.; Mazumdar, S.; Chater, C.; Lin, H.-C.; Coe, R.A.; Kretzschmar, T.; et al. CRISPR-Cas9 and CRISPR-Cpf1 Mediated Targeting of a Stomatal Developmental Gene EPFL9 in Rice. *Plant Cell Rep.* **2017**, *36*, 745–757. [CrossRef]
217. Lee, K.; Zhang, Y.; Kleinstiver, B.P.; Guo, J.A.; Aryee, M.J.; Miller, J.; Malzahn, A.; Zarecor, S.; Lawrence-Dill, C.J.; Joung, J.K.; et al. Activities and Specificities of CRISPR/Cas9 and Cas12a Nucleases for Targeted Mutagenesis in Maize. *Plant Biotechnol. J.* **2019**, *17*, 362–372. [CrossRef]
218. Singh, D.; Chaudhary, P.; Taunk, J.; Kumar Singh, C.; Sharma, S.; Singh, V.J.; Singh, D.; Chinnusamy, V.; Yadav, R.; Pal, M. Plant Epigenomics for Extenuation of Abiotic Stresses: Challenges and Future Perspectives. *J. Exp. Bot.* **2021**, *72*, 6836–6855. [CrossRef]

219. Lloyd, J.P.B.; Lister, R. Epigenome Plasticity in Plants. *Nat. Rev. Genet.* **2022**, *23*, 55–68. [CrossRef]
220. Zhang, Y.; Andrews, H.; Eglitis-Sexton, J.; Godwin, I.; Tanurdžić, M.; Crisp, P.A. Epigenome Guided Crop Improvement: Current Progress and Future Opportunities. *Emerg. Top. Life Sci.* **2022**, *6*, 141–151. [CrossRef] [PubMed]
221. Perrone, A.; Martinelli, F. Plant Stress Biology in Epigenomic Era. *Plant Sci. Int. J. Exp. Plant Biol.* **2020**, *294*, 110376. [CrossRef]
222. Kakoulidou, I.; Avramidou, E.V.; Baránek, M.; Brunel-Muguet, S.; Farrona, S.; Johannes, F.; Kaiserli, E.; Lieberman-Lazarovich, M.; Martinelli, F.; Mladenov, V.; et al. Epigenetics for Crop Improvement in Times of Global Change. *Biology* **2021**, *10*, 766. [CrossRef]
223. Samantara, K.; Shiv, A.; de Sousa, L.L.; Sandhu, K.S.; Priyadarshini, P.; Mohapatra, S.R. A Comprehensive Review on Epigenetic Mechanisms and Application of Epigenetic Modifications for Crop Improvement. *Environ. Exp. Bot.* **2021**, *188*, 104479. [CrossRef]
224. Crisp, P.A.; Bhatnagar-Mathur, P.; Hundleby, P.; Godwin, I.D.; Waterhouse, P.M.; Hickey, L.T. Beyond the Gene: Epigenetic and Cis-Regulatory Targets Offer New Breeding Potential for the Future. *Curr. Opin. Biotechnol.* **2022**, *73*, 88–94. [CrossRef] [PubMed]
225. Kim, E.Y.; Kim, K.D.; Cho, J. Harnessing Epigenetic Variability for Crop Improvement: Current Status and Future Prospects. *Genes Genom.* **2022**, *44*, 259–266. [CrossRef] [PubMed]
226. Villagómez-Aranda, A.L.; Feregrino-Pérez, A.A.; García-Ortega, L.F.; González-Chavira, M.M.; Torres-Pacheco, I.; Guevara-González, R.G. Activating Stress Memory: Eustressors as Potential Tools for Plant Breeding. *Plant Cell Rep.* **2022**, *41*, 1481–1498. [CrossRef]
227. Gahlaut, V.; Zinta, G.; Jaiswal, V.; Kumar, S. Quantitative Epigenetics: A New Avenue for Crop Improvement. *Epigenomes* **2020**, *4*, 25. [CrossRef]
228. Agius, D.R.; Kapazoglou, A.; Avramidou, E.; Baranek, M.; Carneros, E.; Caro, E.; Castiglione, S.; Cicatelli, A.; Radanovic, A.; Ebejer, J.-P.; et al. Exploring the Crop Epigenome: A Comparison of DNA Methylation Profiling Techniques. *Front. Plant Sci.* **2023**, *14*, 1181039. [CrossRef]
229. Wang, J.; Meng, X.; Dobrovolskaya, O.B.; Orlov, Y.L.; Chen, M. Non-Coding RNAs and Their Roles in Stress Response in Plants. *Genom. Proteom. Bioinform.* **2017**, *15*, 301–312. [CrossRef]
230. Dong, Q.; Hu, B.; Zhang, C. microRNAs and Their Roles in Plant Development. *Front. Plant Sci.* **2022**, *13*, 824240. [CrossRef]
231. Wu, L.; Liu, S.; Qi, H.; Cai, H.; Xu, M. Research Progress on Plant Long Non-Coding RNA. *Plants* **2020**, *9*, 408. [CrossRef] [PubMed]
232. Zenda, T.; Wang, N.; Dong, A.; Zhou, Y.; Duan, H. Reproductive-Stage Heat Stress in Cereals: Impact, Plant Responses and Strategies for Tolerance Improvement. *Int. J. Mol. Sci.* **2022**, *23*, 6929. [CrossRef]
233. Balyan, S.; Kumar, M.; Mutum, R.D.; Raghuvanshi, U.; Agarwal, P.; Mathur, S.; Raghuvanshi, S. Identification of miRNA-Mediated Drought Responsive Multi-Tiered Regulatory Network in Drought Tolerant Rice, Nagina 22. *Sci. Rep.* **2017**, *7*, 15446. [CrossRef] [PubMed]
234. Ahmed, W.; Xia, Y.; Zhang, H.; Li, R.; Bai, G.; Siddique, K.H.M.; Guo, P. Identification of Conserved and Novel miRNAs Responsive to Heat Stress in Flowering Chinese Cabbage Using High-Throughput Sequencing. *Sci. Rep.* **2019**, *9*, 14922. [CrossRef]
235. Lu, Q.; Xu, Q.; Guo, F.; Lv, Y.; Song, C.; Feng, M.; Yu, J.; Zhang, D.; Cang, J. Identification and Characterization of Long Non-Coding RNAs as Competing Endogenous RNAs in the Cold Stress Response of *Triticum aestivum*. *Plant Biol.* **2020**, *22*, 635–645. [CrossRef]
236. Zhao, Y.; Wei, Q.; Chen, T.; Xu, L.; Liu, J.; Zhang, X.; Han, G.; Ma, Q. Identification and Characterization of Heat-Responsive miRNAs and Their Regulatory Network in Maize. *Plant Growth Regul.* **2022**, *96*, 195–208. [CrossRef]
237. Xu, J.; Hou, Q.-M.; Khare, T.; Verma, S.K.; Kumar, V. Exploring miRNAs for Developing Climate-Resilient Crops: A Perspective Review. *Sci. Total Environ.* **2019**, *653*, 91–104. [CrossRef]
238. Tirnaz, S.; Batley, J. Epigenetics: Potentials and Challenges in Crop Breeding. *Mol. Plant* **2019**, *12*, 1309–1311. [CrossRef]
239. Sun, C.; Ali, K.; Yan, K.; Fiaz, S.; Dormatey, R.; Bi, Z.; Bai, J. Exploration of Epigenetics for Improvement of Drought and Other Stress Resistance in Crops: A Review. *Plants* **2021**, *10*, 1226. [CrossRef]
240. Miryeganeh, M.; Saze, H. Epigenetic Inheritance and Plant Evolution. *Popul. Ecol.* **2020**, *62*, 17–27. [CrossRef]
241. Ahanger, M.A.; Akram, N.A.; Ashraf, M.; Alyemeni, M.N.; Wijaya, L.; Ahmad, P. Plant Responses to Environmental Stresses-from Gene to Biotechnology. *AoB Plants* **2017**, *9*, plx025. [CrossRef]
242. Choudhary, M.; Wani, S.H.; Kumar, P.; Bagaria, P.K.; Rakshit, S.; Roorkiwal, M.; Varshney, R.K. QTLian Breeding for Climate Resilience in Cereals: Progress and Prospects. *Funct. Integr. Genom.* **2019**, *19*, 685–701. [CrossRef] [PubMed]
243. Husaini, A.M. High-Value Pleiotropic Genes for Developing Multiple Stress-Tolerant Biofortified Crops for 21st-Century Challenges. *Heredity* **2022**, *128*, 460–472. [CrossRef]
244. Li, Z.; Zhong, F.; Guo, J.; Chen, Z.; Song, J.; Zhang, Y. Improving Wheat Salt Tolerance for Saline Agriculture. *J. Agric. Food Chem.* **2022**, *70*, 14989–15006. [CrossRef]
245. Varshney, R.K.; Bohra, A.; Roorkiwal, M.; Barmukh, R.; Cowling, W.; Chitikineni, A.; Lam, H.-M.; Hickey, L.T.; Croser, J.; Edwards, D.; et al. Rapid Delivery Systems for Future Food Security. *Nat. Biotechnol.* **2021**, *39*, 1179–1181. [CrossRef] [PubMed]
246. Hickey, L.T.; Hafeez, A.N.; Robinson, H.; Jackson, S.A.; Leal-Bertioli, S.C.M.; Tester, M.; Gao, C.; Godwin, I.D.; Hayes, B.J.; Wulff, B.B.H. Breeding Crops to Feed 10 Billion. *Nat. Biotechnol.* **2019**, *37*, 744–754. [CrossRef]

247. Xu, Y.; Liu, X.; Fu, J.; Wang, H.; Wang, J.; Huang, C.; Prasanna, B.M.; Olsen, M.S.; Wang, G.; Zhang, A. Enhancing Genetic Gain through Genomic Selection: From Livestock to Plants. *Plant Commun.* **2020**, *1*, 100005. [CrossRef]
248. He, T.; Li, C. Harness the Power of Genomic Selection and the Potential of Germplasm in Crop Breeding for Global Food Security in the Era with Rapid Climate Change. *Crop J.* **2020**, *8*, 688–700. [CrossRef]
249. Crain, J.; DeHaan, L.; Poland, J. Genomic Prediction Enables Rapid Selection of High-Performing Genets in an Intermediate Wheatgrass Breeding Program. *Plant Genome* **2021**, *14*, e20080. [CrossRef]
250. Watson, A.; Ghosh, S.; Williams, M.J.; Cuddy, W.S.; Simmonds, J.; Rey, M.-D.; Asyraf Md Hatta, M.; Hinchliffe, A.; Steed, A.; Reynolds, D.; et al. Speed Breeding Is a Powerful Tool to Accelerate Crop Research and Breeding. *Nat. Plants* **2018**, *4*, 23–29. [CrossRef] [PubMed]
251. Bhat, J.A.; Yu, D.; Bohra, A.; Ganie, S.A.; Varshney, R.K. Features and Applications of Haplotypes in Crop Breeding. *Commun. Biol.* **2021**, *4*, 1266. [CrossRef] [PubMed]
252. Varshney, R.K.; Bohra, A.; Roorkiwal, M.; Barmukh, R.; Cowling, W.A.; Chitikineni, A.; Lam, H.-M.; Hickey, L.T.; Croser, J.S.; Bayer, P.E.; et al. Fast-Forward Breeding for a Food-Secure World. *Trends Genet. TIG* **2021**, *37*, 1124–1136. [CrossRef]
253. Xu, Y.; Zhang, X.; Li, H.; Zheng, H.; Zhang, J.; Olsen, M.S.; Varshney, R.K.; Prasanna, B.M.; Qian, Q. Smart Breeding Driven by Big Data, Artificial Intelligence, and Integrated Genomic-Enviromic Prediction. *Mol. Plant* **2022**, *15*, 1664–1695. [CrossRef]
254. Crossa, J.; Pérez-Rodríguez, P.; Cuevas, J.; Montesinos-López, O.; Jarquín, D.; de Los Campos, G.; Burgueño, J.; González-Camacho, J.M.; Pérez-Elizalde, S.; Beyene, Y.; et al. Genomic Selection in Plant Breeding: Methods, Models, and Perspectives. *Trends Plant Sci.* **2017**, *22*, 961–975. [CrossRef]
255. Voss-Fels, K.P.; Cooper, M.; Hayes, B.J. Accelerating Crop Genetic Gains with Genomic Selection. *TAG Theor. Appl. Genet.* **2019**, *132*, 669–686. [CrossRef]
256. Budhlakoti, N.; Kushwaha, A.K.; Rai, A.; Chaturvedi, K.K.; Kumar, A.; Pradhan, A.K.; Kumar, U.; Kumar, R.R.; Juliana, P.; Mishra, D.C.; et al. Genomic Selection: A Tool for Accelerating the Efficiency of Molecular Breeding for Development of Climate-Resilient Crops. *Front. Genet.* **2022**, *13*, 832153. [CrossRef]
257. Meuwissen, T.H.; Hayes, B.J.; Goddard, M.E. Prediction of Total Genetic Value Using Genome-Wide Dense Marker Maps. *Genetics* **2001**, *157*, 1819–1829. [CrossRef] [PubMed]
258. Bernardo, R. Molecular Markers and Selection for Complex Traits in Plants: Learning from the Last 20 Years. *Crop Sci.* **2008**, *48*, 1649–1664. [CrossRef]
259. Wang, X.; Xu, Y.; Hu, Z.; Xu, C. Genomic Selection Methods for Crop Improvement: Current Status and Prospects. *Crop J.* **2018**, *6*, 330–340. [CrossRef]
260. Qian, L.; Hickey, L.T.; Stahl, A.; Werner, C.R.; Hayes, B.; Snowdon, R.J.; Voss-Fels, K.P. Exploring and Harnessing Haplotype Diversity to Improve Yield Stability in Crops. *Front. Plant Sci.* **2017**, *8*, 1534. [CrossRef] [PubMed]
261. Thomas, W.J.W.; Zhang, Y.; Amas, J.C.; Cantila, A.Y.; Zandberg, J.D.; Harvie, S.L.; Batley, J. Innovative Advances in Plant Genotyping. In *Plant Genotyping: Methods and Protocols*; Shavrukov, Y., Ed.; Methods in Molecular Biology; Springer US: New York, NY, USA, 2023; pp. 451–465, ISBN 978-1-07-163024-2.
262. Samantara, K.; Bohra, A.; Mohapatra, S.R.; Prihatini, R.; Asibe, F.; Singh, L.; Reyes, V.P.; Tiwari, A.; Maurya, A.K.; Croser, J.S.; et al. Breeding More Crops in Less Time: A Perspective on Speed Breeding. *Biology* **2022**, *11*, 275. [CrossRef] [PubMed]
263. Samineni, S.; Sen, M.; Sajja, S.B.; Gaur, P.M. Rapid Generation Advance (RGA) in Chickpea to Produce up to Seven Generations per Year and Enable Speed Breeding. *Crop J.* **2020**, *8*, 164–169. [CrossRef]
264. Swami, P.; Deswal, K.; Rana, V.; Yadav, D.; Munjal, R. Chapter 3—Speed Breeding—A Powerful Tool to Breed More Crops in Less Time Accelerating Crop Research. In *Abiotic Stresses in Wheat*; Khan, M.K., Pandey, A., Hamurcu, M., Gupta, O.P., Gezgin, S., Eds.; Academic Press: Cambridge, MA, USA, 2023; pp. 33–49, ISBN 978-0-323-95368-9.
265. Sinha, P.; Singh, V.K.; Saxena, R.K.; Khan, A.W.; Abbai, R.; Chitikineni, A.; Desai, A.; Molla, J.; Upadhyaya, H.D.; Kumar, A.; et al. Superior Haplotypes for Haplotype-Based Breeding for Drought Tolerance in Pigeonpea (*Cajanus cajan* L.). *Plant Biotechnol. J.* **2020**, *18*, 2482–2490. [CrossRef]
266. Brinton, J.; Ramirez-Gonzalez, R.H.; Simmonds, J.; Wingen, L.; Orford, S.; Griffiths, S.; Haberer, G.; Spannagl, M.; Walkowiak, S.; Pozniak, C.; et al. A Haplotype-Led Approach to Increase the Precision of Wheat Breeding. *Commun. Biol.* **2020**, *3*, 712. [CrossRef]
267. Crossa, J.; Fritsche-Neto, R.; Montesinos-Lopez, O.A.; Costa-Neto, G.; Dreisigacker, S.; Montesinos-Lopez, A.; Bentley, A.R. The Modern Plant Breeding Triangle: Optimizing the Use of Genomics, Phenomics, and Enviromics Data. *Front. Plant Sci.* **2021**, *12*, 651480. [CrossRef]
268. Friesen, M.L.; Porter, S.S.; Stark, S.C.; von Wettberg, E.J.; Sachs, J.L.; Martinez-Romero, E. Microbially Mediated Plant Functional Traits. *Annu. Rev. Ecol. Evol. Syst.* **2011**, *42*, 23–46. [CrossRef]
269. Durán, P.; Thiergart, T.; Garrido-Oter, R.; Agler, M.; Kemen, E.; Schulze-Lefert, P.; Hacquard, S. Microbial Interkingdom Interactions in Roots Promote Arabidopsis Survival. *Cell* **2018**, *175*, 973–983.e14. [CrossRef] [PubMed]
270. Trivedi, P.; Leach, J.E.; Tringe, S.G.; Sa, T.; Singh, B.K. Plant–Microbiome Interactions: From Community Assembly to Plant Health. *Nat. Rev. Microbiol.* **2020**, *18*, 607–621. [CrossRef]

271. Konwar, N.; Singh, K.N.; Narzary, D. Plant–Microbe Interactions in Combating Abiotic Stresses. In *Plant Stress: Challenges and Management in the New Decade*; Roy, S., Mathur, P., Chakraborty, A.P., Saha, S.P., Eds.; Advances in Science, Technology & Innovation; Springer International Publishing: Cham, Switzerland, 2022; pp. 217–234, ISBN 978-3-030-95365-2.
272. Rodriguez, P.A.; Rothballer, M.; Chowdhury, S.P.; Nussbaumer, T.; Gutjahr, C.; Falter-Braun, P. Systems Biology of Plant-Microbiome Interactions. *Mol. Plant* **2019**, *12*, 804–821. [CrossRef] [PubMed]
273. Cohen, A.C.; Bottini, R.; Pontin, M.; Berli, F.J.; Moreno, D.; Boccanlandro, H.; Travaglia, C.N.; Piccoli, P.N. Azospirillum Brasilense Ameliorates the Response of Arabidopsis Thaliana to Drought Mainly via Enhancement of ABA Levels. *Physiol. Plant.* **2015**, *153*, 79–90. [CrossRef] [PubMed]
274. Tiwari, S.; Prasad, V.; Chauhan, P.S.; Lata, C. Bacillus Amyloliquefaciens Confers Tolerance to Various Abiotic Stresses and Modulates Plant Response to Phytohormones through Osmoprotection and Gene Expression Regulation in Rice. *Front. Plant Sci.* **2017**, *8*, 1510. [CrossRef]
275. Ahmed, V.; Verma, M.K.; Gupta, S.; Mandhan, V.; Chauhan, N.S. Metagenomic Profiling of Soil Microbes to Mine Salt Stress Tolerance Genes. *Front. Microbiol.* **2018**, *9*, 159. [CrossRef]
276. Kamran, M.; Imran, Q.M.; Ahmed, M.B.; Falak, N.; Khatoon, A.; Yun, B.-W. Endophyte-Mediated Stress Tolerance in Plants: A Sustainable Strategy to Enhance Resilience and Assist Crop Improvement. *Cells* **2022**, *11*, 3292. [CrossRef]
277. Santos, L.F.; Olivares, F.L. Plant Microbiome Structure and Benefits for Sustainable Agriculture. *Curr. Plant Biol.* **2021**, *26*, 100198. [CrossRef]
278. Omae, N.; Tsuda, K. Plant-Microbiota Interactions in Abiotic Stress Environments. *Mol. Plant-Microbe Interact. MPMI* **2022**, *35*, 511–526. [CrossRef]
279. Chouhan, G.K.; Verma, J.P.; Jaiswal, D.K.; Mukherjee, A.; Singh, S.; de Araujo Pereira, A.P.; Liu, H.; Abd Allah, E.F.; Singh, B.K. Phytomicrobiome for Promoting Sustainable Agriculture and Food Security: Opportunities, Challenges, and Solutions. *Microbiol. Res.* **2021**, *248*, 126763. [CrossRef]
280. Liu, H.; Brettell, L.E.; Qiu, Z.; Singh, B.K. Microbiome-Mediated Stress Resistance in Plants. *Trends Plant Sci.* **2020**, *25*, 733–743. [CrossRef]
281. Singh, B.K.; Trivedi, P.; Egidi, E.; Macdonald, C.A.; Delgado-Baquerizo, M. Crop Microbiome and Sustainable Agriculture. *Nat. Rev. Microbiol.* **2020**, *18*, 601–602. [CrossRef]
282. Mahmud, A.A.; Upadhyay, S.K.; Srivastava, A.K.; Bhojiya, A.A. Biofertilizers: A Nexus between Soil Fertility and Crop Productivity under Abiotic Stress. *Curr. Res. Environ. Sustain.* **2021**, *3*, 100063. [CrossRef]
283. Fadiji, A.E.; Babalola, O.O. Metagenomics Methods for the Study of Plant-Associated Microbial Communities: A Review. *J. Microbiol. Methods* **2020**, *170*, 105860. [CrossRef]
284. Liu, Y.-X.; Qin, Y.; Chen, T.; Lu, M.; Qian, X.; Guo, X.; Bai, Y. A Practical Guide to Amplicon and Metagenomic Analysis of Microbiome Data. *Protein Cell* **2021**, *12*, 315–330. [CrossRef] [PubMed]
285. Jeong, J.-J.; Sang, M.K.; Pathiraja, D.; Park, B.; Choi, I.-G.; Kim, K.D. Draft Genome Sequence of Phosphate-Solubilizing Chryseobacterium Sp. Strain ISE14, a Biocontrol and Plant Growth-Promoting Rhizobacterium Isolated from Cucumber. *Genome Announc.* **2018**, *6*, e00612-18. [CrossRef] [PubMed]
286. Ormeño-Orrillo, E.; Aguilar-Cuba, Y.; Zúñiga-Dávila, D. Draft Genome Sequence of Rhizobium Sophoriradicis H4, a Nitrogen-Fixing Bacterium Associated with the Leguminous Plant Phaseolus Vulgaris on the Coast of Peru. *Genome Announc.* **2018**, *6*, e00241-18. [CrossRef] [PubMed]
287. Raza, A. Metabolomics: A Systems Biology Approach for Enhancing Heat Stress Tolerance in Plants. *Plant Cell Rep.* **2022**, *41*, 741–763. [CrossRef] [PubMed]
288. Alternate Routes to Gene Functions. *Nat. Plants* **2024**, *10*, 1605–1606. [CrossRef] [PubMed]
289. Nguyen, Q.-H.; Nguyen, H.; Oh, E.C.; Nguyen, T. Current Approaches and Outstanding Challenges of Functional Annotation of Metabolites: A Comprehensive Review. *Brief. Bioinform.* **2024**, *25*, bbae498. [CrossRef]
290. Fan, B.-L.; Chen, L.-H.; Chen, L.-L.; Guo, H. Integrative Multi-Omics Approaches for Identifying and Characterizing Biological Elements in Crop Traits: Current Progress and Future Prospects. *Int. J. Mol. Sci.* **2025**, *26*, 1466. [CrossRef]
291. Rane, J.; Singh, A.K.; Kumar, M.; Boraiah, K.M.; Meena, K.K.; Pradhan, A.; Prasad, P.V.V. The Adaptation and Tolerance of Major Cereals and Legumes to Important Abiotic Stresses. *Int. J. Mol. Sci.* **2021**, *22*, 12970. [CrossRef]
292. Zenda, T.; Wang, N.; Yan, X.; Dong, A.; Yang, Q.; Zhong, Y.; Duan, H. Opportunities and Avenues for Achieving Crop Climate Resilience. *Environ. Exp. Bot.* **2023**, *213*, 105414. [CrossRef]
293. Chen, V.; Yang, M.; Cui, W.; Kim, J.S.; Talwalkar, A.; Ma, J. Applying Interpretable Machine Learning in Computational Biology—Pitfalls, Recommendations and Opportunities for New Developments. *Nat. Methods* **2024**, *21*, 1454–1461. [CrossRef]
294. Silva, T.N.; Thomas, J.B.; Dahlberg, J.; Rhee, S.Y.; Mortimer, J.C. Progress and Challenges in Sorghum Biotechnology, a Multipurpose Feedstock for the Bioeconomy. *J. Exp. Bot.* **2022**, *73*, 646–664. [CrossRef] [PubMed]
295. Duan, L.; Ouyang, K.; Xu, X.; Xu, L.; Wen, C.; Zhou, X.; Qin, Z.; Xu, Z.; Sun, W.; Liang, Y. Nanoparticle Delivery of CRISPR/Cas9 for Genome Editing. *Front. Genet.* **2021**, *12*, 673286. [CrossRef] [PubMed]

296. Siddique, A.B.; Shabala, S.; Li, C.; Chen, Z.-H.; Varshney, R.K.; Zhao, C.; Zhou, M. Reducing Heat Stress Damage in Cereal Crops through Agronomic Management and Breeding Strategies. *Plant Stress* **2025**, *16*, 100888. [CrossRef]
297. Sargent, D.; Conaty, W.C.; Tissue, D.T.; Sharwood, R.E. Synthetic Biology and Opportunities within Agricultural Crops. *J. Sustain. Agric. Environ.* **2022**, *1*, 89–107. [CrossRef]
298. Liu, S.; Zenda, T.; Tian, Z.; Huang, Z. Metabolic Pathways Engineering for Drought or/and Heat Tolerance in Cereals. *Front. Plant Sci.* **2023**, *14*, 1111875. [CrossRef]
299. Yaschenko, A.E.; Alonso, J.M.; Stepanova, A.N. Arabidopsis as a Model for Translational Research. *Plant Cell* **2025**, *37*, koae065. [CrossRef] [PubMed]
300. Uauy, C.; Nelissen, H.; Chan, R.L.; Napier, J.A.; Seung, D.; Liu, L.; McKim, S.M. Challenges of Translating Arabidopsis Insights into Crops. *Plant Cell* **2025**, *37*, koaf059. [CrossRef]
301. Zandalinas, S.I.; Mittler, R. Plant Responses to Multifactorial Stress Combination. *New Phytol.* **2022**, *234*, 1161–1167. [CrossRef]
302. Horváth, Á.; Berki, Z.; Balla, K.; Bányai, J.; Mayer, M.; Cseh, A.; Kiss, T.; Karsai, I. Field versus Controlled Environmental Experiments to Evaluate the Heat Stress Response of Barley (*Hordeum vulgare* L.). *Environ. Exp. Bot.* **2024**, *228*, 106038. [CrossRef]
303. de Souza, L.P.; Borghi, M.; Fernie, A. Plant Single-Cell Metabolomics-Challenges and Perspectives. *Int. J. Mol. Sci.* **2020**, *21*, 8987. [CrossRef] [PubMed]
304. Mo, Y.; Jiao, Y. Advances and Applications of Single-Cell Omics Technologies in Plant Research. *Plant J. Cell Mol. Biol.* **2022**, *110*, 1551–1563. [CrossRef] [PubMed]
305. Yu, T.; Scolnick, J. Complex Biological Questions Being Addressed Using Single Cell Sequencing Technologies. *SLAS Technol.* **2022**, *27*, 143–149. [CrossRef]
306. Zhao, D.; Zhu, X.; Zhou, H.; Sun, N.; Wang, T.; Bi, C.; Zhang, X. CRISPR-Based Metabolic Pathway Engineering. *Metab. Eng.* **2021**, *63*, 148–159. [CrossRef]
307. Guo, Y.; Hao, D.; Wang, X.; Wang, H.; Wu, Z.; Yang, P.; Zhang, B. Comparative Transcriptomics Reveals Key Genes Contributing to the Differences in Drought Tolerance among Three Cultivars of Foxtail Millet (*Setaria italica*). *Plant Growth Regul.* **2023**, *99*, 45–64. [CrossRef]
308. Kaur, S.; Seem, K.; Duhan, N.; Kumar, S.; Kaundal, R.; Mohapatra, T. Transcriptome and Physio-Biochemical Profiling Reveals Differential Responses of Rice Cultivars at Reproductive-Stage Drought Stress. *Int. J. Mol. Sci.* **2023**, *24*, 1002. [CrossRef]
309. Wang, Y.; Guo, H.; Wu, X.; Wang, J.; Li, H.; Zhang, R. Transcriptomic and Physiological Responses of Contrasting Maize Genotypes to Drought Stress. *Front. Plant Sci.* **2022**, *13*, 928897. [CrossRef]
310. Wu, G.; Sun, X.; Sun, Q.; Kang, X.; Wang, J.; He, X.; Liu, W.; Xu, D.; Dai, X.; Ma, W. Genetic Variation in Wheat Root Transcriptome Responses to Salinity: A Comparative Study of Tolerant and Sensitive Genotypes. *Int. J. Mol. Sci.* **2025**, *26*, 331. [CrossRef]
311. Wang, W.; Zou, J.; Zhang, Y.; Niu, L.; Yu, L.; Wang, Z.; Wang, F.; Zhang, S.; Yang, X. Salinity Stress Phenotyping of Wheat Germplasm under Multiple Growth Conditions and Transcriptomic Analysis of Two Wheat Varieties Contrasting in Their Salinity Stress Tolerance. *Plant Growth Regul.* **2025**, *105*, 739–757. [CrossRef]
312. Alyahya, N.; Taybi, T. Comparative Transcriptomic Profiling Reveals Differentially Expressed Genes and Important Related Metabolic Pathways in Shoots and Roots of a Saudi Wheat Cultivar (Najran) under Salinity Stress. *Front. Plant Sci.* **2023**, *14*, 1225541. [CrossRef] [PubMed]
313. Deng, Y.; Zhu, Z.; Chen, J.; Kuang, L.; Yan, T.; Wu, D.; Gao, F. Comparative Transcriptomics of Indica and Japonica Rice Roots under Heat Stress Reveals the Crucial Role of OsMAPK3 in Heat Response. *Plant Physiol. Biochem.* **2025**, *221*, 109668. [CrossRef] [PubMed]
314. Cai, H.; Wang, H.; Zhou, L.; Li, B.; Zhang, S.; He, Y.; Guo, Y.; You, A.; Jiao, C.; Xu, Y. Time-Series Transcriptomic Analysis of Contrasting Rice Materials under Heat Stress Reveals a Faster Response in the Tolerant Cultivar. *Int. J. Mol. Sci.* **2023**, *24*, 9408. [CrossRef]
315. Azameti, M.K.; Ranjan, A.; Singh, P.; Gaikwad, K.; Singh, A.K.; Dalal, M.; Arora, A.; Rai, V.; Padaria, J.C. Transcriptome Profiling Reveals the Genes and Pathways Involved in Thermo-Tolerance in Wheat (*Triticum aestivum* L.) Genotype Raj 3765. *Sci. Rep.* **2022**, *12*, 14831. [CrossRef]
316. Konstantinov, D.K.; Zubairova, U.S.; Ermakov, A.A.; Doroshkov, A.V. Comparative Transcriptome Profiling of a Resistant vs Susceptible Bread Wheat (*Triticum aestivum* L.) Cultivar in Response to Water Deficit and Cold Stress. *PeerJ* **2021**, *9*, e11428. [CrossRef]
317. Thapa, R.; Tabien, R.E.; Johnson, C.D.; Septiningsih, E.M. Comparative Transcriptomic Analysis of Germinating Rice Seedlings to Individual and Combined Anaerobic and Cold Stress. *BMC Genom.* **2023**, *24*, 185. [CrossRef]
318. Zhang, P.; Wu, X.; Chen, Y.; Ji, G.; Ma, X.; Zhang, Y.; Xiang, J.; Wang, Y.; Wang, Z.; Li, L. Comparative Transcriptome Combined with Morphophysiological Analyses Revealed Carotenoid Biosynthesis for Differential Chilling Tolerance in Two Contrasting Rice (*Oryza sativa* L.) Genotypes. *Rice* **2023**, *16*, 52. [CrossRef]

319. Zhang, T.; Xiao, J.; Zhao, Y.; Zhang, Y.; Jie, Y.; Shen, D.; Yue, C.; Huang, J.; Hua, Y.; Zhou, T. Comparative Physiological and Transcriptomic Analyses Reveal Ascorbate and Glutathione Coregulation of Cadmium Toxicity Resistance in Wheat Genotypes. *BMC Plant Biol.* **2021**, *21*, 459. [CrossRef]
320. Luo, D.; Li, Q.; Pang, F.; Zhang, W.; Li, Y.; Xing, Y.; Dong, D. Commonalities and Specificities in Wheat (*Triticum aestivum* L.) Responses to Aluminum Toxicity and Low Phosphorus Revealed by Transcriptomics and Targeted Metabolomics. *Int. J. Mol. Sci.* **2024**, *25*, 9273. [CrossRef] [PubMed]
321. Gao, R.; Zhou, L.; Guo, G.; Li, Y.; Chen, Z.; Lu, R.; Liu, C.; Chen, J. Comparative Analysis of Root Transcriptome of High-NUE Mutant and Wild-Type Barley under Low-Nitrogen Conditions. *Agronomy* **2023**, *13*, 806. [CrossRef]
322. Li, J.; Zenda, T.; Liu, S.; Dong, A.; Wang, Y.; Liu, X.; Wang, N.; Duan, H. Integrated Transcriptomic and Proteomic Analyses of Low-Nitrogen-Stress Tolerance and Function Analysis of ZmGST42 Gene in Maize. *Antioxidants* **2023**, *12*, 1831. [CrossRef]
323. Wang, G.; Wang, J.; Yao, L.; Li, B.; Ma, X.; Si, E.; Yang, K.; Li, C.; Shang, X.; Meng, Y. Transcriptome and Metabolome Reveal the Molecular Mechanism of Barley Genotypes Underlying the Response to Low Nitrogen and Resupply. *Int. J. Mol. Sci.* **2023**, *24*, 4706. [CrossRef]
324. Wu, J.; Chen, L.; Yang, Z.; Lu, J.; Yang, J.; Li, N.; Shi, H. Transcriptomics Uncovers Pathways Mediating Low-Nitrogen Stress Tolerance in Two Foxtail Millet Varieties. *Agriculture* **2025**, *15*, 628. [CrossRef]
325. Guo, H.; Tao, W.; Gao, H.; Chen, L.; Zhong, X.; Tang, M.; Gao, G.; Liang, T.; Zhang, X. Physiological Traits, Gene Expression Responses, and Proteomics of Rice Varieties Varying in Heat Stress Tolerance at the Flowering Stage. *Front. Plant Sci.* **2024**, *15*, 1489331. [CrossRef]
326. Kumar, R.; Ghatak, A.; Goyal, I.; Sarkar, N.K.; Weckwerth, W.; Grover, A.; Chaturvedi, P. Heat-Induced Proteomic Changes in Anthers of Contrasting Rice Genotypes under Variable Stress Regimes. *Front. Plant Sci.* **2023**, *13*, 1083971. [CrossRef] [PubMed]
327. Zeng, W.; Peng, Y.; Zhao, X.; Wu, B.; Chen, F.; Ren, B.; Zhuang, Z.; Gao, Q.; Ding, Y. Comparative Proteomics Analysis of the Seedling Root Response of Drought-Sensitive and Drought-Tolerant Maize Varieties to Drought Stress. *Int. J. Mol. Sci.* **2019**, *20*, 2793. [CrossRef]
328. Jha, S.; Maity, S.; Singh, J.; Chouhan, C.; Tak, N.; Ambatipudi, K. Integrated Physiological and Comparative Proteomics Analysis of Contrasting Genotypes of Pearl Millet Reveals Underlying Salt-responsive Mechanisms. *Physiol. Plant.* **2022**, *174*, e13605. [CrossRef] [PubMed]
329. Li, H.; Yang, M.; Zhao, C.; Wang, Y.; Zhang, R. Physiological and Proteomic Analyses Revealed the Response Mechanisms of Two Different Drought-Resistant Maize Varieties. *BMC Plant Biol.* **2021**, *21*, 513. [CrossRef] [PubMed]
330. Ren, W.; Shi, Z.; Zhou, M.; Zhao, B.; Li, H.; Wang, J.; Liu, Y.; Zhao, J. iTRAQ-Based Quantitative Proteomic Analysis Provides Insight into the Drought-Stress Response in Maize Seedlings. *Sci. Rep.* **2022**, *12*, 9520. [CrossRef] [PubMed]
331. Prathap, V.; Kumar, S.; Tyagi, A. Comparative Proteome Analysis of Phosphorus-Responsive Genotypes Reveals the Proteins Differentially Expressed under Phosphorous Starvation Stress in Rice. *Int. J. Biol. Macromol.* **2023**, *234*, 123760. [CrossRef]
332. Wang, Y.; Wang, N.; Liu, S.; Dong, A.; Zenda, T.; Liu, X.; Li, J.; Duan, H. Comparative Proteomic Analysis of Two Contrasting Maize Hybrids' Responses to Low Nitrogen Stress at the Twelve Leaf Stage and Function Verification of ZmTGA Gene. *Genes* **2022**, *13*, 670. [CrossRef]

MDPI AG
Grosspeteranlage 5
4052 Basel
Switzerland
Tel.: +41 61 683 77 34

Plants Editorial Office
E-mail: plants@mdpi.com
www.mdpi.com/journal/plants

www.ingramcontent.com/pod-product-compliance
Lightning Source LLC
Chambersburg PA
CBHW061617220926
51850CB00005B/527
* 9 7 8 3 7 2 5 8 7 5 9 1 7 *